Energy Harvesting Properties
of Electrospun Nanofibers
(Second Edition)

Online at: https://doi.org/10.1088/978-0-7503-5487-5

Energy Harvesting Properties of Electrospun Nanofibers (Second Edition)

Edited by
Jian Fang
College of Textile and Clothing Engineering, Soochow University, Suzhou, People's Republic of China

Tong Lin
School of Textile Science and Engineering, State Key Laboratory of Separation Membranes and Membrane Processes, Tiangong University, Tianjin, People's Republic of China

IOP Publishing, Bristol, UK

ISBN   978-0-7503-5487-5 (ebook)
ISBN   978-0-7503-5485-1 (print)
ISBN   978-0-7503-5488-2 (myPrint)
ISBN   978-0-7503-5486-8 (mobi)

DOI   10.1088/978-0-7503-5487-5

Version: 20250301

IOP ebooks

British Library Cataloguing-in-Publication Data: A catalogue record for this book is available from the British Library.

Published by IOP Publishing, wholly owned by The Institute of Physics, London

IOP Publishing, No.2 The Distillery, Glassfields, Avon Street, Bristol, BS2 0GR, UK

US Office: IOP Publishing, Inc., 190 North Independence Mall West, Suite 601, Philadelphia, PA 19106, USA

Contents

5   Hi-performance piezoelectric nanofiber via advancing β-crystalline phase   5-1

Fatemeh Mokhtari, Pejman Heidarian, Russell J Varley and Joselito M Razal

6   Acoustoelectric energy conversion of nanofibrous materials   6-1

Chenhong Lang, Jingye Jin, Jian Fang and Tong Lin

9 Electrospinning of functional nanofibers: a pathway to flexible piezoelectric and triboelectric wearable devices   **9-1**

Shayan Abrishami, Armineh Shirali, Mahshid Sadeghi,
Amin Sarmadi and Roohollah Bagherzadeh

Preface

The field of energy harvesting is experiencing rapid growth and innovation. Following the publication of our first edition of *Energy Harvesting Properties of Electrospun Nanofibers*, we have been gratified by the enthusiastic reception from our peers in research, engineering, and academia worldwide. At the same time, the application of energy harvesting technologies using various electrospun nanofibers has grown rapidly over the past three years. This growth is particularly pronounced due to the increasing need for flexible and breathable technologies that are essential to the rapid advances in human–machine interfaces, humanoid robotics, and point-of-care wearable devices.

Electrospinning of nanofibers has been instrumental in these advances, marking significant progress in the design of nanofibers and the architecture of fiber assemblies, as well as the fabrication and use of nanofiber devices. Recognizing the need to provide our readers with the latest advances in piezoelectric and triboelectric nanofibers, we felt it was time to update the content of our previous work to keep pace with the rapid technological advances.

In this second edition, we have worked with many of the original authors to update their outstanding chapters. In addition, we have enlisted the expertise of prominent researchers to further enrich this edition. The book now consists of ten chapters. Chapter 1 provides an overview of the electrospinning process, highlighting recent technological advances. Chapter 2 focuses on the quality control of nanofibers by electrospinning bias polarity for electronic applications. Chapter 3 provides an update on the fundamentals of piezoelectricity and the fabrication of piezoelectric materials by electrospinning. Chapter 4 reviews the characterization methods for piezoelectric nanofibers. Chapter 5 presents the latest strategies for controlling the β-crystal phase in polyvinylidene fluoride nanofibers. Chapter 6 updates the applications of electrospun nanofibers in acoustoelectric conversion. Chapter 7 presents the novel research on the use of electrospun polyacrylonitrile nanofibers for high-temperature piezoelectric sensing. Chapter 8 explores the emerging application of electrospun nanofibers in self-powered electronic skin. Chapter 9 provides an update on recent developments in flexible and stretchable nanofiber devices for energy harvesting and sensing. Chapter 10 reviews the use of composite electrospun nanofibers for piezoelectric and triboelectric energy harvesting.

We anticipate that this updated edition, with its expanded scope and wealth of inspiring technological innovations, will prove to be an even more valuable resource for students, researchers, and engineers in the field. Our sincere thanks go to all the contributors for their invaluable insights. Our thanks also go to the editors at IOP Publishing: Ms Caroline Mitchell, Ms Isabelle Defillion, and Ms Betty Barber, for their expert editing, exceptional support, and patience throughout the process. Their invaluable assistance was essential in bringing this book to fruition.

Jian Fang and Tong Lin
September 2024

Acknowledgments

The Editors acknowledge the generous support of the Natural Science Foundation of China for its general fund (projects No. 52173059, No. 52273253, and No. 52373103), the Natural Science Foundation of Jiangsu Higher Education Institutions through the Major Basic Research Project (21KJA540002), and the Hebei Industrial Technology Research Institute of Membranes, Cangzhou Institute of Tiangong University through Science and Technology Program of Cangzhou (No. TGCYY-F-0205).

Editor biographies

Jian Fang

Professor Jian Fang is a professor at Soochow University's College of Textile and Clothing Engineering. He has published over 150 journal articles, 3 books and 6 book chapters. He currently serves as the Director of the Key Laboratory of Smart Textile and Apparel Flexible Devices and the Young Editor of *Advanced Fiber Materials*, *eScience*, *Nano-Micro Letters* and *Engineering*.

Tong Lin

Professor Tong Lin is a professor at Tiangong University, specializing in electrospinning, nanofibers, functional fibers and electro-active polymers, with special emphasis on their applications in environmental and energy-related fields. He has published more than 340 peer-reviewed articles in scientific journals, 40 books and book chapters, and more than 90 other papers.

List of contributors

Shayan Abrishami
School of Engineering, Edith Cowan University, Joondalup, WA 6027, Australia and
Advanced Fibrous Materials LAB, Textile Engineering Department, Institute for Advanced Textile Materials and Technologies (ATMT), Amirkabir University of Technology, Tehran, Iran

S Anandhan
Department of Metallurgical and Materials Engineering, National Institute of Technology Karnataka, Surathkal, Mangaluru 575025, India

Anand Babu
Quantum Materials and Devices Unit, Institute of Nano Science and Technology, Knowledge City, Sector 81, Mohali 140306, India

Roohollah Bagherzadeh
Advanced Fibrous Materials LAB, Textile Engineering Department, Institute for Advanced Textile Materials and Technologies (ATMT), Amirkabir University of Technology, Tehran, Iran

Bin Ding
Innovation Center for Textile Science and Technology, Donghua University, Shanghai 200051, China

Govind S Ekbote
Department of Metallurgical and Materials Engineering, National Institute of Technology Karnataka, Surathkal, Mangaluru 575025, India

Pejman Heidarian
Carbon Nexus at the Institute for Frontier Materials, Deakin University, Waurn Ponds, Victoria 3216, Australia

Jingye Jin
College of Textile Science and Engineering (International Institute of Silk), Zhejiang Sci-Tech University, Hangzhou, Zhejiang 310018, China and
Key Laboratory of Intelligent Textile and Flexible Interconnection of Zhejiang Province, Zhejiang Sci-Tech University, Hangzhou, Zhejiang 310018, China

Xin Jin
School of Materials Science and Engineering, State Key Laboratory of Separation Membranes and Membrane Processes, Tiangong University, Tianjin 300387, China

Chenhong Lang
College of Textile Science and Engineering (International Institute of Silk), Zhejiang Sci-Tech University, Hangzhou, Zhejiang 310018, China

and
Key Laboratory of Intelligent Textile and Flexible Interconnection of Zhejiang
Province, Zhejiang Sci-Tech University, Hangzhou, Zhejiang 310018, China

Zhaoling Li
Shanghai Frontiers Science Center of Advanced Textiles, College of Textiles,
Donghua University, Shanghai 201620, China
and
Innovation Center for Textile Science and Technology, Donghua University,
Shanghai 200051, China

Dipankar Mandal
Quantum Materials and Devices Unit, Institute of Nano Science and Technology,
Knowledge City, Sector 81, Mohali 140306, India

Fatemeh Mokhtari
Carbon Nexus at the Institute for Frontier Materials, Deakin University, Waurn
Ponds, Victoria 3216, Australia

Remya Nair
Kuwait College of Science and Technology (KCST), Doha District 13133,
Kuwait
School of Applied Sciences, Suresh Gyan Vihar University Jaipur, India

Haitao Niu
College of Textiles and Clothing, Qingdao University, Qingdao, China

Zhan Qu
College of Textile and Clothing Engineering, Soochow University, Suzhou,
Jiangsu 215021, China

Joselito M Razal
Institute for Frontier Materials, Deakin University, Waurn Ponds, Victoria 3216,
Australia

Mahshid Sadeghi
Department of biomedical Engineering, Tarbiat Modares University, Tehran, Iran

Amin Sarmadi
School of Engineering, Edith Cowan University, Joondalup, WA 6027, Australia

T Sathies
Central Research Facility, National Institute of Technology Karnataka,
Surathkal, Mangaluru 575025, India

Nader Shehata
Department of Engineering Mathematics and Physics, Faculty of Engineering,
Alexandria University, Alexandria 21544, Egypt
and
Center of Smart Materials, Nanotechnology and Photonics (CSMNP), Smart CI
Research Center, Alexandria University, Alexandria 21544, Egypt

and
Kuwait College of Science and Technology (KCST), Doha District 13133, Kuwait
and
USTAR Bioinnovations Center, Faculty of Science, Utah State University, Logan, UT 84341, USA
and
School of Engineering, Ulster University, Belfast BT15 1AP, Northern Ireland, UK

Armineh Shirali
Advanced Fibrous Materials LAB, Textile Engineering Department, Institute for Advanced Textile Materials and Technologies (ATMT), Amirkabir University of Technology, Tehran, Iran

Junzhu Tao
School of Materials Science and Engineering, State Key Laboratory of Separation Membranes and Membrane Processes, Tiangong University, Tianjin 300387, China

Md Arif Saleh Tasin
College of Textiles and Clothing, Qingdao University, Qingdao, China

Russell J Varley
Carbon Nexus at the Institute for Frontier Materials, Deakin University, Waurn Ponds, Victoria 3216, Australia

Faqiang Wang
Shanghai Frontiers Science Center of Advanced Textiles, College of Textiles, Donghua University, Shanghai 201620, China

Hongxia Wang
School of Textile Science and Engineering, Tiangong University, Tianjin 300387, China

Wenyu Wang
School of Textile Science and Engineering, Tiangong University, Tianjin 300387, China

Jianyong Yu
Innovation Center for Textile Science and Technology, Donghua University, Shanghai 200051, China

Xuekai Zheng
School of Materials Science and Engineering, State Key Laboratory of Separation Membranes and Membrane Processes, Tiangong University, Tianjin 300387, China

Hua Zhou
College of Textiles and Clothing, Qingdao University, Qingdao, China

IOP Publishing

Energy Harvesting Properties of Electrospun Nanofibers (Second Edition)

Jian Fang and Tong Lin

Chapter 1

Electrospinning: an advanced nanofiber-making technology

Haitao Niu, Md Arif Saleh Tasin and Hua Zhou

The electrospinning process has been regarded as one of the most facile and versatile techniques for preparing nanoscale fiber materials. Electrospinning technology has undergone enormous progress since its appearance as early as 1930s. It has evolved into numerous new techniques and diverse methods of making continuous fibers from viscoelastic fluid exploiting the basic electrohydrodynamic phenomenon, and has resulted in the invention of innovative electrospinning apparatuses. This chapter briefly reviews the recent improvements in the electrospinning technologies and summarizes the state-of-the-art electrospinning setups. It will be valuable resource for researchers in the electrospinning field and engineers in related areas.

1.1 Introduction of electrospinning

Nanomaterials are materials that have at least one dimension below 100 nm, e.g. nanoparticles, nanorods, nanowires, nanotubes and nanosheets. Nanomaterials have attracted considerable attention in the past decades owing to their excellent properties, outstanding performance and high application potential. As the dimensions of materials decrease, the percentage of atoms exposing to the surface and the surface-to-volume ratio grow remarkably [1]. For a nanoparticle with a diameter of 100 nm, only less than 0.2% of atoms are on the surface, while 10% of the atoms are on the surface for a 10 nm nanoparticle and around 90% of atoms are distributed on the surface of a 2 nm nanoparticle [2]. The atoms on the nanomaterial surface have more dangling bonds, which make them very active and they tend to bond with adjacent molecules. As a result, these nanomaterials, in comparison with their bulk counterparts, usually exhibit distinctive properties, e.g. higher chemical activity, lower melting point, higher phase transition pressure and solubility [3].

doi:10.1088/978-0-7503-5487-5ch1

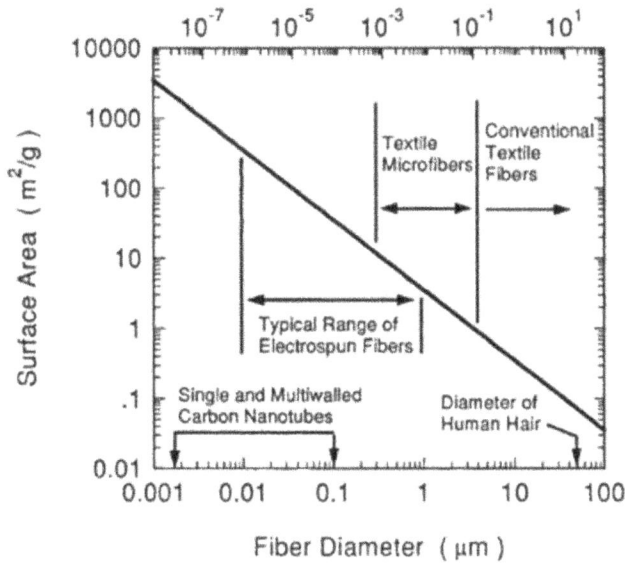

Figure 1.1. Relationship between specific surface area and diameter of different fibers. Reproduced from [11], Copyright (2001), with permission from Elsevier.

In the fiber and textile industry, fibers a with diameter of less than 1 μm (1000 nm) are normally regarded as nanofibers. In comparison with conventional microfibers, these nanofibers have significantly enlarged surface area, for instance, the surface area of electrospun nanofibers is dozens times of conventional fibers (figure 1.1). Accordingly, the pore size reduces and pore volume increases in nanofibrous materials. As a result, nanofibers show outstanding performances in many different areas, e.g. air/liquid filtration, energy generation and storage, biomedical and tissue engineering, environmental protection, sensor and catalyst, smart textiles, drug delivery and nanocomposite [4–10].

Nanofibers can be fabricated by numerous approaches, such as phase separation [12], self-assembly [13, 14], template synthesis [15], melt-blowing [16], flash spinning [17], bi-component spinning [18] and electrospinning [19, 20]. Among all these existing nanofiber-fabricating approaches, electrospinning technology is one of the most investigated and widely used, because of its high efficiency, cost-effectiveness and great adaptability. In particular, electrospinning technology is easy to be scaled up and has high commercialization potential. It can facilely prepare large length-to-diameter nanofibers with tailored fiber structures, e.g. Janus [21], core–sheath [22], hollow [23], multi-channeled [24], tri-layer core–shell [22], side-by-side [25], porous [26], etc. Moreover, small molecular substances or nanoparticles could be easily blended or embedded in electrospun nanofibers during the electrospinning process with additional or enhanced properties.

1.1.1 Electrospinning history

In as early as 1600, William Gilbert first reported the observation about a spherical water drop on a dry surface being drawn and deformed into a cone under the effect of electrostatic force [27]. Nearly 300 years after Gilbert's observation, Formhal, in the 1930s, patented setups to produce continuous fine fibers governed by electrostatic force [28–30], which are regarded as the real beginning of electrospinning technology. In the 1960s, Geffrey Taylor investigated the shape of the cone formed by a fluid droplet under the action of electric field and reported the existence of a conical angle of 49.3° [31]. This angle was later named the 'Taylor cone' and it has been extensively used to explain electrospinning and electrospray processes. In addition, Taylor [32] proposed that a jet in the parallel electric field experienced two critical instabilities: Rayleigh instability and bending instability. His works greatly improved the understanding of electrospinning and promote its development. Henceforward, there were no noteworthy improvements on electrospinning technology until Doshi and Reneker [19, 33] reported their work on using this technique to fabricate nano-structured materials. Since then, electrospinning technology and electrospun nanofibers have begun to draw more and more interest from both academic and industrial fields. Tremendous efforts have been devoted to development of the electrospinning technology, exploiting the fiber-making mechanism, nanofiber characterization, and discovering novel applications. Large-scale production and commercialization of electrospun nanofibers have been accomplished. Electrospinning technology and process have made fast advancement in recent years to meet the ever-increasing demands on nanofibers for various applications.

1.1.2 Basic apparatus

The conventional electrospinning setup (figure 1.2) is comprised of a capillary nozzle connected to a high voltage DC power supply, a grounded collector and a solution reservoir to supply the spinning solution [34]. This type of electrospinning setup works based on the capillary effect that the solution is transported to the tip of a thin nozzle and jet initiation happens at the nozzle tip. In the past over two decades, many advancements have been made to control nanofiber collection, and various

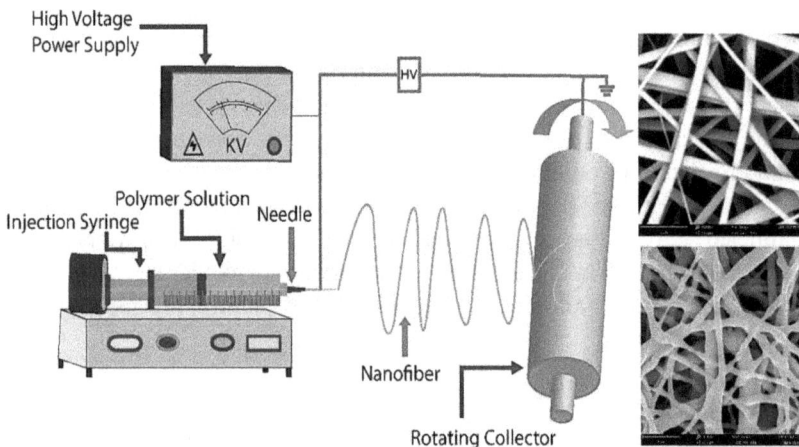

Figure 1.2. Illustration of the typical electrospinning process.

fiber generating designs including needleless, near-field, melt electrospinning, yarn electrospinning, multi-component electrospinning, e.g. have been developed. An overview of the advanced nanofiber-making technologies and state-of-the-art progress are reviewed in this chapter.

A typical electrospinning process can be briefly described as follow: spinning fluid is fed to the capillary tube (usually a syringe needle) from a fluid reservoir. When a high voltage is applied to the needle (typically around 10–50 kV), a high electric field is formed between the needle and the collector (grounded or oppositely charged), which applies electrostatic force to the fluid droplet. Electrospinning fluid is described as 'leaky dielectric' that has sufficient conductivity for the induced charges to quickly accumulate on the free surface in a short time scale or acts as a dielectric [35]. The repulsion between charges on the free surface of fluid droplet works against surface tension and fluid viscosity to deform the droplet into a cone shape (Taylor cone) with escalating applied voltage (high electric field intensity) [31]. When the applied voltage exceeds a critical value, electrostatic force can overcome fluid surface tension and jet initiation happens from the vertex of the 'Taylor cone'. The generated fluid jets fly to the grounded collector under the action of electrostatic force. During the flying process, solvent evaporation from the fluid jet (or cooling downing during the melt electrospinning process) results in dry nanofibers depositing on the collector. Although the electrospinning process is relatively easy to implement, it is very complex considering the co-existence and combined action of Coulombic force, gravity, fluid surface tension and viscosity during fiber formation.

1.2 Electrospinning basis

1.2.1 Mechanism of the electrospinning process

The fiber formation process in electrospinning can be divided into three stages: jet initiation, jet whipping, jet solidification and deposition [36]. As the first stage of electrospinning, jet initiation has been widely studied [37–39]. In the first milliseconds, the solution droplet begins to transform into a conical shape by the high electrostatic force. The round solution droplet tip becomes more and more sharp. Finally, a jet is emitted from the vertex of the cone. Then the cone shape gradually changes back to the rounded shape, indicating the system transit from the jet initiation stage to the electrospinning stage, which lasts as long as the solution consumed by the jet emitting can be replenished promptly.

After ejection, the solution jet experiences a short section of straight jet (figure 1.3), which may extend from few millimeters to several centimeters away from the nozzle tip along its axis direction. This ejected solution jet carries away electrical charges in the form of uncompensated ions from the nozzle. Attributed to the effect of charge repulsion in the solution jets, free charges migrate radially onto the jet surface to satisfy the equilibrium condition. As a result, electrostatic force induced by the electric field applies to the jet surface. The charged fluid jet accelerates under the action of electric field, accompanied by the thinning of the fluid jet [38].

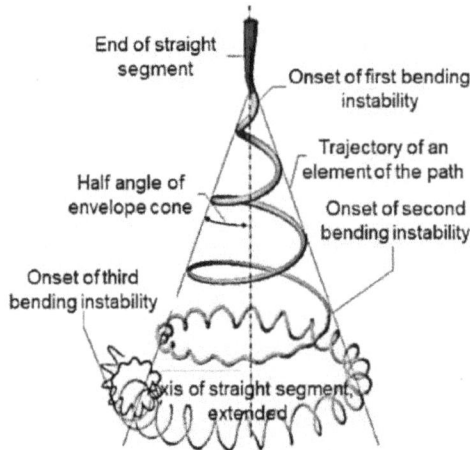

Figure 1.3. Diagram illustrating the typical instant position of an electrospinning jet with three sequential electrical bends instability. Reproduced from [38], Copyright (2008), with permission from Elsevier.

Figure 1.4. Digital photo of an electrospinning solution jet, including stable jet and jet whipping stage. Reproduced from [46], Copyright (2007), with permission from Elsevier.

After the initial steady stage, the jet enters an instability stage (whipping, as shown in figure 1.4) under the influence of the charges carried by the jet, which may involve bending, winding, spiraling and looping movements. Many theoretical models have been proposed to describe this jet instability and there is a prevalent belief that the jet is continuously elongated and becomes longer and thinner with continual stretching [40–44]. The existence of jet instability is due to the action of axisymmetric and non-axisymmetric instabilities caused by the perturbations of surface charges. The axisymmetric instability derives from perturbation of the surface charges along the jet axis direction, which makes different segments of jet

under different strengths of electrostatic force, resulting in uneven jet. The non-axisymmetric instability derives from perturbation of surface charges around circumference of the jet, which induces localized torque around the jet that accounts for whipping motion. In the condition that non-axisymmetric instability plays the major role of instability during the electrospinning process, the jet is likely to be stretched uniformly [45]. During the flying process, the solvent in the jet evaporates leaving dry fibers, which fly to the collector under electrostatic force and deposit as a form of fibrous nonwoven.

1.2.2 Effects of electrospinning parameters

Electrospinning process, fiber morphology, fibrous structure, and fiber production rate are governed by a number of parameters, e.g. applied voltage, flow rate, nozzle diameter, collecting distance, solution properties (polymer molecular weight, concentration, electrical conductivity, surface tension, solvent properties etc), and ambient conditions (temperature, humidity, etc) (figure 1.5). Although, many of these electrospinning parameters are interdependent and there are interactions between them, we can derive a general trend of influences of some parameters on the electrospinning process and fiber properties.

1.2.2.1 Applied voltage
The applied high voltage is an essential factor in electrospinning, without it jet initiation does not happen. When the applied voltage is low, the electrostatic force is insufficient to overcome surface tension of the solution droplet, as a result, no jet is stretched out and dripping happens. With the increasing applied voltage, the electrostatic force increases, and eventually leads to jet initiation and electrospinning process starts. Generally speaking, the fiber diameter decreases with increasing voltage attributable to the growing stretching force [47, 48]. Applied voltage plays a far more important role in needleless electrospinning than in the nozzle electrospinning. It can decrease the fiber diameter and increase the fiber production rate,

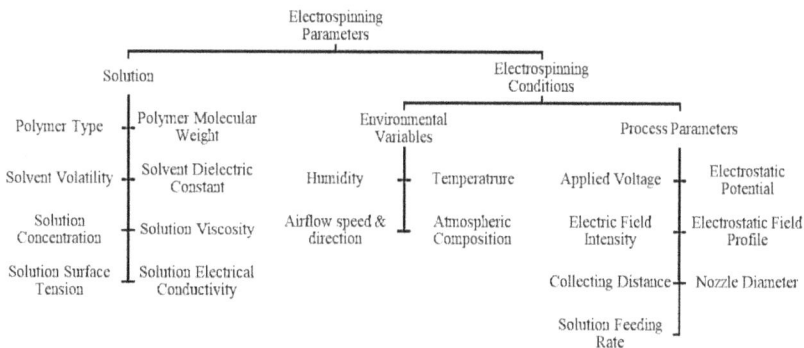

Figure 1.5. Important electrospinning parameters and conditions.

simultaneously, because high voltage can increase the number of jets emitted from the free solution surface of spinneret and escalate jet flying speed [49, 50].

1.2.2.2 Collecting distance

In general, an optimum distance for electrospinning should be long enough for fiber stretching and solvent evaporation. Under the circumstance that the applied voltage is constant, the change of distance between nozzle and collector will affect electric field intensity and fiber diameter accordingly. Another important influence of changing distance is on fiber morphology and structure. The solvent in the solution jet needs to have sufficient time to evaporate and turn the jets into dry nanofibers. When the collecting distance is very short, interconnected nanofibers are often collected [49–51]. This phenomenon happens more frequently in needleless electrospinning due to instantaneous generation of a large number of solution jets. It has also been found that interconnected nanofibers collected at a short distance could benefit the improvement of mechanical strength of nanofiber mats with improved and durable performances in the energy harvesting application [51].

1.2.2.3 Flow rate

The solution flow rate in the nozzle electrospinning has a direct effect on the fiber production rate, large flow rate results in high fiber production rate and small flow rate results in low fiber production rate. However, a large flow rate will generally produce coarser nanofibers [52], because more solution is drawn out from the nozzle tip at the same time. It has been proposed that four electrospinning phenomena may happen with the increasing flow rate, discontinuous, continuous (stable), intermittent, and dripping. When the solution flow rate exceeds a critical value, the congestion of excess solution at the nozzle tip can affect jet the formation process leading to the formation of unstable jet, beaded fibers, or other defects, such as branching, splitting and flattened fibers [53].

1.2.2.4 Solution properties

Polymer molecular weight has substantial effect on electrospinning processes. In general, increasing the molecular weight can enable a polymer solution to be electrospun into uniform fibers at relative low concentrations because high molecular weight induces a large degree of chain entanglement [54]. It was found that polyvinyl alcohol (PVA) solutions produced from beaded fiber, uniform fiber, to coarser un-uniform fibers with the increasing PVA concentration [55]. In another study, the polyamide (PA6) solution produced droplets, merged fibers, smooth fibers at 5, 15, 25 wt% concentrations, respectively [56]. When the solution concentration is very low, electrospray happens instead of electrospinning. At low solution concentrations, electrospinning usually produces defective fibers (discontinuous, merged or beaded) because the surface tension of solution overcomes the viscoelastic force and electrostatic drawing force. When the solution concentration is appropriately high, the chain entanglement of macromolecules is sufficient to overcome surface tension to generate fibers, in addition, the fiber diameter increases with the rising concentration as there is more solid content in the solution jet [54, 57]. However, a too high concentration solution makes electrospinning difficult due to the high viscoelasticity [58], especial for needleless electrospinning that stops jet production at high concentration solutions. In addition, the critical voltage for electrospinning may go up with the rising solution concentration [59]. The minimum solution concentration to produce smooth nanofibers is dependent on the polymer type, polymer molecular weight and solvent used.

The addition of surfactant and salt alters the surface tension and electrical conductivity of polymer solutions, and their influences on the electrospinning process may vary at different conditions. It has been reported that the addition of lithium chloride (LiCl), sodium nitrate (NaNO$_3$), sodium chloride (NaCl) and calcium chloride (CaCl$_2$) salts can increase solution conductivity, as a result, reducing the PAN fiber diameter and the reduction rate is proportional to solution conductivity [60]. In contrast, the addition of salt in the PA-6 solution was found to increase fiber diameter with increasing amounts of salt, which was attributed to the growing viscoelastic force within the solution jet [54]. The addition of surfactant can reduce solution surface tension, as a result, producing thinner nanofibers with improved fiber uniformity [61, 62]. The introducing of surfactant (e.g. dodecylbenzene sulfonic acid, tetrabutylammonium chloride) in the solution could also produce nanofibers with special morphologies, e.g. tree-like [63] and nano-net [64].

1.2.2.5 Temperature and humidity of environment

Electrospun nanofibers normally have smaller fiber diameters with the rising environment temperature during electrospinning because of the declined solution viscosity and surface tension [65]. When using low temperature (200 K–220 K) to prepare poly(lactic acid-co-glycolic acid) nanofibers, the porosity of electrospun nanofibers can be improved by four times because the ice crystals formed at low temperature served as a removable void template to create additional pores [66]. Humidity can also affect fiber diameter, structure and morphology [65, 67, 68]. Pelipenko *et al* [69] described the impact of humidity on PVA nanofiber diameter, it reduced from 667 ± 83 nm to 161 ± 42 nm when the relative humidity increased from 4% to 70%. In contrast, when the relative humidity was raised from 25% to 30%, the fiber diameter of silk fibroin (SF) nanofiber grew from 30 to 120 nm. In a study of electrospinning with PVA, HA, PEO, and CS polymers, it was demonstrated that higher RH level resulted in slower solvent evaporation and produced thinner nanofibers, whereas lower RH benefited solvent evaporation and produced thicker nanofibers [70]. It has been reported that polystyrene nanofibers prepared at high humidity environment form porous structure due to the breath figure formation effect, because water condensed on the fiber from gaseous phase owing to the evaporative cooling of the jet [68].

In the following section of this chapter, state-of-the-art progress of electrospinning techniques and apparatuses will be summarized (figure 1.6), covering nozzle electrospinning techniques (e.g. multi-component electrospinning, multi-nozzle electrospinning, near-field electrospinning and melt electrospinning), needleless electrospinning techniques, and methods of controlling fiber deposition (e.g. patterned nanofiber mat, aligned nanofibers, nanofiber yarns).

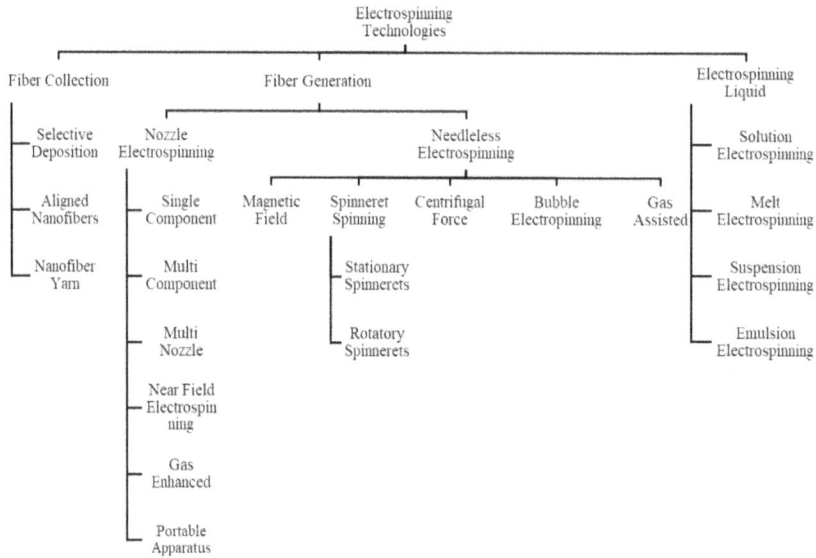

Figure 1.6. Summary of different electrospinning techniques.

1.3 Nozzle electrospinning

In a nozzle spinneret-based electrospinning process, a spinning solution is transported to the jet initiation position through a capillary channel or multiple channels. This kind of electrospinning technique has many unique features, e.g. enabling incessant solution feeding and continuous electrospinning, enclosed system avoiding unnecessary solvent evaporation and maintaining the solution stability, tunable capillary diameter with easily controlled fiber diameter, producing uniform nanofibers with narrow fiber diameter distribution.

1.3.1 Single-component electrospinning

Single-component electrospinning refers to the circumstance that spinning solution is supplied through a single capillary nozzle, and jet initiation occurs at the nozzle tip when high voltage is applied. Although positive high voltage is generally used in electrospinning, negative high voltage has also been applied for making electrospun nanofibers. Tong *et al* investigated the effect of high voltage (HV) polarity on fiber quality of poly(hydroxybutyrate-co-hydroxyvalerate) (PHBV) nanofibers. They noticed that the PHBV fiber diameter increased with the rising applied voltage for positive voltage electrospinning but decreased for negative electrospinning, however, water contact angle, tensile strength and stiffness of the nanofibers were barely affected by the HV polarity [61]. In another work of electrospinning zein/soy protein isolate (95/5), the diameter of positive electrospun nanofibers was smaller than that of negative electrospinning. With the increasing applied voltage, positive electrospinning produced thinner nanofibers and negative electrospinning did not show any obvious effect on the fiber diameter [71].

In addition to the conventional setup that high voltage is connected to the nozzle spinneret, high voltage can also be connected to the collector to perform electro-spinning. It is working based on the inducting effect from the high potential applied collector. It has been found that when high voltage is connected to the collector and nozzle is grounded, electrospinning proceeds successfully but requires higher critical voltage to start the spinning process, also resulting in enlarged fiber diameter and low fiber production rate [72]. Direct current (DC) voltage is usually used for electrospinning that the high voltage power supply has either positive high voltage output or negative output. An alternating current (AC) voltage can not only implement electrospray/electrospinning, but also has better controllability on fiber deposition. It has been found that AC electrospinning of poly(ethylene oxide) (PEO) significantly subsided the whipping phenomenon with a high degree of fiber alignment [73]. Unlike the conventional DC electrospinning, in which a funnel shaped nanofiber mesh is formed, a visible thread can be observed emerging downstream from the needle in the AC electrospinning process. And the fibrous thread is not attracted by the grounded electrode but can be easily deflected away and drawn [74].

There are another two electrospinning techniques, emulsion and suspension electrospinning [75]. Sanders *et al* demonstrated the preparation of polymer fibers by emulsion electrospinning of poly (ethylene-co-vinyl acetate) with dichlorome-thane mixed with bovine serum albumin into aqueous phosphate buffer saline [76]. Zhao *et al* have fabricated the high purity membranes of polytetrafluoroethylene/ polyethene oxide (PTFE/PEO) by emulsion electrospinning using a high temper-ature sintering method, which were subsequently employed across a polyamide 6 (PA6) membrane constructing vertically contact-mode triboelectric nanogenerators (TENGs) [77]. Ru *et al* used PLGA nanofiber prepared by the suspension electro-spinning method to manufacture a scaffold. The keratinocyte cells sown on this scaffold can suspend within the cell culture medium, allowing them to be exposed to air needed for skin tissue creation [78]. By suspension electrospinning, Gonzalez *et al* studied how the morphology of the MMA/BA copolymer with the PVA polymer template was impacted by the molar mass, particle ratio, and size distribution [79].

1.3.2 Multi-component electrospinning

In the textile industry, a single fiber with two components distributed inside in the radial direction, for example core–sheath and islands-in-the-sea structures, can find many unique applications. Although it is possible to obtain core–sheath structured nanofibers induced by phase separation [80] or emulsion electrospinning [81, 82], the normal single component electrospinning nozzle usually produces nanofiber without obvious core–sheath structure. Attributable to the fact that two solutions inside the nozzle (diameter < 1 mm) have small Reynolds numbers, they can be identified as laminar flows and flow independently without mixing, as a result, they can maintain their initial injection state [83]. Because of this, it is possible to conduct bi-component electrospinning and fabricate nanofibers with tunable composition in the radial direction [84].

In a core–sheath electrospinning spinneret, an inner nozzle of small diameter and an outer nozzle of large diameter are positioned axially with the inner nozzle slightly protruding out of the outer nozzle. During the electrospinning process, the sheath solution is fed into the outer nozzle and the core solution is fed into the inner nozzle. At the nozzle tip, the outer solution forms a thin sheath that encloses the inner solution. Under the action of electrostatic force, both solutions are pulled into a compound Taylor cone of core–sheath structure. Then the core–sheath solution jet is stretched into thin core–sheath fibers by the electric field and deposited on the collector [84–86]. When the inner solution is replaced with a liquid (e.g. octane), hollow nanofibers (figures 1.7(a) and (b)) can be obtained after removing the liquid from resulted in the core–sheath nanofibers [85]. Benefiting from the capability of core–sheath electrospinning technique, many un-spinnable materials, e.g. polydimethylsiloxane (PDMS) [87] and medicine [88], can be successfully processed into ultrathin fibers.

When two components distribute inside the bi-component electrospinning nozzle in a side-by-side way, the fabricated nanofibers have a side-by-side structure [89, 90]. Interestingly, self-crimped polyacrylonitrile nanofibers (figure 1.8) were obtained

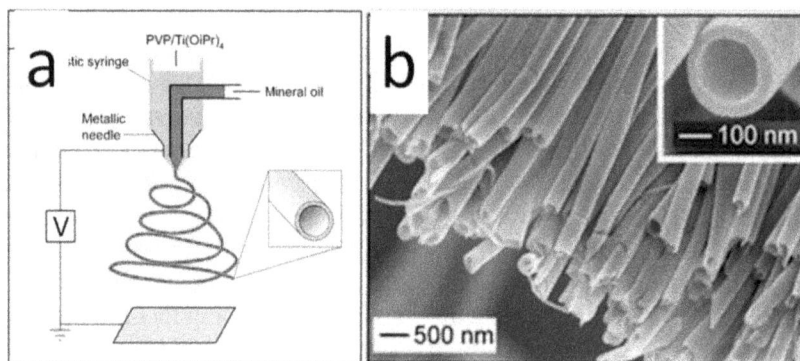

Figure 1.7. (a) Core–sheath electrospinning setup and (b) SEM images of electrospun hollow nanofibers. Reproduced from [85] with permission from Royal Society of Chemistry.

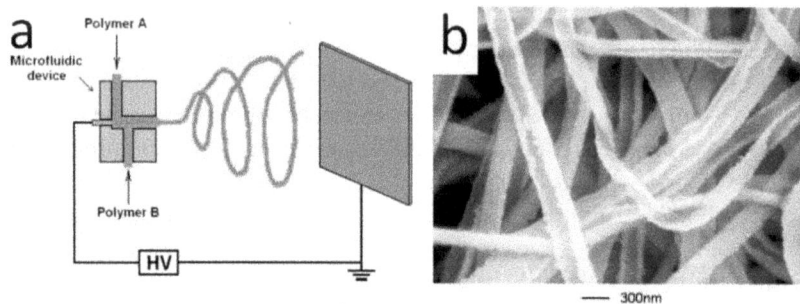

Figure 1.8. (a) Side-by-side electrospinning setup, (b) SEM image of side-by-side bi-component nanofibers with one component removed. Reproduced with permission from [92] John Wiley & Sons. Copyright © 2005 WILEY-VCH Verlag GmbH & Co. KGaA, Weinheim.

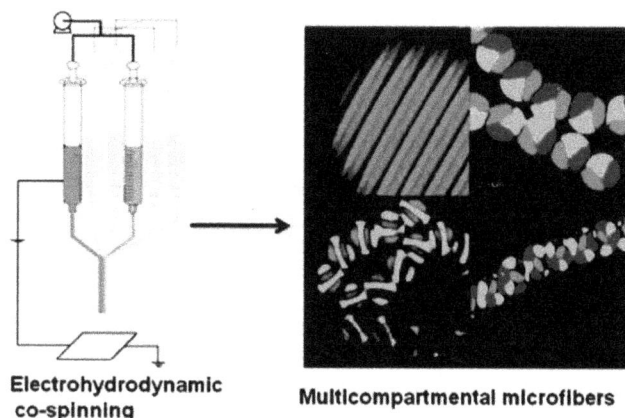

Figure 1.9. Schematic drawing of a multi-compartmental electrospinning setup and prepared fibers. Reproduced from [93] with permission from American Chemical Society.

when the polyurethane component in polyacrylonitrile/polyurethane side-by-side nanofibers were removed, which could provide an efficient way to tailor the structure of electrospun nanofibers [91].

Multi-component nanofibers can (figure 1.9) be obtained via simply increasing the number of components inside the spinneret [93]. However, further increasing the component number to produce islands-in-the-sea structure may be difficult, because the diameter of nozzle is extremely small (< 1 mm) that makes it impractical to manufacture a more complex spinneret and also demand large pressure to feed all the components evenly.

1.3.3 Multi-nozzle and porous spinneret

The single nozzle electrospinning setup generally has a low nanofiber productivity (<0.3 g h^{-1} nozzle), which is far below the requirement for nanofiber industrial production. A direct and practical way of improving nanofiber production rate is to increase the number of nozzles for electrospinning, namely multi-nozzle electrospinning [94–97]. However, there is strong electrostatic repulsion between the adjacent jets in multi-nozzle electrospinning, which can easily affect the electrospinning process, causing solution dripping, poor fiber quality and fibrous membrane unevenness (figure 1.10(a)). Different nozzle position arrangements have been developed to reduce electrostatic repulsion and improve the multi-nozzle electrospinning performances [98]. The application of auxiliary electrode can interfere with the electric field in multi-nozzle electrospinning and improve its fiber distribution (figures 1.10(b) and (c)), as a result improving the electrospinning performances [99–101]. The presence of the external electrode could shrink the fiber deposition area and improve the fiber production rate. In addition, the multi-nozzle stage holding numerous nozzles has been applied to improve the fiber deposition uniformity by stage movement [102].

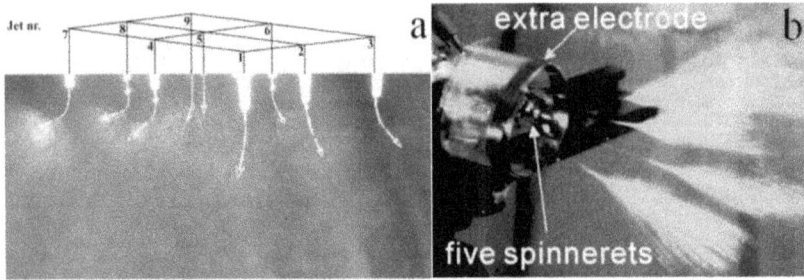

Figure 1.10. (a) A multi-nozzle electrospinning setup. Reproduced from [98], copyright (2005), with permission from Elsevier. (b) Cylindrical auxiliary electrode assisted multi-nozzle electrospinning. Reproduced from [101], copyright (2006), with permission from Elsevier.

Figure 1.11. Horizontal tube electrospinning. Reproduced from [103], copyright (2008) with permission from Elsevier.

Using a cylindrical tube with channels throughout the tube wall is an effective way to increase the solution channel number, thus increasing the fiber production rate [103, 104]. This is still based on the capillary effect that transports the spinning solution from inside of the tube to the outside and the electrospinning happens on the tube surface. Tube electrospinning exhibits improved electrospinning productivity in comparison with the single needle electrospinning, attributable to numerous solution channels (figure 1.11). Although it is possible to improve the nanofiber production rate easily by increasong the tube length and number of channels, the gap between adjacent channels should not be too small because strong jet interference can even result in a nanofiber belt instead of fiber web [103].

Many efforts have been devoted to improve the multi-nozzle spinneret design, for example, Jiang *et al* [105] used the ANSYS software to optimize electric field and developed a unique spinneret structure with nozzles organized in an arc array that generated more uniform electrical field than linear arranged nozzles, as shown in figure 1.12. In electrospinning, the ejected solution jet carries a large amount of charges, which drive the jet stretching and fiber deposition on the collector. The interference among solution jets in multi-jet electrospinning cannot be completely eliminated. In addition, solution and electrospinning conditions (e.g. solution type, solution concentration, applied voltage) have influences on the electrostatic

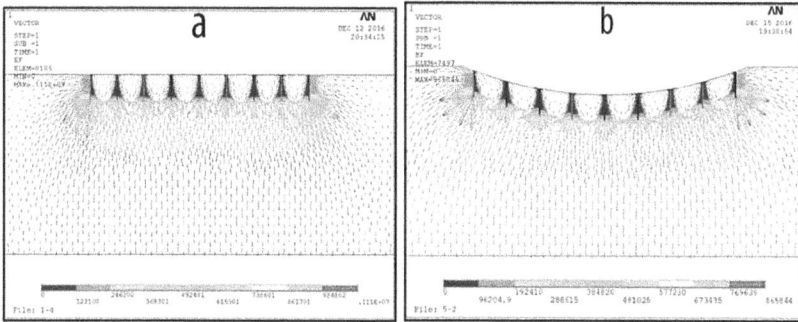

Figure 1.12. A nozzle array with two alternative configurations is simulated in an electrical field: (a) linear array and (b) arc array. Reproduced with permission from MDPI [105] CC BY 4.0.

repulsion between ejected jets, so it is difficult to find an optimized distance between solution channels to be adaptable to all the electrospinning circumstances. If not resolved successfully, the electrostatic repulsion problem will be a serious barrier to the industrialization of multi-jet electrospinning technique.

1.3.4 Near-field electrospinning

Near-field electrospinning can be regarded as a technique of integrating nano-lithography [106, 107] and electrospinning, a good example of interdisciplinary technological convergence. In a normal electrospinning system, the collecting distance is usually over 5 cm and the whipping stability is evident. Near-field electrospinning working based on much lower applied voltage can precisely control electrospun fiber deposition [108, 109]. In near-field electrospinning, the jet whipping instability is eliminated or greatly restricted due to a very short collecting distance (<5 cm). When a collector is controlled by a computer program with precisely positioned movement, the collected fibers can form predesigned patterns. Near-field electrospinning also has a few features, such as it produces fibers with a much larger diameter than that in conventional electrospinning due to insufficient jet stretching, and it has a small nanofiber production rate. The differences between near-field electrospinning and conventional electrospinning are compared in table 1.1.

Figure 1.13(a) illustrates a near-field electrospinning process and the SEM image in figure 1.13(b) shows a collected nanofiber. The nanofibers fabricated by near-field electrospinning have narrower diameter distribution than those from the conventional electrospinning process. This technique can greatly expand the application fields of electrospun nanofiber with ordered structures, e.g. nano-generator, tissue engineering, wearable sensors, and microelectromechanical systems (MEMS) [112–116].

Because of the precise fiber deposition in near-field electrospinning, it is feasible to fabricate patterned 3D fibrous structures [114]. These fibers prepared by near-field electrospinning show precise deposition and they have formed well-organized 3D fibrous structures at microscale (figure 1.14) [117].

Table 1.1. Comparison of conventional electrospinning and near-field electrospinning [110, 111].

	Conventional electrospinning	Near-field electrospinning
Material	Solution, polymer melt	Solution, polymer melt
Applied voltage (kV)	10–50	0.2–12
Collecting distance (mm)	50–500	0.5–50
Feeding rate (ml·h^{-1})	0.1–1500	0.03–2
Fiber diameter (μm)	0.01–10	0.05–200
Advantages	Simple setup, thin fibers, high productivity	Low voltage, 3D structure, precisely controlled fiber deposition
Disadvantages or challenges	Unable to create 3D structures, collect randomly distributed fibers	Low productivity, complicated technique

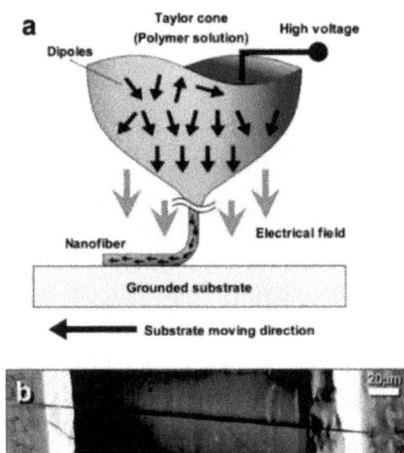

Figure 1.13. (a) Schematic diagram of near-field electrospinning process and (b) SEM image of a single polyvinylidene fluoride nanofiber formed across two electrodes. Reproduced from [112], copyright (2010), with permission from American Chemical Society.

1.3.5 Gas enhanced electrospinning

Applying auxiliary gas to an electrospinning process can diminish the influence of Coulombic repulsion force on multi-nozzle electrospinning [118]. As a result, gas enhanced electrospinning can produce thinner nanofibers [119–121], improve nanofiber collection [122–124], or increase the nanofiber production rate [125]. It has been found that the application of nitrogen gas containing solvent vapour can help to eliminate the whipping motion of the solution jet and lead to the collection of highly aligned nanofibers on the fast-rotating cylindrical collector [122]. In another work, Zs (Zetta spinning) electrospinning system was developed that can process both

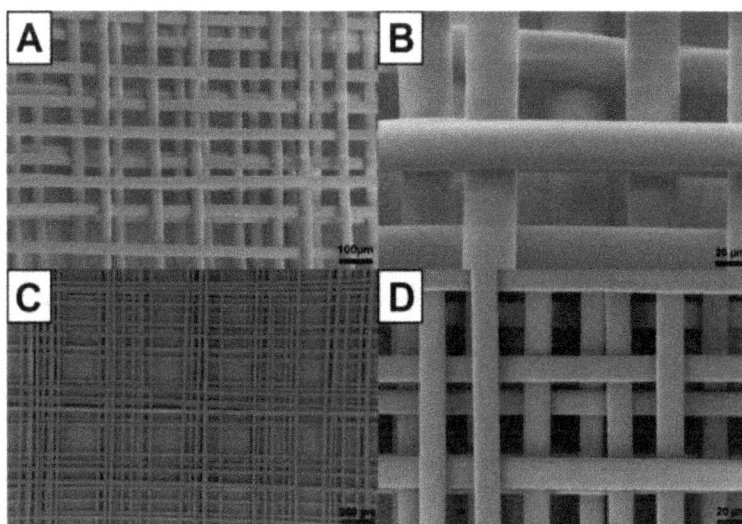

Figure 1.14. SEM images of deposited poly(2-ethyl-2-oxazoline) fibrous structures. Reproduced from [117], copyright (2014), with permission from Elsevier.

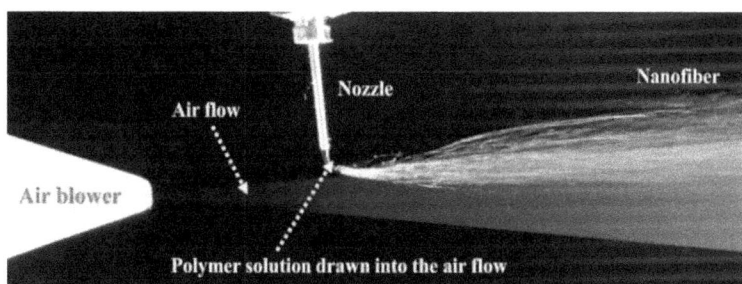

Figure 1.15. Zetta electrospinning process. Reproduced from [125], copyright (2016), with permission from American Chemical Society.

polymer solution and polymer melt into nanofibers. Figure 1.15 shows that Zetta electrospinning uses airflow to enhance electrospinning with impressively improved nanofiber production rate [125].

When preparing thick nanofiber membranes by extending the electrospinning time, it is very easy to build up electrostatic charges on the collector because polymer nanofibers are electrically non-conductive. These accumulated electrostatic charges can weaken the electric field, prevent fiber deposition and result in coarser fibers. Solvent accumulation inside the nanofibrous membrane is another challenge in electrospinning a polymer solution for a long period of time. In this regard, gas enhanced electrospinning can efficiently solve these problems [126]. Airflow has also

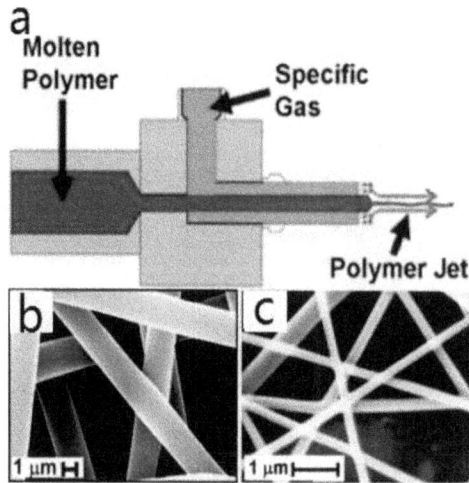

Figure 1.16. (a) Schematic diagram of gas enhanced melt electrospinning, SEM images of fabricated polylactic acid fibers, (b) without gas-assisted system and (c) with gas-assisted system. Reproduced from [119], copyright (2010), with permission from Elsevier.

been applied to assist the melt electrospinning process (figure 1.16), resulting in 10% thinner polylactic acid fibers than un-assisted melt electrospinning [119].

1.3.6 Melt electrospinning

Melt electrospinning technology has been reported in as early as 1936 in a patent filed by Charles Norton from the Massachusetts Institute of Technology [127]. While solution electrospinning uses flammable, toxic solvents, its operation can cause environmental issues and hazard risks, the melt electrospinning could play a more important role in nanofiber production. However, in spite of the early appearance, melt electrospinning had not drawn as much attention as solution electrospinning in the earlier years, for the reasons that it needs to maintain elevated temperature during the electrospinning process, and polymer melts have low conductivity and high viscosity. In recent years, many advances have been made in the heating method of melt electrospinning, e.g. electrical heating [128, 129], heating air [130, 131], circulating fluid [132] and laser heating [133, 134]. Melt electrospinning has shown many merits that solution electrospinning doesn't have, e.g. ability to process insoluble polymers or polymers (e.g. polyethylene, polypropylene) dissoluble only in highly toxic solvents into ultrathin fibers, solvent-free end product, low impact on the environment, which make it favourable for various applications, e.g. filtration [135], sensor [136], textiles [137], especially biomedical applications [132, 138–141]. Figure 1.17 illustrates a melting electrospinning process to prepare poly-hydroxymethylglycolide-co-ε-caprolactone-based fiber scaffolds for the potential cardiac tissue engineering application [141].

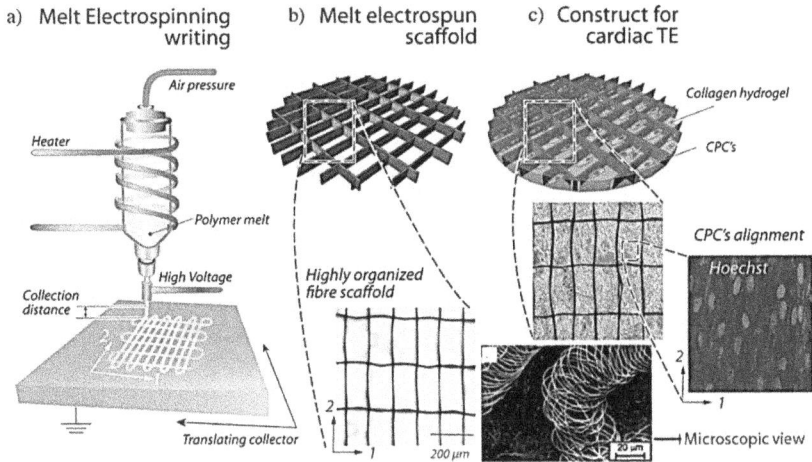

a) Melt Electrospinning writing
b) Melt electrospun scaffold
c) Construct for cardiac TE

Figure 1.17. A Schematic diagram of a melt electrospinning procedure, fiber morphology and tissue scaffold application. Reproduced from [141] John Wiley & Sons. Copyright 2017 The Authors. Published by WILEY-VCH Verlag GmbH & Co. KGaA, Weinheim.

1.4 Needleless electrospinning

In spite of great efforts to improve electrospinning productivity, conventional capillary electrospinning can only produce a very limited quantity of nanofibers, being unable to meet the ever-increasing demand for industrial applications. Needleless electrospinning offers a solution to this problem. Needleless electrospinning (also referred to as free surface electrospinning) is a special type of electrospinning technique. Instead of forming a jet from a capillary tip, needleless electrospinning forms a large number of solution jets directly from an open liquid surface. The needleless electrospinning concept appeared as early as 2004 [142], when Yarin and Zussman reported upward needleless electrospinning of nanofibers from a two-layer liquid system. Since then, this technology has attracted tremendous interest attributable to its ability for large-scale nanofiber production. During a needleless electrospinning process, numerous jets are generated instantaneously from the spinneret surface exempt from the influence of capillary effect [143]. The electrospinning liquid (electrically conductive) on the spinneret surface self-organizes on a mesoscopic scale to form waves, and jet initiation happens from wave crests when the applied voltage exceeds a critical value. The electric field intensity profile around the spinneret and in the electrospinning zone plays a far more important role in needleless electrospinning than in needle electrospinning, in terms of jet initiation, jet stretching, nanofiber productivity and morphology. Most of the works on needleless electrospinning have been focusing on understanding how spinneret structure affects electrospinning process and nanofiber production. It has been established that large curvature can generate a high-intensity electric field, which increases the fiber production rate. Based on their motility, needleless electrospinning spinnerets can be classified into three categories: stationary needleless spinneret, linear-moving needleless spinneret and rotary needleless spinneret. In addition to these three types of needleless spinnerets, there are some other needleless electrospinning designs working on different jet-initiating mechanisms.

1.4.1 Stationary needleless spinnerets

Stationary needleless spinnerets refer to those spinnerets that don't move during the electrospinning process while the spinning solution is fed onto them or the spinnerets work in a batch mode. This kind of electrospinning setup is relatively simple and the electrospinning processes can be implemented with minimum effort. Spinnerets in this category include, wire, twisted wire, conical wire, bowl, sharp edge, cylinder, slit, curved slot, stepped pyramid and cleft (shown in table 1.2).

It is worth introducing the linear needleless spinneret, the jet initiation sites on linear needleless spinnerets are distributed in a parallel way to the linear direction of spinnerets, and a typical design is the wire electrospinning technique. Although the wire spinneret remains stationary, there is a solution feeding device moving linearly along the wire to feed the spinning solution evenly along the wire ensuring

Table 1.2. A summary of stationary needleless electrospinning spinnerets and their spinning process.

Spinneret	Fiber generation	Fiber generation area	References
Cleft	(Reproduced with permission from AIP Publishing)	Cleft surface	[143]
Conical coil	(Reproduced with permission from John Wiley & Sons. Copyright 2009 Society of Plastics Engineers)	Coil wire surface and gap	[50]

(Continued)

Table 1.2. (*Continued*)

Spinneret	Fiber generation	Fiber generation area	References
Plate	**Edge-plate geometry** (Copyright (2010), with permission from Elsevier)	Plate edge	[144]
Bowl	 (Copyright IOP Publishing. All rights reserved)	Bow edge	[145]
Slit	 (Copyright (2015) CC BY 4.0)	Slit	[146–149]
Stepped pyramid	 (Copyright (2013), with permission from Elsevier)	Pyramid edge	[150, 151]

Twisted wire	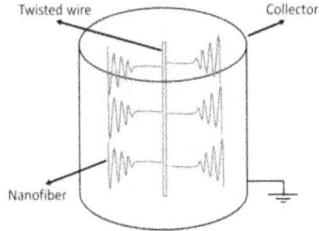	Wire surface	[152]
	(Copyright IOP Publishing. All rights reserved)		
Curved slot	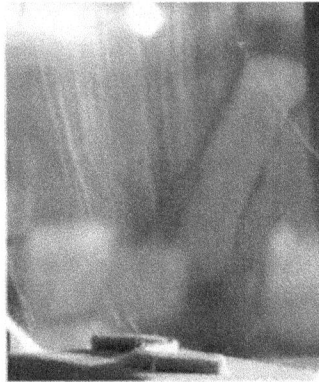	Slot	[153]
	(Reproduced with permission from Springer Nature)		
Cylinder spinneret	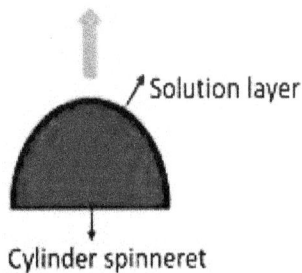	Cylinder surface	[154]
	(Reproduced with permission, copyright (2012) Haitao Niu and Tong Lin)		

continuous electrospinning. With the help of this technology, nanofibrous substances can be produced at uniformity and capacity adequate enough for development on an industrial level [155]. In another design, a horizontal bead chain was used to produce nanofibers and the chain moved in parallel to the flat collector. The beads, as an auxiliary structure, can centralize the electric field and improve electrospinning performances [156].

1.4.2 Rotatory needleless spinnerets

During the electrospinning processes of using stationary spinnerets, there is no effective control over the solution distribution on the spinneret surface. As a result, electrospinning conditions and fibrous membrane uniformity may vary with electrospinning continuing. When rotatory needleless spinnerets are used for electrospinning, the spinnerets can spread the solution evenly onto their surface through rotation and ensure continuous electrospinning. The solution layer thickness can be easily regulated by the rotating speed with improved electrospinning stability and nanofiber uniformity. Jet initiation in the needleless electrospinning process can be summarized in four stages: (1) a thin solution layer is formed on the spinneret surface because of the spinneret rotation; (2) rotation causes perturbations on the solution layer inducing the formation of conical spikes; (3) when high voltage is applied, these spikes centralize electric force intensifying perturbation to form 'Taylor cones'; (4) solution jets are stretched out from 'Taylor cones', resulting in nanofiber formation. Table 1.3 lists the common rotatory spinnerets and fiber generating sites, including cylinder (roller), disc, ball, coil, cone and wire frame.

Table 1.3. A summary of rotatory needleless electrospinning spinneret and their spinning process.

Spinneret	Fiber generation	Fiber generation area	References
Cylinder roller	(Reproduced with permission from John Wiley & Sons, copyright (2009) Wiley Periodicals, Inc.)	Cylinder surface	[49, 157]
Disc	(Reproduced with permission from John Wiley & Sons, copyright (2009) Wiley Periodicals, Inc.)	Disc rim	[49]

Ball	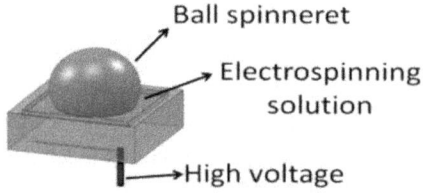	Ball surface	https://sncfibers.com [154]
	(Reproduced with permission, copyright (2012) Haitao Niu and Tong Lin)		
Coil	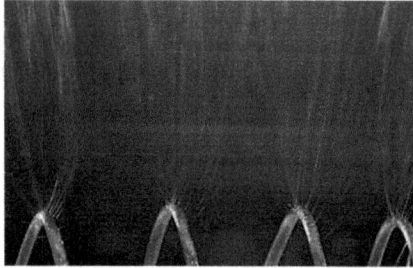	Coil wire surface	[158–160]
	(Reproduced with permission from John Wiley & Sons, copyright (2010) Wiley-VCH Verlag GmbH & Co. KGaA, Weinheim)		
Cone	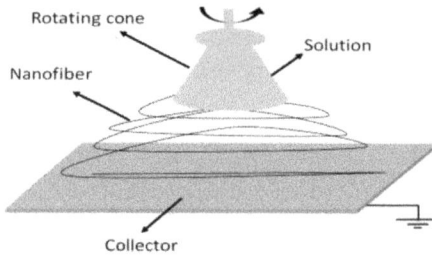	Cone edge	[161]
Wire frame	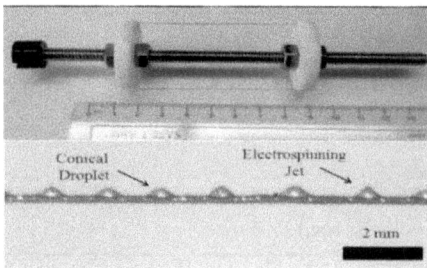	Wire surface	[162–164]
	(Copyright (2012), reproduced with permission from Elsevier)		

(*Continued*)

Table 1.3. (*Continued*)

Spinneret	Fiber generation	Fiber generation area	References
Beaded chain	Beaded chain spinneret · Electrospinning solution · High voltage (Reproduced with permission from Wiley, CC BY 3.0)	Beaded chain	[154]
Cylinder	Solution distributor, Positive pole, Collector 6, 4, 7, High-voltage electronic generator, Polymer solution container, 2, Negative pole 5, Pump 1 (Reproduced with permission from John Wiley & Sons, copyright (2010) Society of Plastics Engineers)	Cylinder surface	[165]

With the help of finite element analysis (FEA), the electric field intensity profile in electrospinning area can be simulated and used for optimizing spinneret design [49, 158, 159]. It has been found that the auxiliary structure on the primary spinneret structure can centralize the electric field around the auxiliary structure with increased intensity. Therefore, auxiliary structures were introduced onto common needleless spinnerets to enhance electric field strength and improve electrospinning performances, e.g. Von Koch curve fractal structure [166], needles on disk or helix slice [167, 168], barbed roller [169], probed cylinder [170] and threaded rod [171]. Wang *et al* identified that coil electrospinning employing a rotating helical coil can generate a strong electric field across the coil wire, as a result, producing a large number of nanofibers from the wire surface [159]. In another work, by winding a two-level copper wire coil around the fundamental helical coil structure of the spinneret, Niu *et al* created a unique needleless electrospinning spinneret and observed that the secondary coil exhibited a small effect on the fiber diameter while increasing electrospinning yields by approximately around 170% and decreasing jet initiation voltage by around 4 kV [160].

With the fast advances in electrospinning technology, commercialized needleless electrospinning technologies are already on the market (e.g. Elmarco, Fanavaran Nano-meghyas company, Revolution Fibers, SPUR company, Shanghai Yuntong Nanomaterials Technology Co. Ltd, Stellenbosch Nanofiber Company, INOVENSO), they produce nanofibers based on different mechanisms but all have a large-scale nanofiber production capacity, as shown in table 1.4.

Table 1.4. Commercialized needleless electrospinning machines.

Technology	Figure	Company name	References
Wire electrode	(Copyright: Elmarco s.r.o., Nanospider Technology, available at: http://www. elmarco.com/)	Elmarco	[155]
Spraying nozzle	(Reproduced with permission)	Fanavaran Nano-meghyas	[172]
Ball spinneret	(Reproduced with permission, SNC BEST™ Ball Electrospinning Technology - Bridging the Gap Between Nanofiber Innovation and Industrial Scale Materials)	Stellenbosch Nanofiber Company	[173]

(Continued)

Table 1.4. (*Continued*)

Technology	Figure	Company name	References
Coil spinneret	 (CC BY 3.0)	Deakin University	[159]
Cylindrical drum	 (Reprinted with permission, SKE Research Equipment, Italy - http://www.ske.it)	SKE Research Equipment	[174]
Sonic electrospinning	 (Reprinted with permission, copyright NanoLayr Ltd http://www.nanolayr.com)	Nanolayr (Revolution Fibers)	[175]

1.4.3 Magnetic field-assisted needleless electrospinning

Apart from the above-mentioned needleless spinnerets that rely on spinnerets with specific geometric shapes to conduct electrospinning, there are also a number of needleless electrospinning technologies that utilize additional forces such as gas blowing, magnetic field or centrifugal force to implement electrospinning. Yarin and Zussman [142] reported an electrospinning technique that used a magnetic field to initiate the jet formation (figure 1.18). The spinning liquid comprised two layers, bottom ferromagnetic fluid layer and the top polymer solution layer. When an external magnetic field was applied to the ferromagnetic fluid and electric field was applied to the polymer solution, the ferromagnetic fluid triggered the formation of

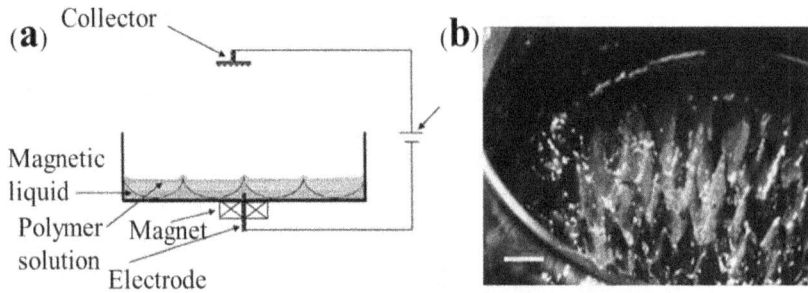

Figure 1.18. (a) Schematic drawing of magnetic field-assisted needleless electrospinning. (b) Spikes formed on the silicone oil-based magnetic fluid under the action of a permanent magnet. Reproduced from [142], copyright 2004, with permission from Elsevier.

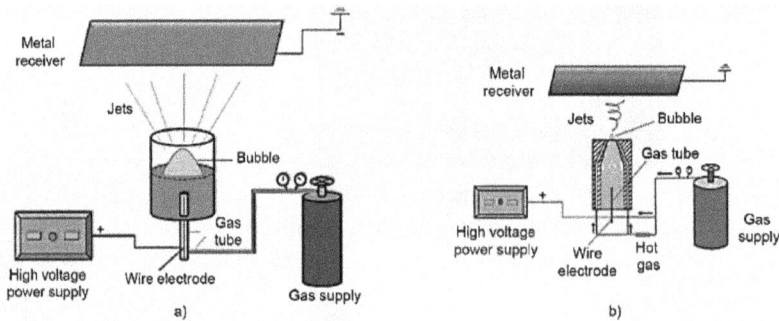

Figure 1.19. Schematic diagram of bubble electrospinning setup. (A) Bubble electrospinning and (B) blown bubble electrospinning. Reproduced with permission from Thermal Science [180].

steady vertical spikes, which perturbed the interlayer interface and solution layer accordingly. When the applied voltage was high enough, solution jets were stretched out from the spikes.

1.4.4 Bubble needleless electrospinning

Because of low cost and high safety, airflow has been used to assist jet initiation in electrospinning, as shown in figure 1.19. A gas-jet electrospinning technique (also referred to as bubble electrospinning) was developed in 2007 [176], it used gas to create bubbles on the liquid surface, which increased surface curvature and facilitated jet initiation. Since the introduction of this, gas initiated electrospinning has attracted great interest. In nozzle electrospinning [177] and needleless electro-spinning [178], high speed gas can improve the nanofiber production rate, due to the effect that airflow could enhance solution jet stretching facilitating jet initiation. In addition to the gas-assisted solution electrospinning, melt electrospinning can also benefit from additional airflow [119, 179].

1.4.5 Centrifugal force assisted needleless electrospinning

Centrifugal spinning has been developed for many years, it was originally extensively used for producing glass fibers [181]. Recently, centrifugal spinning has been applied to assist electrospinning to prepare nanofibers, as shown in figure 1.20 [182–184]. Many parameters in centrifugal electrospinning, e.g. voltage, spinneret rotation speed, solution feed rate, the distance between spinning head and collector and solution concentration, can affect the spinning process and nanofiber quality. The combination of centrifugal force and electric field makes it very effective to fabricate aligned nanofibers as well as randomly distributed nanofibers (figure 1.20) [185–187]. The combination of centrifugal spinning and electrospinning has many advantages [187, 188]:

- For high concentration solution or polymer melt, it is hard to make jet initiation in normal electrospinning because of the high viscosity, however centrifugal force can easily transport these fluids and initiate jet ejection.
- Large nanofiber production rate.
- Small fiber diameter.
- Low jet initiation voltage.

1.4.6 Air enhanced needleless electrospinning

The applied voltage needed to start jet initiation and electrospinning in needleless electrospinning is usually much higher than that in conventional needle electrospinning, because it requires much higher electrostatic force to stretch out a solution jet from the small curvature surface of needleless spinneret [49]. In a recent work of electroaerodynamic-field-aided needleless electrospinning, two auxiliary forces of additional electric field and airflow were used to assist electrospinning process (figure 1.21). The high-intensity electric field was generated between the slot and the inductive electrode (5 cm distance). In this way, electrospinning runs at a voltage equivalent to that in needle electrospinning (e.g. 10–30 kV) [190]. The auxiliary airflow diverted the nanofibers away from the inductive electrodes and directly to the collector.

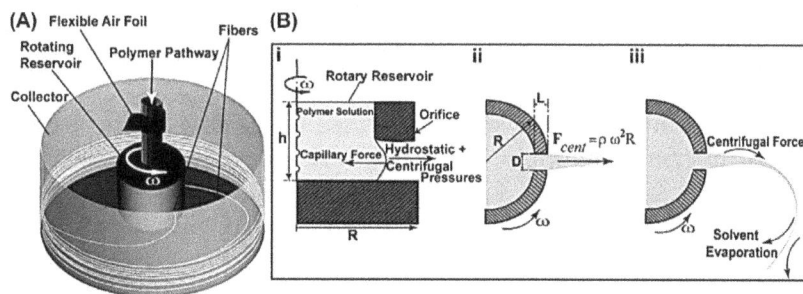

Figure 1.20. (a) Schematic diagram of centrifugal electrospinning process. (b) The mechanism of nanoscale fiber production by centrifugal spinning. (i) Jet ignition, (ii) jet extension and (iii) solvent evaporation. Reproduced from [189], copyright (2019), with permission from Elsevier.

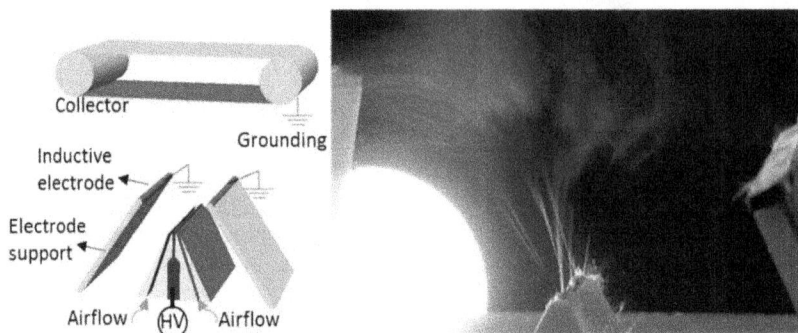

Figure 1.21. Schematic drawing of electroaerodynamic needleless spinning setup and its electrospinning process. Reproduced with permission from IOP Publishing [190], copyright IOP Publishing, all rights reserved.

1.5 Nanofiber collection

Electrospun nanofibers are generally collected as randomly distributed nanofiber mats, the morphology of which is similar to a nonwoven fibrous mat. In lots of literature, such a fibrous structure was described as a fiber web, fiber sheet, nonwoven or membrane. To meet many specific applications, nanofibers are also collected into different forms besides the nonwoven structure. Numerous setups have been developed to manipulate nanofiber deposition. As a result, aligned nanofibers, nanofiber yarn and, 3D nanofiber structures have been achieved.

1.5.1 Selective nanofiber deposition

Due to the large number of electrostatic charges carried by nanofibers, there is significant Coulombic repulsion force between these fibers. As a result, the nanofiber deposition area on the collector is usually large with low fiber collection efficiency. Auxiliary electric field could improve the control over nanofiber deposition [99, 100]. Charged rings have been used to restrain nanofiber disposition in a small area [191], the schematic diagram of this electrospinning setup is shown in figure 1.22. When an auxiliary electrode was used in electrospinning to control fiber deposition, nano-fibers in a 2D pattern or 3D structure could be fabricated [192, 193]. During the electrospinning process, the Taylor cone was stabilized and the jet whipping was converged sufficiently [192]. This method could provide a practical strategy for the fabrication of nanofiber with elaborate structures.

In addition, when a patterned substrate is used to collect nanofibers, the electric field doesn't distribute uniformly on the substrate, the collected nanofibers show predesigned arrangement on the substrate [194–196]. Figure 1.23(a) shows that the tips on the substrate can centralize the electric field, and as a result, more nanofibers are drawn to these tips and collected PEO nanofibers show a patterned structure (figure 1.23(b)).

Electrospinning nanofibers onto columnar shapes collector, instead of flat collector as usual, tubular structured nanofibers can be obtained [197, 198], which will have promising potential in biomedical and industrial applications. Figure 1.24

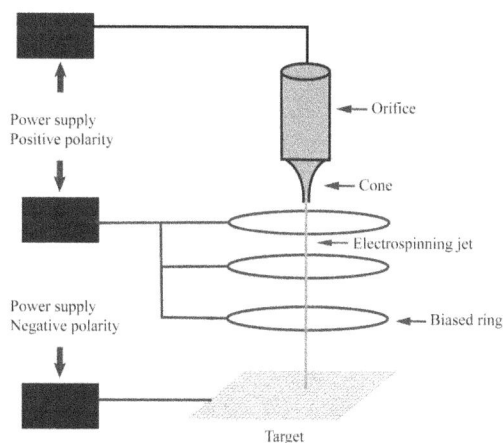

Figure 1.22. Auxiliary electric field-assisted fiber deposition. Reproduced from [191], copyright (2001), with permission from Elsevier.

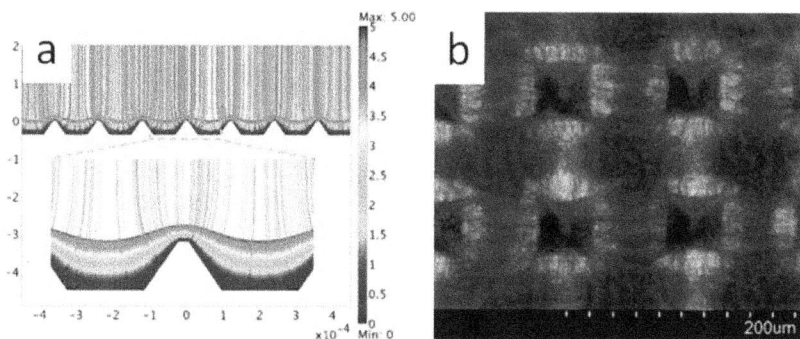

Figure 1.23. (a) Simulation of electric potential and electrical field distribution on the substrate. (b) SEM image of polyethylene oxide fiber deposited on a patterned substrate for 5 min. Reproduced with permission from [194], copyright (2009), American Chemical Society.

shows complex nanofiber tubes collected on columnar collectors. Furthermore, some revolutionary methods, e.g. multi-layer electrospinning [199–201], liquid and template-assisted electrospinning [202, 203], centrifugal electrospinning [204], gas foaming [205], cell sheet and origami engineering [206, 207], short nanofibers assembling into 3D nanofibers [208], electrospray [209] have been applied to prepare 3D nanofiber structure [208].

1.5.2 Aligned nanofibers

When nanofibers are collected uni-directionally, the tensile strength of nanofiber mats in the fiber-aligned direction could be improved significantly. Collecting nanofibers using a high speed rotating drum collector or disc is an efficient way to obtain aligned nanofibers [122, 210, 211]. Aligned nanofibers can also be collected using the stationary parallel electrode collector [212–214] or tip collector [215].

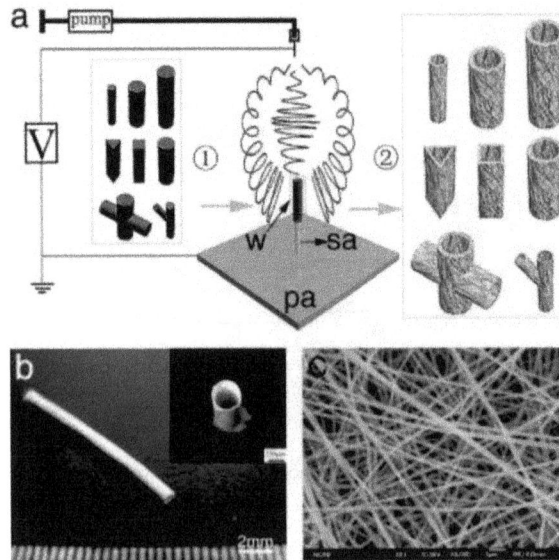

Figure 1.24. (a) Schematic illustration of columnar collectors for fabricating fibrous tubes. (①: columnar collectors and ②: fibrous tubes), (b) fabricated fibrous tube (diameter = 500 μm, inset is the cross-section image), and (c) SEM image of nanofibers. Reproduced with permission from [198], copyright (2008), American Chemical Society.

Figure 1.25. SEM image of aligned nanofibers collected across two parallel Si substrate. Reproduced with permission from [212], copyright (2003), American Chemical Society.

Figure 1.25 shows aligned short nanofibers collected by two parallel Si substrates. When using two magnet bars as the collector, it is also possible to collect aligned wavy polymeric nanofibers [216].

In another work, an electrode with counter polarity voltage was used to govern the fiber deposition [217]. Fiber placement and alignment on both microscale and nanoscale can be archived through controlling the shape and magnitude of the electric field of the counter electrode. This technology demonstrates the ability of making nanofiber membranes with tailored porosity (figure 1.26).

Figure 1.26. A straight fiber being collected with auxiliary counter electrode and SEM image of collected fibers. Reproduced with permission from [217], copyright (2008), American Chemical Society.

1.5.3 Nanofiber yarns

The first attempt in electrospinning nanofiber yarns, back in the 1930s, contains a fiber spinning wheel and fiber yarn collecting device [218]. Recently, the production of uniaxial nanofiber bundles or twisted nanofiber yarns by electrospinning has drawn increasing interest because yarns can be woven or knitted into 1D fabrics, 2D fabrics or 3D structures with tailored structure, mechanical strength, large specific surface area and high porosity. Nanofibers collected in a yarn form can find applications in the traditional textile industry and creating possibilities for a greater number of new applications. In particular, the large-scale production of nanofiber yarns is becoming more and more imperative. The most popular setups for nanofiber yarn/bundle production are listed in table 1.5, and they can be divided into following categories:

- Collect aligned nanofibers in a short length (e.g. two rotating disc, high speed collector) and then twist them into twisted nanofiber yarns [219–222];
- Collect short twisted yarns directly using two rotating tube collectors [223];
- Collect continuous nanofiber bundles with the help of airflow [224, 225];
- Use water bath as the collector to obtain continuous nanofiber bundles [226–228];
- Use an auxiliary electrode to govern nanofiber collection for obtaining nanofiber bundle [229–231];
- Apply both positive and negative potentials in electrospinning to improve nanofiber alignment and collect continuous nanofiber bundles [232–234];
- Use a rotating funnel as the collector to obtain continuously twisted nanofiber yarns directly [235–240];
- Wrap electrospun nanofibers around conventional filaments or yarns to obtain composite nanofiber yarns [241, 242];
- Use AC potential to electrospin nanofibers instead of DC potential, collect continuous nanofiber bundles [243].

In addition to the solution electrospinning, direct yarn production has also been realized in melt electrospinning. Polypropylene nanofiber yarn was continuously

Table 1.5. A summary of nanofiber yarn electrospinning setup, mechanism and features.

Electrospinning setup	Mechanism	Continuous production (Y/N)	Collected form	References
 (Reproduced with permission from Elsevier, copyright (2005))	Two separated rotating disc	N	Twisted yarn	[219]
 (Reproduced with permission from Elsevier, copyright (2005))	Water bath collection	Y	Nanofiber bundle	[226, 227]
 (Reproduced with permission from Elsevier, copyright (2006))	Opposite polarity electrospinning	Y	Nanofiber bundle	[232–234]

(*Continued*)

Table 1.5. (*Continued*)

Electrospinning setup	Mechanism	Continuous production (Y/N)	Collected form	References
 (Reproduced with permission from Elsevier, copyright (2007))	Water bath collection	Y	Nanofiber bundle	[228]
 (Reproduced with permission from Elsevier, copyright (2008))	Grounded bar induced collection	Y	Nanofiber bundle	[229, 230]

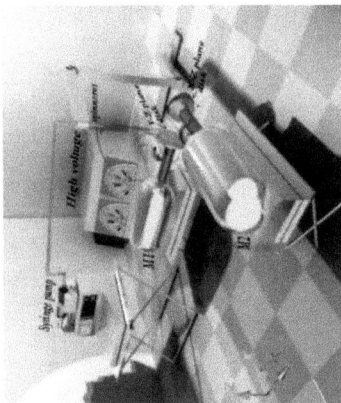 (Reproduced with permission from Elsevier, copyright (2008))	Two rotating disc in perpendicular position	Y	Nanofiber bundle	[231]
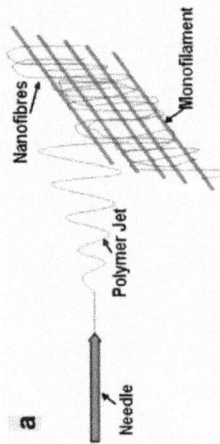 (Reproduced with permission from Elsevier, copyright (2010))	Parallel filament collection	Y	Nanofiber covered filaments	[241]
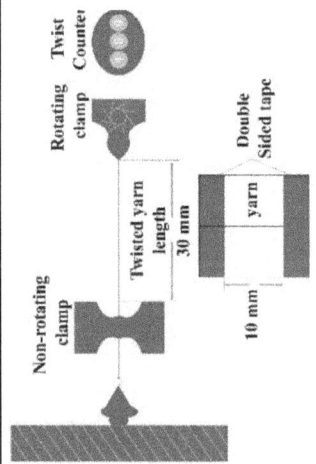	Short aligned nanofiber strip	N	Twisted yarn	[223–225]

(Continued)

Table 1.5. (*Continued*)

Electrospinning setup	Mechanism	Continuous production (Y/N)	Collected form	References
 (Reproduced with permission from Elsevier, copyright (2013))	Two rotating tubes and a winding tube	Y	Twisted yarn	[223]
 (Reproduced with permission from Elsevier, copyright (2011))	Opposite polarity electrospinning, rotating funnel collection	Y	Twisted yarn	[235–238]

Nanofiber covered yarns	Y	Needleless electrospinning, rotating collection	[242]

Twisted nanofiber yarn	Y	Opposite charge to form fibrous web, rotating ring collector	[240]

Twisted nanofiber yarn	Y	A rotating disk with a winding device, roller collector	[234]

(Continued)

Table 1.5. (*Continued*)

Electrospinning setup	Mechanism	Continuous production (Y/N)	Collected form	References
(Reproduced with permission from Elsevier, copyright (2004))	Two metal disks are positioned vertically	Y	Twisted nanofiber yarn	[244]
(Reproduced with permission from Elsevier, copyright (2008))	placing a stainless-steel annular electrode to a hollow aluminium plate	Y	Nanofiber bundle	[245]

Twisted nanofiber yarn [246]

A metal rod containing a sharp point and a hollow metal hemispheric

Y

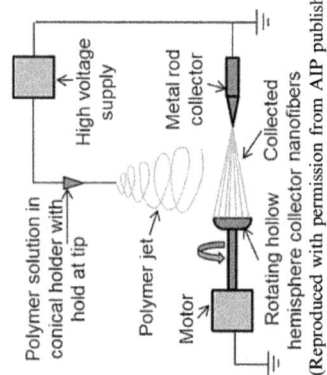

(Reproduced with permission from AIP publishing)

Nanofiber bundle [247]

A flowing water bath, electro-wet spinning

Y

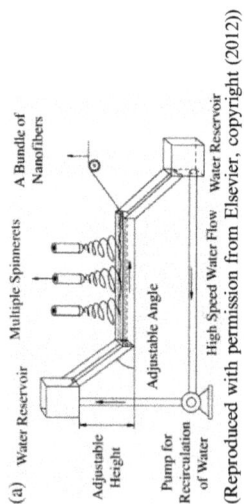

(Reproduced with permission from Elsevier, copyright (2012))

manufactured by a melt electrospinning method, wherein suction airflow was used to facilitate the formation of aligned nanofiber strand. The twist of yarn strand was realized through a tailored rotating collector (figure 1.27) [225].

AC voltage has the potential of minimizing jet whipping instability and enhancing fiber alignment. A method named biased AC electrospinning used amplified AC voltage to perform electrospinning with improved fiber alignment. When electro-spinning process is driven by DC voltage, it has a high level of whipping instability due to action of repulsive Coulombic force. On the other hand, AC power carries both positive and negative charges, both attractive and repulsive Coulombic forces take effect during electrospinning and lead to a weakening of overall whipping instability [248]. In another interesting work, AC electrospinning was used to generate nanofiber bundles directly in the absence of a collector [243]. Smoke-like nanofibers were generated from a rod electrode. Attributable to the existence of both positive changes and negative changes, they are easy to manipulate and can be wound into nanofiber yarns effectively (figure 1.28).

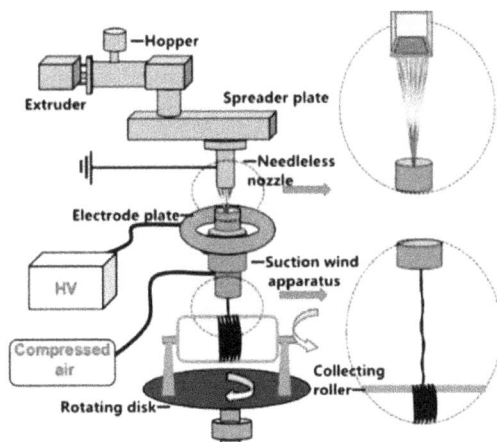

Figure 1.27. A schematic diagram of melt electrospinning. Reproduced with permission from [225] John Wiley & Sons. Copyright 2017 Wiley Periodicals, Inc.

Figure 1.28. (a) Schematic diagram of the AC electrospinning setup. Reproduced with permission from [249], copyright (2022), with permission from Elsevier. (b) generated a compact nanofiber plume, and (c) SEM image of nanofiber yarn. Reproduced with permission from [250] CC BY 4.0.

1.6 Summary and outlook

The science and engineering of nanofiber electrospinning technologies have advanced rapidly in the recent years. Many successful attempts to improve electrospinning yield, nanofiber and nanofiber membrane quality, controlled nanofiber deposition and secondary nanofibrous structure, have contributed to the exciting development of various advanced electrospinning technologies. Commercial production of electrospun nanofibers has also been realized. In spite of the enormous amount of efforts made in electrospinning technology, there are still some major challenges in this field, e.g. small production rate of nanofiber yarn, high electrical hazard risk associated with needleless electrospinning, large fiber diameter in melt electrospinning. The future research work on electrospinning should primarily focus on the large-scale production of high quality nanofibers with improved safety and efficiency.

1. Near-field electrospinning has the ability of producing delicate fiber structures in 2D or 3D arrangements, however, the fiber diameter is much larger than that of conventional needle electrospun nanofibers and fiber production rate is much lower as well. An important trend for near-field electrospinning is to become more productive, e.g. multi-jet synchronous printing, and to have the ability of fabricating nanofiber-based real structures.

2. The current yarn electrospinning technology is either based on needle electrospinning that has low nanofiber production rate or has low yarn production efficiency in needleless electrospinning. The large-scale production of nanofiber yarn can greatly expand the application of electrospun nanofibers.

3. The needleless electrospinning technologies generally require high voltage to initiate jet ejection and ensure continuous electrospinning. The high voltage can lead to electrical discharge, even fire/explosion hazards. The utilization of auxiliary fields, e.g. magnetic field, airflow field, centrifugal force, will facilitate electrospinning process effectively, e.g. reducing critical voltage, thinning fiber diameter, improving nanofiber production rate, which will be a promising direction in the future for energy-efficient, safe production of nanofibers on a large scale.

4. Melt electrospinning is an environmentally friendly process and melt electrospun fibers are of great importance in the applications in biomedical, filtration and textile areas. Due to the deficiency of free charge in polymer melt, the fiber diameter of melt electrospinning is normally at micrometer scale. The utilization of airflow or centrifugal force can effectively improve melt jet stretching and produce thin melt electrospun fibers. The production of thinner fibers in melt electrospinning will be an important research direction in the future.

References

[1] Chen X *et al* 2012 Nanomaterials for renewable energy production and storage *Chem. Soc. Rev.* **41** 7909–37

[2] Burda C *et al* 2005 Chemistry and properties of nanocrystals of different shapes *Chem. Rev.* **105** 1025–102

[3] Alivisatos A P 1996 Perspectives on the physical chemistry of semiconductor nanocrystals *J. Phys. Chem.* **100** 13226–39

[4] Chinnappan A *et al* 2017 An overview of electrospun nanofibers and their application in energy storage, sensors and wearable/flexible electronics *J. Mater. Chem.* C **5** 12657–73

[5] Ramakrishna S *et al* 2006 Electrospun nanofibers: solving global issues *Mater. Today* **9** 40–50

[6] Suja P S *et al* 2017 Electrospun nanofibrous membranes for water purification *Polym. Rev.* **57** 467–504

[7] Zhu M *et al* 2016 Electrospun nanofibers membranes for effective air filtration *Macromol. Mater. Eng.* **302** 1600353

[8] Chen S *et al* 2017 Recent advances in electrospun nanofibers for wound healing *Nanomedicine* **12** 1335–52

[9] Sill T J and von Recum H A 2008 Electrospinning: applications in drug delivery and tissue engineering *Biomaterials* **29** 1989–2006

[10] Li D and Xia Y 2004 Electrospinning of nanofibers: reinventing the wheel? *Adv. Mater.* **16** 1151–70

[11] Gibson P, Schreuder-Gibson H and Rivin D 2001 Transport properties of porous membranes based on electrospun nanofibers *Colloids Surf., A* **187–188** 469–81

[12] Ma P X and Zhang R 1999 Synthetic nano-scale fibrous extracellular matrix *J. Biomed. Mater. Res.* **46** 60–72

[13] Liu G, Qiao L and Guo A 1996 Diblock copolymer nanofibers *Macromolecules* **29** 5508–10

[14] Hartgerink J D, Beniash E and Stupp S I 2001 Self-assembly and mineralization of peptide-amphiphile nanofibers *Science* **294** 1684–8

[15] Martin C R 1996 Membrane-based synthesis of nanomaterials *Chem. Mater.* **8** 1739–46

[16] Ward G F 2001 Meltblown nanofibres for nonwoven filtration applications *Filtr. Sep.* **38** 42–3

[17] Weinberg M, Dee G and Harding T 2006 Flash spun web containing sub-micron filaments and process for forming same *U.S. Patent* 20060135020

[18] Fedorova N and Pourdeyhimi B 2007 High strength nylon micro—and nanofiber based nonwovens via spunbonding *J. Appl. Polym. Sci.* **104** 3434–42

[19] Doshi J and Reneker D H 1995 Electrospinning process and applications of electrospun fibers *J. Electrostat.* **35** 151–60

[20] Srinivasan G and Reneker D H 1995 Structure and morphology of small diameter electrospun aramid fibers *Polym. Int.* **36** 195–201

[21] He H *et al* 2020 Analysis and prediction of the diameter and orientation of AC electrospun nanofibers by response surface methodology *Mater. Des.* **194** 108902

[22] Wang M *et al* 2020 Electrospun environment remediation nanofibers using unspinnable liquids as the sheath fluids: a review *Polymers (Basel)* **12** 103

[23] El-Toni A M *et al* 2016 Design, synthesis and applications of core–shell, hollow core, and nanorattle multifunctional nanostructures *Nanoscale* **8** 2510–31

[24] Yu D G *et al* 2020 Multifluid electrospinning for the generation of complex nanostructures *Wiley Interdiscip. Rev. Nanomed. Nanobiotechnol.* **12** e1601

[25] Cai M *et al* 2019 Efficient synthesis of PVDF/PI side-by-side bicomponent nanofiber membrane with enhanced mechanical strength and good thermal stability *Nanomaterials* **9** 39

[26] Yan L *et al* 2021 Porous Janus materials with unique asymmetries and functionality *Mater. Today* **51** 626–47

[27] William G 1600 *de Magnete* (Trans. By P. E. Mottelay) (London: Peter Short)

[28] Formhals A 1934 Process and apparatus for preparing artificial threads *U.S. Patent* 1,975,504

[29] Formhals A 1943 Production of artifical fibers from fiber forming liquids *U.S. Patent* 2,323,025

[30] Formhals A 1944 Method and apparatus for spinning *U.S. Patent* 2,349,950

[31] Taylor G 1964 Disintegration of water drops in an electric field *Proc. R Soc. Lond. Ser A, Math. Phys. Sci.* **280** 383–97

[32] Taylor G 1969 Electrically driven jets *Proc. R Soc. Lond. Ser A Math. Phys. Sci.* **313** 453–75

[33] Kowalewski T A, Hiller W J and Behnia M 1993 An experimental study of evaporating small diameter jets *Phys. Fluids* A **5** 1883–90

[34] Huang Z *et al* 2003 A review on polymer nanofibers by electrospinning and their applications in nanocomposites *Compos. Sci. Technol.* **63** 2223–53

[35] Sir G and Taylor F R S 1966 Studies in electrohydrodynamics. I. The circulation produced in a drop by an electric field *Proc. R Soc. Lond. Ser A Math. Phys. Sci.* **291** 159

[36] Subbiah T *et al* 2005 Electrospinning of nanofibers *J. Appl. Polym. Sci.* **96** 557–69

[37] Rutledge G C and Fridrikh S V 2007 Formation of fibers by electrospinning *Adv. Drug Deliv. Rev.* **59** 1384–91

[38] Reneker D H and Yarin A L 2008 Electrospinning jets and polymer nanofibers *Polymer* **49** 2387–425

[39] Fong H and Reneker D H 1999 Electrospinning and the formation of nanofibers *Structure Formation in Polymeric Fibers* ed D R Salem (Hanser) ch 6 225–46

[40] Reneker D H *et al* 2000 Bending instability of electrically charged liquid jets of polymer solutions in electrospinning *J. Appl. Phys.* **87** 4531–47

[41] Yarin A L, Koombhongse S and Reneker D H 2001 Bending instability in electrospinning of nanofibers *J. Appl. Phys.* **89** 3018–26

[42] Hohman M M *et al* 2001 Electrospinning and electrically forced jets. I. Stability theory *Phys. Fluids* **13** 2201–20

[43] Hohman M M *et al* 2001 Electrospinning and electrically forced jets. II. Applications *Phys. Fluids* **13** 2221–36

[44] Yarin A L, Koombhongse S and Reneker D H 2001 Taylor cone and jetting from liquid droplets in electrospinning of nanofibers *J. Appl. Phys.* **90** 4836–46

[45] Lin T *et al* 2004 The charge effect of cationic surfactants on the elimination of fibre beads in the electrospinning of polystyrene *Nanotechnology* **15** 1375–81

[46] Han T, Reneker D H and Yarin A L 2007 Buckling of jets in electrospinning *Polymer* **48** 6064–76

[47] Ribeiro C *et al* 2010 Influence of processing conditions on polymorphism and nanofiber morphology of electroactive poly(vinylidene fluoride) electrospun membranes *Soft Mater.* **8** 274–87

[48] Beachley V and Wen X 2009 Effect of electrospinning parameters on the nanofiber diameter and length *Mater. Sci. Eng.* C **29** 663–8

[49] Haitao N, Tong L and Xungai W 2009 Needleless electrospinning. I. A comparison of cylinder and disk nozzles *J. Appl. Polym. Sci.* **114** 3524–30

[50] Wang X *et al* 2009 Needleless electrospinning of nanofibers with a conical wire coil *Polym. Eng. Sci.* **49** 1582–6

[51] Shao H et al 2015 Robust mechanical-to-electrical energy conversion from short-distance electrospun poly(vinylidene fluoride) fiber webs ACS Appl. Mater. Interfaces 7 22551–7

[52] Xiao-Peng T et al 2014 Effect of flow rate on diameter of electrospun nanoporous fibers Therm. Sci. 18 1447–9

[53] Barua B and Saha M C 2015 Investigation on jet stability, fiber diameter, and tensile properties of electrospun polyacrylonitrile nanofibrous yarns J. Appl. Polym. Sci. 132 41918

[54] Mit-uppatham C, Nithitanakul M and Supaphol P 2004 Ultrafine electrospun polyamide-6 fibers: effect of solution conditions on morphology and average fiber diameter Macromol. Chem. Phys. 205 2327–38

[55] Koski A, Yim K and Shivkumar S 2004 Effect of molecular weight on fibrous PVA produced by electrospinning Mater. Lett. 58 493–7

[56] Hekmati A H et al 2013 Effect of needle length, electrospinning distance, and solution concentration on morphological properties of polyamide-6 electrospun nanowebs Textile Res. J. 83 1452–66

[57] Biber E et al 2010 Effects of electrospinning process parameters on nanofibers obtained from Nylon 6 and poly (ethylene-n-butyl acrylate-maleic anhydride) elastomer blends using Johnson SB statistical distribution function Appl. Phys. A 99 477–87

[58] Demir M M et al 2002 Electrospinning of polyurethane fibers Polymer 43 3303–9

[59] Ding B et al 2002 Preparation and characterization of nanoscaled poly(vinyl alcohol) fibers via electrospinning Fibers Polym. 3 73–9

[60] Qin X-H et al 2006 Effect of different salts on electrospinning of polyacrylonitrile (PAN) polymer solution J. Appl. Polym. Sci. 103 3865–70

[61] Ho-Wang T and Min W 2010 Electrospinning of fibrous polymer scaffolds using positive voltage or negative voltage: a comparative study Biomed. Mater. 5 054110

[62] Tong L et al 2004 The charge effect of cationic surfactants on the elimination of fibre beads in the electrospinning of polystyrene Nanotechnology 15 1375

[63] Li Z et al 2016 Fabrication of a polyvinylidene fluoride tree-like nanofiber web for ultra high performance air filtration RSC Adv. 6 91243–9

[64] Yang S et al 2011 Controllable fabrication of soap-bubble-like structured polyacrylic acid nano-nets via electro-netting Nanoscale 3 564–8

[65] Hardick O, Stevens B and Bracewell D G 2011 Nanofibre fabrication in a temperature and humidity controlled environment for improved fibre consistency J. Mater. Sci. 46 3890–8

[66] Simonet M et al 2007 Ultraporous 3D polymer meshes by low-temperature electrospinning: use of ice crystals as a removable void template Polym. Eng. Sci. 47 2020–6

[67] De Vrieze S et al 2008 The effect of temperature and humidity on electrospinning J. Mater. Sci. 44 1357

[68] Fashandi H and Karimi M 2012 Pore formation in polystyrene fiber by superimposing temperature and relative humidity of electrospinning atmosphere Polymer 53 5832–49

[69] Pelipenko J et al 2013 The impact of relative humidity during electrospinning on the morphology and mechanical properties of nanofibers Int. J. Pharm. 456 125–34

[70] Szewczyk P K and Stachewicz U 2020 The impact of relative humidity on electrospun polymer fibers: from structural changes to fiber morphology Adv. Colloid Interface Sci. 286 102315

[71] Phiriyawirut M et al 2008 Morphology of electrospun mats of soy protein isolate and its blend Adv. Mater. Res. 55 733–6

[72] Ali U, Wang X and Lin T 2012 Effect of nozzle polarity and connection on electrospinning of polyacrylonitrile nanofibers *J. Text. Inst.* **103** 1160–8

[73] Kessick R, Fenn J and Tepper G 2004 The use of AC potentials in electrospraying and electrospinning processes *Polymer* **45** 2981–4

[74] Maheshwari S and Chang H 2009 Assembly of multi-stranded nanofiber threads through AC electrospinning *Adv. Mater.* **21** 349–54

[75] Agarwal S and Greiner A 2011 On the way to clean and safe electrospinning—green electrospinning: emulsion and suspension electrospinning *Polym. Adv. Technol.* **22** 372–8

[76] Sanders E H *et al* 2003 Two-phase electrospinning from a single electrified jet: micro-encapsulation of aqueous reservoirs in poly(ethylene-co-vinyl acetate) fibers *Macromolecules* **36** 3803–5

[77] Zhao P *et al* 2018 Emulsion electrospinning of polytetrafluoroethylene (PTFE) nanofibrous membranes for high-performance triboelectric nanogenerators *ACS Appl. Mater. Interfaces* **10** 5880–91

[78] Ru C *et al* 2015 Suspended, shrinkage-free, electrospun PLGA nanofibrous scaffold for skin tissue engineering *ACS Appl. Mater. Interfaces* **7** 10872–7

[79] Gonzalez E *et al* 2021 Green electrospinning of polymer latexes: a systematic study of the effect of latex properties on fiber morphology *Nanomaterials (Basel)* **11** 706

[80] Chen G *et al* 2015 Core–shell structure PEO/CS nanofibers based on electric field induced phase separation via electrospinning and its application *J. Polym. Sci., Part A: Polym. Chem.* **53** 2298–311

[81] Yang Y *et al* 2008 Release pattern and structural integrity of lysozyme encapsulated in core–sheath structured poly(dl-lactide) ultrafine fibers prepared by emulsion electrospinning *Eur. J. Pharm. Biopharm.* **69** 106–16

[82] Yang Y *et al* 2007 Structural stability and release profiles of proteins from core–shell poly (DL-lactide) ultrafine fibers prepared by emulsion electrospinning *J. Biomed. Mater. Res.* A **86A** 374–85

[83] George M C and Braun P V 2009 Multicompartmental materials by electrohydrodynamic cojetting *Angew. Chem. Int. Ed.* **48** 8606–9

[84] Sun Z *et al* 2003 Compound core–shell polymer nanofibers by co-electrospinning *Adv. Mater.* **15** 1929–32

[85] McCann J T, Li D and Xia Y N 2005 Electrospinning of nanofibers with core-sheath, hollow, or porous structures *J. Mater. Chem.* **15** 735–8

[86] Zussman E *et al* 2006 Electrospun polyaniline/poly(methyl methacrylate)-derived turbostratic carbon micro-/nanotubes *Adv. Mater.* **18** 348–53

[87] Niu H *et al* 2014 Ultrafine PDMS fibers: preparation from *in situ* curing-electrospinning and mechanical characterization *RSC Adv.* **4** 11782–7

[88] Yang C *et al* 2016 Electrospun pH-sensitive core–shell polymer nanocomposites fabricated using a tri-axial process *Acta Biomater.* **35** 77–86

[89] Wilkes P G G L 2003 Some investigations on the fiber formation by utilizing a side-by-side bicomponent electrospinning approach *Polymer* **44** 6353–9

[90] Gupta P and Wilkes G L 2003 Some investigations on the fiber formation by utilizing a side-by-side bicomponent electrospinning approach *Polymer* **44** 6353–9

[91] Cai J *et al* 2016 High-performance supercapacitor electrode from cellulose-derived, inter-bonded carbon nanofibers *J. Power Sources* **324** 302–8

[92] Lin T, Wang H and Wang X 2005 Self-crimping bicomponent nanofibers electrospun from polyacrylonitrile and elastomeric polyurethane *Adv. Mater.* **17** 2699–703

[93] Bhaskar S and Lahann J 2009 Microstructured materials based on multicompartmental fibers *JACS* **131** 6650–1

[94] Varesano A, Carletto R A and Mazzuchetti G 2009 Experimental investigations on the multi-jet electrospinning process *J. Mater. Process. Technol.* **209** 5178–85

[95] Yang E, Shi J and Xue Y 2010 Influence of electric field interference on double nozzles electrospinning *J. Appl. Polym. Sci.* **116** 3688–92

[96] Ding B *et al* 2004 Fabrication of blend biodegradable nanofibrous nonwoven mats via multi-jet electrospinning *Polymer* **45** 1895–902

[97] Ying Y *et al* 2008 Electrospun uniform fibres with a special regular hexagon distributed multi-needles system *J. Phys. Conf. Ser.* **142** 012027

[98] Theron S A *et al* 2005 Multiple jets in electrospinning: experiment and modeling *Polymer* **46** 2889–99

[99] Yang Y *et al* 2010 A shield ring enhanced equilateral hexagon distributed multi-needle electrospinning spinneret *IEEE Trans. Dielectr. Electr. Insul.* **17** 1592–601

[100] Zheng Y and Zeng Y 2014 Electric field analysis of spinneret design for multihole electrospinning system *J. Mater. Sci.* **49** 1964–72

[101] Kim G, Cho Y-S and Kim W D 2006 Stability analysis for multi-jets electrospinning process modified with a cylindrical electrode *Eur. Polym. J.* **42** 2031–8

[102] Bioinicia 2018 Multi-nozzle electrospinning machines https://bioinicia.com/multi-nozzle-electrospinning-machines/

[103] Varabhas J S, Chase G G and Reneker D H 2008 Electrospun nanofibers from a porous hollow tube *Polymer* **49** 4226–9

[104] Dosunmu O O *et al* 2006 Electrospinning of polymer nanofibres from multiple jets on a porous tubular surface *Nanotechnology* **17** 1123

[105] Jiang J *et al* 2019 Arced multi-nozzle electrospinning spinneret for high-throughput production of nanofibers *Micromachines (Basel)* **11** 27

[106] Piner R D *et al* 1999 'Dip-Pen' nanolithography *Science* **283** 661

[107] Huo F *et al* 2008 Polymer pen lithography *Science* **321** 1658

[108] Sun D *et al* 2006 Near-field electrospinning *Nano Lett.* **6** 839–42

[109] Hellmann C *et al* 2009 High precision deposition electrospinning of nanofibers and nanofiber nonwovens *Polymer* **50** 1197–205

[110] He X-X *et al* 2017 Near-field electrospinning: progress and applications *J. Phys. Chem.* C **121** 8663–78

[111] Nazemi M M, Khodabandeh A and Hadjizadeh A 2022 Near-field electrospinning: crucial parameters, challenges, and applications *ACS Appl. Bio. Mater.* **5** 394–412

[112] Chang C *et al* 2010 Direct-write piezoelectric polymeric nanogenerator with high energy conversion efficiency *Nano Lett.* **10** 726–31

[113] Fuh Y K, Chen S Z and He Z Y 2013 Direct-write, highly aligned chitosan-poly(ethylene oxide) nanofiber patterns for cell morphology and spreading control *Nanoscale Res. Lett.* **8** 97

[114] Brown T D, Dalton P D and Hutmacher D W 2011 Direct writing by way of melt electrospinning *Adv. Mater.* **23** 5651–7

[115] Liu Z H *et al* 2015 Crystallization and mechanical behavior of the ferroelectric polymer nonwoven fiber fabrics for highly durable wearable sensor applications *Appl. Surf. Sci.* **346** 291–301

[116] SangHoon L *et al* 2007 Chip-to-chip fluidic connectors via near-field electrospinning *2007 IEEE 20th Int. Conf. on Micro Electro Mechanical Systems (MEMS)*

[117] Hochleitner G *et al* 2014 High definition fibrous poly(2-ethyl-2-oxazoline) scaffolds through melt electrospinning writing *Polymer* **55** 5017–23

[118] Wojasiński M, Goławski J and Ciach T 2017 Blow-assisted multi-jet electrospinning of poly-L-lactic acid nanofibers *J. Polym. Res.* **24** 76

[119] Zhmayev E, Cho D and Joo Y L 2010 Nanofibers from gas-assisted polymer melt electrospinning *Polymer* **51** 4140–4

[120] Hsiao H-Y *et al* 2011 Effect of air blowing on the morphology and nanofiber properties of blowing-assisted electrospun polycarbonates *J. Appl. Polym. Sci.* **124** 4904–14

[121] An S *et al* 2014 Supersonically blown ultrathin thorny devil nanofibers for efficient air cooling *ACS Appl. Mater. Interfaces* **6** 13657–66

[122] Kiselev P and Rosell-Llompart J 2012 Highly aligned electrospun nanofibers by elimination of the whipping motion *J. Appl. Polym. Sci.* **125** 2433–41

[123] Varesano A, Montarsolo A and Tonin C 2007 Crimped polymer nanofibres by air-driven electrospinning *Eur. Polym. J.* **43** 2792–8

[124] Chen C *et al* 2016 Use of electrospinning and dynamic air focusing to create three-dimensional cell culture scaffolds in microfluidic devices *Analyst* **141** 5311–20

[125] Tanioka A and Takahashi M 2016 Highly productive systems of nanofibers for novel applications *Ind. Eng. Chem. Res.* **55** 3759–64

[126] Kim G H and Yoon H 2008 A direct-electrospinning process by combined electric field and air-blowing system for nanofibrous wound-dressings *Appl. Phys.* A **90** 389–94

[127] Norton C L 1936 Method of and apparatus for producing fibrous or filamentary material *US Patent* US677277A

[128] Lyons J, Li C and Ko F 2004 Melt-electrospinning part I: processing parameters and geometric properties *Polymer* **45** 7597–603

[129] Hunley M T *et al* 2008 Taking advantage of tailored electrostatics and complementary hydrogen bonding in the design of nanostructures for biomedical applications *Macromol. Symp.* **270** 1–7

[130] Dalton P D *et al* 2006 Direct *in vitro* electrospinning with polymer melts *Biomacromolecules* **7** 686–90

[131] Qin C-C *et al* 2015 Melt electrospinning of poly(lactic acid) and polycaprolactone microfibers by using a hand-operated Wimshurst generator *Nanoscale* **7** 16611–5

[132] Kim S J *et al* 2010 Fabrication and characterization of 3-dimensional PLGA nanofiber/microfiber composite scaffolds *Polymer* **51** 1320–7

[133] Li X *et al* 2012 Preparation and characterization of PLLA/nHA nonwoven mats via laser melt electrospinning *Mater. Lett.* **73** 103–6

[134] Ogata N *et al* 2007 Poly(lactide) nanofibers produced by a melt-electrospinning system with a laser melting device *J. Appl. Polym. Sci.* **104** 1640–5

[135] Li X *et al* 2014 Effect of oriented fiber membrane fabricated via needleless melt electrospinning on water filtration efficiency *Desalination* **344** 266–73

[136] Kim J *et al* 2014 Preparation and gas-sensing properties of pitch-based carbon fiber prepared using a melt-electrospinning method *Res. Chem. Intermed.* **40** 2571–81

[137] Lee S and Kay Obendorf S 2006 Developing protective textile materials as barriers to liquid penetration using melt-electrospinning *J. Appl. Polym. Sci.* **102** 3430–7

[138] Pham Q P, Sharma U and Mikos A G 2006 Electrospinning of polymeric nanofibers for tissue engineering applications: a review *Tissue Eng.* **12** 1197–211

[139] Brown T D *et al* 2012 Design and fabrication of tubular scaffolds via direct writing in a melt electrospinning mode *Biointerphases* **7** 13

[140] Hacker C *et al* 2013 Functionally modified, melt-electrospun thermoplastic polyurethane mats for wound-dressing applications *J. Appl. Polym. Sci.* **131** 40132

[141] Castilho M *et al* 2017 Melt electrospinning writing of poly-hydroxymethylglycolide-co-ε-caprolactone-based scaffolds for cardiac tissue engineering *Adv. Healthcare Mater.* **6** 1700311

[142] Yarin A L and Zussman E 2004 Upward needleless electrospinning of multiple nanofibers *Polymer* **45** 2977–80

[143] Lukas D, Sarkar A and Pokorny P 2008 Self-organization of jets in electrospinning from free liquid surface: a generalized approach *J. Appl. Phys.* **103** 084309

[144] Thoppey N M *et al* 2010 Unconfined fluid electrospun into high quality nanofibers from a plate edge *Polymer* **51** 4928–36

[145] Thoppey N M *et al* 2011 Edge electrospinning for high throughput production of quality nanofibers *Nanotechnology* **22** 345301

[146] Sharma U *et al* 2012 Electrospinning process for manufacture of multi-layered structures *U.S. Patent* 2012/0193836 Al

[147] Yan X *et al* 2015 Slit-surface electrospinning: a novel process developed for high-throughput fabrication of core-sheath fibers *PLoS One* **10** e0125407

[148] Yan X *et al* 2012 High-throughput needleless electrospinning of core-sheath fibers *Fiber Soc. Conf. Proc.* 100

[149] Ucar N and U M 2013 Design of a novel nozzle prototype for increased productivity and improved coating quality during electrospinning *Tekst. Konfeksiyon* **23** 199

[150] Jiang G, Zhang S and Qin X 2013 High throughput of quality nanofibers via one stepped pyramid-shaped spinneret *Mater. Lett.* **106** 56–8

[151] Jiang G and Qin X 2014 An improved free surface electrospinning for high throughput manufacturing of core–shell nanofibers *Mater. Lett.* **128** 259–62

[152] Holopainen J *et al* 2015 Needleless electrospinning with twisted wire spinneret *Nanotechnology* **26** 025301

[153] Yan G *et al* 2017 Curved convex slot: an effective needleless electrospinning spinneret with low solvent evaporation loss *J. Mater. Sci.* **52** 11749

[154] Niu H and Lin T 2012 Fiber generators in needleless electrospinning *J. Nanomater.* **2012** 725950

[155] Elmacro 2023 Patented needle-free Nanospider™ technology [cited 2023; https://elmarco.com/nanospider]

[156] Liu S L *et al* 2014 Needleless electrospinning for large scale production of ultrathin polymer fibres *Mater. Res. Innov.* **18** S4-833–7

[157] Jirsak O *et al* 2009 Method of nanofibres production from a polymer solution using electrostatic spinning and a device for carrying out the method *US Patent* US10/570,806

[158] Niu H, Wang X and Lin T 2012 Upward needleless electrospinning of nanofibers *J. Eng. Fibers Fabrics* **7** 17–22

[159] Wang X *et al* 2012 Needleless electrospinning of uniform nanofibers using spiral coil spinnerets *J. Nanomater.* **2012** 785920

[160] Niu H *et al* 2018 Enhancement of coil electrospinning using two-level coil structure *Ind. Eng. Chem. Res.* **57** 15473–8

[161] Lu B *et al* 2010 Superhigh-throughput needleless electrospinning using a rotary cone as spinneret *Small* **6** 1612–6

[162] Bhattacharyya I *et al* 2016 Free surface electrospinning of aqueous polymer solutions from a wire electrode *Chem. Eng. J.* **289** 203–11

[163] Forward K M and Rutledge G C 2012 Free surface electrospinning from a wire electrode *Chem. Eng. J.* **183** 492–503

[164] Forward K M, Flores A and Rutledge G C 2013 Production of core/shell fibers by electrospinning from a free surface *Chem. Eng. Sci.* **104** 250–9

[165] Tang S, Zeng Y and Wang X 2010 Splashing needleless electrospinning of nanofibers *Polym. Eng. Sci.* **50** 2252–7

[166] Yang W *et al* 2016 Optimal spinneret layout in Von Koch curves of fractal theory based needleless electrospinning process *AIP Adv.* **6** 065223

[167] Liu Z, Chen R and He J 2016 Active generation of multiple jets for producing nanofibres with high quality and high throughput *Mater. Des.* **94** 496–501

[168] Liu Z, Ang K K J and He J 2017 Needle-disk electrospinning inspired by natural point discharge *J. Mater. Sci.* **52** 1823–30

[169] Cengiz-Çallıoğlu F 2014 Dextran nanofiber production by needleless electrospinning process *e-Polymers* **14** 5–13

[170] Moon S, Gil M and Lee K J 2017 Syringeless electrospinning toward versatile fabrication of nanofiber web *Sci. Rep.* **7** 41424

[171] Zheng G *et al* 2018 Self-cleaning threaded rod spinneret for high-efficiency needleless electrospinning *Appl. Phys.* A **124** 473

[172] Nano-meghyas F 2023 Fanavaran nano-meghyas—products https://fnm.ir/Product

[173] Company S N 2023 SNC ball electrospinning technology https://sncfibers.com/snc-best

[174] Equipment S R 2022 EF500 needleless system https://ske.it/products/ef500-needleless-system/

[175] Nanolayr R F 2021 Nanofibre manufacturing on a huge scale https://nanolayr.com/nanofibre-manufacturing/

[176] Liu Y and He J H 2007 Bubble electrospinning for mass production of nanofibers *Int. J. Nonlinear Sci. Numer. Simul.* **8** 393–6

[177] Um I C *et al* 2004 Electro-spinning and electro-blowing of hyaluronic acid *Biomacromolecules* **5** 1428–36

[178] Wang X, Lin T and Wang X 2014 Use of airflow to improve the nanofibrous structure and quality of nanofibers from needleless electrospinning *J. Ind. Text.* **45** 310–20

[179] Zhmayev E, Cho D and Joo Y L 2010 Modeling of melt electrospinning for semi-crystalline polymers *Polymer* **51** 274–90

[180] He J-H *et al* 2012 Review on fiber morphology obtained by bubble electrospinning and blown bubble spinning *Therm. Sci.* **16** 1263–79

[181] Jones F R and Huff N T 2009 The structure and properties of glass fibres *Handbook of Textile Fibre Structure* ed S J Eichhorn *et al* (Woodhead Publishing) ch 9 pp 307–52

[182] Badrossamay M R *et al* 2010 Nanofiber assembly by rotary jet-spinning *Nano Lett.* **10** 2257–61

[183] Sarkar K *et al* 2010 Electrospinning to Forcespinning™ *Mater. Today* **13** 12–4

[184] Weitz R T *et al* 2008 Polymer nanofibers via nozzle-free centrifugal spinning *Nano Lett.* **8** 1187–91

[185] Liu S L, L Y Z, Zhang Z H, Zhang H D, Sun B and Zhang J C 2013 Assembly of oriented ultrafine polymer fibers by centrifugal electrospinning *J. Nanomater.* **1** 713275

[186] Dabirian F *et al* 2011 A comparative study of jet formation and nanofiber alignment in electrospinning and electrocentrifugal spinning systems *J. Electrostat.* **69** 540–6

[187] Edmondson D *et al* 2012 Centrifugal electrospinning of highly aligned polymer nanofibers over a large area *J. Mater. Chem.* **22** 18646–52

[188] Dabirian F, Ravandi S A H and Pishevar A R 2010 Investigation of parameters affecting PAN nanofiber production using electrical and centrifugal forces as a novel method *Curr. Nanosci.* **6** 545–52

[189] Chen C, Dirican M and Zhang X 2019 Centrifugal spinning—high rate production of nanofibers *Electrospinning: Nanofabrication and Applications* ed B Ding, X Wang and J Yu (William Andrew Publishing) ch 10 pp 321–38

[190] Guilong Y *et al* 2018 Electro-aerodynamic field aided needleless electrospinning *Nanotechnology* **29** 235302

[191] Deitzel J M *et al* 2001 Controlled deposition of electrospun poly(ethylene oxide) fibers *Polymer* **42** 8163–70

[192] Kim G H 2006 Electrospinning process using field-controllable electrodes *J. Polym. Sci., Part B: Polym. Phys.* **44** 1426–33

[193] Kim H-Y *et al* 2010 Nanopottery: coiling of electrospun polymer nanofibers *Nano Lett.* **10** 2138–40

[194] Ding Z, Salim A and Ziaie B 2009 Selective nanofiber deposition through field-enhanced electrospinning *Langmuir* **25** 9648–52

[195] Salim A, Son C and Ziaie B 2008 Selective nanofiber deposition via electrodynamic focusing *Nanotechnology* **19** 375303

[196] Park S M *et al* 2018 Direct fabrication of spatially patterned or aligned electrospun nanofiber mats on dielectric polymer surfaces *Chem. Eng. J.* **335** 712–9

[197] Thomas V, Zhang X and Vohra Y K 2009 A biomimetic tubular scaffold with spatially designed nanofibers of protein/PDS® bio-blends *Biotechnol. Bioeng.* **104** 1025–33

[198] Zhang D M and Chang J 2008 Electrospinning of three-dimensional nanofibrous tubes with controllable architectures *Nano Lett.* **8** 3283–7

[199] Chainani A *et al* 2013 Multilayered electrospun scaffolds for tendon tissue engineering *Tissue Eng. Part* A **19** 2594–604

[200] Shim I K *et al* 2009 Chitosan nano-/microfibrous double-layered membrane with rolled-up three-dimensional structures for chondrocyte cultivation *J. Biomed. Mater. Res.* A **90** 595–602

[201] Sadeghi A *et al* 2022 Multilayered 3-D nanofibrous scaffold with chondroitin sulfate sustained release as dermal substitute *Int. J. Biol. Macromol.* **206** 718–29

[202] Sun B *et al* 2014 Fabrication and characterization of mineralized P(LLA-CL)/SF three-dimensional nanoyarn scaffolds *Iran. Polym. J.* **24** 29–40

[203] Sun B *et al* 2015 Fabrication and characterization of mineralized P(LLA-CL)/SF three-dimensional nanoyarn scaffolds *Iran. Polym. J.* **24** 29–40

[204] Wang L *et al* 2017 Multi-compartment centrifugal electrospinning based composite fibers *Chem. Eng. J.* **330** 541–9

[205] Jiang J *et al* 2015 Expanding two-dimensional electrospun nanofiber membranes in the third dimension by a modified gas-foaming technique *ACS Biomater Sci Eng* **1** 991–1001

[206] Lee S J *et al* 2015 Characterization and preparation of bio-tubular scaffolds for fabricating artificial vascular grafts by combining electrospinning and a 3D printing system *Phys. Chem. Chem. Phys.* **17** 2996–9

[207] Song J *et al* 2018 Origami meets electrospinning: a new strategy for 3D nanofiber scaffolds *Bio-Des. Manuf.* **1** 254–64

[208] Chen Y *et al* 2020 Advanced fabrication for electrospun three-dimensional nanofiber aerogels and scaffolds *Bioact. Mater.* **5** 963–79

[209] John J V *et al* 2019 Tethering peptides onto biomimetic and injectable nanofiber microspheres to direct cellular response *Nanomed. Nanotechnol. Biol. Med.* **22** 102081

[210] Katta P, Alessandro M, Ramsier R D and Chase G G 2004 Continuous electrospinning of aligned polymer nanofibers onto a wire drum collector *Nano Lett.* **4** 2215–8

[211] Theron A, Zussman E and Yarin A L 2001 Electrostatic field-assisted alignment of electrospun nanofibres *Nanotechnology* **12** 384–90

[212] Li D, Wang Y and Xia Y 2003 Electrospinning of polymeric and ceramic nanofibers as uniaxially aligned arrays *Nano Lett.* **3** 1167–71

[213] Li D, Y W and Xia Y 2004 Electrospinning nanofibers as uniaxially aligned arrays and layer-by-layer stacked films *Adv. Mater.* **16** 361–6

[214] Yang D *et al* 2007 Fabrication of aligned fibrous arrays by magnetic electrospinning *Adv. Mater.* **19** 3702–6

[215] Rafique J *et al* 2007 Electrospinning highly aligned long polymer nanofibers on large scale by using a tip collector *Appl. Phys. Lett.* **91** 063126

[216] Liu Y *et al* 2010 Magnetic-field-assisted electrospinning of aligned straight and wavy polymeric nanofibers *Adv. Mater.* **22** 2454–7

[217] Carnell L S *et al* 2008 Aligned mats from electrospun single fibers *Macromolecules* **41** 5345–9

[218] Anton F 1934 Process and apparatus for preparing artificial threads *US Patent* US500283A

[219] Dalton P D, Klee D and Möller M 2005 Electrospinning with dual collection rings *Polymer* **46** 611–4

[220] Nakashima R *et al* 2011 Mechanical properties of poly(vinylidene fluoride) nanofiber filaments prepared by electrospinning and twisting *Adv. Polym. Tech.* **32** E44–52

[221] Zhou Y *et al* 2012 Strip twisted electrospun nanofiber yarns: structural effects on tensile properties *J. Mater. Res.* **27** 537–44

[222] Chawla S, Naraghi M and Davoudi A 2013 Effect of twist and porosity on the electrical conductivity of carbon nanofiber yarns *Nanotechnology* **24** 255708

[223] Yan H, Liu L and Zhang Z 2011 Continually fabricating staple yarns with aligned electrospun polyacrylonitrile nanofibers *Mater. Lett.* **65** 2419–21

[224] Ko F *et al* 2003 Electrospinning of continuous carbon nanotube-filled nanofiber yarns *Adv. Mater.* **15** 1161–5

[225] Ma X *et al* 2017 Continuous manufacturing of nanofiber yarn with the assistance of suction wind and rotating collection via needleless melt electrospinning *J. Appl. Polym. Sci.* **134** 44820

[226] Smit E, Büttner U and Sanderson R D 2005 Continuous yarns from electrospun fibers *Polymer* **46** 2419–23

[227] Tian L, Yan T and Pan Z 2015 Fabrication of continuous electrospun nanofiber yarns with direct 3D processability by plying and twisting *J. Mater. Sci.* **50** 7137–48

[228] Teo W-E *et al* 2007 A dynamic liquid support system for continuous electrospun yarn fabrication *Polymer* **48** 3400–5

[229] Dabirian F, Hosseini Y and Ravandi S A H 2007 Manipulation of the electric field of electrospinning system to produce polyacrylonitrile nanofiber yarn *J. Textile Inst.* **98** 237–41

[230] Wang X *et al* 2008 Continuous polymer nanofiber yarns prepared by self-bundling electrospinning method *Polymer* **49** 2755–61

[231] Bazbouz M B and Stylios G K 2008 Novel mechanism for spinning continuous twisted composite nanofiber yarns *Eur. Polym. J.* **44** 1–12

[232] Huan Pan L L, Long H and Xiaojie C 2006 Continuous aligned polymer fibers produced by a modified electrospinning method *Polymer* **47** 4901–4

[233] Hajiani F, Jeddi A A and Gharehaghaji A A 2012 An investigation on the effects of twist on geometry of the electrospinning triangle and polyamide 66 nanofiber yarn strength *Fibers Polym.* **13** 244–52

[234] Maleki H *et al* 2013 Influence of the solvent type on the morphology and mechanical properties of electrospun PLLA yarns *Biofabrication* **5** 035014

[235] Ali U *et al* 2012 Direct electrospinning of highly twisted, continuous nanofiber yarns *J. Text. Inst.* **103** 80–8

[236] Wu S-H and Qin X-H 2013 Uniaxially aligned polyacrylonitrile nanofiber yarns prepared by a novel modified electrospinning method *Mater. Lett.* **106** 204–7

[237] He J *et al* 2013 Continuous twisted nanofiber yarns fabricated by double conjugate electrospinning *Fibers Polym.* **14** 1857–63

[238] He J-X *et al* 2013 Fabrication of continuous nanofiber yarn using novel multi-nozzle bubble electrospinning *Polym. Int.* **63** 1288–94

[239] Afifi A M *et al* 2010 Electrospinning of continuous aligning yarns with a 'funnel' target *Macromol. Mater. Eng.* **295** 660–5

[240] Shuakat M N and Lin T 2016 Direct electrospinning of nanofibre yarns using a rotating ring collector *J. Text. Inst.* **107** 791–9

[241] Zhou F-L, Gong R-H and Porat I 2010 Nano-coated hybrid yarns using electrospinning *Surf. Coat. Technol.* **204** 3459–63

[242] Niu H *et al* 2014 Composite yarns fabricated from continuous needleless electrospun nanofibers *Polym. Eng. Sci.* **54** 1495–502

[243] Pokorny P *et al* 2014 Effective AC needleless and collectorless electrospinning for yarn production *Phys. Chem. Chem. Phys.* **16** 26816–22

[244] Wu S *et al* 2022 State-of-the-art review of advanced electrospun nanofiber yarn-based textiles for biomedical applications *Appl. Mater Today* **27** 101473

[245] Liu C-K *et al* 2008 Preparation of short submicron-fiber yarn by an annular collector through electrospinning *Mater. Lett.* **62** 4467–9

[246] Lotus A *et al* 2008 Electrical, structural, and chemical properties of semiconducting metal oxide nanofiber yarns *J. Appl. Phys.* **103** 024910

[247] Liu J *et al* 2012 Structure and thermo-chemical properties of continuous bundles of aligned and stretched electrospun polyacrylonitrile precursor nanofibers collected in a flowing water bath *Carbon* **50** 1262–70

[248] Sarkar S, Deevi S and Tepper G 2007 Biased AC electrospinning of aligned polymer nanofibers *Macromol. Rapid Commun.* **28** 1034–9

[249] Sivan M *et al* 2022 AC electrospinning: impact of high voltage and solvent on the electrospinnability and productivity of polycaprolactone electrospun nanofibrous scaffolds *Mater. Today Chem.* **26** 101025

[250] Valtera J *et al* 2019 Fabrication of dual-functional composite yarns with a nanofibrous envelope using high throughput AC needleless and collectorless electrospinning *Sci. Rep.* **9** 1801

IOP Publishing

Energy Harvesting Properties of Electrospun Nanofibers (Second Edition)

Jian Fang and Tong Lin

Chapter 2

Bias-controlled electrospinning for improved performance of electronic devices

Anand Babu and Dipankar Mandal

Electrospinning is a versatile technique used to produce nanofibers from a polymer solution or melt. The process involves the application of an electric field to a polymer droplet, causing the droplet to stretch and produce the nanofibers with the desired diameter, roughness, alignment, and porosity. The first electrospinning technique was used by Sir Geffery Taylor in 1967 [1–5].

2.1 Electrospinning setup

The basic setup for electrospinning includes a syringe with a metallic needle, a high-voltage power supply, and a collector. The polymer solution is loaded into the syringe and a high voltage is applied between the needle and the collector. As a result, several key forces come into play (figure 2.1).

Fiber production through electrospinning involves the following steps [6–12].

2.1.1 Electrostatic force (F_e)

The electrostatic force is the driving force in electrospinning and can be calculated using Coulomb's law.

$$F_e = \frac{kQ_1Q_2}{r^2}$$

F_e is the electrostatic force, k is Coulomb's constant, Q_1 and Q_2 are the charges on the polymer droplet and the collector, respectively, r is the distance between the droplet and the collector.

doi:10.1088/978-0-7503-5487-5ch2

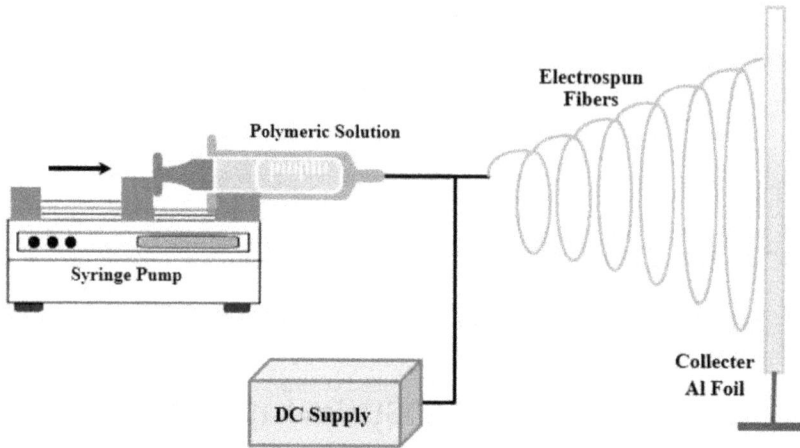

Figure 2.1. Modern electrospinning setup. Reproduced with permission from [5] Copyright 2016 lifescience Global. CC BY 3.0.

2.1.2 Surface tension (F_t)

The surface tension of the polymer solution tries to minimise the surface area of the droplet, opposing the electrostatic force. It can be calculated using the Young–Laplace equation.

$$F_t = 2\pi r\gamma$$

F_e is the surface tension force, r is the radius of the droplet, γ is the surface tension of the polymer solution.

2.1.3 Viscous drag force (F_v)

It encounters resistance because of its viscosity. The viscous drag force is given by the following:

$$F_v = \eta A v$$

F_v is the viscous drag force, η is the viscosity of the polymer solution, A is the cross-sectional area of the fiber, v is the velocity of the polymer solution.

2.1.4 Taylor cone formation

Taylor cone formation is a critical step in the electrospinning process, as it marks the initiation of nanofiber production. The Taylor cone is a cone-shaped structure that forms at the tip of the needle when the electrostatic force applied to a polymer solution overcomes the surface tension. This event marks the onset of the electrospinning process. When the electrostatic force overcomes the surface tension, a cone-like shape called the Taylor cone forms at the tip of the needle. The critical voltage required for Taylor cone formation can be calculated using the following equation:

$$V_c = \frac{2\gamma r}{k}$$

where V_c is the critical voltage, γ is the surface tension, r is the radius of the needle, k is the electric field enhancement factor, dependent on the geometry.

2.1.5 Taylor cone formation process

2.1.5.1 Initially, the polymer solution is subjected to an electric field due to the applied voltage.

2.1.5.2 As the strength of the electric field increases, it exerts an electrostatic force on the surface of the droplet. When this electrostatic force overcomes the surface tension, the droplet starts elongating into a cone-like shape.

2.1.5.3 The elongation continues until the electrostatic force, surface tension, and viscous forces reach equilibrium, forming a stable Taylor cone.

2.1.5.4 Once the Taylor cone is formed, a jet of polymer solution is ejected from its apex. This jet undergoes further stretching and thinning as it travels towards the collector.

2.1.5.5 During flight, solvent evaporation occurs, leading to solidification of the nanofiber.

2.2 Effect of different collectors on the properties of the fibers

The choice of the collector in the electrospinning process can have a significant impact on the properties of the nanofibers produced. Different collectors can influence various aspects of fiber morphology, alignment, and overall characteristics. Here is a discussion of the effects of different collectors on the properties of nanofibers (figure 2.2). There are several types of collectors used in electrospinning, each with its influence on fiber properties [13–18]:

2.2.1 Flat plate collector

A flat plate collector is a common choice where the collector surface is flat and stationary. It is straightforward to use and typically results in randomly orientated nanofibers. Flat plate collectors typically result in randomly orientated fibers. The lack of controlled alignment is advantageous for applications such as tissue engineering, where a three-dimensional network of fibers is desirable. Fibers collected on a flat plate tend to form denser mats with lower porosity.

2.2.2 Rotating drum collector

A rotating drum collector is a cylindrical collector that rotates during electrospinning. This collector type can induce the alignment of fibers along the axial direction of the drum, resulting in highly aligned and oriented nanofibers. Rotating drum collectors promote fiber alignment along the axis of rotation. This alignment is useful in applications such as filtration, where a highly aligned structure improves efficiency. These collectors can produce fibers with varying density and porosity depending on the drum or collector design.

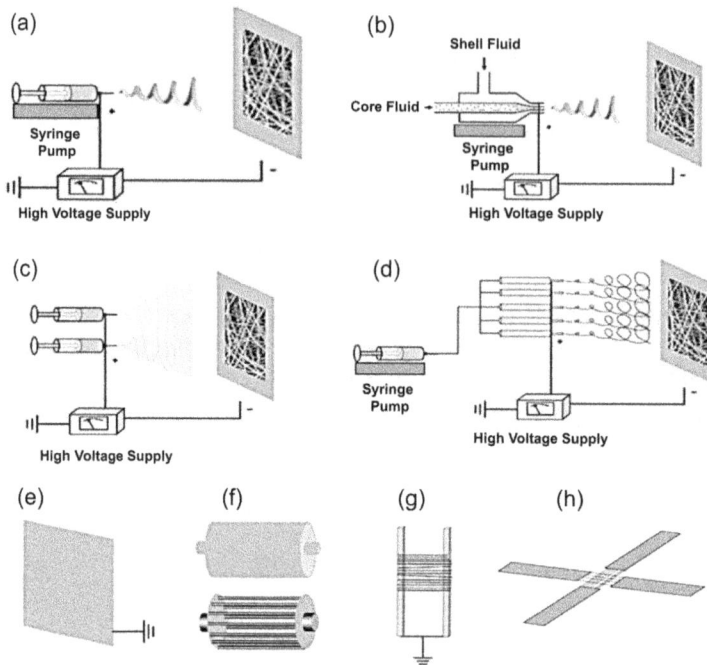

Figure 2.2. Various electrospinning techniques include (a) basic electrospinning, (b) coaxial electrospinning, (c) side-by-side electrospinning, and (d) multiple-jet electrospinning. Additionally, various collector types are employed, such as (e) metallic plate, (f) drum, (g) parallel electrode, and (h) array of counter electrodes. Reproduced with permission from [11] Copyright 2022 MDPI. CC BY 4.0.

2.2.3 Cylindrical collector

Similar to the rotating drum, a cylindrical collector is a static cylinder. This produces aligned fibers, but the alignment is usually not as pronounced as with a rotating drum. Like rotating drum collectors, cylindrical collectors can induce fiber alignment, but the degree of alignment may vary depending on the design. They are useful when precise alignment is not critical.

2.2.4 Conductive grid collector

A conductive grid collector is a grid or mesh structure placed between the needle and a grounded plate. It can help collect nanofibers in a more organized manner, creating a mesh-like structure. Grid collectors produce nanofiber meshes with an ordered structure. Grid spacing and geometry is adjusted to control the mesh properties. Grid collectors create nanofiber meshes with controlled pores and porosity, making them suitable for applications where controlled filtration or permeability is required.

2.2.5 Basket or frame collector

These collectors consist of three-dimensional structures or frames that are used to collect nanofibers in specific configurations. They are useful for applications requiring three-dimensional scaffolds. These collectors create three-dimensional nanofiber scaffolds, allowing for the design of complex structures for tissue engineering and other applications. These collectors form porous three-dimensional structures with high surface area, which is valuable for tissue engineering and scaffolding applications.

Basic electrospinning is the conventional method where a single nozzle is used to generate nanofibers from a polymer solution. This technique is straightforward and widely applicable, providing a simple means of producing nanofibers with diameters ranging from a few nanometers to micrometers (figure 2.2(a)). Coaxial electrospinning involves the use of two concentric nozzles, allowing the simultaneous electrospinning of core–shell structured fibers. This technique enables the encapsulation of different materials within the fibers, providing enhanced functionalities and controlled release properties (figure 2.2(b)). Side-by-side electrospinning utilizes two separate nozzles positioned adjacent to each other, enabling the simultaneous electrospinning of two different polymer solutions. This technique is valuable for creating composite fibers with distinct properties from each component (figure 2.2(c)). Multiple-jet electrospinning involves the simultaneous operation of multiple nozzles, enhancing production throughput and allowing for the fabrication of complex structures. This technique is advantageous for scaling up the electrospinning process (figure 2.2(d)). Metallic plate collectors are commonly used for basic electrospinning. The nanofibers are randomly deposited on the collector, forming a non-woven mat. This collector type is suitable for applications where fiber alignment is not critical (figure 2.2(e)). Drum collectors rotate during the electrospinning process, resulting in aligned nanofiber structures. This controlled alignment is particularly beneficial for applications such as tissue engineering, where scaffold architecture plays a crucial role (figure 2.2(f)).

Parallel electrode collectors consist of two parallel plates, providing a controlled environment for the deposition of nanofibers. This collector type is suitable for applications requiring uniformity and precise control over fiber orientation (figure 2.2(g)). The array of counter electrodes involves the use of multiple electrodes arranged in an array, enabling the deposition of nanofibers on individual electrodes. This collector type is advantageous for high-throughput applications, allowing for the simultaneous electrospinning of multiple fibers (figure 2.2(h)).

2.3 Effect of the bias polarity on nanofibers

Bias polarity, or the direction of the electric field, plays a significant role in the electrospinning process and has specific effects on the properties of the resulting fibers (figure 2.3). Let us explore in detail how changing the bias polarity influences various aspects of electrospun fiber properties [19–22]

The following properties have been tuned by using the unlike bias polarity in electrospinning.

2.3.1 Fiber morphology

Positive bias (cations repelled): When positive bias is applied, it repels positively charged ions in the polymer solution, promoting the formation of thinner and more

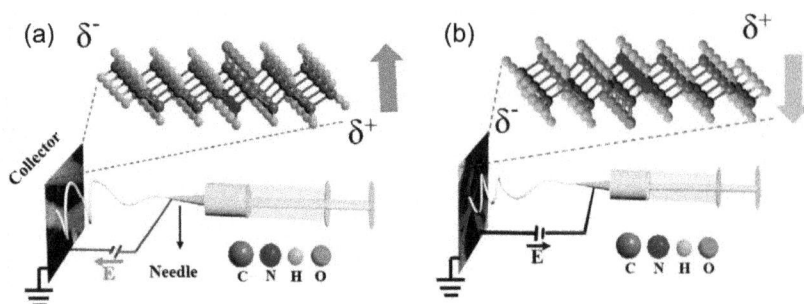

Figure 2.3. (a) +ve bias polarity, and (b) −ve bias polarity in the electrospinning setup. Reproduced from [25] with permission from Sage Publications.

uniform fibers. Electrostatic repulsion assists in the stretching and thinning of the polymer jet, resulting in finer fibers.

Negative bias (cations attracted): In contrast, a negative bias attracts positively charged ions in the solution, which can disrupt the formation of continuous fibers. Instead, it leads to the generation of electrosprayed droplets or particles, making it challenging to produce fibers under negative bias.

2.3.2 Fiber diameter

Positive bias: Positive bias generally leads to smaller fiber diameters due to the repulsion of positively charged polymer ions. Finer fibers often have a higher surface area, making them suitable for applications such as filtration, tissue engineering, and drug delivery.

Negative bias: If negative bias is used for electrospraying, it produces smaller droplets or particles with diameters, which is controlled by adjusting the magnitude of the negative bias.

2.3.3 Fiber alignment and orientation

Positive biases (figure 2.3(a)): Electrospinning under positive bias typically results in random fiber orientation as the fibers are ejected from the Taylor cone in various directions.

Negative bias (figure 2.3(b)): When a negative bias is applied, it enhances fiber alignment. Electrostatic forces tend to align the fibers along the direction of the electric field, leading to improved fiber orientation.

2.3.4 Surface charge and porosity

Positive bias: Fibers produced under positive bias are often positively charged. This positively charged surface affects the interactions of the fibers with negatively charged entities, which is relevant in applications where surface charge is crucial.

Negative bias: Fibers produced under negative bias may acquire a negative charge. This charge influences interactions with positively charged species, which is useful in applications such as drug delivery.

2.3.5 Bead formation

Positive bias: Under certain conditions, positive bias leads to the formation of beads along the fibers due to instability in the jetting process. Beads affect the mechanical properties of the fibers, which is undesirable in some applications.

Negative bias: Electrospraying under negative bias is more prone to bead formation, resulting in the production of smaller droplets or particles.

2.3.6 Material composition

Bias polarity also influences the incorporation of additives, nanoparticles, or drugs into fibers through electrostatic interactions. The choice of bias polarity affects the distribution and concentration of these additives within the fibers.

Bias polarity is one of several key parameters (including voltage, solution properties, and collector geometry) that is optimized to customize the electrospinning process. Figure 2.4 provides a comprehensive exploration of the nuanced effects resulting from the manipulation of various electrospinning parameters on the distinctive properties exhibited by nanofibers. In this visual representation, detailed SEM micrographs are presented, showcasing the morphology of figures 2.4(a) and (b) impeccably smooth fibers, as well as the intriguing characteristics of figures 2.4(c) and (d) porous polycaprolactone (PCL) (+) and PCL (−) respectively. Furthermore, figure 2.4(e) captures the visual essence of a PCL film, elucidating its unique features. For a more quantitative insight, figure 2.4(f) presents a histogram that meticulously outlines the distribution of PCL fiber diameters, offering a comprehensive view of the dimensional landscape within the nanofiber ensemble.

2.3.7 Alignment of ferroelectric dipoles

When ferroelectric electrospinning materials, such as certain piezoelectric polymers or ceramics, the presence of ferroelectricity introduces unique properties and behaviours into the electrospinning process. Ferroelectric materials have spontaneous electric dipoles, which are reversed by an applied electric field.

Ferroelectric materials possess spontaneous electric dipoles, which are naturally aligned along a specific crystallographic axis when the material is in its ferroelectric phase. In the electrospinning process, the application of an external electric field influences the alignment of these dipoles. By adjusting the direction and magnitude of the applied electric field during electrospinning, it is possible to control the orientation of the ferroelectric dipoles in the resulting fibers. This orientation control is crucial to achieving specific material properties and performance characteristics. Piezoelectric behaviour: ferroelectric materials often exhibit piezoelectric properties, meaning that they generate an electric charge when subjected to mechanical stress. By aligning the ferroelectric dipoles during electrospinning, you enhance the piezoelectric response of the resulting fibers. These piezoelectric fibers are used in sensors, actuators, and energy harvesting devices. Properly aligned dipoles in electrospun fibers also enhance their ferroelectric properties, which involves the ability to switch

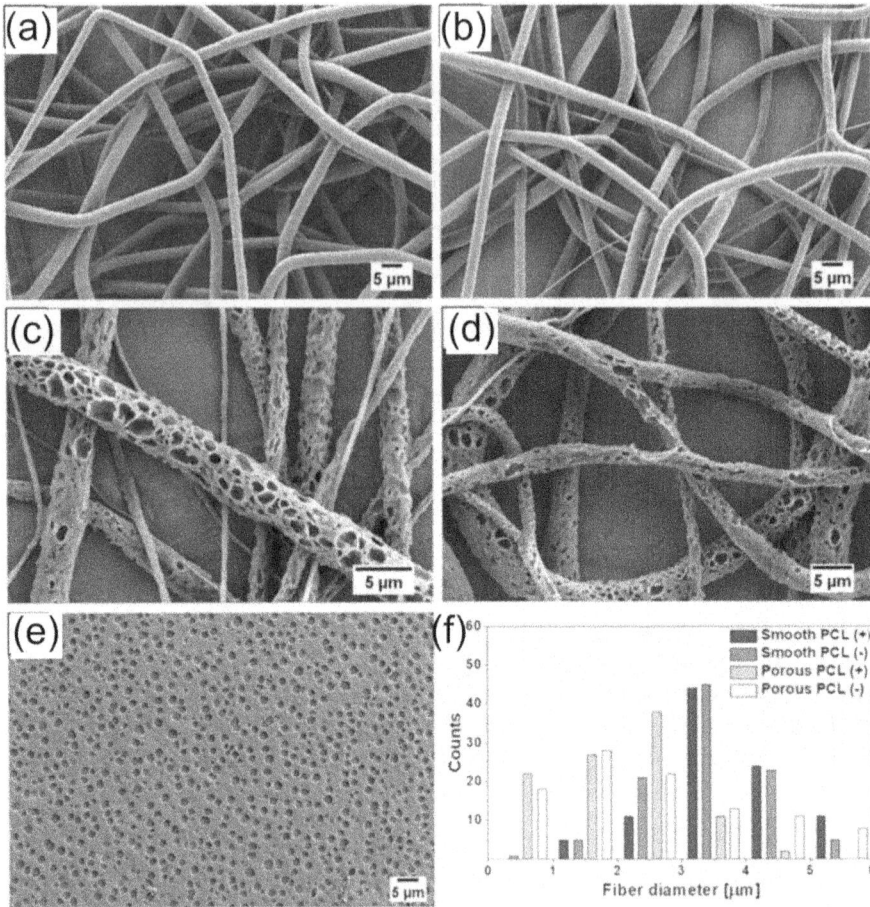

Figure 2.4. Illustrates the impact of various electrospinning parameters on the various properties of nanofibers. The SEM micrographs show smooth fibers (a) and (b), (c) and (d) depict porous PCL (+) and PCL (−), (e) presents a PCL film, and (f) a histogram depicting the distribution of PCL fiber diameters. Reproduced with permission from [64] Copyright 2020 Elsevier. CC BY 4.0.

between polarisation states in response to an external electric field. This property is valuable in applications related to memory devices and nonvolatile storage [23–28].

2.4 Characterisation techniques

Characterizing nanofibers produced under different bias polarities is essential to understand their properties and performance in various applications. Several characterization techniques are employed to evaluate nanofibers, and the choice of techniques depends on the specific properties of interest. Here are some common characterization techniques for nanofibers produced under different bias polarities [29–37].

2.4.1 Scanning electron microscopy (SEM)

SEM is a widely used technique to visualise the morphology and surface topography of nanofibers. It provides high-resolution images, allowing for the assessment of factors such as fiber diameter, fiber alignment, and the presence of beads or defects. SEM helps to compare the change in morphology of nanofibers produced under different bias polarities (both positive and negative bias).

2.4.2 Atomic force microscopy (AFM)

AFM allows for the nanoscale topographical and mechanical characterisation of nanofibers. It measures surface roughness, stiffness, and adhesion properties. AFM is valuable for understanding the mechanical properties of nanofibers produced under different bias polarities. AFM-based Kelvin probe force microscopy (KPFM) is one of the potential tools to probe the surface potential of materials at the nanoscale. The surface potential of different materials under positive and negative electrospinning is illustrated in figure 2.5. Figure 2.5(a) intricately unveils the nuanced interplay between surface potential and diverse biases of polycaprolactone (PC) in a spectrum of

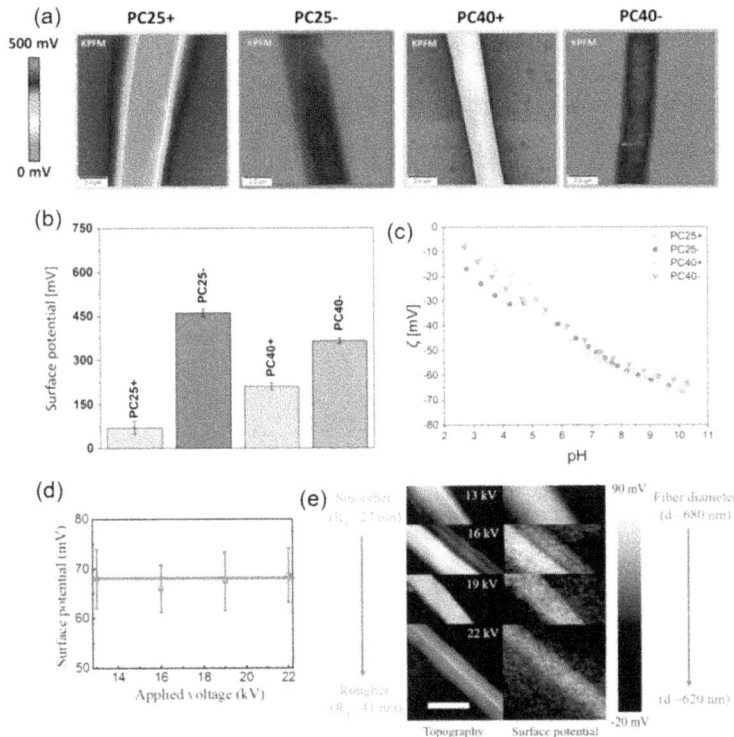

Figure 2.5. Effect of voltage on the (a) surface potential and (b) statistical potential distribution at different humidities and (c) zeta potential of polycaprolactone, (d) Surface potential distribution at different voltage and (e) KPFM images of nylon-11 nanofibers. (c) Reproduced with permission from [28] Copyright 2021 American Chemical Society. Reproduced with permission from [29] John Wiley & Sons. Copyright (2022) Wiley-VCH GmbH.

humidity levels. Figures 2.5(b) and (c) explore the dynamic shifts in zeta potential, offering a comprehensive exploration of the electrochemical landscape.

Figure 2.5(d) delves deeper, elucidating the impact of increasing voltage during electrospinning process, while figure 2.5(e) meticulously captures the absence of perceptible changes in the surface potential, thus enriching our understanding of electrospinning dynamics under varying voltage conditions.

2.4.3 X-ray diffraction (XRD)

X-ray diffraction (XRD) is used to determine the crystallographic structure and orientation of nanofibers. It helps to identify changes in the crystalline structure induced by variations in bias polarity. XRD provides information on the degree of crystallinity and the presence of different polymorphs. Figure 2.6(a) illustrates the x-ray diffraction (XRD) of positively and negatively poled electrospinning nanofibers.

2.4.4 Fourier transform infrared spectroscopy (FTIR)

FTIR is a spectroscopic technique used to identify chemical functional groups and bonding within nanofibers. It helps assess the chemical composition and molecular structure of nanofibers produced under different bias polarities, which influence their properties and applications [38–47].

2.4.5 Differential scanning calorimetry (DSC)

The DSC measures the heat flow associated with phase transitions and is used to assess the thermal properties of nanofibers. It reveals changes in melting points, glass transition temperatures, and crystallinity caused by bias-polarity variations. Changes in differential scanning calorimetry (DSC) of alternatively biased nanofibers are shown in figure 2.6(b) as well as a crystallinity table containing the values (figure 2.6(c)).

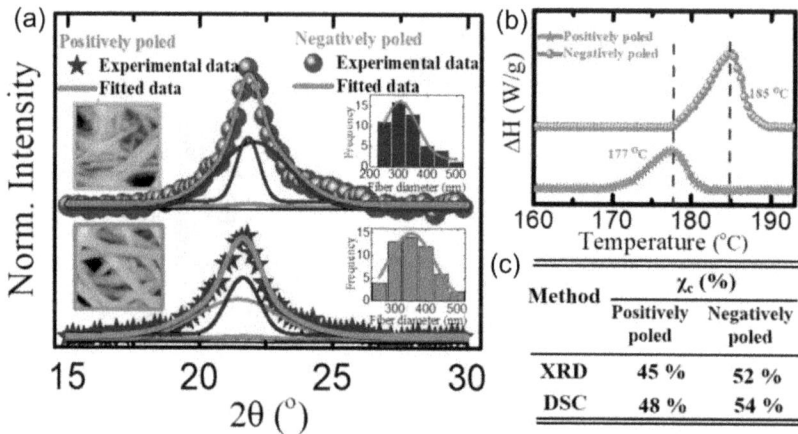

Figure 2.6. Characterization of the produced nanofibers: (a) XRD pattern, (b) DSC, and (c) Summary table of the crystallinity values of positively and negatively biased electrospinning nanofibers. Reproduced with permission from [27], copyright 2022 AIP Publishing.

2.4.6 Mechanical testing

The mechanical properties of nanofibers, such as tensile strength, elasticity, and Young's modulus, are evaluated through mechanical testing methods such as tensile testing or nanoindentation. These tests help determine how bias polarity influences the mechanical behavior of nanofibers.

2.4.7 Energy harvesting performance tests

For energy harvesting applications, the characterization involves testing the energy conversion efficiency and output of nanofiber-based devices under different bias polarities and mechanical or environmental conditions [48–53].

Figures 2.8(a) and (b) represent the SEM images and the diameter distribution of positively poled and negatively poled nanofibers, respectively. Figures 2.8(c) and (d) show the statistical distribution of fibers.

The presented SEM images in figures 2.7(a) and (b) depict nanofibers subjected to positive and negative biases, respectively, revealing distinct surface morphologies influenced by the applied electrical charge. The images suggest that the electrostatic

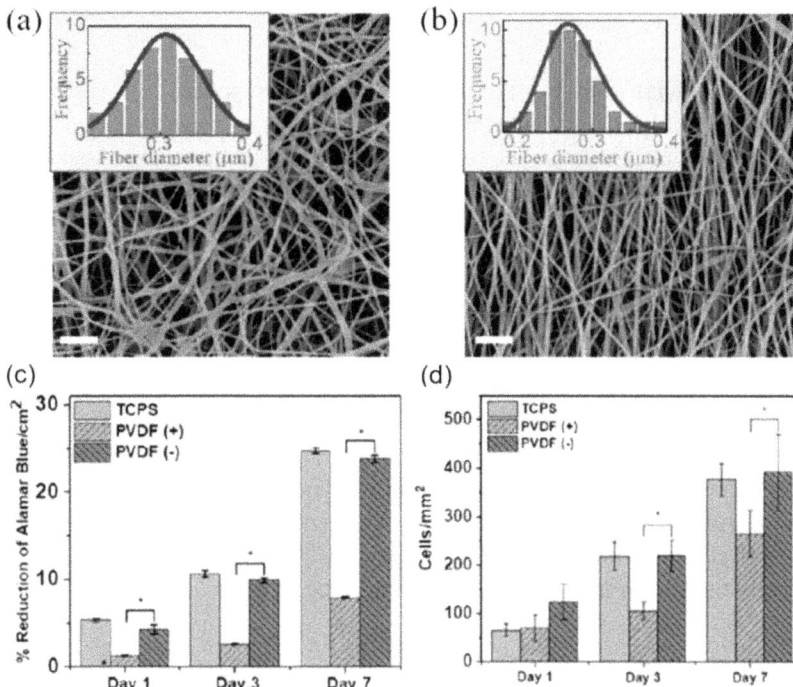

Figure 2.7. Scanning electron microscopy (SEM) images of the (a) positively biased and (b) negatively biased nanofibers. (c) Quantification of cell proliferation and growth on different fiber substrates. (c) Histograms represent cell proliferation measured by the Alamar Blue assay at 1, 3, and 7 days in culture. (d) Analysis of cell density per mm2 in TCPS control, polyvinylidene fluoride (PVDF) (+), and PVDF (−) fibers over the same culture period. Asterisks (*) indicate significant differences between PVDF (+) and PVDF (−) samples, determined using the Tukey test ($p < 0.05$). Reproduced with permission from [27], copyright 2022 AIP Publishing.

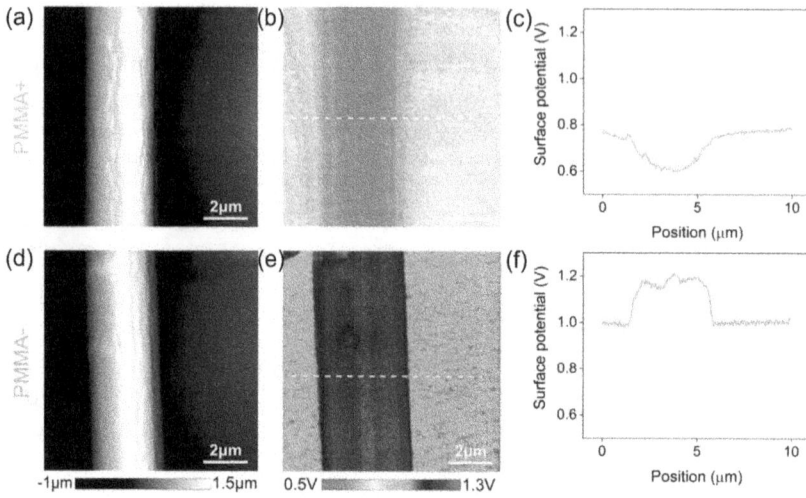

Figure 2.8. AFM data depicting PMMA fibers. (a) Topographic representation. (b) Visualisation of the surface potential. The dashed line indicates the position of the line scan. (c) Line profile along the *x*-axis illustrating the surface potential of PMMA+ fibers. (d) Topographic view. The dashed line denotes the line scan location. (f) Profile of the *x*-axis line detailing PMMA− fibers. Reproduced with permission from [30] Copyright 2019 Elsevier. CC BY 4.0.

forces play a crucial role in shaping the nanofibers, potentially affecting their physicochemical properties and interactions with biological entities.

To gain further insights into the biological response to these nanofiber substrates, figure 2.7(c) provides a quantitative assessment of cell proliferation and growth. The Alamar Blue assay, a widely used indicator of cellular metabolic activity, was employed to measure cell proliferation at 1, 3, and 7 days in culture. The histograms indicate varying levels of cell proliferation on different fiber substrates, emphasizing the temporal dynamics of cellular responses.

In figure 2.7(d), the analysis of cell density per mm^2 is presented over the same culture period for three different conditions: tissue culture polystyrene (TCPS) control, positively biased PVDF (+) fibers, and negatively biased PVDF (−) fibers. The data reveal noteworthy distinctions in cell density, suggesting that the electrical biasing of PVDF nanofibers significantly influences cell adhesion and proliferation. The asterisks (*) highlight statistically significant differences between positively and negatively biased PVDF samples, as determined by the Tukey test ($p < 0.05$), underscoring the importance of electrical polarization in modulating cellular responses.

2.5 Literature review

The impact of electrical polarity on the electrospinning process is explored. The study reveals that electrical polarity influences the process dynamics, mechanical properties, and microstructure of electrospun fibers. Specifically, the length of the straight jet is determined by the electric field strength, while the fiber deposition radius remains independent of the electric field strength and polarity. A negatively charged spinneret results in longer straight jets and wider fibers, whereas a positively charged setup leads to fibers with enhanced mechanical properties and higher crystallinity. The investigation focuses on the effect of positive or negative voltage on the mechanical performance of polymethyl methacrylate (PMMA) fiber mats [24].

The surface potential and roughness of PCL fibers play a crucial role in influencing cell adhesion and collagen mineralization. A higher surface potential is found to enhance collagen mineralization after 7 days, and negative zeta potential promotes calcium mineralization for tissue formation. Control of the surface potential through voltage polarity is highlighted, and it is demonstrated that porous scaffolds exhibit higher cell adhesion and capture percentage. Electrospun PCL scaffolds enhance cell-material interaction and cellular activity. The study emphasizes the role of surface potential in regulating cell adhesion and signaling for tissue regeneration [59].

In surface potential tailoring of PMMA fibers by electrospinning for enhancement, the focus is on modifying the surface potential of PMMA fibers through voltage polarity during the electrospinning process. Research demonstrates that electrospun PMMA fibers exhibit enhanced triboelectric performance compared to spin-coated films. Positive voltage polarity imparts a more tribonegative character to the fibers, while negative polarity results in a higher tribopositive character. The study employs x-ray photoelectron spectroscopy (XPS) to analyse the surface chemistry of electrospun PMMA fibers, revealing the effective control of the position of oxygen in the ester group.

The tailor-made adjustment of surface potential enhances the triboelectric energy harvesting capability of PMMA fibers, with potential applications in the development of wearable and implantable medical devices. Figure 2.8 presents a detailed analysis of electrospun PMMA nanofibers, encompassing morphology, surface potential, and line profiling. In figure 2.8(a), PMMA+ exhibits a distinct nanofiber arrangement influenced by the electrospinning process, resulting in a well-defined morphology. Figure 2.8(b) provides the surface potential map for PMMA+, revealing electrostatic properties with colour-coded potential levels. Understanding surface potential is crucial for its impact on nanofiber interactions in various applications. Figure 2.8(c) complements the analysis with line profiling of PMMA+, offering a quantitative assessment of surface potential variations in the nanofiber diameter. Figures 2.8(d)–(f) compare morphology, surface potential, and line profiling for PMMA−. Differences between PMMA+ and PMMA− provide insights into how electrospinning variations influence nanofiber characteristics. The morphology of PMMA− (figure 2.8(d)) highlights distinctive features compared to PMMA+, essential for tailoring the properties for specific applications such as filtration or tissue engineering. Figure 2.9(e)'s surface potential map for PMMA− enables a direct comparison with PMMA+, indicating potential variations with implications for adhesion, sensing, or controlled release applications [30].

Babu *et al* have proposed the fabrication of a single-material-based triboelectric nanogenerator (S-TENG). The surface potential of nylon-11 nanofibers in S-TENG is tuned by electrospinning, resulting in ultra-high mechanoacoustic sensitivity and a wide frequency response range. The S-TENG is showcased as an ultra-high-sensitivity acoustic sensor capable of discriminating different instrumental sounds

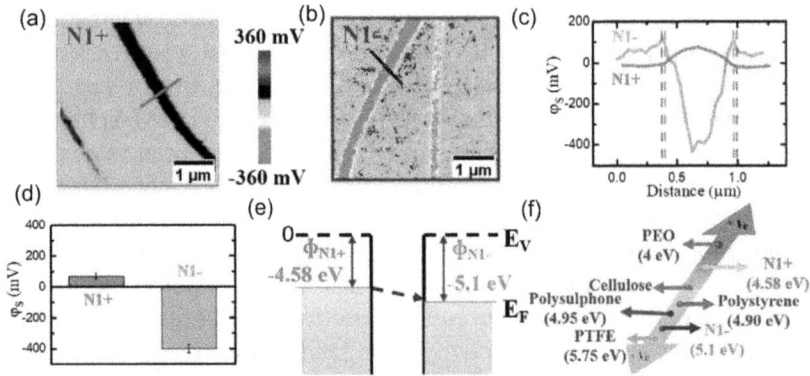

Figure 2.9. Surface potential images of (a) N1+ and (b) N1 measured by kelvin probe force microscopy (KPFM). (c) Profile of the x-axis line of N1+ and N1−. (d) Histogram representing the surface potential of N1+ and N1− (mean ± SD; $N = 8$). (e) Band diagram of N1+ and N1. (f) The triboelectric series consists of N1+ and N1− along with some reference materials. Reproduced with permission from [29] Copyright 2022 Wiley-VCH.

and human voice words. The research suggests potential applications in the evaluation of diseases related to speech signals in the healthcare sector. Figure 2.9 demonstrates the surface potential changes in the nylon-11 nanofibers (figures 2.9 (a)–(d), respectively). Change in the work function due to changes in the bias voltage (figure 2.9(e)) and change in position in the triboelectric series (figure 2.9(f)) [29].

Ura *et al* explore how adjusting electrical polarity and humidity improve the water collection efficiency of electrospun meshes. The study reveals that PC meshes with negative electrical polarity and 40% humidity exhibit a higher water collection rate. The electrospinning process at different relative humidities influences the fiber diameter. The control of voltage polarity and humidity is shown to be crucial to enhance water collection capabilities. The research emphasises the significance of surface potential and morphology adjustments in a single-step fiber production process for efficient water harvesting from fog [32].

Urabanek *et al* demonstrated the influence of charge polarity on PCL/CHT fiber morphology and properties. Negative polarity is found to increase PCL-chitosan interactions and reduce fiber wettability, resulting in lower PCL crystallinity.

The study employs SEM imaging to reveal differences in fiber morphology based on charge polarity, and negative polarity is shown to improve the efficiency of surface modification by chondroitin sulfate. The research highlights the importance of charge polarity in controlling the organization of polymer functional groups and its impact on the electrospinning process [23].

Cell proliferation and collagen mineralization on electrospun. This article discusses the use of positive and negative voltage polarities during electrospinning to control the surface potential on polymer fibers used as scaffolds in tissue engineering applications. PVDF scaffolds with a higher surface potential (−95 mV) are observed to show increased cell proliferation, cell and filopodia, and enhanced collagen production. Two types of PVDF scaffolds, PVDF(+) and PVDF(−), are investigated, with PVDF(−)

scaffolds demonstrating stronger cell adhesion, proliferation, and greater potential for bone regeneration. The article emphasizes the tailored adjustment of the fibrous scaffold potential of the cell membrane potential for fast mineralization. It presents similar findings regarding the surface potential control on polymer fibers used as scaffolds in tissue engineering applications. The study, which focuses on PVDF scaffolds with a higher surface potential (-95 mV), reports increased cell proliferation, cell shaping of cells and filopodia shaping, and enhanced collagen production. Adjustment in the surface potential of scaffolds is demonstrated to enhance collagen mineralization without additional biochemical modifications. Similarly to the previous article, PVDF($-$) scaffolds show stronger cell adhesion and proliferation, with greater potential for bone regeneration. Research highlights the tailoring of fibrous scaffolds to the potential of cell membranes for fast mineralization [64].

2.6 Applications across diverse domains

Electrospinning fibers are promising in diverse applications, from energy harvesting to self-powered sensor applications. Owing to the functionality of the air permeable, flexible, and easy to produce, it has provided applications in numerous dimensions. We will discuss applications in different categories [54–62].

2.6.1 Energy harvesting applications

Nanofibers have garnered significant attention in recent years as promising materials for energy harvesting applications. Energy harvesting refers to the process of converting ambient energy sources, such as mechanical vibrations, thermal gradients, or solar radiation, into usable electrical energy. Here, we will discuss how bias polarity, different collectors, and other factors can influence the energy harvesting properties of nanofibers.

2.6.1.1 Piezoelectric effect
Piezoelectric materials generate electrical charges when subjected to mechanical stress or vibration, making them suitable for mechanical energy harvesting. Nanofibers are engineered to possess piezoelectric properties, and their energy-harvesting capabilities are influenced by various factors:

Positive bias: When a positive bias to piezoelectric nanofibers, it improves the alignment of ferroelectric dipoles, which may lead to an increased piezoelectric response under mechanical stress.

Negative bias: A negative bias may have a different effect depending on the specific properties of the material. It influences the alignment of dipoles in the opposite direction, potentially altering the piezoelectric response.

Rotating drum collector: Using a rotating drum collector during electrospinning aligns the piezoelectric nanofibers in a specific direction, enhancing their response to mechanical vibrations along that axis.

Flat plate collector: A flat plate collector results in randomly orientated nanofibers, which have isotropic piezoelectric properties, making them suitable for harvesting energy from vibrations in multiple directions.

2.6.1.2 Triboelectric effect

Triboelectric nanogenerators (TENGs) are devices that harvest energy through the contact and separation of materials with different electronegativities. Nanofibers are integrated into TENGs to enhance their performance:

Bias polarity in TENGs: In TENGs, the choice of materials with different electronegativities plays a more critical role than bias polarity. The relative electronegativities of the materials in contact determine the triboelectric effect.

Collector design in TENGs: TENGs often use specific collector designs with one electrode serving as a stationary part and another as a moving part to generate friction. The choice of materials and their interaction with the nanofibers affect TENG performance. The piezoresponse phase and amplitude characteristics of the electrospinning fibers under both positive and negative bias conditions are examined, as illustrated in figures 2.10(a) and (b) for positively biased fibers, and figures 2.10(c) and (d) for negatively biased fibers.

2.6.1.3 Photovoltaic effect

Nanofibers are used in photovoltaic applications to convert solar radiation into electrical energy. Although bias polarity and collectors may not have a direct impact on photovoltaic properties, their morphology and alignment can influence efficiency:

Collector surface texture: The choice of collector surface texture affects the alignment and packing density of photovoltaic nanofibers. A structured collector surface promotes better light absorption and electron transport.

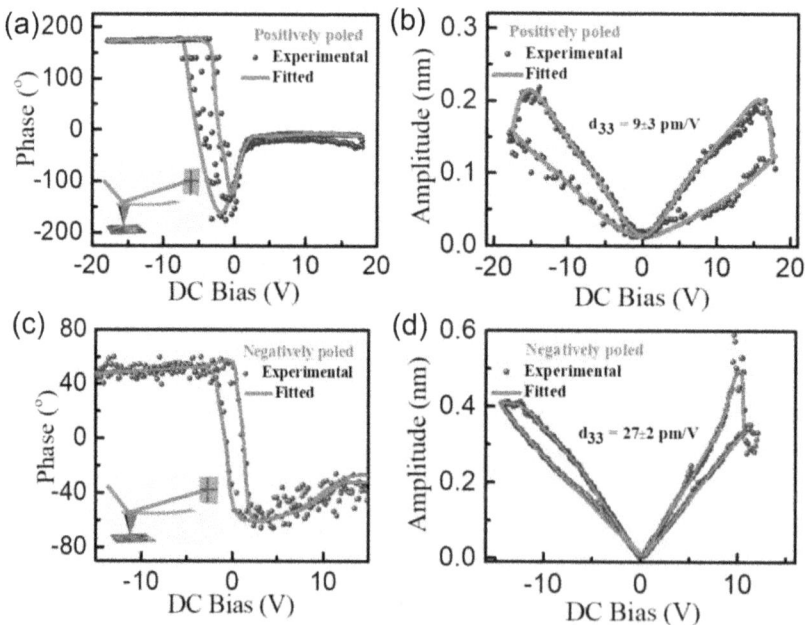

Figure 2.10. Piezoelectric force microscopy measurement of positively biased (a) and (b) and negatively biased (c) and (d) produced nanofibers. Reproduced with permission from [27], copyright 2022 AIP Publishing.

2.6.1.4 Thermal effect

Thermoelectric nanofibers generate electricity from temperature gradients. Bias polarity and collector design do not directly affect their performance, but other factors like material choice and nanofiber alignment are crucial.

Thermal conductivity: The choice of materials in thermal nanofibers influences their thermoelectric efficiency. Materials with low thermal conductivity and high electrical conductivity are preferred.

Alignment: Proper alignment of the thermoelectric nanofibers along the direction of the temperature gradient is essential for maximizing the thermoelectric efficiency.

2.6.2 Self-powered sensors applications

Self-powered sensors based on electrospun nanofibers have found diverse applications in various fields. The inherent piezoelectricity of these nanofibers, coupled with bias polarity changes, contributes to the development of innovative and efficient sensing mechanisms. This section explores two key sensing mechanisms: piezoelectric sensing and chemical sensing [63–72].

2.6.2.1 Electromechanical sensing

The electromechanical properties of electrospun nanofibers play a pivotal role in the development of self-powered sensors. This section delves into specific applications, such as:

Pressure sensors for prosthetics

Electrospun nanofiber-based pressure sensors, responsive to bias polarity changes, offer a promising solution for prosthetics. Sensors detect pressure variations, providing valuable feedback for improved comfort and functionality in prosthetic limbs.

Impact sensors for structural health monitoring

In structural health monitoring, the use of electrospun nanofiber-based impact sensors is instrumental. The sensors, driven by changes in bias polarity, can effectively detect and analyze impacts on structures, aiding in the early identification of potential problems.

2.6.2.2 Chemical sensing

Electrospun nanofibers are designed for chemical sensing applications. Sensitivity to bias polarity changes enhances their utility in self-powered sensors for detecting specific gases or substances. Applications explored include the following.

Manufacturing industry

In manufacturing, self-powered sensors that utilize electrospun nanofibers are used for chemical sensing. Sensors dynamically respond to changes in bias polarity, making them valuable for detecting and monitoring chemical processes in manufacturing environments.

Agriculture

The agricultural sector can benefit from electrospun nanofiber-based sensors designed for chemical sensing. These sensors, which are responsive to bias polarity changes, can detect and monitor specific chemicals relevant to soil health and crop management.

2.6.3 Healthcare applications

The applications of self-powered sensors span across diverse domains, showcasing the versatility and adaptability of electrospun nanofiber-based sensing technologies [73–81].

2.6.3.1 Physiological monitoring

Beyond wound healing applications, electrospun nanofiber-based sensors find utility in real-time physiological monitoring. In particular, they contribute to the following.

Continuous glucose monitoring for diabetic patients

The inherent sensitivity of electrospun nanofibers concerning bias polarity enables the development of noninvasive, self-powered sensors for continuous glucose monitoring. This application provides diabetic patients with a convenient and effective tool for managing their health.

2.6.3.2 Drug release systems

Electrospun nanofibers are engineered for drug delivery systems, with different bias polarity; influencing encapsulation and release rates. This section explores the following.

Targeted drug delivery in cancer treatment and chronic disease management

The precise control offered by electrospun nanofiber-based sensors makes them promising candidates for targeted drug delivery. In cancer treatment and chronic disease management, these sensors can enhance therapeutic efficacy while minimising side effects.

2.6.4 Environmental and aerospace monitoring

Environmental monitoring stands to benefit significantly from the use of self-powered sensors based on electrospun nanofibers.

2.6.4.1 Monitoring of air and water quality

The capacity of electrospun nanofibers to dynamically respond to changes in bias polarity positions them as ideal candidates for transforming air quality monitoring. Some specific applications encompass:

Detection of particulate matter, volatile organic compounds (VOCs), and airborne contaminants

Self-powered sensors efficiently detect particulate matter, VOCs, and other airborne contaminants. This capability improves the accuracy and sensitivity of air quality monitoring systems.

In monitoring water quality, electrospun nanofiber-based sensors offer distinct advantages.

Sensitivity and durability in detecting contaminants

The enhanced sensitivity of electrospun nanofibers to variations in bias polarity increases their efficacy in detecting contaminants in water sources. This positions them as invaluable instruments for protecting drinking water quality and actively contributing to environmental conservation initiatives. The aerospace industry can leverage self-powered sensors based on electrospun nanofibers for various applications [82–88].

2.6.4.2 Structural health monitoring

Structural health monitoring is crucial to ensuring the safety and reliability of aircraft components. Electrospun nanofiber-based sensors, responsive to bias polarity changes, offer the following.

Early detection of wear, fatigue, or damage

By enhancing sensor sensitivity, bias polarity changes contribute to the early detection of wear, fatigue, or damage to critical aircraft components. This proactive approach helps with maintenance and ensures the structural integrity of aerospace systems.

In agriculture, self-powered sensors can play a pivotal role in optimising crop yields and promoting sustainable practices.

2.6.4.3 Soil health monitoring

Self-powered sensors based on electrospun nanofibers are well-suited for real-time soil health monitoring.

Real-time data on soil conditions

Electrospun nanofibers, with their sensitivity to different bias polarities, provide real-time data on soil conditions. This information is invaluable for implementing precision agriculture practices and optimizing crop yields. The integration of electrospun nanofibers into self-powered sensors opens up a wide array of possibilities across these diverse domains, showcasing their potential to revolutionize various industries. Figure 2.11 represents the various aspects of electrospinning fibers for energy harvesting applications.

2.6.5 AI/ML integration in electrospun nanofiber applications

2.6.5.1 Predictive maintenance

Incorporating machine learning algorithms into systems utilising electrospun nanofiber sensors enables predictive maintenance (figure 2.12). Through the examination of historical data related to structural health, chemical exposure, or soil conditions, artificial intelligence (AI) can anticipate the need for maintenance or intervention, effectively minimising downtime and lowering overall maintenance costs. Intelligent wearable sensors, which incorporate electrospun nanofibers, are revolutionising

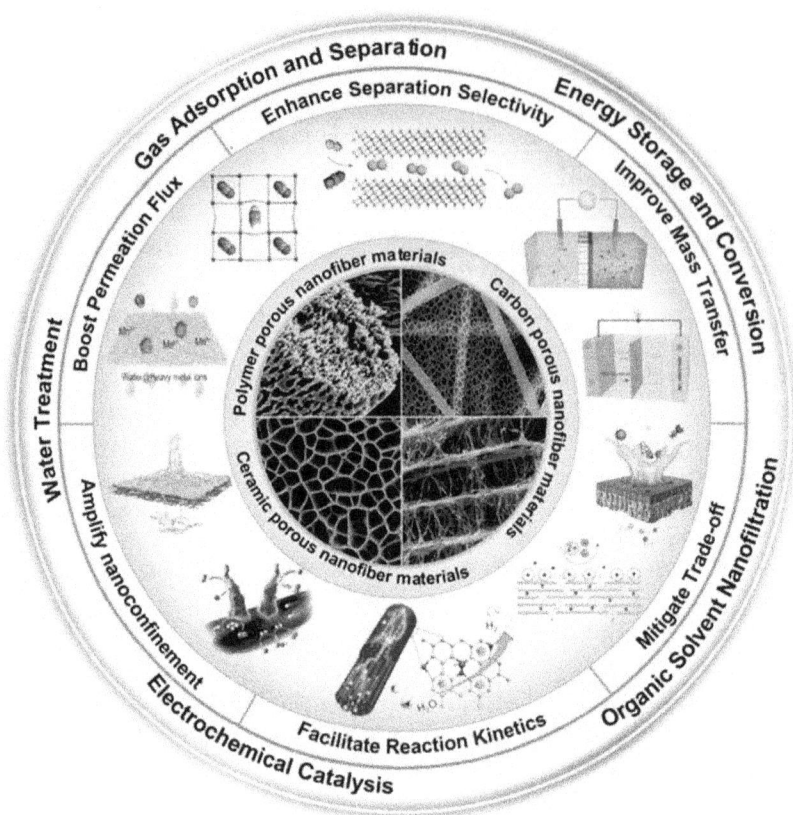

Figure 2.11. Various aspects of tuning electrospinning nanofibers for energy harvesting applications. Reproduced from [82], copyright (2023), with permission from Elsevier.

structural health monitoring. These sensors, when integrated with AI algorithms, continuously assess the structural integrity of various components in real time. The archival data obtained from wearable sensors plays a crucial role in empowering machine learning models to forecast potential issues or structural fatigue. This, in turn, facilitates prompt and targeted maintenance interventions. Such intelligent wearables find applications in construction, civil engineering, and even personal safety equipment.

2.6.5.2 Pattern recognition and smart wearables

AI can enhance the capabilities of electrospun nanofiber-based sensors by providing advanced pattern recognition. This is particularly useful in applications such as air quality monitoring, where AI algorithms can identify specific pollutant patterns, aiding in source identification and targeted pollution control measures.

Wearable devices incorporating electrospun nanofiber sensors can be worn by individuals for personal air quality monitoring. AI-driven pattern recognition algorithms analyse sensor data, identifying trends and specific pollutant patterns

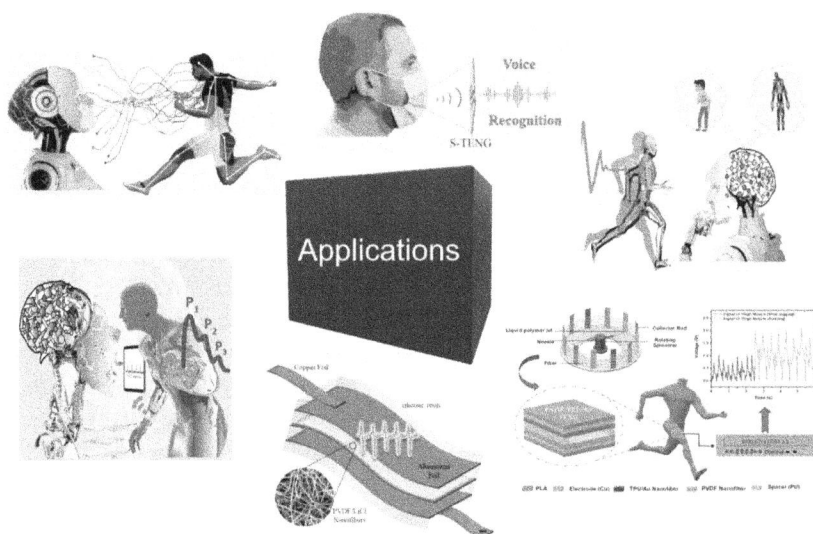

Figure 2.12. Diverse applications of the sensors fabricated from electrospinning fibers. Reproduced from [97] with permission from Springer Nature; [29] John Wiley & Sons, copyright (2022) Wiley-VCH GmbH; [93] John Wiley & Sons, copyright (2023) Wiley-VCH GmbH; reprinted with permission from [58], copyright (2021), American Chemical Society; reproduced from [77] CC BY 4.0; [96] John Wiley & Sons, copyright (2023) Wiley-VCH GmbH.

in the wearer's environment. This information not only provides real-time feedback to the individual, but also contributes to larger datasets used for urban planning and pollution control initiatives. The integration of AI ensures accurate and context-sensitive analysis of air quality patterns, making the wearables more effective in providing relevant information to users.

In healthcare applications, AI algorithms can analyse continuous physiological monitoring data from electrospun nanofiber sensors. This allows for personalised medicine approaches, where treatment plans can be tailored based on individual patient responses, optimising therapeutic outcomes and minimising side effects.

2.6.5.3 Early assessment of disease biomarkers

Intelligent wearable devices that incorporate electrospun nanofiber sensors are becoming an integral part of personalised healthcare. These wearables, equipped with AI algorithms, continuously monitor physiological parameters such as heart rate, blood glucose levels, and other vital signs.

Nanofiber sensors provide accurate and reliable data, while AI analyses this information to create personalised health profiles for individuals. Machine learning models predict health trends, enabling early detection of potential problems and allowing healthcare providers to customise treatment plans for patients. Electrospun nanofibers are engineered to serve as drug delivery systems in wearable devices [88–94]. By integrating AI, these wearables can adapt drug release profiles based on real-time physiological data. For example, an insulin delivery wearable for diabetic patients could use AI to adjust the insulin release rates according to the wearer's glucose levels, ensuring optimal blood sugar control. This intelligent drug delivery approach improves the effectiveness of treatments and minimizes the risk of side effects [95–100].

2.7 Conclusion and future prospective

This chapter has unveiled the fascinating interplay between bias polarity and fiber properties in electrospinning, showcasing its potential as a potent tool for crafting advanced sensor materials. By meticulously adjusting the bias, we have achieved exquisite control over fiber diameter, morphology, surface texture, and functional group orientation. This fine-tuned control paves the way for the development of highly sensitive, selective, and responsive sensors, marking a significant leap forward in sensor technology. However, the implications of this technique extend far beyond sensor applications. Tailoring fiber properties through bias polarity modulation holds immense promise for diverse fields, including energy storage, filtration, drug delivery, and even tissue engineering. As we delve deeper into this realm, we can envision a future where electrospun materials play a pivotal role in shaping cutting-edge technologies and revolutionising various industries.

The future of bias polarity in electrospinning is brimming with exciting possibilities. To fully unlock its potential, we must delve further into optimising the electrospinning process. Precise control over voltage, flow rate, and solution properties will enable even more refined manipulation of fiber characteristics and sensor performance, pushing the boundaries of what is achievable. Additionally, exploring the frontiers of co-electrospinning and nanoparticle integration opens doors to the development of multifunctional sensors capable of detecting multiple parameters simultaneously. This leap in complexity would usher in a new era of advanced sensing, with applications in environmental monitoring, healthcare diagnostics, and even security systems. Scaling up electrospinning for large-scale sensor production and seamless integration into existing devices presents an essential challenge. Addressing this hurdle will be crucial for translating laboratory break-throughs into real-world solutions. Additionally, investigating the use of biocom-patible and biodegradable materials holds immense promise for creating implantable medical devices and environmentally friendly sensors, paving the way for a more sustainable future. As we continue to unravel the secrets of bias polarity in electrospinning, we stand at the threshold of a transformative era in materials science and engineering. By embracing these exciting possibilities, we can forge a path toward groundbreaking technologies that benefit humanity in countless ways.

References

[1] Rutledge G C and Fridrikh S V 2007 Formation of fibers by electrospinning *Adv. Drug Deliv. Rev.* **59** 1384–91
[2] Greiner A and Wendorff J H 2007 Electrospinning: a fascinating method for the preparation of ultrathin fibers *Angew. Chem. Int. Ed.* **46** 5670–703

[3] Yoon J, Yang S, Lee S and Yu R 2018 Recent progress in coaxial electrospinning: new parameters, various structures, and wide applications *Adv. Mater.* **30** 1704765

[4] Taylor G I 1969 Electrically driven jets *Proc. R. Soc. Lond. A Math. Phys. Sci.* **313** 453–75

[5] Alharbi A, Alarifi I M, Khan W S and Asmatulu R 2016 Highly hydrophilic electrospun polyacrylonitrile/polyvinylpyrrolidone nanofibers incorporated with gentamicin as filter medium for dam water and wastewater treatment *J. Memb. Separ. Technol.* **5** 38–56

[6] Ji D, Lin Y, Guo X, Ramasubramanian B, Wang R, Radacsi N, Jose R, Qin X and Ramakrishna S 2024 Electrospinning of nanofibers *Nat. Rev. Methods Primers* **4** 1–21

[7] Xue J, Wu T and Dai Y 2019 Electrospinning and electrospun nanofibers: methods, materials, and applications *Chem. Rev.* **119** 5298–415

[8] Li D and Xia Y 2004 Electrospinning of nanofibers: reinventing the wheel? *Adv. Mater.* **16** 1151–70

[9] Shin Y M, Hohman M M, Brenner M P and Rutledge G C 2001 Electrospinning: a whipping fluid jet generates submicron polymer fibers *Appl. Phys. Lett.* **78** 1149–51

[10] Shin Y M, Hohman M M, Brenner M P and Rutledge G C 2001 Experimental characterization of electrospinning: The electrically forced jet and instabilities *Polymer* **42** 9955–67

[11] Chinnappan B A, Krishnaswamy M, Xu H and Hoque E 2022 Electrospinning of biomedical nanofibers/nanomembranes: effects of process parameters *Polymers* **14** 3719

[12] Yarin L, Koombhongse S and Reneker D H 2001 Bending instability in electrospinning of nanofibers *J. Appl. Phys.* **89** 3018–26

[13] Hohman M M, Shin M M, Rutledge G C and Brenner M P 2001 Electrospinning and electrically forced jets. I. Stability theory *Phys. Fluids* **13** 2201–20

[14] Reneker D H, Yarin A L, Fong H and Koombhongse S 2000 Bending instability of electrically charged liquid jets of polymer solutions in electrospinning *J. Appl. Phys.* **87** 4531–47

[15] Hohman M M, Shin M M, Rutledge G C and Brenner M P 2001 Electrospinning and electrically forced jets. II. Applications *Phys. Fluids* **13** 2221–36

[16] Dai H, Gong J, Kim H and Lee D 2002 A novel method for preparing ultra-fine alumina-borate oxide fibers via an electrospinning technique *Nanotechnol* **13** 674–7

[17] Edmondson D, Cooper A, Jana S, Wood D and Zhang M 2012 Centrifugal electrospinning of highly aligned polymer nanofibers over a large area *J. Mater. Chem.* **22** 18646–52

[18] Park S M and Kim D S 2015 Electrolyte-assisted electrospinning for a self-assembled, free-standing nanofiber membrane on a curved surface *Adv. Mater.* **27** 1682–7

[19] Li D, Wang Y and Xia Y 2003 Electrospinning of polymeric and ceramic nanofibers as uniaxially aligned arrays *Nano Lett.* **3** 1167–71

[20] Luo C, Stoyanov S D, Stride E, Pelan E and Edirisinghe M 2012 Electrospinning versus fiber production methods: from specifics to technological convergence *Chem. Soc. Rev.* **41** 4708–35

[21] Jian S, Zhu J, Jiang S, Chen H, Song Y, Duan G, Zhang Y and Hou H 2018 Nanofibers with diameter below one nanometer from electrospinning *RSC Adv.* **8** 4794–802

[22] Fong H, Chun I and Reneker D H 1999 Beaded nanofibers formed during electrospinning *Polymer* **40** 4585–92

[23] Urbanek O, Sajkiewicz P and Pierini F 2017 The effect of polarity in the electrospinning process on PCL/chitosan nanofibers' structure, properties and efficiency of surface modification *Polymer* **124** 168–75

[24] Ura D P *et al* 2019 The role of electrical polarity in electrospinning and on the mechanical and structural properties of as-spun fibers *Materials* **13** 4169

[25] Kilic A, Oruc F and Demir A 2008 Effects of polarity on electrospinning process *Textile Res. J.* **78** 532–9

[26] Ura D P and Stachewicz U 2022 The significance of electrical polarity in electrospinning: a nanoscale approach for the enhancement of the polymer fibers' properties *Macromol. Mater. Eng.* **307** 2100843

[27] Babu A, Gupta V and Mandal D 2022 Negatively bias-driven enhancement in piezo response for self-powered biomedical and facial expression sensor *Appl. Phys. Lett.* **120** 093701

[28] Tong H and Wang M 2013 A novel technique for the fabrication of 3D nanofibrous scaffolds using simultaneous positive voltage electrospinning and negative voltage electrospinning *Mater. Lett.* **94** 116–20

[29] Babu A, Malik P, Das N and Mandal D 2022 Surface potential tuned single active material comprised triboelectric nanogenerator for a high-performance voice recognition sensor *Small* **18** 2201331

[30] Busolo T, Ura D P, Kim S K, Marzec M M, Bernasik A, Stachewicz U and Kar-Narayan S 2019 Surface potential tailoring of PMMA fibers by electrospinning for enhanced triboelectric performance *Nano Energy* **57** 500–6

[31] Metwally S, Karbowniczek J E, Szewczyk P K, Marzec M M, Gruszczyński A, Bernasik A and Stachewicz U 2018 Single-step approach to tailor surface chemistry and potential on electrospun PCL fibers for tissue engineering application *Adv. Mater. Interfaces* **6** 1801211

[32] Ura D P, Knapczyk-Korczak J, Szewczyk P K, Sroczyk E A, Busolo T, Marzec M M, Bernasik A, Kar-Narayan S and Stachewicz U 2021 Surface potential-driven water harvesting from fog *ACS Nano* **15** 8848–59

[33] Szewczyk P K, Ura D P, Metwally S, Gajek M, Marzec M M, Bernasik A and Stachewicz U 2018 Roughness and fiber fraction dominated wetting of electrospun fiber-based porous meshes *Polymers* **11** 34

[34] Krysiak Z J, Gawlik M Z, Kaniuk Ł and Stachewicz U 2019 Hierarchical composite meshes of electrospun PS microfibers with PA6 nanofibers for regenerative medicine *Materials* **13** 1974

[35] Peng L, Jiang S, Seuß M, Fery A, Lang G, Scheibel T and Agarwal S 2015 Two-in-one composite fibers with side-by-side arrangement of silk fibroin and poly(L-lactide) by electrospinning *Macromol. Mater. Eng.* **301** 48–55

[36] Arinstein A, Burman M, Gendelman O and Zussman E 2006 Effect of supramolecular structure on polymer nanofiber elasticity *Nat. Nanotechnol.* **2** 59–62

[37] Stachewicz U, Szewczyk P K, Kruk A, Barber A H and Czyrska-Filemonowicz A 2019 Pore shape and size dependence on cell growth into electrospun fiber scaffolds for tissue engineering: 2D and 3D analyses using SEM and FIB-SEM tomography *Mater. Sci. Eng.* C **95** 397–408

[38] Duan G *et al* 2015 Ultralight, soft polymer sponges by self-assembly of short electrospun fibers in colloidal dispersions *Adv. Funct. Mater.* **25** 2850–6

[39] Liu Y, Zhang X, Xia Y and Yang H 2010 Magnetic-field-assisted electrospinning of aligned straight and wavy polymeric nanofibers *Adv. Mater.* **22** 2454–7

[40] Jesse S, Baddorf A P and Kalinin S V 2006 Switching spectroscopy piezoresponse force microscopy of ferroelectric materials *Appl. Phys. Lett.* **88** 062908

[41] Li M, Long Y-Z, Yang D, Sun J, Yin H, Zhao Z, Kong W, Jiang X and Fan Z 2011 Fabrication of one-dimensional superfine polymer fibers by double-spinning *J. Mater. Chem.* **21** 13159–62

[42] Kong L and Ziegler G R 2014 Rheological aspects in fabricating pullulan fibers by electrowet-spinning *Food Hydrocoll* **38** 220–6

[43] Wu J and Hong Y 2016 Enhancing cell infiltration of electrospun fibrous scaffolds in tissue regeneration *Bioact. Mater.* **1** 56–64

[44] Hong S and Kim G 2011 Fabrication of size-controlled three-dimensional structures consisting of electrohydrodynamically produced polycaprolactone micro/nanofibers *Appl. Phys. A: Mater. Sci. Process.* **103** 1009–14

[45] Teo W-E, Gopal R, Ramaseshan R, Fujihara K and Ramakrishna S 2007 A dynamic liquid support system for continuous electrospun yarn fabrication *Polymer* **48** 3400–5

[46] Wang L, Wu Y, Guo B and Ma P X 2015 Nanofiber yarn/hydrogel core–shell scaffolds mimicking native skeletal muscle tissue for guiding 3D myoblast alignment, elongation, and differentiation *ACS Nano* **9** 9167–79

[47] Liu J, Chen G, Gao H, Zhang L, Ma S, Liang J and Fong H 2012 Structure and thermochemical properties of continuous bundles of aligned and stretched electrospun polyacrylonitrile precursor nanofibers collected in a flowing water bath *Carbon* **50** 1262–70

[48] Park S M, Eom S, Kim W and Kim D S 2018 Role of grounded liquid collectors in precise patterning of electrospun nanofiber mats *Langmuir* **34** 284–90

[49] Peng S, Li L, Kong Yoong Lee J, Tian L, Srinivasan M, Adams S and Ramakrishna S 2016 Electrospun carbon nanofibers and their hybrid composites as advanced materials for energy conversion and storage *Nano Energy* **22** 361–95

[50] Wu M, Wang Q, Li K, Wu Y and Liu H 2012 Optimization of stabilization conditions for electrospun polyacrylonitrile nanofibers *Polym. Degrad. Stab.* **97** 1511–9

[51] Gergin İ, Ismar E and Sarac A S 2017 Oxidative stabilization of polyacrylonitrile nanofibers and carbon nanofibers containing graphene oxide (GO): a spectroscopic and electrochemical study *Beilstein J. Nanotechnol.* **8** 1616–28

[52] Zhang B, Yu Y, Xu Z L, Abouali S, Akbari M, He Y-B, Kang F and Kim J-K 2014 Correlation between atomic structure and electrochemical performance of anodes made from electrospun carbon nanofiber films *Adv. Energy Mater.* **4** 1301448

[53] Zhou Z, Lai C, Zhang L, Qian Y, Hou H, Reneker D H and Fong H 2009 Development of carbon nanofibers from aligned electrospun polyacrylonitrile nanofiber bundles and characterization of their microstructural, electrical, and mechanical properties *Polymer* **50** 2999–3006

[54] Yang K S, Edie D D, Lim D Y, Kim Y M and Choi Y O 2003 Preparation of carbon fiber web from electrostatic spinning of PMDA-ODA poly(amic acid) solution *Carbon* **41** 2039–46

[55] Heim D, Keerl M and Scheibel T 2009 Spider silk: from soluble protein to extraordinary fiber *Angew. Chem. Int. Ed.* **48** 3584–96

[56] Andersson M, Jia Q, Abella A, Lee X-Y, Landreh M, Purhonen P, Hebert H, Tenje M, Robinson C V, Meng Q *et al* 2017 Biomimetic spinning of artificial spider silk from a chimeric minispidroin *Nat. Chem. Biol.* **13** 262–4

[57] Woodings C (ed) 2001 *Regenerated Cellulose Fibers* (Woodhead) ch 5

[58] Abir S S H, Sadaf M U K, Saha S K, Touhami A, Lozano K and Uddin M J 2021 Nanofiber-based substrate for a triboelectric nanogenerator: high-performance flexible energy fiber mats *ACS Appl. Mater. Interfaces* **13** 6040–60412

[59] Szewczyk P K, Metwally S, Karbowniczek J, Marzec M, Stodolak-Zych E, Gruszczyński A and Stachewicz U 2018 Surface-potential-controlled cell proliferation and collagen mineralization on electrospun polyvinylidene fluoride (PVDF) fiber scaffolds for bone regeneration *ACS Biomater. Sci. Eng.* **5** 582–93

[60] McIntyre J E 2005 *Synthetic Fibers: Nylon, Polyester, Acrylic, Polyolefin* (Woodhead) ch 1

[61] Gupta V and Kothari V (ed) 1997 *Manufactured Fiber Technology* (Chapman & Hall) ch 1

[62] Boys C V 1887 On the production, properties, and some suggested uses of the finest threads *Proc. Phys. Soc. London* **9** 8–19

[63] Dzenis Y 2004 Spinning continuous fibers for nanotechnology *Science* **304** 1917–9

[64] Metwally S, Ferraris S, Spriano S, Krysiak Z J, Kaniuk Ł, Marzec M M, Kim S K, Szewczyk P K, Gruszczyński A, Wytrwal-Sarna M *et al* 2020 Surface potential and roughness controlled cell adhesion and collagen formation in electrospun PCL fibers for bone regeneration *Mater. Des.* **194** 108915

[65] Gray S A 1731 Letter concerning the electricity of water, from Mr Stephen Gray to Cromwell Mortimer *Philos. Trans. R. Soc. London* **37** 227–60

[66] Nollet J A X 1748 Part of a letter from Abbe Nollet, of the Royal Academy of Sciences at Paris, to Martin Folkes, concerning electricity *Philos. Trans.* **45** 187–94

[67] Rayleigh L 1882 On the equilibrium of liquid conducting masses charged with electricity *Philos. Mag.* **14** 184–6

[68] Cooley J F 1902 Apparatus for electrically dispersing fluids *U.S. Patents* 692,631

[69] Morton W J 1902 Method of dispersing fluid *U.S. Patents* 705,691

[70] Formhals 1944 Method and apparatus for spinning *U.S. Patents* 2,349,950

[71] Taylor G I 1964 Disintegration of water drops in an electric field *Proc. R. Soc. London* A **280** 383–97

[72] Taylor G I 1966 The force exerted by an electric field on a long cylindrical conductor *Proc. R. Soc. London, Ser.* A **291** 145–58

[73] Taylor G I 1969 Electrically driven jets *Proc. R. Soc. London, Ser.* A **313** 453–75

[74] Martin G E and Cockshott I D 1977 Fibrillar lining for prosthetic device *U.S. Patents* 4,044,404

[75] Larsen G, Velarde-Ortiz R, Minchow K, Barrero A and Loscertales I G 2003 A method for making inorganic and hybrid (organic/inorganic) fibers and vesicles with diameters in the submicrometer and micrometer range via sol–gel chemistry and electrically forced liquid jets *J. Am. Chem. Soc.* **125** 1154–5

[76] Theron E, Zussman E and Yarin A 2001 Electrostatic field-assisted alignment of electrospun nanofibers *Nanotechnology* **12** 384–90

[77] Mokhtari F, Shamshirsaz M, Latifi M and Foroughi J 2020 Nanofibers-based piezoelectric energy harvester for self-powered wearable technologies *Polymers* **12** 2697

[78] Dersch R, Liu T, Schaper A, Greiner A and Wendorff J 2003 Electrospun nanofibers: Internal structure and intrinsic orientation *J. Polym. Sci. A: Polym. Chem.* **41** 545–53

[79] Sun Z, Zussman E, Yarin A L, Wendorff J H and Greiner A 2003 Compound core–shell polymer nanofibers by co-electrospinning *Adv. Mater.* **15** 1929–32

[80] Smit E, Buttner U and Sanderson R D 2005 Continuous yarns from electrospun fibers *Polymer* **46** 2419–23

[81] Xue J, Xie J, Liu W and Xia Y 2017 Electrospun nanofibers: new concepts, materials, and applications *Acc. Chem. Res.* **50** 1976–87

[82] Zhang F, Si Y, Yu J and Ding B 2023 Electrospun porous engineered nanofiber materials: a versatile medium for energy and environmental applications *Chem. Eng. J.* **456** 140989

[83] Panwar V, Babu A, Sharma A, Thomas J, Chopra V, Malik P and Rajput S 2021 Tunable, conductive, self-healing, adhesive and injectable hydrogels for bioelectronics and tissue regeneration applications *J. Mater. Chem.* B **9** 6260–70

[84] Huang T, Lu M, Yu H, Zhang Q, Wang H and Zhu M 2015 Enhanced power output of a triboelectric nanogenerator composed of electrospun nanofiber mats doped with graphene oxide *Sci. Rep.* **5** 1–8

[85] Mishra H K, Gupta V, Roy K, Babu A, Kumar A and Mandal D 2023 Revisiting of δ-PVDF nanoparticles via phase separation with giant piezoelectric response for the realization of self-powered biomedical sensors *Nano Energy* **95** 107052

[86] Gupta V, Kumar A, Mondal B, Babu A, Ranpariya S, Sinha D K and Mandal D 2023 Machine learning-aided all-organic air-permeable piezoelectric nanogenerator *ACS Sustain. Chem. Eng.* **11** 6173–82

[87] Gupta V, Babu A, Ghosh S K, Mallick Z, Mishra H K, Saini D and Mandal D 2021 Revisiting δ-PVDF based piezoelectric nanogenerator for self-powered pressure mapping sensor *Appl. Phys. Lett.* **119** 25

[88] Busolo T, Szewczyk P K, Nair M, Stachewicz U and Kar-Narayan S 2021 Triboelectric yarns with electrospun functional polymer coatings for highly durable and washable smart textile applications *ACS Appl. Mater. Interfaces* **13** 16876–86

[89] Rahaman M S A and Ismail A F 2007 A review of heat treatment on polyacrylonitrile fiber *Polym. Degrad. Stab.* **92** 1421–32

[90] Dong K, Peng X and Wang Z L 2020 Fiber/fabric-based piezoelectric and triboelectric nanogenerators for flexible/stretchable and wearable electronics and artificial intelligence *Adv. Mater.* **32** 1–43

[91] Babu A, Kassahun G, Dufour I, Mandal D and Thuau D 2024 Self-learning e-skin respirometer for pulmonary disease detection *Adv. Sensor Res.* **3** 2400079

[92] Ye C, Yang S, Ren J, Dong S, Cao L, Pei Y and Ling S 2022 Electroassisted core-spun triboelectric nanogenerator fabrics for intelliSense and artificial intelligence perception *ACS Nano* **16** 4415–25

[93] Babu A, Ranpariya S, Sinha D K and Mandal D 2023 Deep learning enabled perceptive wearable sensor: an interactive gadget for tracking movement disorder *Adv. Mater. Technol.* **8** 1–10

[94] Sun T *et al* 2024 Artificial intelligence meets flexible sensors: emerging smart flexible sensing systems driven by machine learning and artificial synapses *Nano-Micro Lett.* **16** 14

[95] Babu A, Raoul E, Kassahun G, Dufour I, Mandal D and Thuau D 2023 Programmable polymeric-interface for voiceprint biometrics *Adv. Mater. Technol.* **2301551** 1–9

[96] Babu A, Ranpariya S, Sinha D K, Chatterjee A and Mandal D 2023 Deep learning enabled early predicting cardiovascular status using highly sensitive piezoelectric sensor *Adv. Mater. Technol.* **2202021** 1–10

[97] Babu A, Thuau D and Mandal D 2023 AI-enabled wearable sensor for real-time monitored personalized training of sportsperson *MRS Commun.* **13** 1071–5

[98] Zhao S *et al* 2018 All-nanofiber-based ultralight stretchable triboelectric nanogenerator for self-powered wearable electronics *ACS Appl. Energy Mater.* **1** 2326–32

[99] Xu Y, Li W J and Lee K K 2008 *Intelligent Wearable Interfaces* (Wiley-Interscience)

[100] Babu A and Mandal D 2024 Roadmap to human–machine interaction through triboelectric nanogenerator and machine learning convergence *ACS Appl. Energy Mater.* **7** 822–33

IOP Publishing

Energy Harvesting Properties of Electrospun Nanofibers
(Second Edition)

Jian Fang and Tong Lin

Chapter 3

Piezoelectric energy conversion performance of electrospun nanofibers

Zhan Qu, Hongxia Wang and Jian Fang

Piezoelectric materials that can convert mechanical forces or vibrations into electricity have attracted increasing attention due to energy crises and environmental pollution problems. Electrospinning is an effective technique for producing piezo-electric nanofibers for energy conversion purposes. In the electrospinning process, the dipoles in materials tend to be oriented within the high electrical field, resulting in piezoelectricity. Polyvinylidene fluoride (PVDF) is the most widely studied piezoelectric polymer because of its high piezoelectric coefficient and easy electro-spinning preparation. Besides, other electrospun piezoelectric polymers, piezoelec-tric ceramics, and composites with nanofibrous structures have also been prepared for energy harvesting. In this chapter, a detailed introduction of electrospun piezoelectric materials and their piezoelectricity will be provided.

3.1 Introduction

Mechanical-to-electrical energy conversion can be achieved based on different principles such as piezoelectric, triboelectric, electromagnetic effects, and Schottky contact diode. The different principles can lead to great differences in device structure and electric output features. Among them, the piezoelectric effect, which converts mechanical forces or vibration into electric signals, has attracted the most attention owing to the large amount of existing piezoelectric materials around our life.

Piezoelectric materials are conventionally divided into two main classes, inor-ganic ceramics and organic polymers, and piezoelectric polymers are further divided into three subclasses, bulk polymers, composite and void charged polymers (VCP). Most of the piezoelectric materials except VCP have a dense thin film structure. Piezoelectric ceramics normally have higher piezoelectric coefficients and higher

doi:10.1088/978-0-7503-5487-5ch3 3-1

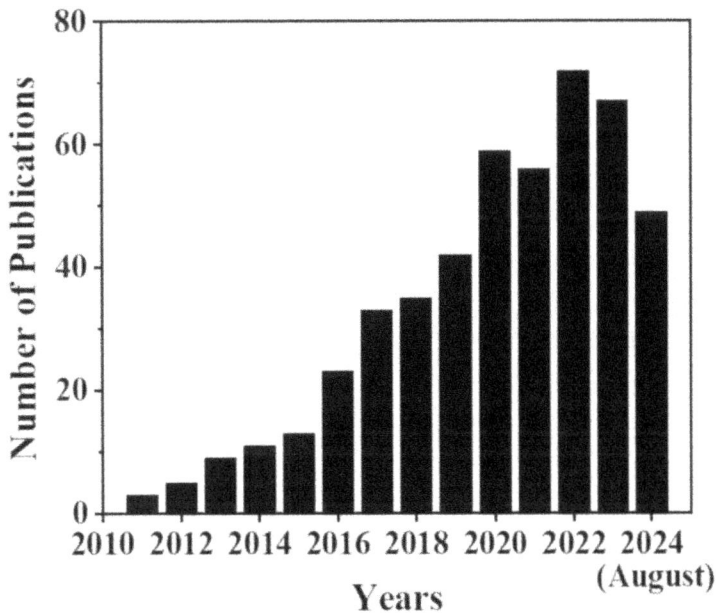

Figure 3.1. Number of scientific publications on electrospun nanofibers for piezoelectric energy conversion application between 2010 and 2024 (Source: Web of Science; August 2024).

piezoelectric response. Piezoelectric polymers with non-toxic, biocompatibility, biodegradability and high-flexibility properties have exhibited great significance in many application fields. The fabrication of piezoelectric films often requires harsh and costly conditions. Recently, electrospinning has been found as a simple and effective tool to process piezoelectric materials into nanofibrous structures, especially for piezoelectric polymers. Piezoelectric nanofibers from electrospinning have shown many unique features apart from high surface area-to-volume ratios, high porosity, good flexibility and light weight [1]. As shown in figure 3.1, since the first piezoelectric energy generator (PENG) prepared by electrospinning was reported in 2009, the research has continuously increased, as indicated by the increased publication numbers. In this chapter, the details of research findings about electrospun nanofibers for piezoelectric energy conversion will be summarized.

3.2 Brief history

The word 'piezo' is derived from the Greek word 'piezoin', which means to press or squeeze, and 'piezoelectricity' means electricity resulting from pressing. Since it was discovered in 1880, piezoelectricity has been developed for over one century. It has two effects, consisting of direct piezoelectric effect and converse piezoelectric effect, as shown in figure 3.2. Applying a force to a piezoelectric material alters the dipole moment, generating electric outputs (direct effect). On the other hand, piezoelectric material deforms in the direction of the poling voltage when it is charged by an external electrical voltage (converse effect). As shown in figure 3.3, the deformation

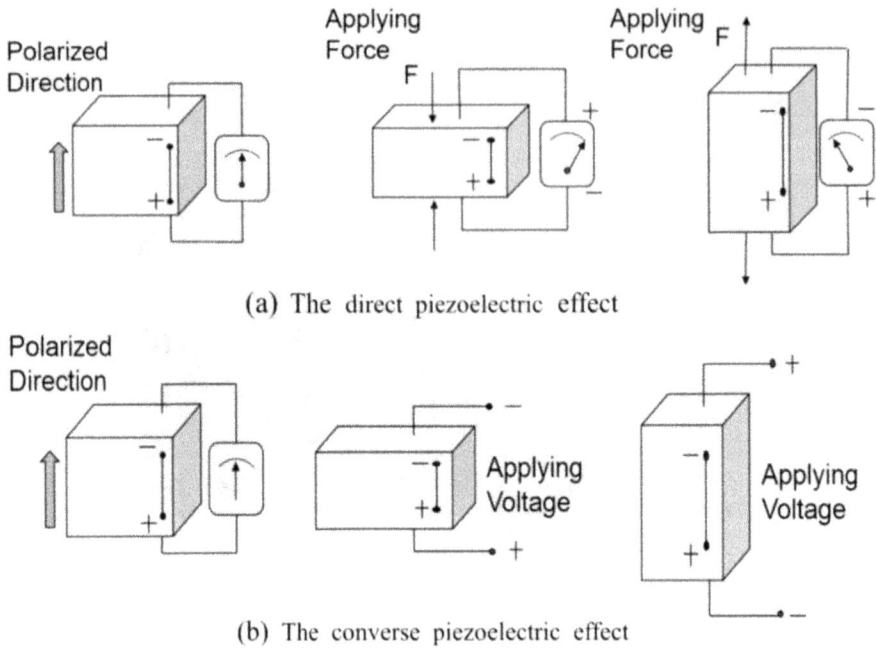

Figure 3.2. A schematic of (a) direct and (b) converse effects of piezoelectric materials. Reproduced from [2], copyright (2017) IOP Publishing Ltd.

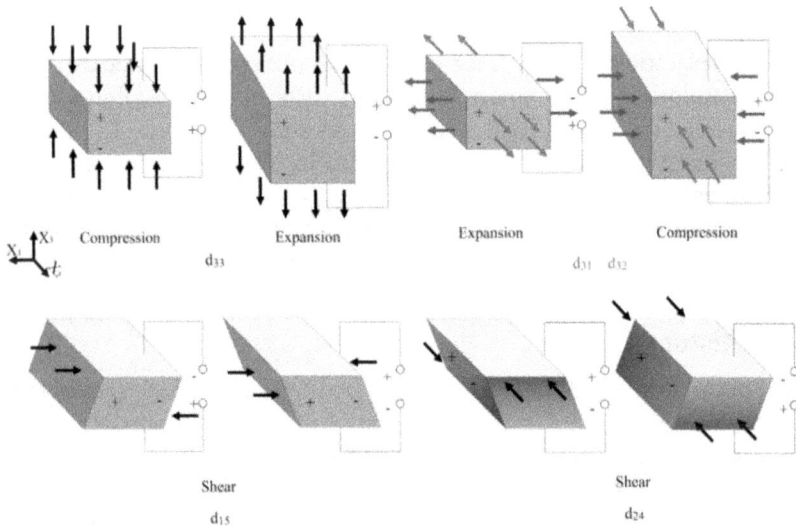

Figure 3.3. Deformation of piezoelectric materials in different modes with corresponding piezoelectric coefficients and polarity of output voltage. Reproduced from [3], copyright (2023), with permission from Elsevier.

of the piezoelectric material under different loading modes and the components of d corresponding to each mode are systematically illustrated. For d_{ij}, the subscript i is the direction of polarization, and the subscript j is the direction of applied force.

Numerous natural materials, such as quartz, DNA, and bone, exhibit piezo-electric properties [4]. Artificial piezoelectric materials that have been developed including lead zirconate titanate (PZT), barium titanate (BaTiO$_3$), zinc oxide (ZnO), and polymers etc. PZT was ranked as having the highest piezoelectric coefficient among the piezoelectric ceramics. However, the toxicity originating from lead has restricted its wide application. Thus, environmentally friendly piezoelectric ceramics, such as BaTiO$_3$ and ZnO, have been developed.

PVDF was the first reported piezoelectric polymer [5]. After that, more piezo-electric polymers, such as PVDF-copolymers, nylon-11, polyurea-9 were discovered [6]. To date, piezoelectric polymers normally show lower piezoelectric coefficients than piezoelectric ceramics, but they are much more flexible and can undergo a higher level of physical deformation. Recently, electrospinning was found to be an effective method to prepare piezoelectric webs from polymers, ceramics, and their combinations.

3.3 Piezoelectricity of PVDF nanofibers

PVDF is a multi-ferric material with superior piezoelectric, pyroelectric and ferro-electric properties [7]. PVDF has a simple monomer structure of $-(C_2H_2F_2)-$, which consists of atoms of carbon, hydrogen and fluorine. In the solid state, PVDF is a semicrystalline polymer that can have five different crystal phases: α, β, γ, δ and ε. These crystal phases contain three different conformations, all-trans (TTT) planar zigzag for β phase, trans-gauche-trans-gauche (TGTG') for α and δ phases, and TTTGTTTG' for γ and ε phases, as shown in figure 3.4. The β crystal phase has the highest electrical dipole moment per unit, about $5-8 \times 10^{-30}$ C · m, which is the main reason for PVDF piezoelectricity.

The α crystal phase shows no piezoelectric property, but it is the most thermodynamically stable phase. Conventionally, piezoelectric PVDF films are prepared by conducting a series of post-treatments to PVDF cast films which are α crystal phase dominated. The post-stretching enables a PVDF film to have β crystal phase and electrical poling at an elevated temperature assists in improvement of aligning the dipole moments in the film.

Electrospinning is able to directly produce PVDF nanofibers with high β crystal phase content without any post-treatments [8]. However, it was not until 2009 that the direct piezoelectric effect was reported on a single electrospun PVDF fiber. Chang *et al* [8] used a near-field electrospinning technique to prepare a single PVDF fiber. The single PVDF fiber device showed piezoelectric property with pulse outputs of 2.5 mV when an external strain of 0.092% was applied. Slightly later, strong piezoelectric property was reported on aligned nanofiber web [9] and randomly orientated PVDF nanofibers webs, i.e., nonwovens [10].

According to fiber structure and electric output feature, the studies on the piezoelectric effect of electrospun PVDF nanofibers can be separated into a few

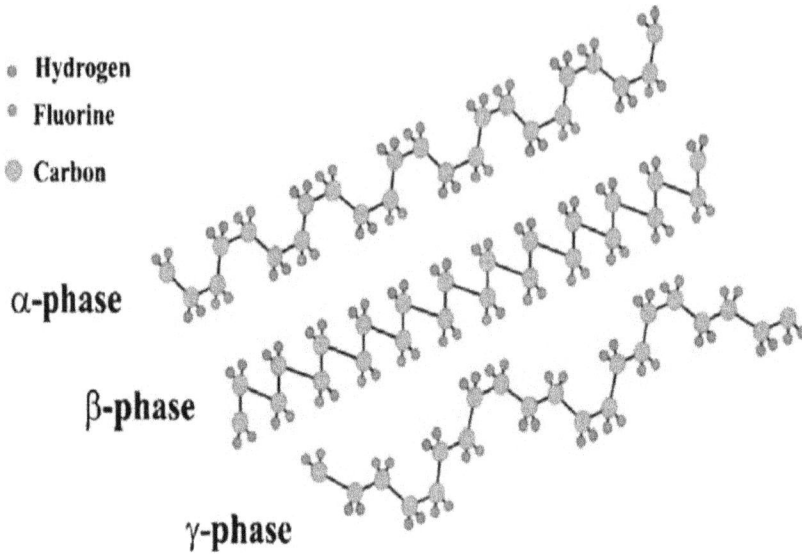

Figure 3.4. Chain conformation for the α, β and γ phases of PVDF. Reprinted from [7], copyright (2014), with permission from Elsevier.

different groups: single fiber/multi-fibers, randomly orientated fiber web (nonwoven) and aligned nanofiber web. A summary of neat PVDF piezoelectric devices are provided, as shown in table 3.1.

3.3.1 Single fiber/multi-fibers

Near-field electrospinning (NFES) is an effective method to produce a single fiber. In the electrospinning process, the stretching forces and strong electric fields (greater than 10^7 V m^{-1}) naturally align dipole moments in the fiber to form polar β phase, determining the polarity of the electrospun nanofiber (figure 3.5). The piezoelectric device fabricated based on a single electrospun PVDF nanofiber with a diameter of 700 nm could produce an electric voltage output up to 2.5 mV. Limited electric outputs were generated by a single fiber. Later, the same group found that the single PVDF nanofiber device could generate the electrical output of 32 mV and 4.5 nA under low stretch strain (around 0.08%) [9]. Figures 3.5(c) and (d) indicate the typical electrical voltage and current outputs. A pulse electrical signal accompanied with an opposite pulse electrical output was generated. The first signal was derived from the deformation of nanofiber, while the subsequent opposite output was attributed to the recovery deformation of the nanofiber due to the release of the force. Guo *et al* [11] combined NEFS and TENG technology to prepare PVDF single fiber (figure 3.6). Under 0.098% strain, the electric energy and the energy conversion efficiency can reach 7.74 pJ and 13.5%, respectively, which is comparable to the fibers produced by the NFES system powered by commercial DC supply. Due to the limitations of electromechanical conversion of single fibers, researchers also

Table 3.1. A summary of piezoelectric electrospun neat PVDF fiber devices.

PVDF	Testing mode	Fiber diameter	Web thickness	Working area (cm^2)	Strain/force/frequency	Output types	Poling	V	I	References
Single fiber	Stretching/releasing	700 nm	—	—	0.092% strain	d_{33}	No	8.5 mV	—	[8]
	Stretching/releasing	500 nm - 6.5 μm	—	—	0.085% strain, 2 Hz	d_{33}	No	32 mV	4.5 nA	[9]
	Bending/releasing	6.11 ± 0.44 μm - 30.61 ± 1.02 μm	—	—	0.098% strain, 0.5 Hz	d_{33} and d_{31}	No	6.1 mV	—	[11]
Multi-fibers	Vibration	0.9-5 μm	~50 parallel fibers	9.75	Wind speed 11.3 m s^{-1}	d_{33} and d_{31}	No	2.5 V	100 nA	[12]
	Vibration	0.9-2.5 μm	~500 fibers	5	Wind speed 19.8 m s^{-1}	d_{33} and d_{31}	No	2.0 V	—	[13]
	Bending/releasing	0.9-2.5 μm	~500 fibers	3.5	0.5% strain, 10 Hz	d_{33} and d_{31}	No	4 V	75 nA	[14]
	Stretch/release		~120 buckled fibers	—	30% strain, 0.5 Hz	d_{33}	No	40 mV	1.2 nA	[15]
	Stretch/release		40 fibers	—	70% strain, 0.8 Hz	d_{33}	No	—	2.8 nA	[16]
	Stretch/release		20-150 fibers	—	160 mm s^{-1}, 120% strain	d_{33} and d_{31}	No	—	20 nA	[17]
	Bending	400 nm	—	—	0.05% strain, 1.67%s^{-1} strain rate	d_{33}	20 kV mm^{-1}, 15 min	20 mV	0.3 nA	[18]
	Compression/decompression	10 μm	3000 fibers	2.8	4 Hz	d_{33}	No	125 mV	208 nA	[19]
	Compression/decompression	500 nm	200 fibers	6.25	4 Hz	d_{33} and d_{31}	No	2 V	200 nA	[20]
	Compression/decompression	700 nm - 2.5 μm	500 fibers	6	0.1% strain, 2 Hz	d_{33} and d_{31}	No	6 V	280 nA	[21]

3-6

Nonwoven web	Compression/decompression	183 ± 37 nm	140 µm	2	5 Hz	d_{33}	No	2.21 V	4.5 µA	[10]
	Compression/decompression	—	20 µm	2.5	0.2 MPa, 3 Hz	d_{33}	No	175 mV	—	[22]
	Drop stainless steel	812 ± 123 nm	310 µm	2	5 g	d_{33}	12.6 kV mm^{-1}, 4 h room temperature	4.0 V	—	[23]
	Compression/decompression	284 ± 70 nm	70 µm	4	10 N, 1 Hz	d_{33}	No	2.2 V	2.3 µA	[24]
	Finger compression/decompression	124 nm	120 µm	2	—	d_{33}	No	1 V	—	[25]
	Vibration	300 ± 70 nm	—	2	1.54 Hz	d_{33}	20 kV mm^{-1}, 30 min	1.3 V	13 nA	[26]
	Compression/decompression	—	—	2.25	4 Hz, 15 N	d_{33}	No	0.97 V	—	[27]
	Periodic bending	500 ± 220 nm	50 µm	3	—	d_{33} and d_{31}	No	2.8 V	1.32 µA	[28]
	Compression/decompression	300 nm	100 µm	4	10 N, 1 Hz	d_{33}	No	1.0 V	1.2 µA	[29]
	Compression/decompression	539 ± 175 nm	100 µm	2	10 N, 5 Hz	d_{33}	No	2.6 V	4.5 µA	[30]
	Compression/decompression	—	—	—	4.44 N, 20 Hz	d_{33}	No	0.28 V	—	[31]
	Sound wave reacted vibration	150–180 nm	—	—	100 Hz 100 dB	d_{33}	No	0.3 V	—	[32]
	Bending/releasing	—	200 µm	8.4	37 N, 1.4 Hz	d_{33} and d_{31}	25 kV mm^{-1}	1.5 V	400 nA	[33]
	Bending/releasing	124 nm	120 µm	16	0.5 Hz	d_{33} and d_{31}	No	0.7 V	—	[34]
	Compression/decompression	~100 nm	260 µm	6	8.3 kPa	d_{33}	No	48 V	6 µA	[35]
	Compression/decompression	1137 ± 200 nm (Smooth fibers)	100 µm	15	5 Hz, 10 N	d_{33}	No	1 V	1.6 µA	[36]
	Compression/decompression	1129 ± 168 nm (Porous fibers)	100 µm	15	5 Hz, 10 N	d_{33}	No	1.3 V	2.1 µA	[36]
	Compression/decompression	1130 ± 158 nm (Wrinkled fibers)	100 µm	15	5 Hz, 10 N	d_{33}	No	1.8 V	3.2 µA	[36]
	Bending	350 ± 23 nm	60 µm	1.2	2 Hz	d_{33} and d_{31}	No	0.5 V	13 nA	[37]

(Continued)

Table 3.1. (*Continued*)

PVDF	Testing mode	Fiber diameter	Web thickness	Working area (cm^2)	Strain/force/ frequency	Output types	Poling	V	I	References
	Compression/ decompression	779 ± 109 nm (Wrinkled fibers, M_w of 180 000)	100 μm	15	5 Hz, 10 N	d_{33}	No	1.97 V	2.7 μA	[38]
	Compression/ decompression	1134 ± 126 nm (Wrinkled fibers, M_w of 275 000)	100 μm	15	5 Hz, 10 N	d_{33}	No	2.55 V	2.98 μA	[38]
	Compression/ decompression	1437 ± 167 nm (Wrinkled fibers, M_w of 530 000)	100 μm	15	5 Hz, 10 N	d_{33}	No	2.92 V	4.1 μA	[38]
	Compression/ decompression	260–280 nm	80 ± 10 μm	1	—	d_{33}	No	0.505 V	—	[39]
	Compression/ decompression	1.26 ± 0.5 μm	100 μm	1.5	1 Hz, 20 N	d_{33}	No	4 V	80 nA	[40]
	Compression/ decompression	250 nm	700 nm	—	362 Hz	d_{33}	No	0.04 V	600 nA	[41]
	Compression/ decompression	30–220 nm	15 μm	9	1 Hz, 40 N	d_{33}	No	124 V	174 nA	[42]
	Bending/releasing	250–400 nm	80 μm	12	—	d_{33} and d_{31}	No	2.8 V	—	[43]
	Sound wave reacted vibration	—	—	100	230 Hz, 98 dB	d_{33}	No	27.54 mV	—	[44]
	Compression/ decompression	500–1000 nm	—	4	1.6 Hz, 9 N	d_{33}	No	32.6 V	0.58 μA	[45]
	Compression/ decompression	312 ± 24 nm	105 ± 21 μm	1	0.5 Hz	d_{33}	30 V μm^{-1}, 6 h	0.8 ± 0.04 V	0.86 ± 0.09 mA	[46]
	Compression/ decompression	328 nm	1000 μm	36	—	d_{33}	2.5 kV cm^{-1}	—	—	[47]
Aligned fiber web	Stretch/release	1.6 μm	100 fibers	—	0.05% Strain, 7 Hz	d_{33}	No	76 mV	39 nA	[48]
	Vibration	Hollow tube web, 13.18 μm outer, 3.26 μm inner	—	5.25	7 Hz	d_{33}	No	71.66 mV	153 nA	[49]

Finger press	0.9–3 µm	400 fibers	1	—	d_{31}	No	4 V	100 nA	[50]
Bending	400 ± 35 nm	60 µm	1.2	2 Hz	d_{33} and d_{31}	No	1.1 V	40 nA	[37]
Compression/ decompression	1052 ± 129 nm (Smooth fibers)	100 µm	15	5 Hz, 10 N	d_{33}	No	1.7 V	2.2 µA	[36]
Compression/ decompression	1038 ± 113 nm (Porous fibers)	100 µm	15	5 Hz, 10 N	d_{33}	No	2.2 V	2.7 µA	[36]
Compression/ decompression	1047 ± 100 nm (Wrinkled fibers)	100 µm	15	5 Hz, 10 N	d_{33}	No	2.8 V	3.9 µA	[36]
Compression/ decompression	—	5 µm	3	0.5 Hz, 1 N	d_{33}	No	0.48 V	2.7 nA	[51]
Bending/releasing	1679 ± 178 nm (Grooved fibers)	100 µm	7.5	1 Hz	d_{33} and d_{31}	No	0.74 V	457 nA	[52]
Compression/ decompression	—	—	4	8 Hz, 12 N	d_{33}	No	2.1 V	—	[53]
Bending/releasing	70 nm	—	3.75	0.5 Hz	d_{33} and d_{31}	No	0.2 V	—	[54]
Bending/releasing	125 µm	—	—	1 Hz, 1.29 N	d_{33} and d_{31}	No	1 V	—	[55]
Compression/ decompression	292.2 nm	120 µm	21	2.6 N	d_{33}	No	84.8 mV	—	[56]

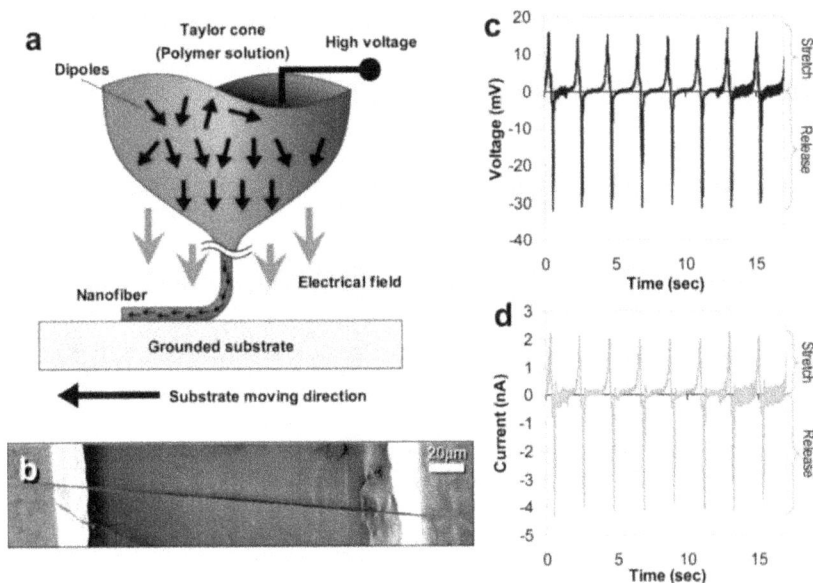

Figure 3.5. (a) Near-field electrospinning combining direct-writing, mechanical stretching and electrical poling to create piezoelectric nanogenerators onto a substrate. (b) SEM image of a single PVDF nanofiber. Output (c) voltage and (d) current measured with respect to time under an applied strain at 2 Hz. Reproduced with permission from [9], copyright (2010), American Chemical Society.

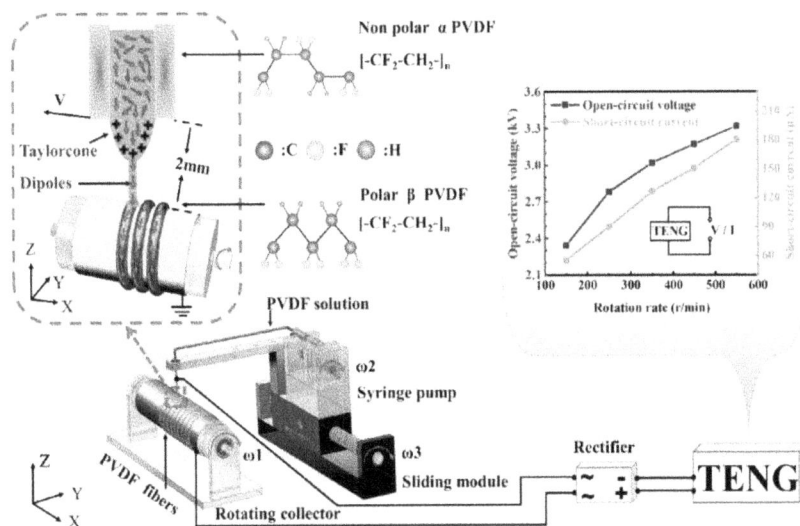

Figure 3.6. Preparation process of PVDF fibers via the NFES system driven by a TENG. Reproduced with permission from [11], copyright (2023), American Chemical Society.

reported that the electrical output enhanced by directly-writing many aligned PVDF nanofibers to form a device.

Fuh et al [12–14] developed highly flexible devices with multiple PVDF fibers for sensor and energy generating applications. The sensor device attached to the flag surface produced maximum outputs of 100 nA current and 2.5 V voltage, respectively, when the wind speed was 11.3 m s^{-1}. For energy harvesting purpose, the output voltage of the device exceed 4 V while the output current reached 75 nA [13, 14]. Later, the same group fabricated a transparent and flexible graphene piezoelectric fiber generator by spinning PVDF piezoelectric fibers onto the chemical vapor deposition (CVD) grown graphene, cyclic stretching at 4 Hz on the device achieved an output voltage/current of approximately 2 V/200 nA [20]. On the basis of this research, Fuh et al made a a hybrid self-powered sensor based on graphene and PVDF nanofibers. This device incorporated both piezoelectric and triboelectric effects for improved output performance, with output voltages, currents and powers up to 6 V, 280 nA and 172 nW. Stretchable nanogenerators (NGs) with highly uniform controllable buckled PVDF fiber arrays were prepared by direct-writing electrospinning [15–17]. As shown in figure 3.7(a), piezoelectric device containing about 120 buckled PVDF fibers. Under periodic stretching (30% strain) and releasing process, the output current and voltage of 1.2 nA and 40 mV were observed, respectively (figure 3.7(b)).

Apart from near-field electrospinning, conventional electrospinning was also used to prepare aligned multiple nanofibers. Benjemin et al [18] prepared an NG by using aligned PVDF nanofibers obtained by using two grounded copper pieces with a 2 cm gap as a fiber collector. They indicated that random dipoles of the polar β crystal phase existed in the nanofibers, however, they did not show any piezoelectric property. To obtain the piezoelectricity, a high electrical field poling (20 kV mm^{-1}) was employed to align the dipoles along the in-plane direction. The device was prepared could generate 20 mV voltage and 0.3 nA current.

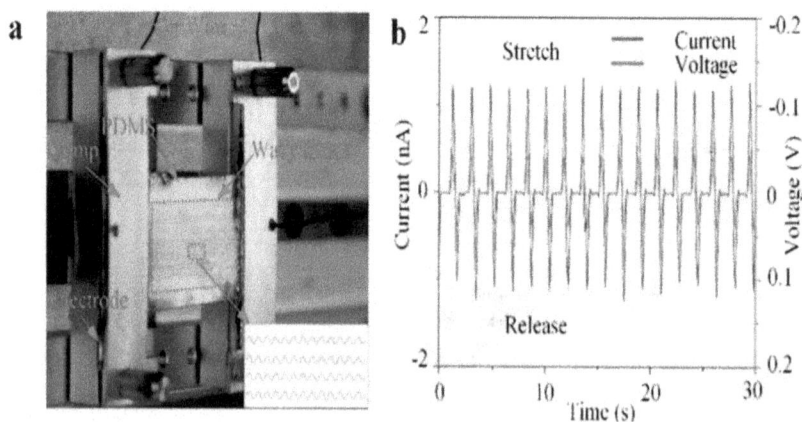

Figure 3.7. (a) Photograph of the tensile test platform. (b) Output current and voltage of the device consisting of 120 PVDF fibers measured under 30% applied strain at 0.5 Hz. Reproduced from [15]. Copyright (2014), with permission from Royal Society of Chemistry.

3.3.2 Nonwoven

Nanofiber nonwovens are predominant products of conventional electrospinning. In 2011, Fang *et al* [10] first reported that randomly oriented PVDF nanofiber webs prepared by conventional electrospinning showed mechanical-to-electrical energy conversion properties without a post-poling process. When an electrospun PVDF fiber nonwoven was sandwiched between two aluminium foils, such a simple device could generate voltage and current of 2.4 V and 4.2 μA upon receiving a compressive impact (figure 3.8).

Lei *et al* [22] compared nanofiber webs produced by a mechanical spinning process without static electric field bias (forcespinning) and an electrospinning technique. It was interesting to find that the two nanofiber webs showed completely difference piezoelectric responses. The electrospun PVDF possesses piezoelectric effect, where the forcespun PVDF does not. The electric field in the electrospinning process performs poling and inducing dipole orientation in PVDF fibers.

The effects of electrospinning parameters and other conditions on β crystal phase content and mechanical-to-electrical energy conversion of randomly-oriented electrospun PVDF nanofiber webs were studied systematically [24–28]. Electrospinning parameters include electrospinning equipment parameters, electrospinning solution parameters, and environmental parameters. Firstly, electrospinning equipment parameters such as voltage, spinning distance, solution feed flow rate, needle size, and collection drum speed influence the stretching process of PVDF fibers, which in

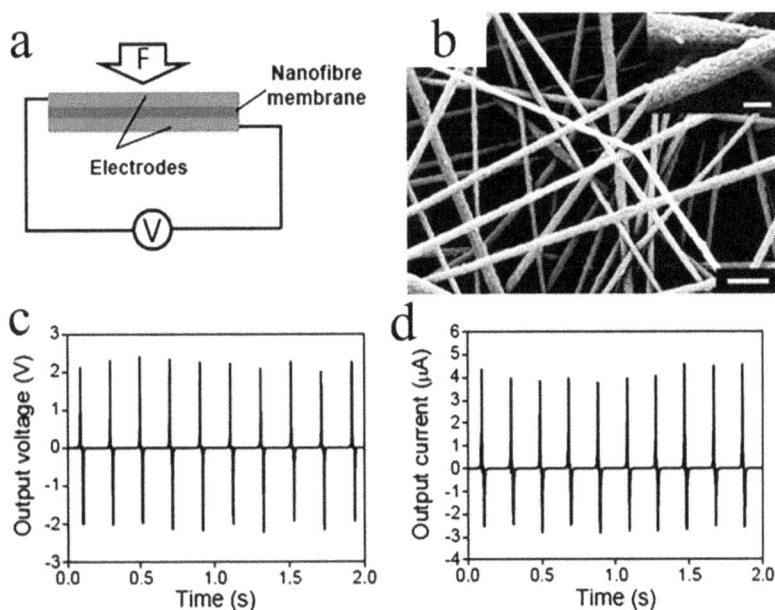

Figure 3.8. (a) Illustration of power generator. (b) SEM images of electrospun PVDF nanofibers (scale bar: 1 mm, inset scale bar: 200 nm). (c) Voltage and (d) current outputs of the same nanofiber device under 5 Hz repeated compressive impacts. (Nanofiber web thickness = 140 mm; working area = 2 cm^2). Reproduced from [10], copyright (2011), with permsision from Royal Society of Chemistry.

turn affects the content and orientation of the β crystal phase [43, 57–59]. Secondly, polymer solution concentration, surface tension, solvent polarity and ratio influence the morphology and surface structure of electrospun PVDF nanofibers, which in turn affects the expression of fiber piezoelectric properties [60–63]. Finally, environmental parameters such as temperature and relative humidity have an impact on the solvent evaporation time during the electrospinning process, which in turn affects the content of β crystal phase [64, 65]. Moreover, it was found that finer uniform PVDF fibers showed higher β crystal phase content hence the energy harvesting devices having higher electrical outputs (figure 3.9). Nanofiber web thickness also plays a role in determining the level of electrical outputs. When the fiber web was thinner than 20 μm, a short circuit occurred occasionally during the compressing process. With increasing the web thickness from 20 to 70 μm, both voltage and current outputs increased. However, electrical outputs reduced with further increasing the web thickness [24]. It is also found that the residue charges contribute to the mechanical-to-electrical energy conversion of electrospun PVDF nanofibers [29]. The electric outputs of the device decreased from 1.0 V and 1.2 μA to 0.45 V and 0.5 μA after the residual charges removal.

Interconnected PVDF fiber webs were obtained by using short-distance electrospinning [66]. Despite the insufficient fiber whipping movement and limited solvent evaporation, the interconnected fiber webs produced had a comparable β crystal phase content and mechanical-to-electrical energy conversion property to those produced by conventional electrospinning. Moreover, the interconnected fiber webs had higher delamination resistance and tensile strength, hence to be more robust to maintain structure integrity during compression/decompression process.

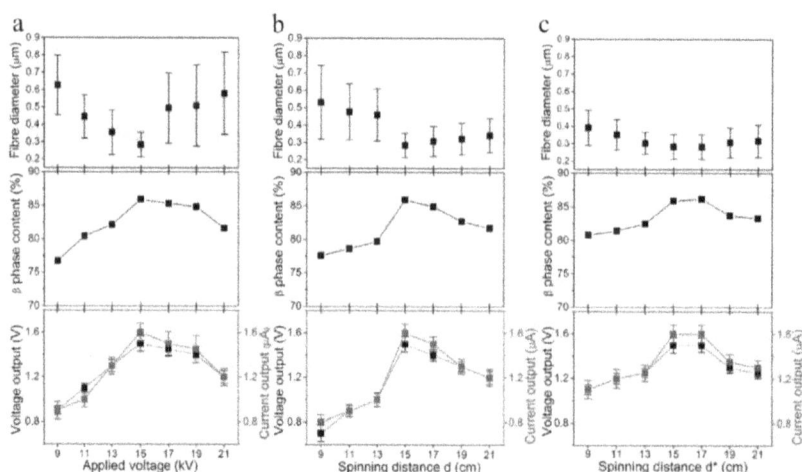

Figure 3.9. Effect of (a) applied voltages, (b) spinning distances, and (c) electric field intensity on PVDF fiber diameters, β crystal phase contents and electrical outputs of PVDF nanofiber webs (PVDF concentration 20%; nanofiber web thickness 100 mm). Reproduced from [24], copyright (2015), with permission from Royal Society of Chemistry.

Fang *et al* [30] examined the effect of electrospinning models on the piezo-electricity of electrospun nanofiber webs. It is reported that PVDF nonwovens prepared by a disk (needleless) electrospinning showed enhanced piezoelectric performance compared to those from conventional needle electrospinning. A device made of needleless spun nanofibers generated 2.6 V and 4.5 μA outputs while the one from needle electrospun nanofibers only produced 2.05 V and 3.12 μA, at the same condition. This was explained by the fact that that the higher applied voltage used in needleless electrospinning leads to higher β crystal phase and better molecular orientation.

Conventional electrospinning was also employed to produce randomly oriented PVDF nanofibers for energy generator and sensor applications [31, 32, 34, 35, 67]. Wang *et al* [31] reported to fabricate piezoelectric force sensors based on random PVDF nanofiber webs. Different PVDF fiber morphologies were obtained by varying the applied voltage and flow rate. As a result, the nanofibers with different β crystal phase contents were obtained. The PVDF nonwoven with higher β crystal phase content exhibited higher sensitivity. Boongik *et al* [32] integrated randomly oriented electrospun PVDF nanofiber web with organic photovoltaic device to fabricate a hybrid energy generator.

Zaarour *et al* [38] prepared PVDF nanofiber webs with different M_w using electrospinning and found that M_w affects the surface structure mechanical properties, crystalline phase and piezoelectric properties, and the piezoelectric properties increased from 1.97 V and 2.7 A to 2.92 V and 4.1 A at 5 Hz and 10 N when M_w was increased from 180 000 to 530 000. Szewczyk *et al* [40] investigated the effects of ambient humidity and positive and negative electrospinning voltage on the piezo-electric properties of PVDF nanofiber webs, and the results showed that higher ambient humidity and negative voltage were beneficial to the formation of β phase, and the PENG fabricated with negative voltage at 60% humidity had the highest d_{33} (5.558 pm·V^{-1}) and power density (0.60 μW·cm^{-2}). As shown in figure 3.10, Lan *et al* [44] were inspired by the human cochlea to design a multichannel piezoacoustic acoustic sensor (MAS) based on electrospun PVDF nanofibers, with different channels recognizing different frequencies of sound. The developed MAS exhibited a 300% response band and 334.6% acoustoelectric output enhancement compared to conventional electrospun piezoelectric membranes. Combined with machine learning, speech recognition can significantly achieve high accuracy rates of up to 100%. Shao *et al* [45] prepared a nonwoven web composed of hollow PVDF fibers by coaxial electrospinning. The study found that PVDF hollow fiber with large inner diameter can effectively enhance its deformation ability, so as to obtain better piezoelectric output performance. The piezoelectric voltage and current of PENG prepared based on this are as high as 32.6 V and 0.58 μA, respectively

In some studies, the electrical poling process was applied to the initial electrospun PVDF webs for getting the device piezoelectricy [23, 26, 33]. Gheibi *et al* [23] indicated that the piezoelectric response of initial electrospun PVDF webs was not desirable and detectable, due to random orientation of fibers causing neutralization of electric dipoles of each other. Thus, a high voltage of 3.9 kV was applied through

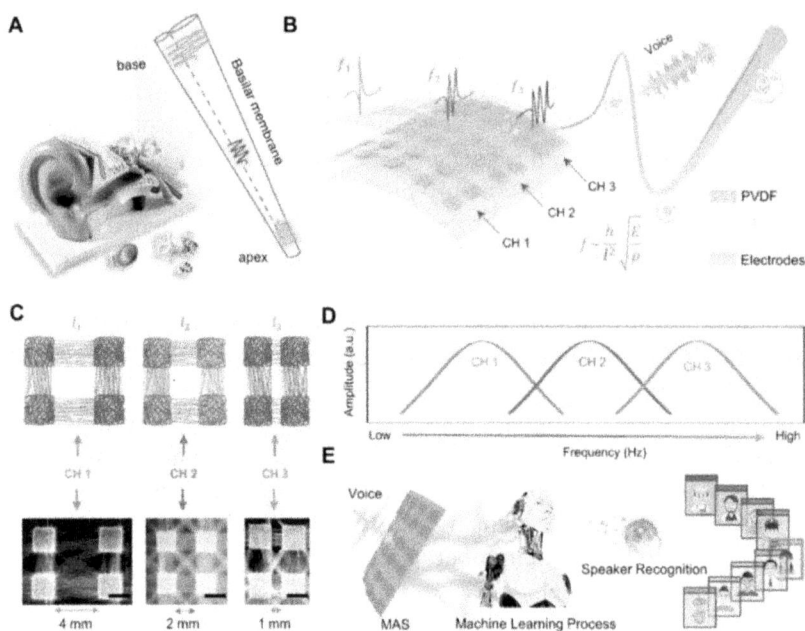

Figure 3.10. (A) Human cochlea with multi-frequency band. (B) MAS based on structured electrospun PVDF nanofibers. (C) Schematic illustration and digital photo of the electrospun membrane with different channels. (D) Schematic description of MAS-based multitunable resonant frequencies broadening piezoelectric response band. (E) The speaker recognition system with deeplearning, in which the multichannel MAS is used as a personalized acoustic interface for voice recognition. Reproduced with permission from [44], copyright (2023), American Chemical Society.

the terminals coming off from the electrode faces for sample with 310 μm thickness. The poling process was carried out for 4 h. The typical voltage output was 4.0 V when a stainless steel weight (5 g) was dropped from 15 cm height onto the sample. Liu *et al* [33] polarized electrospun PVDF webs by applying external current electrical fields of 5 kV. The prepared device generated an output voltage and current up to 1.5 V and 400 nA under periodically bending by a line motor. Kumar *et al* [46] polarized PVDF nanofibers at an electric field of 30 V μm^{-1} and 30 °C for 6 h. The obtained nonwoven webs exhibited higher piezoelectric properties and β-phase content, the output voltage increased from 0.71 ± 0.06 V to 0.8 ± 0.04 V.

3.3.3 Aligned fiber web

The aligned PVDF nanofibers were prepared for piezoelectric energy conversion property and it is found that the electrical outputs of the PENG based on the aligned fiber webs were normally higher than those based on the randomly oriented fiber nonwovens [68].

By modifying the near-field electrospinning setup, various PVDF products resulted, as shown in figure 3.11. Liu *et al* [48] prepared aligned PVDF nonwoven

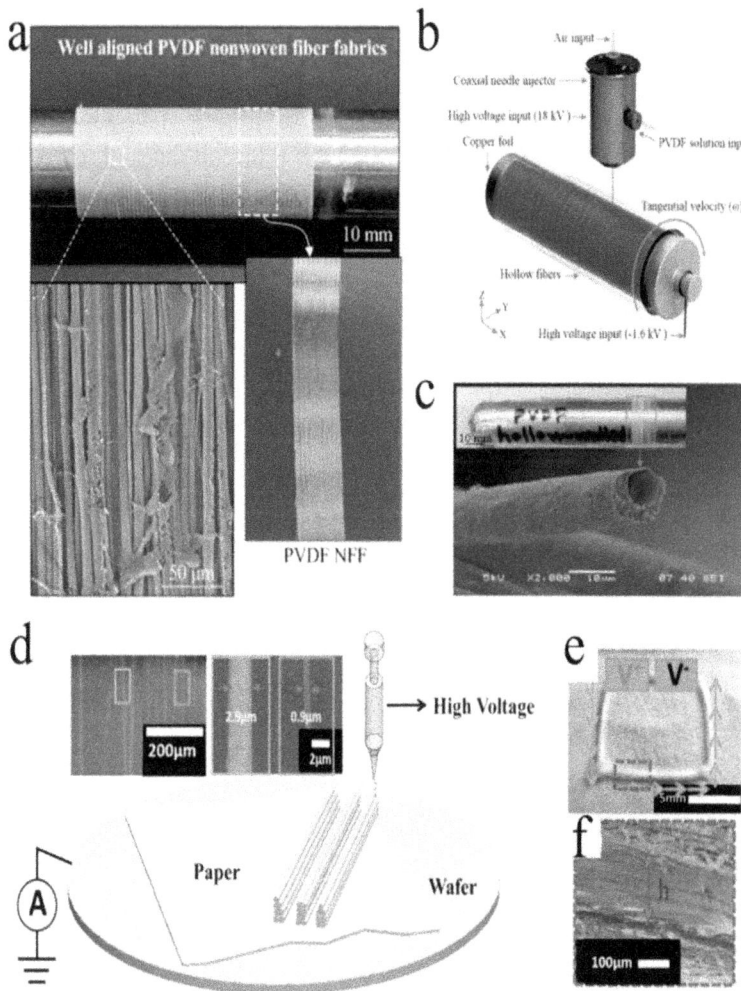

Figure 3.11. (a) The top optical photograph shows well aligned PVDF fibers on a glass tube collector and a piece of PVDF nonwoven fiber fabric is easily extracted from the glass tube collector. The bottom SEM image shows dense deposition of fibers to form the nonwoven fiber webs. (b) The coaxial needle injector electrospinning and (c) SEM image of the cross sections of PVDF fiber tubes. (d) Schematic setup of the sequentially stacked 3D fiber structure via near-field electrospinning process. (e) A photograph of the 3D square structure of area 1×1 cm^2 was constructed on a paper substrate. (f) SEM image showing the area of the square and micro-wall structure made of about 600 layers of electrospun fibers stacked on top of each other, the electrospun and stacked finer bundles is measured to have a width of $\sim$70 µm and height 300 µm. Reproduced from [48], copyright 2014, IOP Publishing Ltd, and reproduced from [50], copyright (2016), with permission from Elsevier.

fiber web energy harvesting devices by near-field electrospinning with a hollow cylindrical collector. A maximum peak voltage and current of 76 mV and 39 nA were obtained under repeated stretching and releasing of the device with a strain of 0.05% at 7 Hz. Pan *et al* [49] fabricated PVDF hollow fiber tubes with the metallic coaxial needle injector. The voltage output generated from the PVDF fiber tubes was about 71.66 mV, which was much higher than that of solid PVDF fibers. 3D architectures of PVDF fibers were also prepared by near-field electrospinning with a programmable x–y translational motion stage [50]. The device with PVDF stacked height of 400 μm (about 800 fibers) generated voltage of 4 V and current of 100 nA, respectively. The electrical outputs increased with the increase of fiber stacked height. Luo *et al* [51] fabricated a PENG consisting of 60 PVDF nanofibers using 3D electrospinning technique, and measured their piezoelectric response to be about 0.48 V and 2.7 nA.

Zhang *et al* [55] encapsulated the arrays of PVDF-NFES fibers in poly(lactic acid) (PLA) to manufacture a self-powered device (figure 3.12(a)). As shown in figure 3.12(b), the authors compared the mechanism of dipole polarization under different forms of electrospinning. During the far-field electrospinning (FFES) process, the shear force oriented the electric dipoles in the direction of PVDF fibers, which resulted in a decrease in the piezoelectricity of the PVDF fibers. Nonetheless, during the NFES process, the vertical electric field polarized the dipoles and oriented them along the thickness direction of the nanofibers, which resulted in an increase in the piezoelectricity of the PVDF fibers. The peak voltage of the PVDF-NFES fibers-based bending sensor was linearly proportional to the

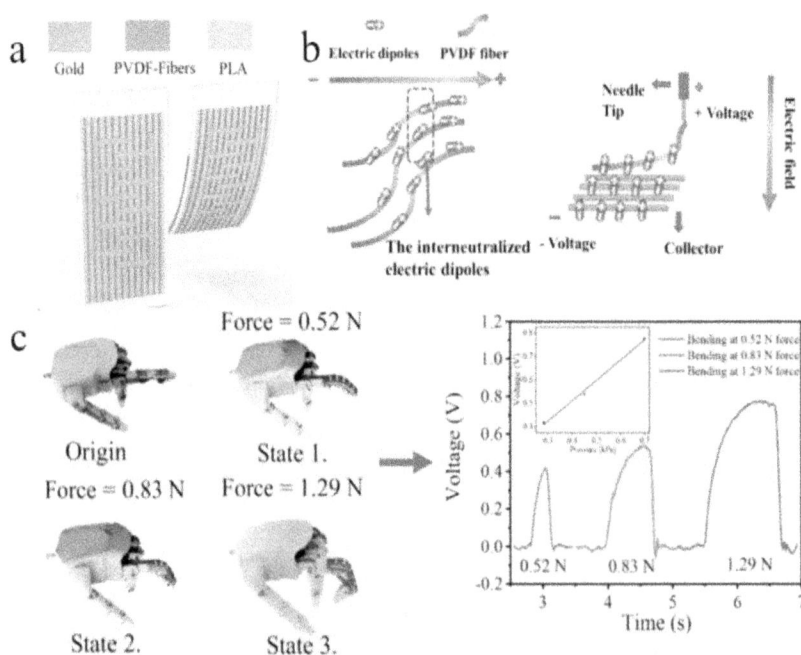

Figure 3.12. (a) Schematic diagram of the bending sensor based on the arrays of PVDF-NFES fibers and PLA. (b) The polarized PVDF fibers and the interaction of electric dipoles with different applied electric fields. (c) Different bending states and corresponding output voltages. Reproduced with permission from [55], copyright (2021), American Chemical Society.

force applied at different bending states. When the bending force was 1.29 N, the maximum output voltage of the sensor was 0.77 V and the sensitivity reached 864.68 mV kPa^{-1}, respectively.

Conventional electrospinning was also used to prepare aligned nanofiber webs. Two types of collectors were reported to prepare aligned fibrous structure: motion-free separated electrodes and high-speed rotating collector. In terms of motion-free separated electrodes, Kang *et al* [37] prepared aligned nanofibers using a metal frame as a collector. Figure 3.13(b) shows the SEM image of prepared PVDF nanofibers. The aligned PVDF nanofibers had a higher content of the β crystal phase as compared to those of randomly orientated fibers. The aligned nanofiber-based piezoelectric device could generate the output voltage and current of 1.1 V and 40 nA, respectively, under bending condition at a frequency of 2 Hz. At the same measurement conditions, it exhibited a two-fold increase in the output voltage and a three-fold increase in the output current as compared to the corresponding values obtained for the device manufactured from randomly oriented nanofibers (figure 3.13).

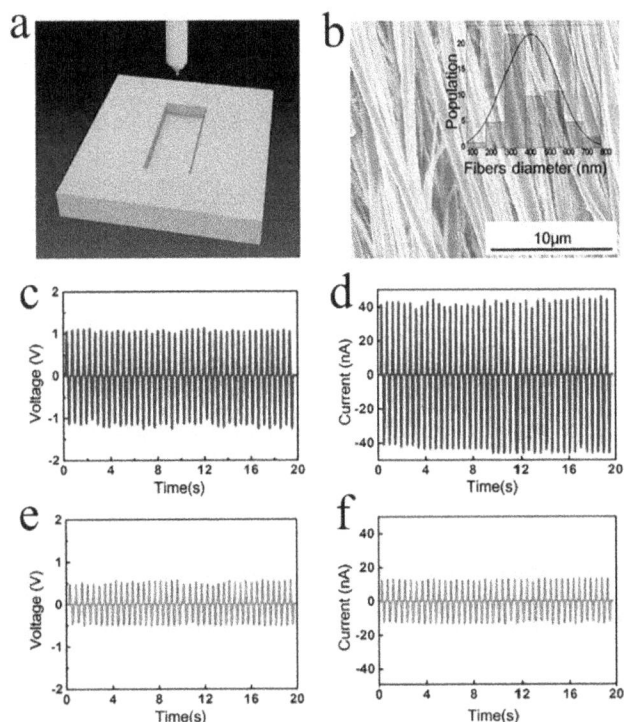

Figure 3.13. (a) A schematic illustration of the electrospinning setup containing a stepped collector, (b) SEM image of the aligned PVDF nanofibers. Output voltage (c) and current (d) piezoelectric device fabricated from the aligned nanofibers. Output voltage (e) and current (f) piezoelectric device fabricated from the randomly oriented nanofibers. Reproduced from [37], copyright IOP Publishing Ltd. All rights reserved.

In terms of high-speed rotating collector, Zaarour *et al* [36] prepared PVDF nanofibers by using a grounded drum collector. Aligned fibers were obtained when the drum rotating speed was 2000 rpm. Fibers with three different morphologies (wrinkled fibers, smooth fibers and porous fibers) were obtained by controlling the relative humidity, as shown in figure 3.14. The highest β crystal phase content was found on wrinkled fibers while the lowest one was shown on porous fibers. Also, the

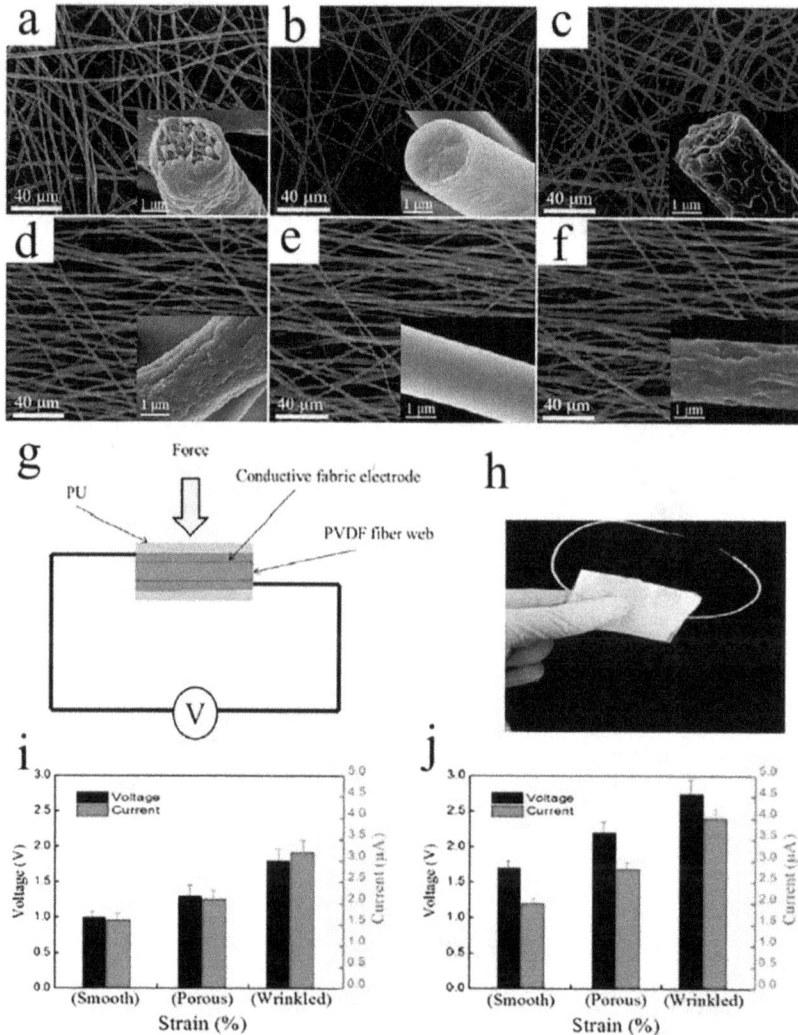

Figure 3.14. The samples prepared by electrospinning of PVDF solutions with different morphologies. (a–c) Randomly oriented fibers, (a) wrinkled, (b) smooth, and (c) porous. (d–f) Aligned fibers, (d) wrinkled, (e) smooth, and (f) porous. (g) Schematic structure of the PENG and (h) photo of the actual device. Summary results of the average voltage and current outputs generated by (i) randomly oriented fiber device and (j) aligned fiber device [36]. Reproduced with permission. Copyright (2018), Wiley Periodicals, Inc.

β crystal phase content of randomly oriented and aligned PVDF fibers at the same morphology was similar.

For comparison, the piezoelectric property of randomly oriented fiber web and aligned fiber web were both tested under compressive impacts. Figures 3.14(g) and (h) show the schematic structure and photo of the actual device. The outputs of randomly oriented fiber web were 1 V and 1.6 μA for the smooth fibers, 1.3 V and 2.1 μA for the porous fibers, and 1.8 V and 3.2 μA for the wrinkled fibers (figure 3.14(i)). The output voltage and current values were enhanced for the aligned fiber webs (figure 3.14(j)).

Zhang et al [52] prepared a piezoelectric nanogenerator based on aligned grooved PVDF fiber webs, the aligned nanofibers had higher β crystal phase content and fewer air gaps between fibers, which facilitateed the increase of piezoelectric outputs. The PENG was mounted on the human elbow and its performance was estimated based on a 90° bend-release of the elbow, the output voltage and current of the device based on the aligned fiber were 0.74 V and 457 nA.

3.3.4 Nanofillers

The nucleation behavior of a polymer depends on the properties of the nucleating agents, such as surface charge, surface area, concentration and dispersion, as well as the interfacial interactions between the nucleating agents with PVDF chains. Nucleating agents reduce the nucleation energy barrier, increase crystallization kinetics and the degree of crystallinity [69]. By introducing appropriate nanofillers into electrospun PVDF, the β crystal phase content in the nanofibers can be significantly increased, which further enhances the mechanical-to-electrical energy conversion property. To get high energy conversion, a variety of nanofillers have been explored, as describe below.

3.3.4.1 Salts

To evaluate the effects of surface charge on nucleation efficiency, various salts have been embedded into electrospun PVDF nanofibers as nucleating agents. Dhakras et al [70] studied the effect of nickel chloride hexahydrate ($NiCl_2 \cdot 6H_2O$) (NC) on the PVDF phase formation. Addition of 0.5 wt.% NC was found to enhance the β crystal phase in the nanofibers by ~30% and the voltage generated from PVDF-salt nanofiber webs was three times higher than that of neat PVDF nanofiber webs. Prasad et al [71] investigated the effect of several chloride salts ($CaCl_2$, $BaCl_2$, NaCl, $SnCl_2$, and $N_2H_6Cl_2$) on the crystallinity and piezoelectricity of PVDF, and found that the PVDF nanofiber webs doped with $CaCl_2$ had the best piezoelectricity and the highest β phase content, and the piezoelectric output voltage increased from 0.48 to 0.89 V.

In another study, 26 different salts were used to dope into PVDF nanofibers for understanding their effect on piezoelectric properties [72]. These salts were divided into three groups. Group I, such as $SnCl_4 \cdot 5H_2O$ and $Co(Ac)_2 \cdot 4H_2O$, had no obvious influence on nanofiber piezoelectric property. Group II, including $CoCl_2 \cdot 6H_2O$ and $FeCl_2 \cdot 4H_2O$ etc, had a significant enhancement in nanofiber piezoelectric performance, while group III ($MgCl_2 \cdot 6H_2O$, $Co(NO_3)_2 \cdot 6H_2O$, etc) was featured to have a modest improvement in piezoelectric property. Furthermore, salts of low concentration induced PVDF transform from amorphous polymer phases to β crystal phase, resulting in further increase of piezoelectric outputs. However, the energy conversion property of the PVDF nanofibers was negatively affected if the salt content was at too high concentration.

3.3.4.2 Ceramic particles

A lot of work has been reported on the incorporation of piezoelectric inorganic particles into electrospun PVDF nanofibers to enhance the piezoelectric property [73–75]. BaTiO$_3$ is a typical piezoelectric ceramic which was added to PVDF nanofibers and with 10 wt.% BaTiO$_3$ nanoparticles the nanofibers showed 1.66 times greater mechanical strength, and the output voltage was found to be 50 V [76]. As shown in figure 3.15, Zhao *et al* [77] composited PVDF with core-double shell PMMA-coated hyperbranched BaTiO$_3$ nanowires to prepare a PENG, which open-circuit voltage and short circuit current reached 3.4 V and 0.32 μA. Athira *et al* [78] fabricated PENGs generating 50 V at 3 N, 4 Hz pressure by sandwiching electro-spun PVDF/BaTiO$_3$ nanofiber webs with ITO coated PET films. The device had been used to monitor the operating status of central processing unit (CPU) fans, hard disk drives and electric sewing machines and to map their mechanical vibrations in real time (figure 3.16).

A vibration energy harvester was prepared by adding piezoelectric (Na$_{0.5}$K$_{0.5}$)NbO$_3$ (NKN) ceramic nanoparticles into PVDF nanofibers via electrospinning [79]. The PVDF/NKN composite nanofiber web reached the highest piezoelectric coefficient when the NKN concentration was 50 vol%. The largest power output obtained was 145 nW cm^{-2} per vibration cycle at 170 Hz. Ji *et al* [80] synthesized composite nanofibers from BNT-ST (0.78Bi$_{0.5}$Na$_{0.5}$TiO$_3$–0.22SrTiO$_3$) ceramic and PVDF, and examined the effect of BNT-ST concentrations on nanofiber

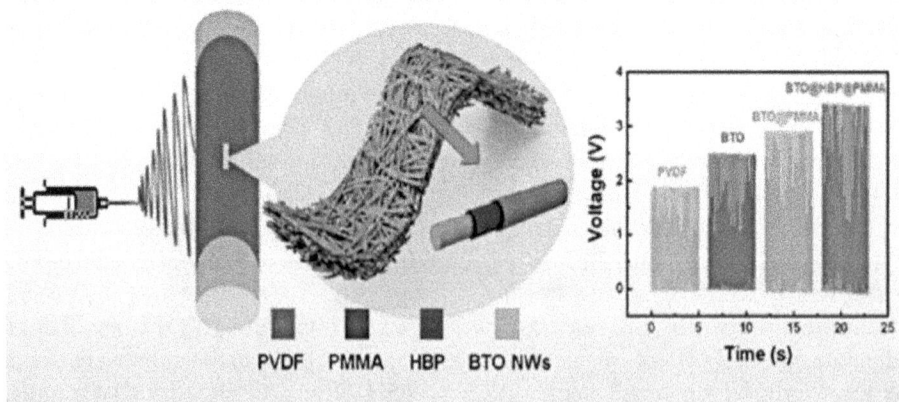

Figure 3.15. The open-circuit voltage and short circuit current of the PENG composed of PVDF and core-double shell PMMA-coated hyperbranched BaTiO$_3$. Reproduced with permission from [77], copyright (2022), American Chemical Society.

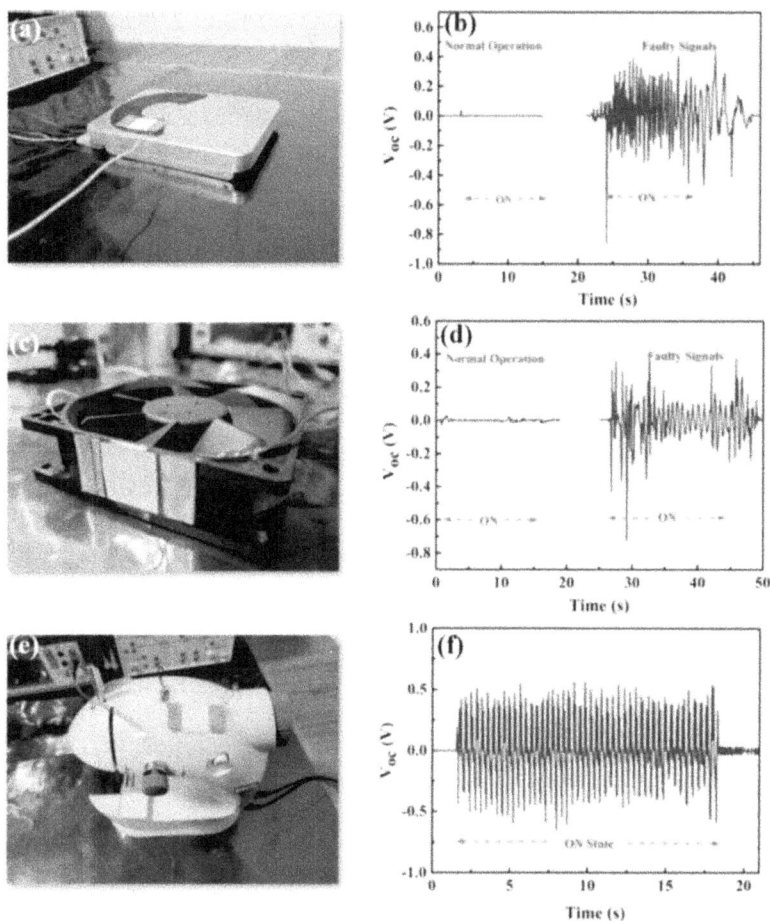

Figure 3.16. Demonstration of the PENG as a self-powered vibrational sensor; (a) photograph and (b) open-circuit voltage of the PENG attached to the hard disk drive (c) photograph and (d) open-circuit voltage of the self-powered vibrational sensor attached to the CPU fan and (e) photograph and (f) open-circuit voltage of the self-powered vibrational sensor attached to the electric sewing machine. Reproduced with permission from [78], copyright (2022), American Chemical Society.

piezoelectricity. BNT-ST has been considered the most promising candidates for lead-free piezoelectric materials, owing to its strong piezoelectric property. It has been found that 60 wt.% BNT-ST content allowed to nanofiber web to produce peak voltage output of 1.31 V under the resonance frequency of 6 MHz.

ZnO nanoparticles were incorporated into PVDF nanofibers to improve the piezoelectric properties. Compared with neat PVDF samples (0.351 V) in their study, under compression, the composite samples with 15 wt.% ZnO nanoparticles generated the higher voltage outputs [81]. ZnO nanorods were also used to increase the piezoelectricity of the PVDF piezoelectric webs, and the samples doped with 6 wt.% ZnO nanorods were able to generate piezoelectric voltages of 14.3 V [82].

3.3.4.3 Nanoclays

Nanoclays are used to reinforce and improve barrier functions of PVDF. Mohammed et al [83] reported that the incorporation of halloysite nanotubes (HNTs) into PVDF not only reduced nanofiber diameter, but also increased β crystal phase content. 10 wt.% was reported to be the optimum HNT concentration to make PVDF nanofibers reach the highest β crystal phase content, from 34% of neat PVDF to 81.1%. Meanwhile, it showed the highest voltage output of 0.955 V under 100 g load, which was three times higher than that of neat PVDF nanofibers. Xin et al [84] reported that 5 wt.% nanoclay greatly improved the β crystal phase content and piezoelectric output of PVDF nanofiber web. Hectorite (HRT) clay was also added into electrospun PVDF to enhance the piezoelectric property [85]. The prepared nanofibers with 0.1 wt.% HRT clay had increased β crystal phase content to 95% and voltage output to 12 V (neat PVDF nanofibers contain 80% β crystal phase and can generate 5 V voltage outputs). Tiwari et al [86] doped 15 wt.% of nanoclay into PVDF by electrospinning to obtain nanofiber webs, and the piezoelectric device made from the webs had an output voltage of up to 70 V and an power density of 68 μW cm^{-2}.

3.3.4.4 Carbon nanotubes and graphene

Multiwalled carbon nanotubes (MWCNTs) and graphene have been used to modify electrospun PVDF nanofibers. Liu et al [87–89] prepared PVDF/MWCNTs composite nanofiber webs by near-field electrospinning and studied their piezoelectricity. During the electrospinning process, MWCNTs interacted with PVDF, resulting in obvious increase in β crystal phase content. When the PVDF nanofiber web was doped with 0.03 wt.% MWCNTs (maximum in this study), it showed the highest piezoelectric outputs about 25 mV and 130 nA under a 15 Hz impact. Hao et al [90] examined the effect of MWCNTs amount in high doping level up to 10 wt.% on the fiber piezoelectricity. It was found that the β crystal phase content reached the highest level when the MWCNTs amount was 5 wt.%. At that time, the generated voltage output was as high as 6 V, which increased by 200% compared with neat PVDF nanofiber webs. Sharafkhani et al [91] achieved preferential orientation of carbon nanotubes (CNTs) along the longitudinal axis of PVDF nanofibers by adjusting the electrostatic spinning conditions, and the weight percentage of well-oriented CNTs increased from 0 to 1.25 wt.%, resulting in a β-phase content of close to 100% and an increase in the output voltage from 4 to 6.8 V. Zeng et al [92] designed a self-powered piezoelectric sensor based on CNT/PVDF composites to monitor ball–shoe interactions with excellent sensitivity of 80 mV N^{-1} and durability of over 15 000 cycles as well as output voltages and currents of 4.9 V and 90 nA.

In another study, the effect of graphene on PVDF piezoelectricity was examined [93]. Electrospun PVDF nanofiber web with just 0.1 wt.% graphene had the β crystal phase content as high as 83%, and voltage of 7.9 V under compression, in comparison to the value of 77% and 3.8 V for neat PVDF nanofiber web. Lee et al [94] prepared a novel piezoelectric fiber sensor based on PVDF-doped graphene using near-field electrostatic spinning, and the maximum output voltage of the PVDF fiber was 56.5 V at 4 wt.% graphene doping ratio, which was 11.54 times higher than that of the pure PVDF sensor. The presence of graphene not only further increased the β phase content in the PVDF nanofibers, but also formed a new conductive network between the PVDF nanofibers, which facilitated electron transport and increased the peak current of the PVDF material.

3.3.4.5 Polymer blending

Some polymers were blended with PVDF to enhance the β crystal phase content and piezoelectricity. Electrospun PVDF nanofibers containing cellulose nanocrystal (CNC) showed an increase in the β crystal phase content. When the CNC content was increased from 0 to 2 wt.% [95, 96], the β crystal phase content increased by 5.7%. Further increasing the CNC content resulted in decrease of β crystal phase content. PVDF fiber matrix contained 2 wt.% CNCs could produce a voltage output of 60 V, which was almost three times the value for neat PVDF web NG [96]. Sengupta et al [97] showed that doping of polypyrrole (PPy), polyaniline (PANI), and L-glutamic acid-modified polyaniline (PANI-LGA/P-LGA) into PVDF was able to improve the electrical conductivity of PVDF, while the β crystal phase content was also significantly increased. Sanchez et al [98] explored the preparation of a spongy electrospun PVDF nanofiber mat with a three-dimensional structure by adding poly(ethylene oxide) (PEO) and lithium chloride to PVDF. It was able to generate an average peak-to-peak voltage of 15.1 V and an instantaneous output power of 40.7 μW cm^{-2} at 69 Hz and 4.1 N impact force.

3.3.4.6 Small molecules

Electron and hole transfer agents, tri-p-tolylamine (TTA) and 2-(4-tert-butylphenyl)-5-(4-biphenylyl)-1,3,4-oxadiazole (Butyl-PBD), were added separately into PVDF nanofibers [99]. When the PVDF nanofibers contain 0.5% TTA or 1% Butyl-PBD, they show highest β phase content and electric outputs (figure 3.17). By applying 10 N compression force at 1 Hz to the fiber webs, the PVDF/TTA device could generate maximum voltage of 2.55 V while PVDF/Butyl-PBD generated 2.65 V. These values increased by 70% and 77%, respectively, compared with those of the neat PVDF nanofiber web.

3.4 Piezoelectricity of other electrospun nanofibers

3.4.1 Polymers

Besides PVDF, some other electrospun nanofibers have also been reported to have piezoelectric properties, such as PVDF-TrFE, PVDF-HFP, poly(γ-benzyl-L-glutamate), poly(L-lactic acid), polyacrylonitrile, collagen, gelatin, vhitin, etc.

3.4.1.1 PVDF copolymers

Copolymerisation is an effective method to tune the polymorph structure and phase transition behavior of PVDF. The comonomer type and composition ratio determine the crystalline structure of the polymer and degree of crystallinity which directly affect the piezoelectric properties. So far, a variety of VDF-copolymers have

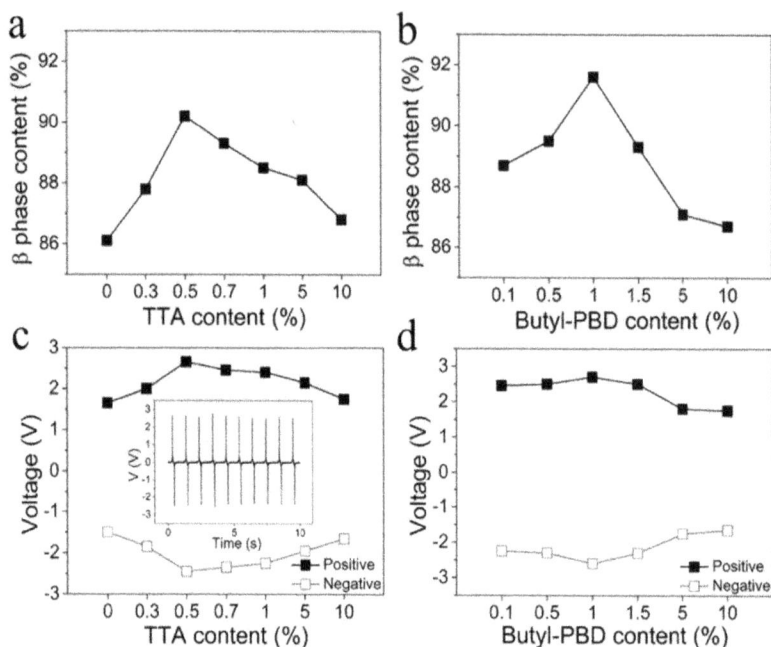

Figure 3.17. Calculated β phase content of PVDF nanofibers with different (a) TTA and (b) Butyl-PBD doping contents. Effect of (c) TTA and (d) Butyl-PBD doping contents on the peak voltage of PVDF nanofibers. Inset in (c): voltage output signals of PVDF nanofibers with 0.5% TTA doping content (web thickness: 100 μm; working area: 4 cm²) [99] John Wiley & Sons. Copyright 2017 WILEY-VCH Verlag GmbH & Co. KGaA Weinheim.

been synthesized by the incorporation of trifluoroethylene (TrFE) and hexafluor-opropene (HFP) and exist high piezoelectricity [73].

PVDF-TrFE, including α (nonpolar), β (polar), γ (polar), and δ (polar) phases, is one of the most studied copolymers. The β crystal phase, having the antipolar arrangement of fluorine and hydrogen atoms with partial negative and positive charges, respectively, is the most piezoelectrically responsive phase. One advantage of PVDF-TrFE over PVDF is that regardless of processing method, it always contains β-crystalline phase. This is attributed to the steric hindrance of TrFE in the copolymer, which forces PVDF into all-trans (TTTT) configuration (β phase). Furthermore, the high degree of crystallinity in the preferred orientation of well-grown crystallites result the higher remnant polarization present compared to PVDF, which indicates the higher efficiency in mechanical-to-electrical energy conversion [100].

PVDF-HFP has lower crystallinity than PVDF due to the presence of the bulky CF_3 groups [101]. PVDF-HFP has an unusual piezoelectric response, i.e., $|d_{31}/d_{33}| > 1$, which is not the case in other piezoelectric polymers [102]. It also has much higher piezoelectric coefficient (d_{31}), making it a promising material in the piezoelectric application field.

Electrospun PVDF-TrFE and PVDF-HFP nanofibers have been reported to have mechanical-to-electrical energy conversion properties [103–107]. Zhang *et al* [54] showed that PVDF-TrFE fiber webs have better mechanical properties, crystallinity and piezoelectric properties than PVDF fiber webs. In addition, the piezoelectric nanogenerator based on PVDF-TrFE fiber webs has a higher electrical output than PVDF due to its high β phase content. Kim *et al* [108] elucidated the role of solvent in crystallization kinetics and fiber formation during electrospinning, and prepared PVDF-TrFE nanofibers with an output voltage of 139.5 V under bending strain after an annealing process at 125 °C–130 °C. Hafner *et al* [109] demonstrated that the piezoelectric properties of the PVDF-TrFE were closely temperature dependent as the temperature was elevated, before phase transition from the ferroelectric to the paraelectric regime. The piezoelectric constant d_{33} was also gradually increased to 150% of the initial state as the temperature was raised from 25 °C to 80 °C. The increase in the piezoelectric response of the polymer was found to be directly related to the temperature dependence of its elastic modulus and polarization. Conte *et al* [110] established the relationship between yield strength, Young's modulus, and piezoelectric output with increasing draw ratio. Based on this, PVDF-HFP was drawn and the piezoelectric voltage output was increased by a factor of 4 compared to the undrawn one when the draw ratio was 3.

3.4.1.2 Poly(γ-benzyl-L-glutamate)

Poly(γ-benzyl-L-glutamate) (PBLG), a liquid crystalline rod like polymer, was reported to have high piezoelectricity when processed into nanofibers. PBLG is a synthetic polypeptide that forms hierarchically ordered structures containing α-helices. The α-helically structured PBLG serves as a rod-like structure in the solid state and common organic solvents. PBLG features a great number of directionally aligned hydrogen bonds along the helical axis that form a macroscopic dipole, which can couple synergistically with external electric field and shear force (figure 3.18) [111].

Pan *et al* [112] used cylindrical near-field electrospinning to produce PBLG piezoelectric fibers showing permanent piezoelectricity. A single PBLG fiber could generate the maximum voltage output of 33.27 mV with an external resistance of 8 MΩ. The corresponding power output reached 138.42 pW. PBLG fibers were further patterned on a cicada wing with interdigitated electrodes for energy harvesting application. The voltage output increased with increasing vibration frequency and it ranged from 7.67 to 14.25 mV when the strains of 0.04%–0.1% at 10–30 Hz were applied. They also reported that the PBLG fibers had much better mechanical properties than PVDF fibers. Nguyen *et al* [113] prepared PBLG nanofibers having the highest piezoelectric coefficients among biocompatible polymers, with piezoelectric charge coefficients in the transverse and longitudinal directions reaching 10.2 and 54 pC/N, respectively. The sensor consisting of PBLG nanofiber webs and PDMS substrate had a peak voltage of 200 mV and a sensitivity of 615 mV N^{-1}.

3.4.1.3 Poly(L-lactic acid)

Poly(L-lactic acid) (PLLA) is a biocompatible, biodegradable and piezoelectric polymer which has shown great potential for applications in tissue engineering scaffolds, energy harvesting and sensors [114]. The piezoelectric property of PLLA

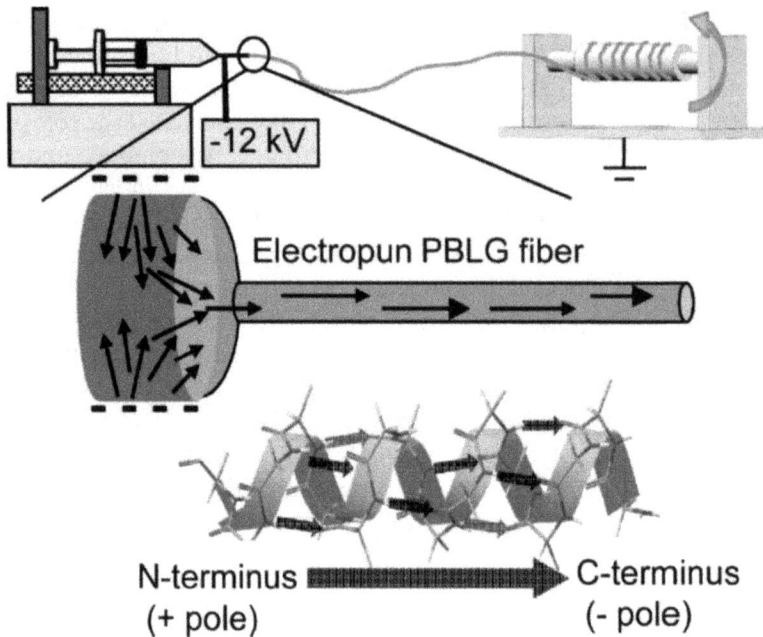

Figure 3.18. Schematic of electrospinning and the PBLG poling process. The α-helical molecular structure of PBLG is shown without side chains for simplicity. Small block arrows in the PBLG helix are individual hydrogen bonds and long arrows under the PBLG helix and in the expanded portion of needle tip area represent macroscopic dipoles of PBLG molecules [111] John Wiley & Sons. Copyright (2011) WILEY-VCH Verlag GmbH & Co. KGaA, Weinheim

fibers is originated from the C=O dipoles in the polymer chain [115]. The C=O dipoles are located circularly around the helical structure of PLLA with an angle of 125° relative to C–O–C main chain. As a result, the dipole component in the perpendicular direction of the polymer chian cancel out and the dipole component along the polymer main chain remains. For these reasons, it is difficult to show piezoelectricity when PLLA is electrically poled along the thickness. With the inherent shear force and electric field force in the same direction, the electrospinning process shows the prospect to align the dipole component along the polymer chain. It results in electrospun PLLA fibers to show piezoelectricity along the d_{33} direction [114].

The shear-piezoelectric behavior of electrospun PLLA nanofiber web sensors were reported as a function of web stacking arrangement including constructive, destructive and multi-layered folding [116]. Comparing to electrospun PVDF nanofiber web sensors, PLLA sensor showed lower voltage outputs in the constructive stacking case while exhibited slightly higher voltage output in the destructive case. With the increase of stacking nanofiber web layer, the voltage output of PLLA device increased.

Sultana et al [117] prepared a flexible and wearable piezoelectric bio-e-skin based on electrospun PLLA nanofibers for non-invasive human physiological signal monitoring. The device had a sensitivity of 22 V N^{-1} stably as well as pressure detection extent down to 18 Pa for detecting light pressure. Zhu et al [118] reported aligned porous PLLA nanofibers on a comb electrode for strain sensing and human joint motion energy harvesting. The device had the output voltage of 0.55 V and current of 230 pA under the strain deformation angle of 28.9°. Xu et al [119] designed a PLLA nanofiber-based PENG with an open-circuit voltage of 7.9 V, a short-circuit current of 286 nA, and an output power density of 1.25 mV cm^{-3}, which were able to exhibit stable piezoelectric signals within 1000 s.

Due to the excellent biocompatibility and biodegradability, as well as the considerable piezoelectricity, PLLA nanofibers are also widely used in the field of biological tissue engineering. Tai et al [120] showed that the transverse and longitudinal piezoelectric voltage outputs of electrostatically spun PLLA nanofibers were significantly dependent on fiber diameter and heat treatment temperature. Specifically, when the temperature is above the glass transition temperature of PLLA, an increase in the heat treatment temperature significantly reduces the transverse voltage output but increases the longitudinal voltage output. This modulation of the piezoelectric properties of PLLA nanofibers was able to significantly affect the differentiation behavior of stem cells, with longitudinal piezoelectricity and shear piezoelectricity enhancing neurogenesis and bone formation, respectively. Xia et al [121] revealed the connection between cell adhesion, scaffold surface morphology and electrical stimulation on the differentiation effect of neural stem cells. PLLA scaffolds modified with PDA were able to achieve significant neural differentiation within seven days under in situ electrical stimulation generated by neural stem cell adhesion.

3.4.1.4 Polyacrylonitrile

Polyacrylonitrile (PAN) is an amorphous polymer comprising a cyano group (–CN) in each repeat unit. The dipole moment of PAN was calculated as nearly 3.5 Debye units, which is much higher than PVDF's (2.1 Debye units) [122]. However, PAN often shows a low piezoelectricity [123].

Wang et al [124] reported that PAN nanofibers prepared by electrospinning can have a piezoelectricity even stronger than PVDF (figure 3.19). The prepared device consisted of an electrospun nanofiber web, two alumimum foils and two transparent polyethylene terephthalate (PET) films. Under the periodically mechanical compression, PAN nanofiber device generated the electric outputs of 2.0 V and 1.2 μA. In the same condition, the device made of the PVDF nanofibers had lower electric outputs. By adjusting the collector drum rotating speed, aligned PAN nanofibers were also prepared. They further discovered that aligned PAN nanofiber devices could generate much larger electrical outputs, up to 6.0 V and 5.1 μA. In solid state, PAN has two typical confirmations of planar zigzag and 3^1-helical. The zigzag confirmation with all-transform (TTT) structure is the reason for PAN's piezoelectricity. High content of the zigzag conformation around 80% exists in electrospun PAN nanofibers, which suggests their higher piezoelectricity.

Shao et al [125] reported an acoustoelectric device based on a PAN electrospun nanofiber membrane, which is capable of generating a voltage of 58 V, a current of 12 μA, and a maximum output power of 210.3 μW at a noise of 117 dB and 100–500

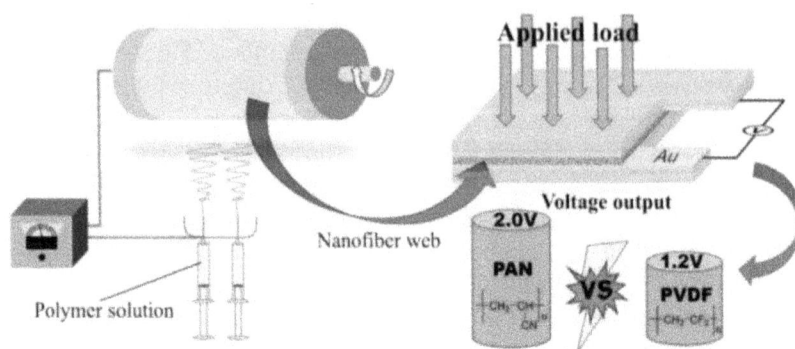

Figure 3.19. Polyacrylonitrile after being electrospun into nanofibrous membranes shows even stronger piezoelectricity than electrospun polyvinylidene fluoride fibrous membranes in the same condition. Reprinted from [124], copyright (2019), with permission from Elsevier.

Hz, respectively. The same research group [126] reported an acoustic sensor based on electrospun PAN nanofibers with high sensitivity, signal-to-noise ratio and fidelity up to 57.2 dB and 0.995, respectively. In the same year, the group [127] also prepared an acoustic sensor with remarkable performance in human voice recognition, the device with a sensitivity of up to 23 401 mV Pa^{-1} for a sound of 70 dB. The same group [128] also heat-stabilized the electrospun PAN fiber membrane to give it a cyclized structure. Devices made from the treated PAN membranes can operate at temperatures up to 550 °C and can produce a voltage of 9.7 V and a maximum power density of 26.4 mV cm^{-2} at 450 °C. Yu *et al* [129] systematically investigated the size-dependent piezoelectric properties of PAN nanofibers, and found that the voltage output and piezoelectric charge constant (d_{33}) increased exponentially with decreasing nanofiber diameter, and d_{33} was further enhanced after annealing of 40 nm nanofibers at 95 °C, reaching as high as 39.0 pm V^{-1}.

3.4.1.5 Collagen, gelatin and chitin

Collagen, gelatin and chitin are reported to show piezoelectric properties and could be used for making pressure sensors. One advantage of using the biological materials is their biocompatibility, allowing for long-term physiological detection without causing any skin infection. Piezoelectricity of collagen, a structure of protein, was reported as early as 1950s, and the piezoelectricity was originated from the oriented crystalline collagen fiber [130]. Gelatin is produced through partial hydrolyzation of collagen. Chitin is one of the polysaccharides containing of β-(1.4) linked N-acetylglucosamine monomers.

Street *et al* [131] reported an electrospun chitin nanofiber web with piezo-electricity and aligned chitin nanofiber webs had a 400% higher piezoelectric response than the random chitin nanofiber webs. The increase in piezoelectricity is a consequence of an increase in α-chitin crystallinity in the nanofibers. They also prepared aligned and random collagen nanofibers as well as gelatin nanofibers but found lower piezoelectric response.

Ghosh *et al* [132] prepared a self-powered pressure sensor based on electrospun gelatin nanofibers for healthcare monitoring. The sensor exhibited high sensitivity in low pressure region, about 0.8 V kPa^{-1}, suitable for physiological signal monitoring. It has high durability over 108 000 cycles piezoelectric output under dynamic tactile stimulus indicating the potential application in the field of implantable and portable biomedical as well as personal electronic devices.

3.4.2 Inorganic materials

Piezoelectric inorganic materials normally have higher piezoelectric coefficient compared to polymer materials. However, they are often rigid and work at a very low strain level. Through processing these piezoelectric inorganic materials into nanofibrous structure, their work durability can be enhanced. The preparation of piezoelectric inorganic nanofibers by electrospinning contains two steps, electrospinning to prepare polymer/inorganic composite nanofibers and calcinating to remove organic components. Unlike the preparation of piezoelectric polymer nanofibers, the electric poling process is always required for making piezoelectric inorganic nanofibers. Various inorganic nanofibers have been prepared by electrospinning technique and their mechanical energy conversion properties are detailed below (table 3.2).

3.4.2.1 Lead zirconate titanate
In 2010, the first inorganic piezoelectric nanofibers prepared by electrospinning was reported, in which lead zirconate titanate (PZT) nanofibers were used [133]. The PVP/PZT composite nanofibers were prepared by electrospinning and then annealed at 650 °C to obtain PZT nanofibers with a perovskite phase. Interdigitated electrodes of fine platinum wires with the distance of 500 μm were used for fabrication of piezoelectric devices and the diameter of the PZT nanofibers was controlled as around 60 nm (figure 3.20). Finally, PZT nanofibers were treated within an electric field of 4 kV mm^{-1} at a temperature over 140 °C for about 24 h. The poled nanofibers had orientated electric dipoles along the electric field with an output voltage of 1.63 V. In another study, PZT nanofibers were prepared by the same process and the effect of poling process on PZT piezoelectricity was studied [134]. The electromechanical coupling effect was increased as high as 3.7 times after 90 min of polarization under 3 kV mm^{-1} electric field. The produced device generated an output voltage of 100 mV in response to the falling ball (8.34 g) impact.

PZT nanowires prepared by electrospinning have also been reported to make energy harvesting device and self-powered devices [135]. PZT precursor salts containing tetrabutyl titanate, zirconium acetylacetonate, lead subacetate and PVP at a specific ratio were used for making electrospinning solution. After electrospinning, PZT nanowires are prepared by calcination the composite fiber at 650 °C for about 3 h. Later on, the nanowires were electrically poled in a 4 kV mm^{-1} electric field at a temperature of 130 °C for 15 min. The device could generate electric outputs of 6 V and 45 nA. He *et al* [141] prepared a flexible sensor array based on PZT nanowires. The sensor array consists of a 3 × 3 sensor unit (75% PZT

Table 3.2. A summary of inorganic nanofiber energy generators.

Inorganic nanofibers	Fiber diameter (nm)	Precursor polymer	Calcination temperature (°C)	Poling voltage (kV mm⁻¹)	Poling temperature/time	Device size (cm²)	Force/strain	Electric outputs	Reference
PZT multi-fibers	60	PVP	650	4	140 °C 24 h	—	Finger pressure	1.63 V	[133]
PZT multi-fibers	50–120	PVP	650	3	90 min	—	8.34 g falling ball impact	0.1 V	[134]
PZT web	370	PVP	650	4	130 °C 15 min	1.2	Bending/releasing	6 V 45 nA	[135]
PZT web	319	PVP	650	2.4	100 °C 15 min	—	0.796 N pressure	14 mV	[136]
PZT web	96	PVP	600	—	—	—	Finger pressure	40 V	[137]
PZT web	450	PVP	550	—	—	3	Pressure, 5 N	4 V	[138]
PZT aligned fiber web	220	PVP	650	7.5	150 °C 30 min	8	Bending strain	V 1.4 µA	[139]
PZT nanowire array	370	PVP	650	5	130 °C 15 min	2.25	0.53 MPa pressure	209 V 53 µA	[140]
PZT nanowire array	300	PVP	800	30	130 °C 1 h	Sensor unit:1 Sensor array:25	Pressure	1.4 V	[141]
PbTiO₃ web	140	PVP	500	0.4	150 °C 24 h	—	~35 g load	1 V	[142]
BiScO₃–PbTiO₃ nanofibers	100	PVP	650	7.5	—	—	Pressure, 5 N	4 V	[143]
Neat ZnO web, doped ZnO web	82.8, 61	PVA	550	—	—	1	1.25 kg load	0.12 V, 0.35 V	[144]
BaTiO₃ random, horizontal, vertical webs	354.1	PVP	1000	5	120 °C 12 h	2.21, 1.19, 1.53	0.002 MPa pressure	0.56 V 57.78 nA, 1.48 V 103.33 nA, 2.67 V 261.4 nA	[145]
BaTiO₃ nanofibers	45	PVP	750	—	—	5.6	0.16% bending strain	2.7 V	[146]
BaTiO₃ nanofibers	100	PVP	850	—	—	—			[147]
BaTiO₃ nanofibers	328	PVP	850	15	100 °C 3 h	1	100 kPa pressure	1.05 V 4.8 nA	[148]

(Continued)

Table 3.2. (*Continued*)

Inorganic nanofibers	Fiber diameter (nm)	Precursor polymer	Calcination temperature (°C)	Poling voltage (kV mm^{-1})	Poling temperature/time	Device size (cm^2)	Force/strain	Electric outputs	Reference
BaTi$_{0.9}$Zr$_{0.1}$O$_3$ nanofibers	300	PVP	800	3	60 °C 60 min	—	Finger pressure	0.26 V	[149]
0.5Ba(Zr$_{0.2}$Ti$_{0.8}$)O$_3$–0.5(Ba$_{0.7}$Ca$_{0.3}$)TiO$_3$ nanofibers	175	PVP	700	1.5	90 °C 2 h	1.2	Stretch/release	3.25 V 55 nA	[150]
(Ba$_{0.85}$Ca$_{0.15}$)(Ti$_{0.9}$Zr$_{0.1}$)O$_3$, (Ba$_{0.85}$Ca$_{0.15}$)(Ti$_{0.9}$Zr$_{0.1}$)O$_3$-Y	300, 350	PVP	700	12.0 kV	25 °C 30 min	—	Finger pressure	1.9 V 48 nA, 3 V 85 nA	[151]
(K,Na)NbO$_3$ nanofibers	100	PVP	700	2.3	—	1	Compression strain 6.0%, 5 Hz	65 mV	[152]
(K,Na)NbO$_3$ nanofibers	50–100	PVP	600	5	Room temperature 3 min	—	—	—	[153]
NaNbO$_3$ nanofibers	78	PVP	700	2	Room temperature	2	Bending	2 V	[154]
0.96(K$_{0.48}$Na$_{0.52}$)(Nb$_{0.95}$Sb$_{0.05}$)O$_3$–0.04Bi$_{0.5}$(Na$_{0.82}$K$_{0.18}$)$_{0.5}$ZrO$_3$ nanofibers	250	PVP	700	5	150 °C 2 h	4	Pressure, 1 Hz	10 V	[155]

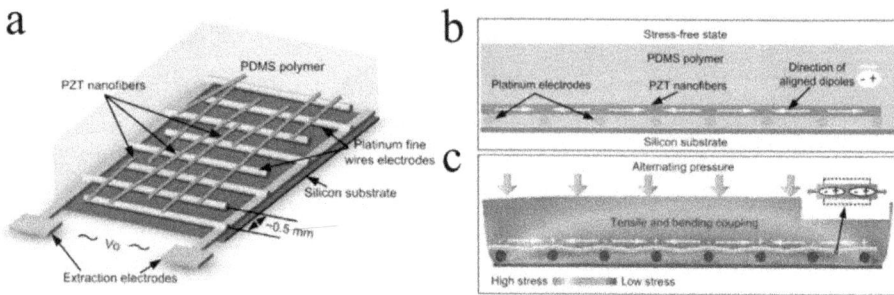

Figure 3.20. (a) Schematic view of the PZT nanofiber generator; (b) cross-sectional view of the polled PZT nanofibers in the generator and (c) schematic view explaining the mechanism of the PZT nanofibers working in the longitudinal mode. Reprinted with permission from [133]. Copyright (2010), American Chemical Society.

nanowires) with a complete bottom electrode and nine top electrodes. The sensor unit has high force-electric conversion capability and high sensitivity, and also shows good linearity and good stability over 1600 cycles. Electrospinning was also used to prepare vertically aligned $Pb(Zr_{0.52}Ti_{0.48})O_3$ (PZT) nanowire arrays [140]. The voltage output generated by the device was as high as 209 V and a current density of 23.5 μA cm^{-2}, which can be directly used to stimulate the frog's sciatic nerve and induce a contraction of a frog's gastrocnemius.

Randomly aligned PZT nanofiber webs prepared by electrospinning have been widely used in energy harvesting and sensing applications [137, 138]. Hayat *et al* [137] deposited PZT nanofiber webs on staggered electrodes of silver wires, and the output voltage of the device could reach 40 V under periodic pressure. Jiang *et al* [138] fabricated pure inorganic PZT electrospun webs by low-temperature calcination method, and decreasing the calcination temperature could effectively improve the flexibility of the fiber webs, with the maximum normal stress and the maximum bending strain not less than 0.18 MPa and 4.2% ± 0.08%, respectively. The device is capable of generating a piezoelectric voltage of 4 V under a pressure of 5 N.

Highly aligned PZT nanofiber webs have also been reported for flexible piezoelectric NGs [139]. The orientation and bending direction of the PZT fibers were found to have a significant effect on the electrical properties of the devices. An assembled NG with an area of 8 cm^2 and a thickness of 80 μm can produce an output voltage of 1.1 V and a current of 1.4 μA under bending strain.

Sutka *et al* [142] produced $PbTiO_3$ nanofibers by electrospinning a solution contained Pb acetate, acetic acid, titanium tetraisopropoxide and PVP, and by subsequently annealing the composite nanofiber web at an elevated temperature (500 °C for 3 h). Finally, the $PbTiO_3$ nanofiber web was packed into a device and poled in 0.4 kV mm^{-1} electric field. It could generate peak voltage output as high as 1 V under compression. Xin *et al* [143] produced $BiScO_3$–$PbTiO_3$ nanofibers and combined them with interdigital electrodes to prepare flexible sensors, the sensors showed an approximately linear output voltage in the pressure range of 0.07–0.55 N in the temperature interval of 160 °C–260 °C, and the sensitivity was still higher than 5.4 V N^{-1} at 220 °C.

3.4.2.2 Lead-free ceramics

Previously, lead oxides, especially lead zirconate titanate (PZT), have been widely used as piezoelectric materials because of their excellent piezoelectric properties. However, their lead contained raises environmental concerns, which hinder the applications in many fields. Thus, developing a lead-free and environmentally friendly material with a piezoelectric coefficient comparable to that of PZT is highly desirable. In the last decade, many lead-free piezoelectric ceramics, such as ZnO, $BaTiO_3$ (BTO), $BaTi_{0.9}Zr_{0.1}O_3$ (BTZO), $0.5Ba(Zr_{0.2}Ti_{0.8})O_3$–$0.5(Ba_{0.7}Ca_{0.3})TiO_3$ (BZTO-BCTO), $(Ba_{0.85}Ca_{0.15})(Ti_{0.9}Zr_{0.1})O_3$-Y (BCTZY), (K,Na)NbO$_3$ (KNN) and $0.96(K_{0.48}Na_{0.52})(Nb_{0.95}Sb_{0.05})O_3$–$0.04Bi_{0.5}(Na_{0.82}K_{0.18})_{0.5}ZrO_3$ (KNNS–BNKZ), have been processed into nanofibers.

ZnO nanofibers and Al doped ZnO nanofibers were fabricated by electrospinning at different solution flow rates of 2, 4 and 6 $\mu l\ min^{-1}$ followed by annealing [144]. A higher solution flow rate resulted in a larger fiber diameter and worse mechanical-to-electrical conversion performance. The maximum power density of neat ZnO was 17.6 nW cm^{-2} and for the doped ZnO nanofibers, the device showed maximum power density of 51.7 nW cm^{-2}.

Yan *et al* [145] studied on the energy conversion properties of $BaTiO_3$ nanofibers with different alignment features: aligned vertically, horizontally and randomly (figure 3.21). The vertically aligned $BaTiO_3$ nanofibers showed the best piezoelectric performance with maximum voltage output of 2.67 V and current of 261.40 nA, under mechanical stress of 0.002 MPa. Comparably, the devices from horizontally aligned nanofibers and random nanofibers generated lower voltage, 1.48 and 0.56 V, and lower current 103.33 and 57.78 nA, respectively. Another report indicated that electrospun $BaTiO_3$ nanofibers with finer diameter exhibited better piezoelectricity [146]. A device contained nanofibers with an average diameter of 45 nm produced a maximum voltage of 7.94 V_{p-p} under a strain of 0.16%. Yan *et al* [148] produced a flexible $BaTiO_3$ nanofiber webs to fabricate piezoelectric sensors by electrospinning, which showed a response time of 80 ms, an open-circuit voltage of 1.05 V, and a short-circuit current of 4.8 nA at a mechanical pressure of 100 kPa.

$BaTi_{0.9}Zr_{0.1}O_3$ (BTZO) nanofibers with diameter nearly 300 nm were prepared by electrospinning [149]. Aligning BTZO nanofibers after poling generated 0.26 V voltage outputs when a finger applied a dynamic load. High piezoelectric coefficient lead-free $0.5Ba(Zr_{0.2}Ti_{0.8})O_3$–$0.5(Ba_{0.7}Ca_{0.3})TiO_3$ (BZTO-BCTO) nanowires were synthesized by electrospinning [150]. By assembling the aligned nanowires into a flexible device, and poled in 1.5 $kV\ mm^{-1}$ electric field, one could harvest weak mechanical energy and generate 3.25 V and 55 nA.

Flexible BCTZY and BCTZ nanofibers were synthesized by electrospinning and the piezoelectric properties of the vertical aligned nanofibers were studied [151]. The introduction of Y^{3+} into BCTZ nanofibers could effectively enhance the nanofiber continuity, flexibility and stability. Moreover, the Curie temperature of the BCTZY nanofibers is about 280 °C, which is much higher than that of about 90 °C for BCTZ nanofibers. Also, the electrical properties such as dielectricity, ferroelectricity and piezo-electricity were all improved after doping with Y^{3+}. The devices made of vertically aligned BCTZY nanofibers showed output of 3.0 V and current of 85 nA by finger tapping.

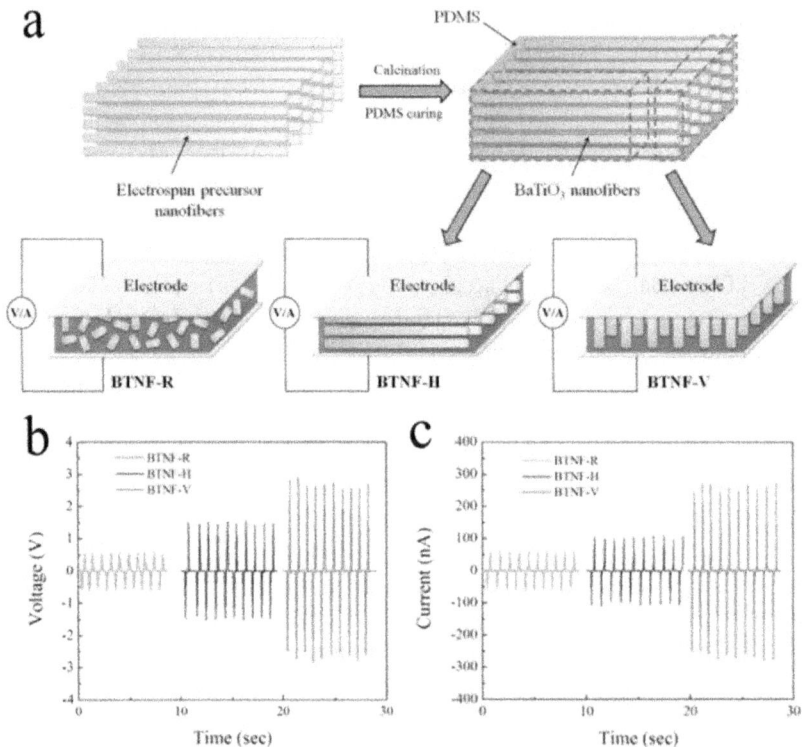

Figure 3.21. (a) Schematic fabrication procedure of the NGs based on BaTiO₃ nanofibers in three kinds of alignment modes. (b) Voltage and (c) current outputs of the BaTiO₃ NGs under a periodic mechanical compression. Reproduced with permission from [145], copyright (2016), American Chemical Society.

Well aligned electrospun (K,Na)NbO₃ (KNN) nanofibers after electric poling showed a fast response to dynamic strain by generating impulsive voltage signals that is dependent on the amplitude of dynamic strain and its frequency [152]. At 1 Hz repeated impact, the voltage output increased from 1 to 40 mV when the strain changed from 1 to 6%. Also, the voltage output showed linear increase when the input frequency increased from 0.2 to 5 Hz. Yousry *et al* [153] fabricated KNN nanofiber webs with randomly aligned structure using electrospinning method. It was found that the crystallization temperature of the KNN gels was 570 °C, while that of the KNN nanofibers was 472 °C, and that the KNN nanofibers were more likely to form (100) oriented perovskite phases. The analysis shows that the high surface-to-volume ratio and the resulting reduced energy barrier due to surface-induced heterogeneous nucleation result in lower crystallization temperatures of KNN nanofibers. The reduction of crystallization temperature and the high crystallinity of (100) orientation can effectively reduce the dielectric loss, increase the electrode polarization, and obtain the high-voltage electrical performance of KNN.

Electrospun piezoelectric NaNbO₃ nanofibers were prepared for humidity sensor application [154]. When the relative humidity changed from 5% to 80%, the peak-to-peak value of voltage generated by the sensor decreased from 0.40 to 0.07 V. This was due to the increased leakage current in the $NaNbO_3$ nanofibers, which was generated due to proton hopping among the H_3O^+ groups in the absorbed H_2O layers under the driving force of the piezoelectric potential. Piezoelectric KNNS–BNKZ nanofibers were fabricated by electrospinning and calcination [155]. The nanofiber device after being poled at 150 °C with an electric field of 5 kV mm^{-1} for 2 h, could generate an output voltage up to 10 V under dynamic pressures.

3.4.3 Composite

A number of nanofillers have been added into electrospun PVDF nanofibers for enhancing the device piezoelectric property. Besides, some nanofillers were also mixed with other piezoelectric polymers to form composite nanofibers for achieving better energy conversion property.

Piezoelectric ceramic PZT particles were doped into PVDF fibers as a piezo-electric reinforcing phase, which can effectively promote the formation of polar β-phase in PVDF [156]. The electrospun fiber webs based on PZT/PVDF composites exhibited excellent flexibility with Young's modulus of 227.2 MPa and elongation of 262.3%. The optimal output voltage of the PENG assembled from the composites was up to 62.0 V with a power of 136.9 μW. As shown in figure 3.22, the PENGs were capable of charging the capacitor rapidly and maintaining stability in a short period of time, and they were able to perform better energy harvesting in different application scenarios, such as finger pressing, fist beating, bending on one side and arm pressing.

Highly sensitive PVDF-HFP/ZnO composite nanofiber piezoelectric sensors were prepared by epitaxial growth of ZnO nanosheets on the surface of electrospun PVDF-HFP nanofibers [157]. It was shown that the sensor achieved optimal pressure sensing performance with a sensitivity of 1.9 V kPa^{-1}, a short response time of 20 ms, and excellent stability under 5000 press-release cycles when the force is in the range of 0.02–0.5 N. The sensor had promising applications in the fields of medical treatment, rehabilitation medicine and motion detection.

Shi et al [158] produced the piezoelectric reinforced phase of $BaTiO_3$ nanowires by hydrothermal synthesis and polymerized a PMMA coating on the surface of the nanowires to form a strong interface between the $BaTiO_3$ nanowires and PVDF-TrFE, which effectively improved the dispersion of the $BaTiO_3$ nanowires in the polymer matrix, thus enhancing the piezoelectric output performance of the composite PENGs. The output voltage and current of the PENGs can reach 12.6 V and 1.30 μA, respectively, and the maximum output power reached 4.25 μW.

Lu et al [159] prepared Tb-modified $(BaCa)(ZrTi)O_3$ (BCZT) particles with 3D structure by freeze-drying method. Composite nanofibers with 3D-Tb-BCZT/PVDF core–shell structure were fabricated by coaxial electrostatic spinning and assembled into PENGs. The PENGs exhibited remarkable piezoelectric outputs of 48.5 V and 3.35 μA. Even after 5000 cycles and 3 months, the prepared PENG exhibited stable output performance and showed good durability and stability.

Apart from adding piezoelectric ceramic materials, materials without piezo-electric property were also added into piezoelectric polymer nanofibers. PVDF-HFP

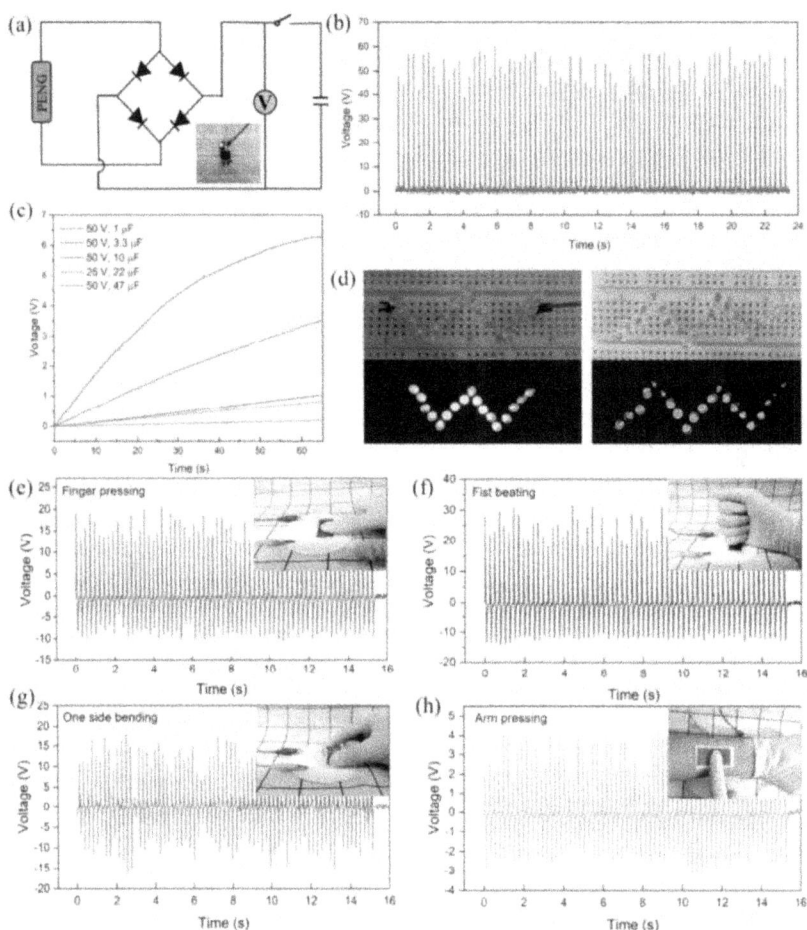

Figure 3.22. (a) Equivalent circuit diagram of the charging commercial capacitor with a charging system inset; (b) rectified voltages of f-P 0.10 fiber-based PENG with a thickness of ~220 μm; (c) charging voltages of various capacitors; (d) green LEDs lit by the rectified voltages of the PENG, output voltages from f-P 0.00 fiber-based PENG under (e) finger pressing, (f) fist beating, (g) one side bending, and (h) pressing on the arm. Reproduced from [156] CC BY 4.0.

nanofibers containing silver nanoparticles were prepared by electrospinning and used for piezoelectric purpose [160]. The content of β phase increased by the addition of silver nanoparticles, resulting in better piezoelectricity. A maximum output voltage of 3.0 V with a current density of 0.9 μA cm^{-2} was achieved under mechanical pressure. Eu^{3+} doped PVDF-HFP/graphene nanofibers were prepared [102]. The addition of Eu^{3+} and graphene sheets were reported to increase the β crystal phase content and crystallinity. The composite nanofibers were able to generate a voltage output of 9 V under finger-touch (pressure 5.6 kPa), whereas a 4 V voltage was generated from the Eu^{3+} doped PVDF-HFP nanofibers. A novel two-dimensional conductive material, MXene, was introduced into PVDF-TrFE to form a nanofiber pressure sensor by electrospinning method is reported [161]. The PENG containing 2.0 wt.% MXene achieved an instantaneous output power density of about 3.64 mW m^{-2} at a pressure of 1 Hz, 20 N. MXene was able to interact with the dipoles of PVDF-TrFE molecular chains and had high electrical conductivity, which increased the polarization of PVDF-TrFE during the electrospinning process. In addition, materials with piezoelectric properties and non-piezoelectric properties can also be doped into piezoelectric polymer nanofibers simultaneously. Su *et al* [162] dispersed polydopamine (PDA) in BaTiO$_3$/PVDF nanofibers, which significantly enhanced the interfacial bonding and connection between the materials and improve the overall mechanical strength and piezoelectric properties (3.02 wt.% PDA doping resulted in a 47% increase in piezoelectric voltage). The PENG based on PDA@ BaTiO$_3$/PVDF nanofibers showed excellent sensitivity (3.95 V N^{-1}) and long-term stability (<3% decrease after 7400 cycles). Later, the same group [163] simultaneously doped Sm-PMN-PT piezoelectric ceramic nanoparticles and MXene with high electrical conductivity into PVDF to prepare composite nanofiber mesh membranes. The introduction of piezoelectric ceramics improved the overall piezoelectric properties, the presentation of MXene effectively enhanced the electrical conductivity and mechanical ductility of the polymer. The combined effect of the two fillers optimized the interfacial coupling mechanism between the materials and improved the piezoelectric response of the composite nanofibers by 160%.

3.5 Piezoelectric–triboelectric hybrid energy generator devices

Hybrid energy generators with both piezoelectric and triboelectric mechanisms can be an effective way to enhance energy conversion performance. So far, there are two ways to fabricate hybrid energy generator devices. One is that using piezoelectric materials as the triboelectric layers, the energy generators combining piezoelectricity and triboelectricity can be fabricated. The other method is by mechanically integrating piezoelectric device unit and triboelectric device unit into one device.

Huang *et al* [164] developed a wearable TENG-based insole composed of electrospun PVDF nanofibers sandwiched between a pair of conducting fabric electrodes (figure 3.23). PVDF nanofibers produced by electrospinning formed high β crystal phase content with strong piezoelectricity. The difference between the PENGs and TENGs is that for PENGs, the top electrode is firmly and directly attached to the PVDF nanofibers without any spacer so there is no relative displacement between the fabric electrode and PVDF nanofibers. The electrode in PENGs is only to collect the piezoelectric charges. In this study, they prepared a PENG with a configuration of electrode/PVDF/electrode and indicated that the direction of the voltage generated was reversed when the direction of dipoles was reversed (figure 3.23). The PVDF PENG could produce the maximum voltage of 4 V. In comparison, the direction of voltage generated by triboelectricity would remain the same. Interestingly, the maximum voltage of the TENG was increased from 200 to 210 V when the voltage direction is the same as that of the voltage generated by piezoelectricity. These indicate that the overall voltage is the sum of the voltages generated by piezoelectricity and triboelectricity. A positive contribution achieved by piezoelectricity is tuning its voltage direction in accordance with that of

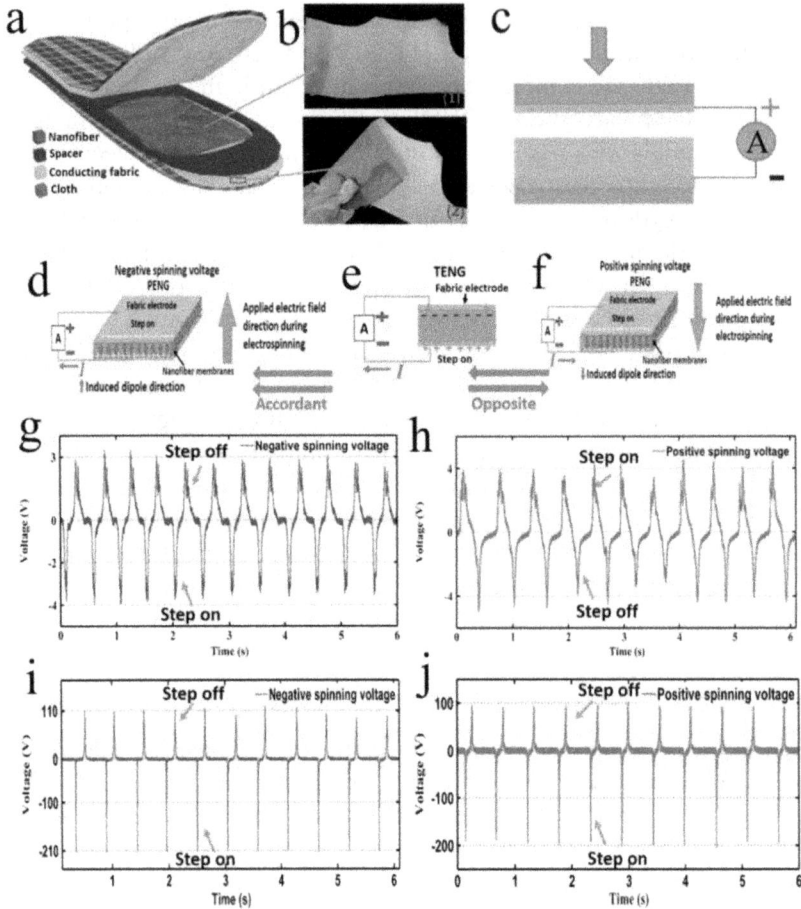

Figure 3.23. (a) Schematic structure illustration of the TENG-based insole; (b) digital photographs of the electrospun PVDF nanofibers on the conducting fabric (1) front side and (2) back side. (c) Schematic illustration of the TENG device. Schematic diagrams of (d) a PENG fabricated with negative spinning voltage; (e) a TENG fabricated by either negative or positive spinning voltage, and (f) a PENG fabricated by positive spinning voltage. Voltage generated by piezoelectricity of PVDF nanofibers fabricated under (g) negative and (h) positive spinning voltage. Voltage generated by triboelectricity of PVDF nanofibers fabricated under (i) negative and (j) positive spinning voltage. Reproduced from [164], copyright (2015), with permission from Elsevier.

the triboelectricity by reversing the dipole direction. The wearable TENG-based insole could be used to convert mechanical energy to light 214 LEDs connected in series simply by stepping force.

An all-fiber hybrid piezoelectric-enhanced triboelectric nanogenerator fabricated by electrospinning silk fibroin and PVDF nanofibers on conductive fabrics was presented in [165]. Due to the large specific surface area of nanofibers and the extraordinary ability of silk fibroin to donate electrons in triboelectrification, the hybrid nanogenerator shows outstanding electrical performance, with the voltage, current and power density of 500 V, 12 μA, and 310 μW cm^{-2}, respectively. The ability of the smart textile device to detect the abrupt motion and falling of a person as well as the high power harvesting make the hybrid generator a promising candidate in the area of wearable real-time health monitoring.

Another type of hybrid generator was developed comprising cascaded piezo-electric and triboelectric units [166]. Electrospun PVDF-TrFE/Ag nanowire nano-fiber web was sandwiched between two fabric electrodes serving as a PENG, while a TENG was formed by a PDMS/graphite composite film and a fabric electrode. Ag nanowires could enhance nanofiber web piezoelectricity, electric conductivity, and dielectric constant. A typical hybrid generator showed a peak voltage output of 247.87 V by its piezoelectric unit and a voltage output of 84.43 V by its triboelectric unit, respectively, under a peak compression force of 1500 N and frequency of 3 Hz. At the same compression condition, the peak voltage reached around 300 V when hybrid generator units were connected in parallel.

A flexible textile-based hybrid nanogenerator was prepared by containing an electrospun $PVDF/CNT/BaTiO_3$ nanofiber nonwoven piezoelectric unit and a triboelectric unit with a microstructured surface configuration [167]. CNT and $BaTiO_3$ were embedded to modify the dielectric constant and improve the piezo-electric performance. The optimized electrospun piezoelectric web in their study consisted of 1 wt.% CNT and 18 wt.% $BaTiO_3$. The piezoelectric NG could generate a voltage of 21.31 V while the voltage ouptut of 157.39 V was achieved for the triboelectric NG. This hybrid NG has low-cost and can be suitable for industrial applications in the wearable device fields.

Wu et al [168] selected silicone rubber as the friction-negative electrode to prepare a hybrid nanogenerator based on BCZT/PVDF-HFP composite nanofibers (BP-based nanofibers). In this work, BCZT nanoparticles were uniformly distributed in PVDF-HFP, which significantly improved the piezoelectric and triboelectric proper-ties of the PTNG, the device reached a maximum power density of 161.7 mW m^{-2} and showed a stable output over 1200 cycles.

Shao et al [169] used polymer blends of PAN and PVDF to prepare electrospun nanofibers based acoustic-electric conversion devices (figure 3.24). The noise-electric conversion principle originates from the piezoelectric effect of the PAN and PVDF nanofibers and the endogenous triboelectric effect within the monolayer fiber membrane. This triboelectric effect comes from the tiny contact separation and friction of PAN and PVDF within the nanofibers. In a noisy environment, a device made of a $3 \times 4 \text{ cm}^2$ PAN-PVDF nanofiber membrane can generate a peak voltage output of up to 94.10 V and an output current of 17.40 μA (14.5 mA m^{-2}) with a power density of 250.1 mW m^{-2} and an energy conversion efficiency of 25.6%. Under the same conditions, the voltage output is 3.9 times higher than that of single-component PAN nanofibers and 4.5 times higher than that of single-component PVDF nanofibers.

Jiang et al [170] doped $Cs_3Bi_2Br_9$ perovskite in PVDF-HFP to design nanofiber-based NGs for energy harvesting using both piezoelectric and triboelectric effects. $Cs_3Bi_2Br_9$ chalcocite acted as a nucleating agent to promote the formation of the polar crystalline phase of PVDF-HFP, and also improved the charge transfer efficiency. The TPENG showed excellent electrical output (400 V, 1.63 μA cm^{-2} and

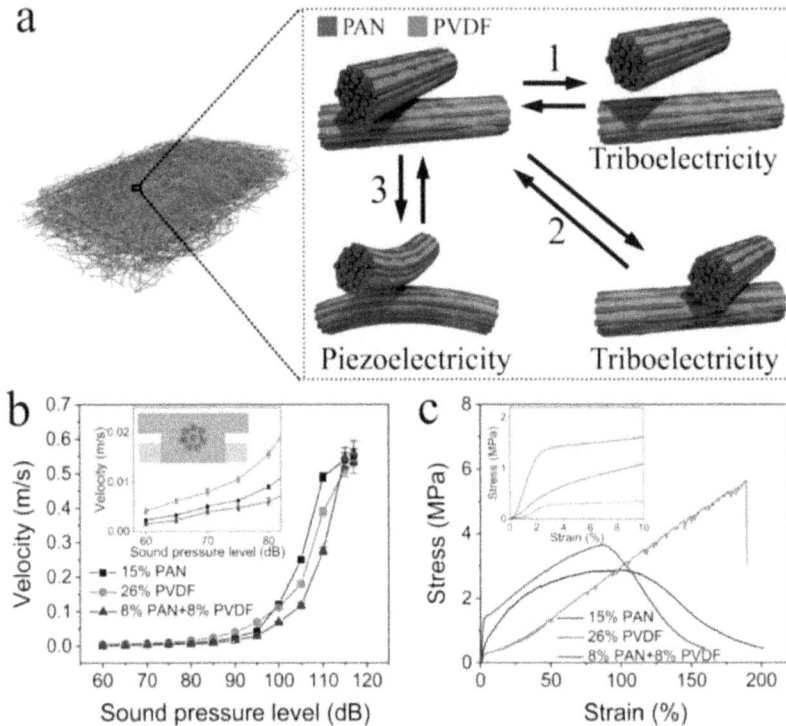

Figure 3.24. (a) A schematic illustration of the proposed charge generation mechanism within the PAN-PVDF nanofiber membrane. (b) Vibration velocity of fiber membranes in the central part (as marked) under different SPLs (250 Hz); Inset: enlarged view of curves at SPLs below 85 dB. (c) Stress–strain curves of fiber membranes; Inset: enlarged view of the curves at the elastic deformation stage (membrane thickness: 40 μm). Reproduced from [169], copyright (2021), with permission from Elsevier.

$2.34\ \mathrm{W\ m^{-2}})$ at 30 N, 5 Hz. The device had excellent stretchability, waterproofness and breathability and could be used as an insole for harvesting energy from walking or running. During pedaling, the PENG and TENG provided high output voltages of 17 V and 273 V, respectively. In addition, the energy harvester survived multiple washing, folding and wrinkling processes without performance degradation and maintained stable power output for up to five months, demonstrating its great potential as a smart textile and wearable power source.

Das *et al* [171] combined PENG and TENG to fabricate hybrid nanogenerators (HNG) based on lithium-modified zinc-titanium oxide (LZTO)/PVDF nanofibers, where Kapton and LZTO were used as negative and positive tribolayers, respectively, for the TENG. HENGs were capable of generating rectified voltages, currents, and power densities of 75 V, 3.2 μA, and 240 μW cm^{-2}, respectively. A wall-mounted punch bag was developed by fixing four HNGs on different sides of the sandbag as shown in figure 3.25, through which different types of punches can be monitored and boxers can use it to self-monitor the type of punches they are delivering.

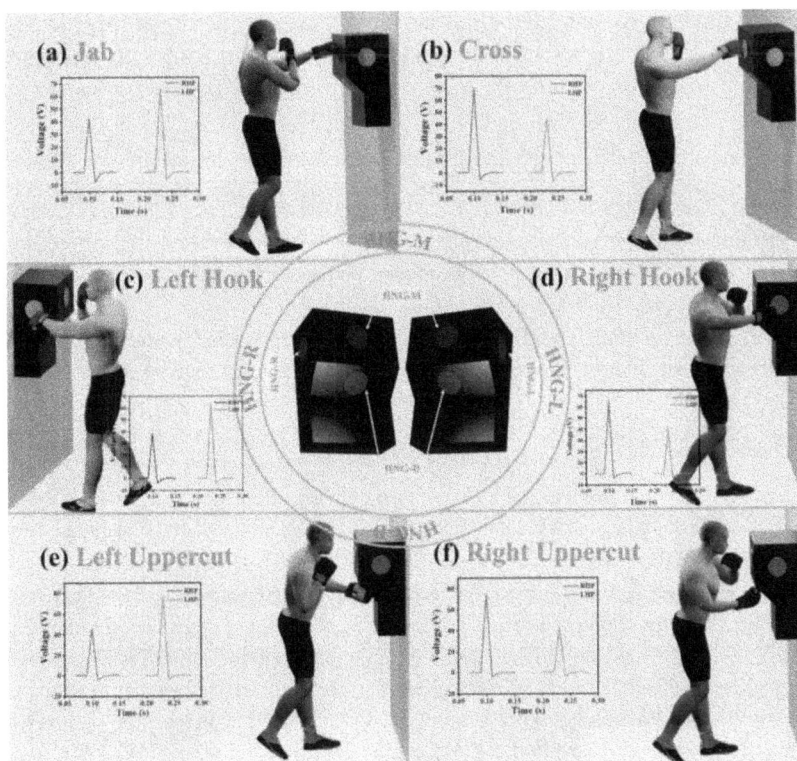

Figure 3.25. Schematic and voltage output of different punches by both a right-handed player and left-handed player, (a) jab; (b) right cross; (c) left hook; (d) right hook; (e) left uppercut; and (f) right uppercut. Reproduced with permission from [171], copyright (2023), American Chemical Society.

3.6 Summary

This chapter has provided a detailed introduction of significant progress in the piezoelectric electrospun nanofibers, their fabrication and applications in mechanical-to-electrical energy conversion. Electrospun nanofibers exhibit large surface area, high porosity, and unique web structure properties. PVDF is a widely studied piezoelectric polymer because of its high piezoelectric coefficient and easy electrospinning preparation. Nanofiber alignment, nanofiller doping and device structure have been employed to enhance electrospun PVDF nanofiber device mechanical-to-electrical energy conversion performance. Besides PVDF, other piezoelectric polymers including PVDF-TrFE, PVDF-HFP, PBLG, PLLA, PAN etc were also fabricated into nanofibers by electrospinning. These materials have shown comparable energy harvesting performance with PVDF, or even much better in some cases. The ceramic piezoelectric nanofibrous materials are mainly studied in the direction of lead-free materials. They normally have a much higher piezoelectric coefficient than the piezoelectric polymers. However, the rigid and fragile feature can limit their wide application in flexible and stretchable mechanical-to-electrical energy conversion devices. Electrospun piezoelectric devices can be integrated with triboelectric devices to enhance the electrical outputs. Many hybrid devices are capable of driving microelectronic devices directly without using any energy storage modules.

References

[1] Dong Z, Kennedy S J and Wu Y 2011 Electrospinning materials for energy-related applications and devices *J. Power Sources* **196** 4886–904

[2] Huo L *et al* 2017 Smart washer—a piezoceramic-based transducer to monitor looseness of bolted connection *Smart Mater. Struct.* **26** 025033

[3] Chen J, Ayranci C and Tang T 2023 Piezoelectric performance of electrospun PVDF and PVDF composite fibers: a review and machine learning-based analysis *Mater. Today Chem.* **30** 101571

[4] Manbachi A and Cobbold R S 2011 Development and application of piezoelectric materials for ultrasound generation and detection *Ultrasound* **19** 187–96

[5] Kawai H 1969 The piezoelectricity of poly (vinylidene fluoride) *Jpn. J. Appl. Phys.* **8** 975

[6] Harrison J and Ounaies Z 2002 Piezoelectric polymers *Encyclopedia of Polymer Science and Technology* (Wiley)

[7] Martins P, Lopes A and Lanceros-Mendez S 2014 Electroactive phases of poly (vinylidene fluoride): determination, processing and applications *Prog. Polym. Sci.* **39** 683–706

[8] Chang C, Fuh Y K and Lin L 2009 A direct-write piezoelectric PVDF nanogenerator *Transducers 2009–2009 Int. Solid-State Sensors, Actuators and Microsystems Conf.* pp 1485–8

[9] Chang C *et al* 2010 Direct-write piezoelectric polymeric nanogenerator with high energy conversion efficiency *Nano Lett.* **10** 726–31

[10] Fang J, Wang X and Lin T 2011 Electrical power generator from randomly oriented electrospun poly(vinylidene fluoride) nanofiber membranes *J. Mater. Chem.* **21** 11088–91

[11] Guo Y C *et al* 2023 Triboelectric nanogenerator-based near-field electrospinning system for optimizing PVDF fibers with high piezoelectric performance *ACS Appl. Mater. Interfaces* **15** 5242–52

[12] Fuh Y *et al* 2014 A highly flexible and substrate-independent self-powered deformation sensor based on massively aligned piezoelectric nano-/microfibers *J. Mater. Chem. A* **2** 16101–6

[13] Fuh Y *et al* 2015 Self-powered sensing elements based on direct-write, highly flexible piezoelectric polymeric nano/microfibers *Nano Energy* **11** 671–7

[14] Fuh Y *et al* 2015 Hybrid energy harvester consisting of piezoelectric fibers with largely enhanced 20 V for wearable and muscle-driven applications *ACS Appl. Mater. Interfaces* **7** 16923–31

[15] Duan Y Q *et al* 2014 Non-wrinkled, highly stretchable piezoelectric devices by electro-hydrodynamic direct-writing *Nanoscale* **6** 3289–95

[16] Ding Y, Duan Y and Huang Y 2015 Electrohydrodynamically printed, flexible energy harvester using *in situ* poled piezoelectric nanofibers *Energy Technol. (Weinheim, Ger.)* **3** 351–8

[17] Duan Y *et al* 2017 Ultra-stretchable piezoelectric nanogenerators via large-scale aligned fractal inspired micro/nanofibers *Polymers (Basel, Switz.)* **9** 714

[18] Hansen B J *et al* 2010 Hybrid nanogenerator for concurrently harvesting biomechanical and biochemical energy *ACS Nano* **4** 3647–52

[19] Hoe Z Y *et al* 2020 Enhancement of PVDF sensing characteristics by retooling the near-field direct-write electrospinning system *Sensors* **20** 4873

[20] Fuh Y K *et al* 2016 A transparent and flexible graphene-piezoelectric fiber generator *Small* **12** 1875–81

[21] Fuh Y K *et al* 2018 A fully packaged self-powered sensor based on near-field electrospun arrays of poly(vinylidene fluoride) nano/micro fibers *Express Polymer Lett.* **12** 136–45

[22] Lei T *et al* 2015 Electrospinning-induced preferred dipole orientation in PVDF fibers *J. Mater. Sci.* **50** 4342–7

[23] Gheibi A *et al* 2014 Electrical power generation from piezoelectric electrospun nanofibers membranes: electrospinning parameters optimization and effect of membranes thickness on output electrical voltage *J. Polym. Res.* **21** 1–14

[24] Shao H *et al* 2015 Effect of electrospinning parameters and polymer concentrations on mechanical-to-electrical energy conversion of randomly-oriented electrospun poly(vinylidene fluoride) nanofiber mats *RSC Adv.* **5** 14345–50

[25] Gheibi A *et al* 2014 Piezoelectric electrospun nanofibrous materials for self-powering wearable electronic textiles applications *J. Polym. Res.* **21** 469

[26] Pan X *et al* 2016 A self-powered vibration sensor based on electrospun poly(vinylidene fluoride) nanofibres with enhanced piezoelectric response *Smart Mater. Struct.* **25** 105010

[27] Hu J *et al* 2018 Effect of electrospinning parameters on piezoelectric properties of electrospun PVDF nanofibrous mats under cyclic compression *J. Text. Inst.* **109** 843–50

[28] Hu P *et al* 2018 Linear dependence between content of effective piezo-phase and mechanical-to-electrical conversion in electrospun poly(vinylidene fluoride) fibrous membrane *Mater. Lett.* **218** 71–5

[29] Shao H *et al* 2017 Effect of static charges on mechanical-to-electrical energy conversion of electrospun PVDF nanofiber mats *Adv. Mater. Lett.* **8** 418–22

[30] Fang J *et al* 2013 Enhanced mechanical energy harvesting using needleless electrospun poly(vinylidene fluoride) nanofiber webs *Energy Environ. Sci.* **6** 2196–202

[31] Wang Y R *et al* 2011 A flexible piezoelectric force sensor based on PVDF fabrics *Smart Mater. Struct.* **20** 045009

[32] Park B *et al* 2013 Highly efficient hybrid energy generator: coupled organic photovoltaic device and randomly oriented electrospun poly(vinylidene fluoride) nanofiber *J. Nanosci. Nanotechnol.* **13** 2236–41

[33] Liu Z *et al* 2017 Flexible piezoelectric nanogenerator in wearable self-powered active sensor for respiration and healthcare monitoring *Semicond. Sci. Technol.* **32** 064004

[34] Zandesh G *et al* 2017 Piezoelectric electrospun nanofibrous energy harvesting devices: Influence of the electrodes position and finite variation of dimensions *J. Ind. Text.* **47** 348–62

[35] Maity K, Mandal D and Mandal D 2018 All-organic high-performance piezoelectric nanogenerator with multilayer assembled electrospun nanofiber mats for self-powered multifunctional sensors *ACS Appl. Mater. Interfaces* **10** 18257–69

[36] Zaarour B *et al* 2019 Enhanced piezoelectric properties of randomly oriented and aligned electrospun PVDF fibers by regulating the surface morphology *J. Appl. Polym. Sci.* **136** 47049

[37] Kang S B *et al* 2017 Enhanced piezoresponse of highly aligned electrospun poly(vinylidene fluoride) nanofibers *Nanotechnology* **28** 395402

[38] Zaarour B, Zhu L and Jin X Y 2019 Controlling the surface structure, mechanical properties, crystallinity, and piezoelectric properties of electrospun PVDF nanofibers by maneuvering molecular weight *Soft Mater.* **17** 181–9

[39] Satthiyaraju M and Ramesh T 2019 Effect of annealing treatment on PVDF nanofibers for mechanical energy harvesting applications *Mater. Res. Express* **6** 105366

[40] Szewczyk P K *et al* 2020 Enhanced piezoelectricity of electrospun polyvinylidene fluoride fibers for energy harvesting *ACS Appl. Mater. Interfaces* **12** 13575–83

[41] Lin Y X *et al* 2021 Studies on the electrostatic effects of stretched PVDF films and nanofibers *Nanoscale Res. Lett.* **16** 79

[42] Tabari R S *et al* 2022 Piezoelectric property of electrospun pvdf nanofibers as linking tips of artificial-hair-cell structures in cochlea *Nanomaterials* **12** 1466

[43] Oflaz K and Ozaytekin I 2022 Analysis of electrospinning and additive effect on beta phase content of electrospun PVDF nanofiber mats for piezoelectric energy harvester nano-generators *Smart Mater. Struct.* **31** 105022

[44] Lan B L *et al* 2023 Multichannel gradient piezoelectric transducer assisted with deep learning for broadband acoustic sensing *ACS Appl. Mater. Interfaces* **15** 12146–53

[45] Shao Z Z *et al* 2023 Simulation guided coaxial electrospinning of polyvinylidene fluoride hollow fibers with tailored piezoelectric performance *Small* **19** 2303285

[46] Kumar R S *et al* 2019 Enhanced piezoelectric properties of polyvinylidene fluoride nanofibers using carbon nanofiber and electrical poling *Mater. Lett.* **255** 126515

[47] Gade H, Bokka S and Chase G G 2021 Polarization treatments of electrospun PVDF fiber mats *Polymer* **212**

[48] Liu Z H *et al* 2014 Direct-write PVDF nonwoven fiber fabric energy harvesters via the hollow cylindrical near-field electrospinning process *Smart Mater. Struct.* **23** 025003

[49] Pan C *et al* 2015 Near-field electrospinning enhances the energy harvesting of hollow PVDF piezoelectric fibers *RSC Adv.* **5** 85073–81

[50] Fuh Y K and Wang B S 2016 Near field sequentially electrospun three-dimensional piezoelectric fibers arrays for self-powered sensors of human gesture recognition *Nano Energy* **30** 677–83

[51] Luo G X *et al* 2020 The radial piezoelectric response from three-dimensional electrospun PVDF micro wall structure *Materials* **13** 1368

[52] Zhang W X *et al* 2020 A comparative study of electrospun polyvinylidene fluoride and poly (vinylidenefluoride-co-trifluoroethylene) fiber webs: mechanical properties, crystallinity, and piezoelectric properties *J. Eng. Fibers Fabrics* **15** 1558925020939290

[53] Kim M *et al* 2020 Development of in-situ poled nanofiber based flexible piezoelectric nanogenerators for self-powered motion monitoring *Appl. Sci.-Basel* **10** 3493

[54] Jin L *et al* 2020 Design of an ultrasensitive flexible bend sensor using a silver-doped oriented poly(vinylidene fluoride) nanofiber web for respiratory monitoring *ACS Appl. Mater. Interfaces* **12** 1359–67

[55] Zhang S Z *et al* 2021 Enhanced piezoelectric performance of various electrospun pvdf nanofibers and related self-powered device applications *ACS Appl. Mater. Interfaces* **13** 32242–50

[56] Azmi S *et al* 2022 Tuning energy harvesting devices with different layout angles to robust the mechanical-to-electrical energy conversion performance *J. Ind. Text.* **51** 9000S–16S

[57] Gee S, Johnson B and Smith A L 2018 Optimizing electrospinning parameters for piezoelectric PVDF nanofiber membranes *J. Membr. Sci.* **563** 804–12

[58] Castkova K *et al* 2020 Structure-properties relationship of electrospun PVDF fibers *Nanomaterials* **10** 1221

[59] He Z C *et al* 2021 Electrospun PVDF nanofibers for piezoelectric applications: a review of the influence of electrospinning parameters on the beta phase and crystallinity enhancement *Polymers* **13** 174

[60] Kalimuldina G *et al* 2020 A review of piezoelectric PVDF film by electrospinning and its applications *Sensors* **20** 5214

[61] Singh R K, Lye S W and Miao J M 2021 Holistic investigation of the electrospinning parameters for high percentage of beta-phase in PVDF nanofibers *Polymer* **214** 123366

[62] Yin J Y *et al* 2022 Effects of solvent and electrospinning parameters on the morphology and piezoelectric properties of PVDF nanofibrous membrane *Nanomaterials* **12** 962

[63] Papez N *et al* 2022 A brief introduction and current state of polyvinylidene fluoride as an energy harvester *Coatings* **12** 1429

[64] Zhang L W *et al* 2023 Recent progress on structure manipulation of poly(vinylidene fluoride)-based ferroelectric polymers for enhanced piezoelectricity and applications *Adv. Funct. Mater.* **33** 2301302

[65] Bai Y B *et al* 2022 Processes of electrospun polyvinylidene fluoride-based nanofibers, their piezoelectric properties, and several fantastic applications *Polymers* **14** 4311

[66] Shao H *et al* 2015 Robust mechanical-to-electrical energy conversion from short-distance electrospun poly(vinylidene fluoride) fiber webs *ACS Appl. Mater. Interfaces* **7** 22551–7

[67] Asadnia M *et al* 2016 From biological cilia to artificial flow sensors: biomimetic soft polymer nanosensors with high sensing performance *Sci. Rep.* **6** 32955

[68] Jiang Y *et al* 2018 Aligned P(VDF-TrFE) nanofibers for enhanced piezoelectric directional strain sensing *Polymers (Basel, Switz.)* **10** 364

[69] Wu Y *et al* 2012 The role of surface charge of nucleation agents on the crystallization behavior of poly(vinylidene fluoride) *J. Phys. Chem.* B **116** 7379–88

[70] Dhakras D *et al* 2012 Enhanced piezoresponse of electrospun PVDF mats with a touch of nickel chloride hexahydrate salt *Nanoscale* **4** 752–6

[71] Prasad G *et al* 2017 Piezoelectric characteristics of electrospun PVDF as a function of phase-separation temperature and metal salt content *Macromol. Res.* **25** 981–8

[72] Yu B *et al* 2017 Enhanced piezoelectric performance of electrospun polyvinylidene fluoride doped with inorganic salts *Macromol. Mater. Eng.* **302** 1700214

[73] Wan C and Bowen C R 2017 Multiscale-structuring of polyvinylidene fluoride for energy harvesting: the impact of molecular-, micro- and macro-structure *J. Mater. Chem.* A **5** 3091–128

[74] Lee C *et al* 2016 Electrospun uniaxially-aligned composite nanofibers as highly-efficient piezoelectric material *Ceram. Int.* **42** 2734–40

[75] Lu X, Qu H and Skorobogatiy M 2017 Piezoelectric microstructured fibers via drawing of multimaterial preforms *Sci. Rep.* **7** 2907

[76] Jiang J *et al* 2020 Flexible piezoelectric pressure tactile sensor based on electrospun batio3/ poly(vinylidene fluoride) nanocomposite membrane *Acs Appl. Mater. Interfaces* **12** 33989–98

[77] Zhao B B *et al* 2022 Piezoelectric nanogenerators based on electrospun PVDF-coated mats composed of multilayer polymer-coated $BaTiO_3$ nanowires *ACS Appl. Nano Mater.* **5** 8417–28

[78] Athira B S *et al* 2022 High-performance flexible piezoelectric nanogenerator based on electrospun $PVDF-BaTiO_3$ nanofibers for self-powered vibration sensing applications *ACS Appl. Mater. Interfaces* **14** 44239–50

[79] Kato M and Kakimoto K-I 2015 Processing and energy-harvesting ability of $(Na,K)NbO_3$ particle-dispersed fibrous polyvinylidene fluoride multilayer composite *Mater. Lett.* **156** 183–6

[80] Ji S H *et al* 2016 Flexible lead-free piezoelectric nanofiber composites based on BNT-ST and PVDF for frequency sensor applications *Sens. Actuators, A* **247** 316–22

[81] Sorayani Bafqi M S, Bagherzadeh R and Latifi M 2015 Fabrication of composite PVDF-ZnO nanofiber mats by electrospinning for energy scavenging application with enhanced efficiency *J. Polym. Res.* **22** 1–9

[82] Kim M and Fan J T 2021 Piezoelectric properties of three types of PVDF and ZnO nanofibrous composites *Adv. Fiber Mater.* **3** 160–71

[83] Khalifa M, Mahendran A and Anandhan S 2016 Probing the synergism of halloysite nanotubes and electrospinning on crystallinity, polymorphism and piezoelectric performance of poly(vinylidene fluoride) *RSC Adv.* **6** 114052–60

[84] Xin Y *et al* 2016 Full-fiber piezoelectric sensor by straight PVDF/nanoclay nanofibers *Mater. Lett.* **164** 136–9

[85] Rahman W *et al* 2018 Enhanced mechanical energy harvesting ability of electrospun poly (vinylidene fluoride)/hectorite clay nanocomposites *AIP Conf. Proc.* **1942** 050081

[86] Tiwari S *et al* 2019 Enhanced piezoelectric response in nanoclay induced electrospun PVDF nanofibers for energy harvesting *Energy* **171** 485–92

[87] Liu Z-H *et al* 2013 Piezoelectricity of well-aligned electrospun fiber composites *IEEE Sens. J.* **13** 4098–103

[88] Liu Z H *et al* 2014 A flexible sensing device based on a PVDF/MWCNT composite nanofiber array with an interdigital electrode *Sens. Actuators A* **211** 78–88

[89] Liu Z H *et al* 2015 Crystallization and mechanical behavior of the ferroelectric polymer nonwoven fiber fabrics for highly durable wearable sensor applications *Appl. Surf. Sci.* **346** 291–301

[90] Yu H *et al* 2013 Enhanced power output of an electrospun PVDF/MWCNTs-based nanogenerator by tuning its conductivity *Nanotechnology* **24** 405401

[91] Sharafkhani S and Kokabi M 2022 Enhanced sensing performance of polyvinylidene fluoride nanofibers containing preferred oriented carbon nanotubes *Adv. Compos. Hybrid Mater.* **5** 3081–93

[92] Zeng W *et al* 2023 Gradient CNT/PVDF piezoelectric composite with enhanced force-electric coupling for soccer training *Nano Res.* **16** 11312–9

[93] Abolhasani M M, Shirvanimoghaddam K and Naebe M 2017 PVDF/graphene composite nanofibers with enhanced piezoelectric performance for development of robust nanogenerators *Compos. Sci. Technol.* **138** 49–56

[94] Lee M C *et al* 2022 Development of piezoelectric silk sensors doped with graphene for biosensing by near-field electrospinning *Sensors* **22** 9131

[95] Fashandi H *et al* 2016 Morphological changes towards enhancing piezoelectric properties of PVDF electrical generators using cellulose nanocrystals *Cellulose (Dordrecht, Neth.)* **23** 3625–37

[96] Fu R *et al* 2017 Improved piezoelectric properties of electrospun poly(vinylidene fluoride) fibers blended with cellulose nanocrystals *Mater. Lett.* **187** 86–8

[97] Sengupta P *et al* 2020 A comparative assessment of poly(vinylidene fluoride)/conducting polymer electrospun nanofiber membranes for biomedical applications *J. Appl. Polym. Sci.* **137** 49115

[98] Sanchez F J D *et al* 2022 Sponge-like piezoelectric micro- and nanofiber structures for mechanical energy harvesting *Nano Energy* **98** 107286

[99] Shao H *et al* 2017 Mechanical energy-to-electricity conversion of electron/hole-transfer agent-doped poly(vinylidene fluoride) nanofiber webs *Macromol. Mater. Eng.* **302** 1600451

[100] Baniasadi M *et al* 2016 Thermo-electromechanical behavior of piezoelectric nanofibers *ACS Appl. Mater. Interfaces* **8** 2540–51

[101] Neese B *et al* 2007 Piezoelectric responses in poly(vinylidene fluoride/hexafluoropropylene) copolymers *Appl. Phys. Lett.* **90** 242917

[102] Adhikary P, Biswas A and Mandal D 2016 Improved sensitivity of wearable nano-generators made of electrospun Eu^{3+} doped P(VDF-HFP)/graphene composite nanofibers for self-powered voice recognition *Nanotechnology* **27** 495501

[103] Beringer L T *et al* 2015 An electrospun PVDF-TrFe fiber sensor platform for biological applications *Sens. Actuators, A* **222** 293–300

[104] Park S *et al* 2016 Flexible and stretchable piezoelectric sensor with thickness-tunable configuration of electrospun nanofiber mat and elastomeric substrates *ACS Appl. Mater. Interfaces* **8** 24773–81

[105] Yang E *et al* 2017 Nanofibrous smart fabrics from twisted yarns of electrospun piezopolymer *ACS Appl. Mater. Interfaces* **9** 24220–9

[106] Kim Y W *et al* 2018 Enhanced piezoelectricity in a robust and harmonious multilayer assembly of electrospun nanofiber mats and microbead-based electrodes *ACS Appl. Mater. Interfaces* **10** 5723–30

[107] Guangyi R *et al* 2013 Flexible pressure sensor based on a poly(VDF-TrFE) nanofiber web *Macromol. Mater. Eng.* **298** 541–6

[108] Kim M, Lee S and Kim Y I 2020 Solvent-controlled crystalline beta-phase formation in electrospun P(VDF-TrFE) fibers for enhanced piezoelectric energy harvesting *APL Mater.* **8** 071109

[109] Hafner J *et al* 2019 Origin of the strong temperature effect on the piezoelectric response of the ferroelectric (co-)polymer P(VDF70-TrFE30) *Polymer* **170** 1–6

[110] Conte A A *et al* 2019 Effects of post-draw processing on the structure and functional properties of electrospun PVDF-HFP nanofibers *Polymer* **171** 192–200

[111] Farrar D *et al* 2011 Permanent polarity and piezoelectricity of electrospun α-helical poly(α-amino acid) fibers *Adv. Mater.* **23** 3954–8

[112] Pan C *et al* 2014 Energy harvesting with piezoelectric poly(γ-benzyl-L-glutamate) fibers prepared through cylindrical near-field electrospinning *RSC Adv.* **4** 21563–70

[113] Nguyen D N and Moon W 2020 Piezoelectric polymer microfiber-based composite for the flexible ultra-sensitive pressure sensor *J. Appl. Polym. Sci.* **137** 48884

[114] Zhao G *et al* 2017 Electrospun poly(l-lactic acid) nanofibers for nanogenerator and diagnostic sensor applications *Macromol. Mater. Eng.* **302** 1600476

[115] Masamichi A *et al* 2012 Film sensor device fabricated by a piezoelectric poly(l-lactic acid) Film *Jpn. J. Appl. Phys.* **51** 09LD14

[116] Lee S J, Arun A P and Kim K J 2015 Piezoelectric properties of electrospun poly(L-lactic acid) nanofiber web *Mater. Lett.* **148** 58–62

[117] Sultana A *et al* 2017 Human skin interactive self-powered wearable piezoelectric bio-e-skin by electrospun poly-L-lactic acid nanofibers for non-invasive physiological signal monitoring *J. Mater. Chem. B* **5** 7352–9

[118] Zhu J, Jia L and Huang R 2017 Electrospinning poly(L-lactic acid) piezoelectric ordered porous nanofibers for strain sensing and energy harvesting *J. Mater. Sci., Mater. Electron.* **28** 12080–5

[119] Xu M H *et al* 2023 Flexible piezoelectric generator based on PLLA/ZnO oriented fibers for wearable self-powered sensing *Compos. Part A-Appl. Sci. Manuf.* **169** 107518

[120] Tai Y Y *et al* 2021 Modulation of piezoelectric properties in electrospun PLLA nanofibers for application-specific self-powered stem cell culture platforms *Nano Energy* **89** 106444

[121] Xia G B *et al* 2022 Piezoelectric charge induced hydrophilic poly(L-lactic acid) nanofiber for electro-topographical stimulation enabling stem cell differentiation and expansion *Nano Energy* **102** 107690

[122] Ueda H and Carr S H 1984 Piezoelectricity in polyacrylonitrile *Polym. J.* **16** 661

[123] Stupp S and Carr S 1978 Spectroscopic analysis of electrically polarized polyacrylonitrile *J. Polym. Sci.: Polym. Phys. Ed.* **16** 13–28

[124] Wang W *et al* 2019 Unexpectedly high piezoelectricity of electrospun polyacrylonitrile nanofiber membranes *Nano Energy* **56** 588–94

[125] Shao H *et al* 2020 Efficient conversion of sound noise into electric energy using electrospun polyacrylonitrile membranes *Nano Energy* **75** 104956

[126] Peng L *et al* 2021 High-precision detection of ordinary sound by electrospun polyacrylonitrile nanofibers *J. Mater. Chem.* C **9** 3477–85

[127] Shao H *et al* 2021 High-performance voice recognition based on piezoelectric polyacrylonitrile nanofibers *Adv. Electron. Mater.* **7** 2100206

[128] Wang W Y *et al* 2021 High-temperature piezoelectric conversion using thermally stabilized electrospun polyacrylonitrile membranes *J. Mater. Chem.* A **9** 20395–404

[129] Yu S *et al* 2022 Maximizing polyacrylonitrile nanofiber piezoelectric properties through the optimization of electrospinning and post-thermal treatment processes *Acs Appl. Polym. Materials* **4** 635–44

[130] Fukada E and Yasuda I 1957 On the piezoelectric effect of bone *J. Phys. Soc. Jpn.* **12** 1158–62

[131] Street R M *et al* 2018 Variable piezoelectricity of electrospun chitin *Carbohydr. Polym.* **195** 218–24

[132] Ghosh S K *et al* 2017 Electrospun gelatin nanofiber based self-powered bio-e-skin for health care monitoring *Nano Energy* **36** 166–75

[133] Chen X *et al* 2010 1.6 V nanogenerator for mechanical energy harvesting using PZT nanofibers *Nano Lett.* **10** 2133–7

[134] Chen X *et al* 2013 Flexible piezoelectric nanofiber composite membranes as high performance acoustic emission sensors *Sens. Actuators* A **199** 372–8

[135] Wu W *et al* 2012 Lead zirconate titanate nanowire textile nanogenerator for wearable energy-harvesting and self-powered devices *ACS Nano* **6** 6231–5

[136] Chamankar N *et al* 2019 Comparing the piezo, pyro and dielectric properties of PZT particles synthesized by sol-gel and electrospinning methods *J. Mater. Sci.-Mater. Electron.* **30** 8721–35

[137] Hayat K *et al* 2020 Fabrication and characterization of Pb(Zr0.5Ti0.5)O-3 nanofibers for nanogenerator applications *J. Mater. Sci.-Mater. Electron* **31** 15859–74

[138] Jiang L L *et al* 2021 Low temperature calcination induced flexibility in purely inorganic lead zirconate titanate and its application in piezoelectric enhanced adsorption *J. Eur. Ceram. Soc.* **41** 7630–8

[139] Lee H *et al* 2019 Pure piezoelectricity generation by a flexible nanogenerator based on lead zirconate titanate nanofibers *ACS Omega* **4** 2610–7

[140] Gu L *et al* 2012 Flexible fiber nanogenerator with 209 V output voltage directly powers a light-emitting diode *Nano Lett.* **13** 91–4

[141] He J *et al* 2020 A high-resolution flexible sensor array based on PZT nanofibers *Nanotechnology* **31** 155503

[142] Sutka A *et al* 2015 Fabrication of lead titanate PbTiO3 nanofiber mats via electrospinning *Int. J. Appl. Ceram. Technol.* **12** E117–21

[143] Xin Y *et al* 2023 $BiScO_3$-$PbTiO_3$ nanofibers piezoelectric sensor for high-temperature pressure and vibration measurements *Measurement* **212** 112694

[144] Suyitno S *et al* 2014 Fabrication and characterization of zinc oxide-based electrospun nanofibers for mechanical energy harvesting *J. Nanotechnol. Eng. Med.* **5** 011002

[145] Yan J and Jeong Y G 2016 High performance flexible piezoelectric nanogenerators based on $BaTiO_3$ nanofibers in different alignment modes *ACS Appl. Mater. Interfaces* **8** 15700–9

[146] Shirazi P *et al* 2017 Size-dependent piezoelectric properties of electrospun $BaTiO_3$ for Enhanced energy harvesting *Adv. Sustain. Syst.* **1** 1–8

[147] Maldonado-Orozco M C *et al* 2019 Absence of ferromagnetism in ferroelectric Mn-doped $BaTiO_3$ nanofibers *J. Am. Ceram. Soc.* **102** 2800–9

[148] Yan J *et al* 2019 Polymer template synthesis of flexible $BaTiO_3$ crystal nanofibers *Adv. Funct. Mater.* **29** 1907919

[149] Xiao X *et al* 2012 Preparation and mechanical energy harvesting of $BaTi_{0.9}Zr_{0.1}O_3$ ceramic nanofibers *Key Eng. Mater.* **512–515** 1359–62

[150] Wu W *et al* 2013 Electrospinning lead-free $0.5Ba(Zr_{0.2}Ti_{0.8})O_3$-$0.5(Ba_{0.7}Ca_{0.3})TiO_3$ nanowires and their application in energy harvesting *J. Mater. Chem.* A **1** 7332–8

[151] Wu Y *et al* 2019 Vertically-aligned lead-free BCTZY nanofibers with enhanced electrical properties for flexible piezoelectric nanogenerators *Appl. Surf. Sci.* **469** 283–91

[152] Wang Z *et al* 2015 K,Na)NbO3 nanofiber-based self-powered sensors for accurate detection of dynamic strain *ACS Appl. Mater. Interfaces* **7** 4921–7

[153] Yousry Y M *et al* 2019 Structure and high performance of lead-free $(K_{0.5}Na_{0.5})NbO_3$ piezoelectric nanofibers with surface-induced crystallization at lowered temperature *ACS Appl. Mater. Interfaces* **11** 23503–11

[154] Gu L, Zhou D and Cao J C 2016 Piezoelectric active humidity sensors based on lead-free $NaNbO_3$ piezoelectric nanofibers *Sensors* **16** 833

[155] Zhu R *et al* 2016 High output power density nanogenerator based on lead-free 0.96 $(K_{0.48}Na_{0.52})(Nb_{0.95}Sb_{0.05})O_3$-$0.04Bi_{0.5}(Na_{0.82}K_{0.18})_{0.5}ZrO_3$ piezoelectric nanofibers *RSC Adv.* **6** 66451–6

[156] Du X X *et al* 2022 Porous, multi-layered piezoelectric composites based on highly oriented PZT/PVDF electrospinning fibers for high-performance piezoelectric nanogenerators *J. Adv. Ceram.* **11** 331–44

[157] Li G-Y *et al* 2022 Hierarchical PVDF-HFP/ZnO composite nanofiber-based highly sensitive piezoelectric sensor for wireless workout monitoring *Adv. Compos. Hybrid Mater.* **5** 766–75

[158] Shi K *et al* 2021 Interface induced performance enhancement in flexible $BaTiO_3$/PVDF-TrFE based piezoelectric nanogenerators *Nano Energy* **80** 105515

[159] Lu H W *et al* 2022 Enhanced output performance of piezoelectric nanogenerators by Tb-modified (BaCa)(ZrTi)O$_3$ and 3D core/shell structure design with PVDF composite spinning for microenergy harvesting *ACS Appl. Mater. Interfaces* **14** 12243–56

[160] Mandal D, Henkel K and Schmeisser D 2014 Improved performance of a polymer nanogenerator based on silver nanoparticles doped electrospun P(VDF-HFP) nanofibers *Phys. Chem. Chem. Phys.* **16** 10403–7

[161] Wang S *et al* 2021 Boosting piezoelectric response of PVDF-TrFE via MXene for self-powered linear pressure sensor *Compos. Sci. Technol.* **202** 108600

[162] Su Y *et al* 2021 Muscle fibers inspired high-performance piezoelectric textiles for wearable physiological monitoring *Adv. Funct. Mater.* **31** 2010962

[163] Su Y *et al* 2022 High-performance piezoelectric composites via beta phase programming *Nat. Commun.* **13** 4867

[164] Huang T *et al* 2015 Human walking-driven wearable all-fiber triboelectric nanogenerator containing electrospun polyvinylidene fluoride piezoelectric nanofibers *Nano Energy* **14** 226–35

[165] Guo Y *et al* 2018 All-fiber hybrid piezoelectric-enhanced triboelectric nanogenerator for wearable gesture monitoring *Nano Energy* **48** 152–60

[166] Chen S *et al* 2017 Quantifying energy harvested from contact-mode hybrid nanogenerators with cascaded piezoelectric and triboelectric units *Adv. Energy Mater.* **7** 1601569

[167] Song J *et al* 2018 Highly flexible, large-area, and facile textile-based hybrid nanogenerator with cascaded piezoelectric and triboelectric units for mechanical energy harvesting *Adv. Mater. Technol.* **3** 1800016

[168] Wu Y *et al* 2019 Flexible composite-nanofiber based piezo-triboelectric nanogenerators for wearable electronics *J. Mater. Chem.* A **7** 13347–55

[169] Shao H *et al* 2021 Single-layer piezoelectric nanofiber membrane with substantially enhanced noise-to-electricity conversion from endogenous triboelectricity *Nano Energy* **89** 106427

[170] Jiang F *et al* 2022 Stretchable, breathable, and stable lead-free perovskite/polymer nanofiber composite for hybrid triboelectric and piezoelectric energy harvesting *Adv. Mater.* **34** 2200042

[171] Das N K, Nanda O P and Badhulika S 2023 Piezo/triboelectric nanogenerator from lithium-modified zinc titanium oxide nanofibers to monitor contact in sports *ACS Appl. Nano Mater.* **6** 1770–82

IOP Publishing

Energy Harvesting Properties of Electrospun Nanofibers
(Second Edition)

Jian Fang and Tong Lin

Chapter 4

Different characterizations and recent applications of piezoelectric nanofibers

Nader Shehata and Remya Nair

Piezoelectric nanofibers membranes/mats have attracted tremendous interest over the last two decades in both research and product tracks. Such nanofibers can convert mechanical excitations into electric signals with an improved efficiency according to higher surface-to-volume ratio through the synthesis of lower nano-dimensions fibers. This chapter introduces an intensive review on the recent progress of piezoelectric nanofibers over the last few years from material up to application levels. Here, we introduce the most important polymers and inorganic additives used to form piezoelectric nanofibers-based-composites, along with fabrication techniques overview. Then, an important section in this chapter is presented about the different piezo analysis techniques with details of the used technical set-ups. Furthermore, different applications have been explained according to the usage of piezoelectric nanofibers such as self-powered units, sensors, and vibrational/acoustic transducers over the last few years. This chapter can be helpful for readers of different disciplines including materials engineering, physics, and electrical engineering who are interested in the application of piezoelectric membranes in energy harvesting and transducers applications.

4.1 Introduction

Nanotechnology is a field of science and technology that deals with the exploitation of the ability to control matters in the range or dimensions between 0.1 and 100 nm. Materials in nanoscale or in general nano-systems are dominant over bulk-sized materials due to many of their fascinating properties such as high surface-to-volume ratio, greater surface area, high reactivity, solubility, etc [1]. A material reduced to nanosize from its bulk size greatly affects its internal structure thereby changing the electrical, optical, and magnetic properties, and the melting point, boiling point and

doi:10.1088/978-0-7503-5487-5ch4

even band gap will be affected [2]. The higher surface area greatly improved accessibility to surface functionality resulting in exceptional mechanical properties with quantum effects begins to dominate over continuum-like behavior [3, 4].

Among different nanostructures, our field of interest mainly focuses on nanofibers and nanofibrous mats due to their many captivating properties such as high porosity, large surface-to-volume ratio, flexibility in surface functionality, good mechanical performances, flexibility, light weight, etc [3–6]. The nanofibers are examples to one-dimensional nanostructure, where one dimension will be greater than 100 nm while the other two dimensions will be less than 100 nm. Some exceptions of nanofibers consideration are accepted within the two nano-dimensions region up to 500 nm [7].

Generally, nanofibers can be classified into different groups based on their physical structure and composition or material content. The physical structure-based classification mainly includes mesoporous, nonporous, core–shell and hollow fibers. From the material perspective, nanofibers can be classified into organic, inorganic, carbon-based and composite types. These nanofibers are developed using several fabrication techniques including spinning and non-spinning techniques and will be discussed in this chapter. The nanofibers generated using these techniques possess several innovating capabilities or properties that makes them inevitable for several important applications including energy generation/storage, sensors, bio-medical applications, supercapacitors, tissue engineering, filters, environmental protection etc. Among these broad applications, this chapter mainly concentrates on both energy harvesting and sensing related application of nanofibers. One of the principal phenomena that both energy harvesting and sensing related applications depend on its piezoelectricity, which is an electromechanical behavior exhibited by some materials to transduce the mechanical excitation into electric energy [8]. This transducing mechanism can offer one of the promising resources for green renewable energy generation/harvesting and consequently contributes to energy resources shortage, especially in the case of global warming and political conflicts. Mechanical vibrations, noise or sound energy, thermal energy etc are in general the natural resources for ambient energy harvesting using piezoelectric materials. Mechanical vibrations or motions in Nature mainly include human body move-ments and some sort of vibrations are visible in industry, buildings, home appliances etc. In addition to this, noise pollution and airflow-based sensing also paves a way towards acoustic energy harvesting and sensing applications. The energy generation from these sources can be utilized for several smart applications and can be considered as an alternative green energy resource. We are interested in the nanostructuring of these piezoelectric materials thereby generating piezoelectric nanofibers. By nanostructuring, the physical dimension is reduced thereby improv-ing the piezoelectric performance. These piezoelectric nanofibers with dominant electroactivity pave a way towards nanogenerators, wearable electronics, self-charging units, sensors, actuators etc. Hence, the main challenge in the literature now is producing highly efficient piezoelectric nanofibers with higher piezosensitivity and how to enhance the piezo performance of the generated nanofibers through material additive or process parameters' control [9, 10].

In this chapter, we introduce a full literature survey about the recent improvement of piezoelectric nanofibers membranes including fabrication processes, materials and additives, detailed characterization techniques including various piezoelectric analysis, along with wide applications of piezoelectric nanofibers including recent progress of triboelectric nanogenerator (TENG)/piezoelectric nanogenerator (PENG), wearable electronics, self-charging units, and others. The target of this chapter is to offer a complete guidance to the interested researchers in the field of piezoelectric nanofibers about the characterizations and recent updates in materials and correlated applications over the last five years.

4.2 Theoretical background

Piezoelectricity is the basic principle utilized in some of the applications of nanofibers. The term piezoelectricity means ability of a material to transform mechanical energy into the form of electrical energy. These types of materials can be used for energy harvesting applications to harness electrical energy from mechanical vibrations or motion. Mechanical vibrations or movements found in Nature are an ambient source of energy and a major focus on piezoelectric effect can be help us to utilize these energy sources for excellent performance. Here, an electrical polarization is developed from the mechanical vibrations due to the structural anisotropy resulting in a net dipole moment greater than zero. Transient deformations occurring in the materials are responsible for this structural anisotropy and it leads to the accumulation of charges inside the material due to the dipole distribution. The application of pressure on such materials will result in a negative charge generation on the expanded side and consequently an equal amount of positive charge will be developed in the compressed side [3].

In this chapter, we are interested in the studies based on one-dimensional material that are piezoelectric because these nanosized materials have exceptional mechanical stability and tunable electrical properties which greatly strengthens the piezoactivity or response compared to a bulky material. The piezoresponse of a nanomaterial is dominant over macro-sized material of the same kind due to its large surface-to-volume ratio, porosity, robustness, and easy reusability with excellent thermal property. Materials showing piezoactivity include crystalline-type, ceramic-type and polymer-type contents. Among these, crystalline-type materials are brittle and ceramic-type materials have toxicity, so polymer-type piezomaterials with remarkable properties are our field of focus. Some of the remarkable properties of polymers are their flexibility, light weight, mechanical stability and elasticity. In addition to pure polymers, nanocomposites prepared from the combination of either crystals or ceramics with pure polymers can greatly enhance the piezoresponse. Quartz and Rochelle salt are some of the examples of crystals that exhibits high piezoelectric behavior. Barium titanate, bismuth titanate, zinc oxide single-crystal etc are some of the ceramics that exhibit good piezo behavior [4].

The piezoactivity of a material can be represented by the direct piezoelectric effect equation [5]

$$D = d \times T + \varepsilon \times E \tag{4.1}$$

Here 'D' represents displacement, 'T' represents the applied mechanical stress, 'ε' represents the material permittivity, the piezoelectric coefficient is represented by 'd' and electric field is notated by 'E'. For an equal amount of applied mechanical stress, the material with higher piezocoefficient value 'd' generates more piezoelectric voltage. The piezoelectric coefficient 'd' is an anisotropic physical quantity and so the constant relates to both the direction of the applied mechanical or electrical force and the directions perpendicular to the applied force. As a result, the constant is usually represented by two subscripts that designate the direction of two related physical quantities: stress and strain. Stress represents the applied force on the material to its surface area and strain represents the ratio of change in length to original length of the element for elasticity [5].

Furthermore, an energy from dielectric materials is generated from the internal elastic energy by the application of an external load on polymeric piezomaterials. The application of an external load results in a mechanical deformation and this creates equal and opposite charges on the surface of the nanofiber membrane. Usually, piezoelectric polymers are anisotropic material and the piezoelectric characteristics are different for each direction and so the corresponding piezoelectric coefficient can be expressed in matrix form as shown below [6],

$$d_{ij} = \begin{bmatrix} 0 & 0 & 0 & 0 & d_{15} & 0 \\ 0 & 0 & 0 & d_{24} & 0 & 0 \\ d_{31} & d_{32} & d_{33} & 0 & 0 & 0 \end{bmatrix} \tag{4.2}$$

The subscript 'i' refers to the electrical value measurement direction or polarization direction and 'j' represents the direction of mechanical action or direction of applied stress. For a rectangular system with xyz axes, the positive polarization direction is made to coincide with z-axis. Subscripts 1, 2 and 3 have been used to mention directions x, y and z, respectively, and the corresponding shear about these axes is represented by 4, 5, and 6 respectively [6]. As a clear definition, the piezoelectric charge constant 'd' quantifies the polarization generated per unit of mechanical stress (T) applied to a piezo material. T_1, T_2, and T_3 represent tensile stresses where deforming force is applied at a right angle to the surface of the material and T_4, T_5, and T_6 represent shear stresses in directions 1, 2, and 3, respectively, where deforming force is applied parallel to the surface of the material. Thus, the total output electrical charge generated on a piezo material in our study can be represented as

$$Q = (d_{31}T_1 + d_{31}T_2 + d_{33}T_3)A \tag{4.3}$$

where 'A' represents the area. The coefficient 'd_{33}' is used when we apply stress only in one direction where 'Q' becomes

$$Q = d_{33}T_3 \tag{4.4}$$

In addition to piezoelectric coefficient, sensitivity is an important parameter that shows the piezoelectric performance of a nanofiber membrane. Sensitivity is defined as the ratio of charge or voltage developed to the applied force that is responsible for the generation of the voltage, and its unit is Volt/Newton [6].

4.3 Materials, fabrication, and characterization

4.3.1 Materials

Piezoelectric properties are exhibited by a group of polymer families and these materials became predominant in our day-to-day life due to their scope in many of our inevitable applications. The class of polymer families that exhibits piezoactivity mainly includes polyureas, fluoropolymers, polysaccharides, polyamides, polypeptides and polyesters with good mechanical resistance, toughness, ductility, viscoelasticity etc. Thus, the material composition associated with the fabrication of piezoelectric nanofibers mainly includes the above-mentioned components and their mixture with the addition of a variety of organic and inorganic particles. The material composition in general includes the solute, solvent and the additives or dopants involved in the solution preparation of polymers as part of fabrication. Some of the material properties associated with the synthesis of nanofiber and solution preparation includes molar mass, viscosity, conductivity etc. All these parameters should be considered while preparing the solution for nanofiber fabrication. Among the various types of material compositions available for the solution preparation, polymers are the most efficient material employed for the fabrication of nanofibers via electrospinning technique. In general, polymers are the ancient source of nanofibers and a set of parameters such as polymer-type, its solvent, polymer solution, its viscosity, environmental conditions such as pressure, temperature and processing parameter determine the structure and morphology of the generated polymeric nanofiber [7].

A polymer can be generally defined as a single large molecule made up of a set of small parts or repeating chemical units, where the repeating chemical units are called monomers. Some of the fascinating properties of the polymers that can be utilized for the generation of nanofibers are their elasticity, low density, easy production methods, resistance towards heat and light, corrosion resistance etc. A wide variety of polymers are available, and they can be classified based on different factors. One of the major classifications of polymers is based on their origin: polymers that are existing in Nature or naturally occurring polymers are natural polymers, and those polymers which are human-made are called synthetic polymers. However, natural polymers are found to be limited in the environment so to meet rising demand, synthetic polymers are synthesized. Cellulose, starch, and chitosan are some of the widely available natural polymers. Polyethylene, polyvinyl alcohol (PVA), and polyethylene oxide (PEO) are some of the widely available synthetic polymers and can be a co-host for piezoelectric polymers. Polymers can be classified into elastomers and fibers based on the molecular forces by which the monomers are linked. If the monomers are linked by weak molecular forces, they are called elastomers, and if there is strong intermolecular force between monomers then the polymers are labelled as fibers. Natural rubber is one of the best examples of elastomers, whereas both nylon 66 and silk form a thin thread-like structure and fall under the category of fiber-like polymer. Here the strong intermolecular force of attraction is either by hydrogen bonding or by dipole–dipole interaction. If the same type of monomer is added on to form a polymer, then that type of polymer is called addition polymer and if two different types of monomers are condensed on to each other by the elimination of some molecules, then that type of polymer is called condensation polymer. Synthetic rubber is an addition polymer but nylon 66 falls under the category of condensation polymer. Based on their response towards heat, polymers are classified as thermosetting and thermoplastic. Thermosetting polymers are linked by irreversible strong chemical bond and are highly resistant towards heat and chemicals. Once they are heated and set into a specific shape, then they cannot be melted or remolded by the application of heat. Polymers such as polyurethane (PU), polyesters, polyamides etc fall under the thermosetting category. But thermoplastic polymers have low melting point and are ideal for applications that use recycled materials. They can be reheated and can be cooled into any desired shape as needed without causing any chemical change. Some common polymers like polyvinylidene difluoride (PVDF), thermoplastic polyurethane (TPU), and polyvinyl chloride fall under the thermoplastic category [8].

Based on a detailed review of polymers, it is found that some of the natural/biopolymers and some synthetic polymers exhibit piezo behavior and this is discussed in detail. Some of the specific features on which a piezoelectric polymer depends for a successful piezoactivity are the presence of permanent molecular dipole and the ability to align or orient them and at the same time to maintain the alignment after achieving it. In addition to these properties, the specific polymer should be strong enough to withstand large strain during the application of mechanical stresses [9]. The biopolymers with piezoactivity mainly include collagen, gelatin, silk, chitin and cellulose-based materials and these are discussed in detail below.

4.3.1.1 Cellulose

The presence of confined charges and aligned dipoles in the cellulose-based materials accounts for their piezoelectric and pyroelectric behavior. Some of the cellulose-based polymers are wood, chitin, starch and amylase. A non-centrosymmetric crystal structure is found in cellulose due to the intra- and intermolecular hydrogen bonds along cellulose chain sides. The cellulose nanofibers are found to be very lengthy, around micrometers, comprising alternate amorphous and crystalline domains arrangement. Cellulose crystals usually consist of longitudinal, transverse and shear piezoelectric constants due to the presence of permanent dipole moment along their axes. When the applied mechanical disturbances and the generated voltage are in the same direction, the mechanical deformation occurring in the material will be maximum and this was explained by the d_{33} longitudinal piezoelectric coefficient. But the homogeneous orientation of randomly oriented cellulose nanocrystals to achieve high d_{33} was practically difficult over a large area and as a result only around 5 pC/N d_{33} value was shown by pure nanocellulose which is not very high compared to some synthetic piezoelectric polymers. But a new work proposed that the electrochemical poling of the cellulose nanofiber can enhance the piezocoefficient almost to be mostly comparable to the synthetic piezopolymer range. The electrochemical poling resulted in an increase of piezocoefficient in the range 35–55 pC/m which is comparable to the piezopolymer PVDF which is having around 44 pC/m piezocoefficient value [10].

4.3.1.2 Silk

Silk is a biopolymer obtained from the cocoon of a silkworm and is comprised of two types of protein called fibroin and sericin. Of these, fibroin is a copolymer with disulphide bridges acting as a link for heavy and light chain proteins. The fibrous fibroin protein has outstanding biological, mechanical and electrical properties and sericin protein holds this fibrous protein. Thus, piezoactivity is shown by the fibroin protein part of silk material which exhibits a semicrystalline nature. Cocoon silk-based nanofiber exhibits a polymorphic crystalline nature with Silk I, II and III polymorphs. Among these, Silk II possess a pleated secondary structure with antiparallel beta sheets having a monoclinic unit cell. Silk I has a metastable helical chain conformation with a orthorhombic unit cell. Both these units possess a lack of symmetry and this accounts for the piezoeffect generation. The silk fibroin can be easily processed into nanofibers by electrospinning process with excellent orientation of the polymer chains. Oriented silk II shows around 1.5 pC/N piezocoefficient value due to the intrinsic beta sheet formation [11]. In addition to cocoon silk, spider silk has also attracted our attention due to its remarkable mechanical properties. This naturally abundant spider-based silk material shows piezoelectric coefficient in the range 0.36 pm V^{-1} and the cocoon silk-based biomaterial exhibits a piezoelectric response of 1 pC/N which is comparable to quartz crystal coefficient. The spider-based silk comprises helical α elastic and β platelet sheet crystalline structure with repeating amino acid pattern units, where they are interconnected by intra- and intermolecular H-bonding which results in the development of electrical dipoles within the lattice crystal. The spider silk-based fiber acts as a highly efficient nanogenerator due to its ability to generate shear piezoelectricity (d_{14}) under the application of tensile stress in a direction parallel to the fiber. The vertical piezoelectric behavior exhibited by the spider silk fibers is due to its complex structure with beta sheet crystals, alpha helix, inter/intra H-bonding, microfibril groups and the carbonyl group contents. Thus, silk in general is an efficient natural polymer that conquers the field of energy harvesting, tactile and sensing applications [12, 13]. The electrospinning of the silk fibroin polymer has resulted in the generation of a longitudinal piezocoefficient (d_{33}) of 38 ± 2 pC/N [14].

4.3.1.3 Collagen

Collagen is an organic piezoelectric polymer that is a functional protein found in mammals which is highly biocompatible, biodegradable and useful for several applications. It exhibits a fiber-like structure and can generate electric potential in response to applied mechanical stress. The structure of collagen protein mainly includes a long sequence of polypeptide chain. The NH and CO groups involved in the chain undergo hydrogen bonding and as a result a number of hydrogen bonds are formed. The piezoelectric coefficient of collagen ranges from 0.2 pC/N to 2 pC/N. In general, a hexagonal crystalline unit is responsible for the piezoactivity of collagen. This material exhibits a piezoelectric effect of around 10% of that exhibited quartz crystal. Hexafluoroisopropanol is used as solvent for collagen for successful solution preparation. Collagen/polycaprolactone (PCL), collaged PEO and collagen compositions are mainly used for nanofiber application [9].

4.3.1.4 Chitin

Chitin is another natural piezopolymer which comes under polysaccharide category. The piezoelectric coefficient of chitin lies in the range 0.2–1.5 pC/N. This material is a main component of cuticles of insects, mollusks and crustaceans. Chitin is a major component of cell walls in fungi and exoskeletons of arthropods. The non-centrosymmetric structure of both alpha and beta polymorphs contributes to the intrinsic polarization of chitin. This natural polymer exhibits polymorphism with semicrystalline nature and ordered crystal structure and can be easily electrospun for various applications [9].

4.3.1.5 Gelatin

The partial hydrolysis of the collagen polymer extracted from the connective tissue of animals generates gelatin, a natural polymer having a good piezocoefficient d_{33} up to 20 pC/N. For practical applications, mechanical properties of gelatin nanofiber have to be enhanced and for that crosslinked gelation nanofibers are produced using electrospinning. This material is very sensitive and reliable to be used within sensors. The supramolecular interactions controlled by the hydrogen bonding and the dipole moment across the peptide chain axis are responsible for the piezo behavior of gelatin [14]. In addition to biopolymers, a large variety of synthetic polymers are synthesized to meet the growing demands. The most common synthetic polymers having piezoactivity mainly include PVDF and its copolymers, odd numbered nylon, poly-L-lactic acid (PLLA), polyacrylonitrile (PAN), poly3-hydroxy butyrate-3 hydroxy valerate (PHBV), polypropylene, PVC and its composites with some additives as organic and inorganic fillers. The specific compositions at optimum rate were performed for the enhancement of the piezo behavior so that the effect can be utilized with maximum efficiency [15–17].

4.3.1.6 PVDF

PVDF finds a major place in the center of scientific and industrial research in polymer science due to many of its fascinating properties such as pyroelectricity, piezoelectricity and ferroelectricity. PVDF is a semicrystalline thermoplastic fluoro-polymer which is highly inert and formed by polymerization of vinylidene difluoride. It has excellent mechanical strength, toughness and flexibility, low flammability, low moisture absorption and high dielectric strength. PVDF has a distinct polymorphic structure generally formed by free radical polymerization of 1-difluoroethylene ($CH_2{=}CF_2$), the monomer synthesized from vinylidene or acetylene. Here, CH_2 group represents the head and CF_2 group represents the tail of the monomer unit and then a dipole moment is generated inside the material due to the electro-negativity difference between fluorine and hydrogen. The molecular chain packing and internal structural transformation resulted in the generation of macroscopic polarization and excellent thermos-mechanical properties which makes them an exceptional material for some basic mechanical to highly sensitive micro/nano electronic applications [18]. PVDF polymer possesses an excellent piezoresponse with a piezocoefficient value in the range 20–40 pC/N. PVDF material is anisotropic in nature and exhibits polymorphic structures with five multiple phases such as α, β,

Y, δ and ε. Among the five crystalline phases, α and β are the most abundant phases with α phase having a TGTG conformation with a monoclinic unit cell. β phase, which is the best polar phase, has all trans (TTTT) conformation with an orthorhombic unit cell and has all dipoles aligned along the identical direction, perpendicular to the chain axes resulting in a net nonzero dipole moment. The crystalline form with nonzero net dipole moment results in spontaneous polarization with an efficient piezoactivity. For the α nonpolar phase, the dipole components are antiparallel to each other resulting in the neutralization of the dipole moment. The third phase called 'Y' phase also possess an orthorhombic unit cell with TTTG-TTTG conformation. The polar and antipolar analogues of the α form are represented by the last two polymorphs δ and ε. Thus beta phase is the only polymorph mainly responsible for the piezoactivity in PVDF and this phase makes PVDF an attractive polymer for a wide range of applications. Thus, the enhancement of piezoresponse can be achieved by increase in β-phase fraction and crystallinity degree. The increase in β-phase fraction can be achieved by the addition of nanofillers or dopants onto the pure PVDF nanomembrane [18] (figure 4.1).

A detailed review based on the piezoresponse enhancement of PVDF nanomembrane by additives is focused on and discussed. ZnO is an inorganic nanofiller that is found to be very efficient in increasing the β-phase content of PVDF electrospun nanofibers. This study revealed the effectiveness of high molecular weight PVDF with ZnO nanoparticles in generating devices with excellent piezoactivity [19]. Another efficient PENG is fabricated by using plasticizer (tetrahydrofuran—THF) treatment to improve the β-phase content. The effect of plasticizer on nanofiber web was positive and this effect greatly enhanced the piezoresponse and this paved a way towards the generation of an outstanding piezo nanogenerator [20]. Another important additive that improved the mechanical stability with elasticity of the PVDF membrane was TPU. This polymer blended with PVDF improved its electroactivity and the blended ratio of 15%–17% TPU generated the best performance with superior mechanical elasticity [21]. A soft nanotechnology approach was performed by the addition of a cationic surfactant, tetra-n-butyl ammonium chloride (TBAC), at a different loading weight (1–5 wt.%) onto the PVDF solution before electrospinning. An improvement in both β-phase content and dielectric properties with reduction in crystallinity is identified and a maximum piezoelectric performance was exhibited by 3 wt.% TBAC/PVDF nanofiber, where a maximum of

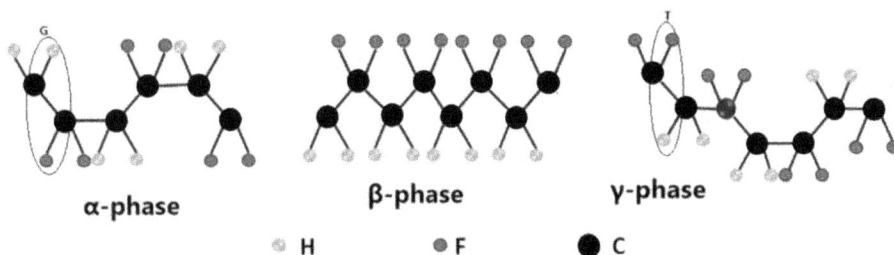

Figure 4.1. Different phases of PVDF polymer.

17.2 V output voltage is generated under an applied force of 5 N. That was around 19.1 times higher than the case of prisitine PVDF nanofiber. Here, an increase of around 89% β-phase content with an overall reduction of 8% crystallinity is obtained. The dielectric constant value exhibited by PVDF increased greatly by the addition of TBAC and a power density of approximate 1.4 μW cm^{-2} was observed under the application of 3 N force [22]. Another study revealed the impact of cobalt doped ZnO nanofiller onto PVDF hexafluoropropylene (HFP) electrospun nanofiber. PVDF-HFP is a fluoro-copolymer of PVDF formed by the emulsion polymerization of (VDF) vinylidene fluoride and HFP. An increase of dielectric constant from 8 to 38 is observed after doping PVDF-HFP with cobalt doped zinc oxide at a concentration of 2 wt.%. An increase in piezo potential is observed from 0.12 to 2.8 V after doping with cobalt doped zinc oxide, which paved a way for the development of an impressive nanogenerator/wearable sensor-based device [23].

PVDF-TrFe is a copolymer of PVDF that offers exceptional piezoelectricity and the effect can be improved again by poling and annealing process [24]. A detailed review based on piezoresponse enhancement of PVDF-TrFe nanomembrane by additives clearly showed the scope of carbon-based additives such as CNT, Gr, Go. In addition to carbon-based additive silver nanowire, silver nanoparticles, SiO$_2$ and BT, BiCl$_3$, LiCl, and PEDOT can enhance the piezoresponse of PVDF [6]. Another study revealed the enhancement of piezoactivity of PVDF by the addition of MWCNT doped with Cu nanoparticles. This prototype efficiently expressed its ability to harvest energy upon integration on a shoe back and while walking at a leisurely pace it generates around 18–20 V potential [25]. PVDF-TrFe, PVDF polyaniline, PVDF/BaTiO$_3$/PDA, PVDF-AU nanocages, PVDF/Dopamine, PVDF/ZnO, P(VDF-TrFe)/AIN are some of the energy harvesting-based compositions that have sensing applications. The electrospinning of PVDF with carbon dots greatly enhanced the piezoresponse and multifluorescent performance of the PVDF [26]. Tri-valent cerium ions (Ce^{3+}) doped electrospun PVDF/graphene composite exhibited a good piezoresponse and acoustic sensing behaviour [27]. High-performance piezoelectric porous multilayered composites are developed by the incorporation of lead zirconate titanate (PZT) in PVDF and the generated electrospun PVDF PZT nanofiber acted as a well-organized piezo nanogenerator [28, 29]. Carbon nanofiber-based PVDF electrospun nanofibers proved to have an outstanding piezoresponse on the basis of a recent study [30]. A hitherto unseen polarisation-locking phenomena of PVDF-TrFe perpendicular to the basal plane of 2D TiC$_2$T$_n$ Mxene nanosheets. The composition generated a piezoelectric charge coefficient d_{33} of −52 pC/N, which is higher than poled PVDF-TrFe having −38pC/n. Here a new basic low-energy input mechanism of fluoropolymers poling those results in a new level of performance is studied [31].

4.3.1.7 More polymers
PAN is an effective semicrystalline piezoelectric polymer with chemical formula (CH$_2$CHCN)$_n$. This polymer can be easily electrospun into nanofibers by adjusting the processing parameters. The electroactive content was further enhanced by

dimensional reduction to approximately 40 nm. Finally, annealing-based post treatment on PAN electrospun fibers greatly improved its piezo performance. The fabricated electrospun nanomembrane showed a piezo charge constant of around 39 pm V^{-1} via diameter reduction of 40 nm and annealing at 95 °C. Thus fine tuning of the piezoelectric electrospun nanofibers was achieved through dimensional confinement effect and annealing in this study [16]. Again, polylactic acid (PLA) is a piezoelectric thermoplastic polyester that lacks ferroelectricity. PLA possess a higher order structure with mixed crystalline state and amorphous nature. It possesses an orthorhombic unit cell structure with a polymer helix. PLA exhibits piezoelectricity even in the amorphous state with partial long-range ordering of polymer chains through any thermo-mechanical work [15]. PLLA is a homopolymer of PLA, having excellent biocompatibility and mechanical stability. PLLA has a scope in the field of tissue engineering. These PLLA scaffolds can be fabricated using electrospinning for sensor-based applications where they are stretched under electric field and thus converting α-phase into β-phase inducing polarization and piezo-electricity. It is a biodegradable polymer having a wide range of scope in the field of biomedical engineering both as *in vivo* and *in vitro* models [32].

PHBV is a natural thermoplastic polyester with aliphatic nature. Some of the attractive properties of PHBV include biocompatibility, biodegradability and semi-crystallinity; one of the applications of PHBV is in the field of wound dressing where it has a great potential in the closure of wounds and healing. One study showed effectiveness of PHBV with curcumin where it can be electrospun easily in the form of nanofibers for wound healing-related applications [17]. Another study revealed the successful formation of radially oriented electrospun nanofiber scaffolds that exhibited enhanced biological properties over randomly oriented fibers with good cell proliferation and alignment. PHBV/SiO$_2$-based superhydrophobic mats containing microfibers and nanofibers with beads fabricated via electrospinning is demonstrated in a study. Chloroform is used as a solvent for pure PHBV where we can develop fibers of desired morphology. The static wetting behavior of the electrospun membranes are improved by the incorporation of fumed SiO$_2$ treated with polydimethylsiloxane (PDMS) due to the enhanced surface roughness [33–35].

4.3.2 Fabrication techniques

The demand for nanofibers has increased considerably over the past few years and this led to the development of a wide variety of fabrication techniques. A brief review about the prominent fabrication techniques is provided in the present chapter. Fabrication techniques mainly include several non-spinning and spinning techniques. Non-spinning techniques mainly include phase separation, template synthesis, drawing techniques, STEP (Spinneret-based tunable engineered parameter method) interfacial polymerization of nanofibers, solution blowing method, freeze drying method etc. Spinning-based fabrication techniques mainly include traditional electrospinning, melt blowing, wet spinning, dry spinning, centrifugal spinning, mechanical/force-based spinning etc. Out of all these spinning methods,

Conventional electrospinning is the most used method for fabrication of nanofibers with controlled experimental parameters [36].

Electrospinning uses strong electric field to stretch the polymer solution into fine nanofibers. This technique is based on the principle of interaction between electrically active surfaces, electrified liquids, and fluid dynamics. This electrohydrodynamic process uses a high voltage power supply which acts as a source for electric field and this high voltage output is connected onto the needle tip of the sample-loaded syringe with needle. Thus, the polymer solution droplet coming out of the needle tip will be affected by the strong electric field and this electric field results in the distortion of the sample droplet, to overcome the surface tension and viscoelastic forces, and this electric force dominates over the viscous forces above certain threshold voltage and finally the ejection of jet and its thinning occurs. The polymer solution ejected out of the syringe with needle is controlled via an injection pump. The negative terminal of the high voltage power supply is connected to the collector which is electrically grounded. As a result, the erupted and stretched filaments after traversing through air are deposited on to the grounded collector electrode in the form of ultrathin fibers after evaporation of the solvent. In addition to the aforementioned mechanisms, modifications on the conventional electrospinning set-up such as co-axial electrospinning, needleless electrospinning, bubble electrospinning, electroblowing, cylindrical porous hollow tube electrospinning, flash spinning, self-bundling electrospinning, charge injection electrospinning etc are some of recent ideas used for the generation of efficient nanofibers. The modifications are done mainly in the experimental set-up including the design of spinneret, collector electrode shape etc.

Melt blowing is a method used for the generation of continuous nanofibers. Hence, the polymer-based sample is melted and loaded inside the matrix. Here, the melted sample comes out through small holes and the ejected solution is directed towards a rotating drum by using hot air. This method results in generation of ultrafine nanofiber on non-woven web where the fibers are solidified during the drawing process. This method lacks the uniformity of the fibers and generated ones are very weak fibers which restricts their usage in many applications. Wet spinning is another ancient method used to fabricate nanofibers. Here the spinneret is submerged in a chemical bath so that the generated fibers are precipitated during the emerging time. This process has no requirement for additional purification. This method is applicable for any polymer that does not melt and dissolve in solvent that is thermally unstable and nonvolatile [37].

Solution blowing or solution blow-spinning method is another method which is highly straightforward and economically beneficent. This method consists of a blow-spinning device where a heated reactor is presented, and the gel solution is fed through heated injection nozzle where finally the polymeric gel is dragged by the accelerated stream of air to the collector electrode. Dry spinning is another technique utilized for the fabrication of nanofibers where the polymer solution is prepared in a volatile solvent and then the solution is extracted through a spinneret having a large number of holes. Heated air is used to evaporate the solvent content so that the nanofibers get solidified [38].

4.3.3 Piezoelectric characterization techniques

4.3.3.1 Materials characterization

To detect the piezoelectric capabilities of the fabricated piezoelectric nanofibers mat, researchers analyze both mechanical–electrical transducing possibilities, along with material characteristics proving the piezoelectric property of the mat. From a material perspective, it is required to test the quality of generated nanofibers with the original practical characteristics of material content and comparing the quality rate obtained during the synthesis process. Since our focus of research is mainly on the energy harvesting application of nanofibers, our characterization part mainly checks the piezoelectric properties of the nanofiber. Demand for characterization techniques is very high and it is improving continuously. Some of the common characterization techniques used for checking the piezoresponse of nanofibers are discussed in the present study. As part of piezo characterization, studies based on Fourier transform infrared spectroscopy (FTIR), d_{33} meter, and x-ray diffraction (XRD) are performed, which provides us with a signature of the beta sheets content and crystallinity of the piezoelectric polymer.

The FTIR-based analysis clearly indicates the functional group present in a sample in the wavenumber range from 400 to 4000 cm^{-1}. Infrared spectroscopy or spectra mainly originate from the vibrational level transition in a molecule in ground state and are represented in the form of absorption or transmission spectra. The IR region is mainly classified into three regions, namely: far IR which is less than 400 cm^{-1}, mid IR representing 400–4000 cm^{-1}, and finally near IR in the range 14 000–4000 cm^{-1}. Out of these three regions, most of the compounds generate their emission and absorption spectra in the mid IR region and thus FTIR mainly deals with the mid IR region. The vibration levels in a molecule are associated with stretching and bending modes. Bending mode is related to bond angle changes and stretching mode generates symmetric and asymmetric mode by varying the bond length. During this process, the frequency of a particular vibration matches with the frequency of applied emission, and this results in the absorption of complete radiation by the sample content that results in the development of a transmittance (or absorbance) plot versus wavenumber. Based on these plots, the chemical structure of the material can be clearly identified, and the beta sheet content can also be calculated. Beer–Lambert's law-based equation is used to calculate the beta sheet content and is given by

$$f(\beta) = \frac{A_\beta}{1.26A_\alpha + A_\beta} \tag{4.5}$$

where 'A_β' stands for the absorbance value at 840 cm^{-1} and 'A_α' indicates the absorbance at 760 cm^{-1} and $f(\beta)$ indicates the beta sheet content percentage. Thus, by applying the absorbance value at 840 and 760 cm^{-1} in Beer–Lambert's law, the beta sheet content in the sample can be easily calculated. This method is used for calculating the beta sheet content of any pure polymer or polymer-based composition which in turn clearly demonstrates the piezoresponse of the composition [39].

XRD spectrometry is a powerful technique used for the analysis of nanofiber membrane structural properties. The phase identification, crystal orientation and crystallinity related information can be easily obtained via XRD spectra. The basic principle behind XRD analysis is the constructive interference between the crystalline sample and monochromatic x-rays. This phenomenon is achieved by the collimation and direction of x-rays onto the nanomaterial sample, and the resultant interaction of x-rays with the sample generates a diffracted ray. A diffraction pattern is finally developed based on the plot obtained for diffracted rays that are scattered at different angles. Usually, the phase associated with a material corresponds to a unique diffraction pattern. The pattern of a diffracted signal is greatly influenced by the imperfection in a sample and thus the crystal structure, and defects associated with a sample can be easily studied using XRD spectra. The crystalline phase analysis of the tested sample can be easily performed based on the comparison of the generated spectra with the standard crystallographic databases such as International Centre for Diffraction Data (ICDD) [39].

Piezoelectric coefficient is another important parameter that is usually calculated for testing the piezoresponse of a nanofiber sample. A d_{33} meter is an instrument that is used to measure the piezoelectric coefficient of piezomaterials. This instrument mainly measures the d_{33} value of ceramic, single-crystal and polymer-based materials. Here the sample is clamped and subjected to a low frequency force. The d_{33} coefficient represents the charge per unit force in the direction of polarization. Here the sample is processed under electrical signals and compared with the built-in reference, thus enabling the system to provide a direct reading of d_{33}. This system requires an electrode area in the sample where a force can be applied, and the generated charge can be collected. Another set-up used a system comprising two devices, one for electro force application called uniaxial test machine, and another one called charge meter to measure charge. The charge meter measures the induced charge developed during the compression loading to ascertain range of force. Based on the charge value and applied load, the d_{33} value can be calculated using equation (4.6).

$$d_{33} = \frac{\text{Induced charge}}{\text{Applied load}} \tag{4.6}$$

4.3.3.2 Mechanical–electrical transducing set-ups

There are different measurements to analyze the piezoresponse of the synthesized piezoelectric nanofibers. The idea of the different piezoelectric analysis is to apply different mechanical excitations and to detect the corresponding electric outcome in the form of electric voltage, electric current, and power density consumed on an external load. The first and traditional measurement is the detection of generated voltage from free-falling masses or what is called in some literature impulse loading [39]. As shown in figure 4.2, the piezoelectric nanofibers mat is sandwiched between two aluminum metal electrodes, and then connected through isolated wires toward a high-impedance oscilloscope (preferred to be a mixed domain one such as Tektronix MDO30xx model), which is used to measure the output peak-to-peak voltage. Different weights in grams are freely thrown from a certain height (mostly 1–10 cm)

Different weights

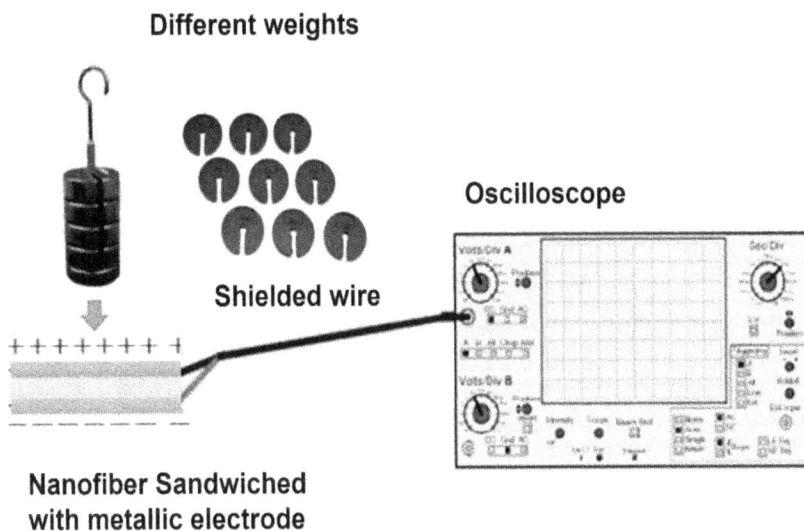

Figure 4.2. The schematic of the impulse loading set-up with different weights.

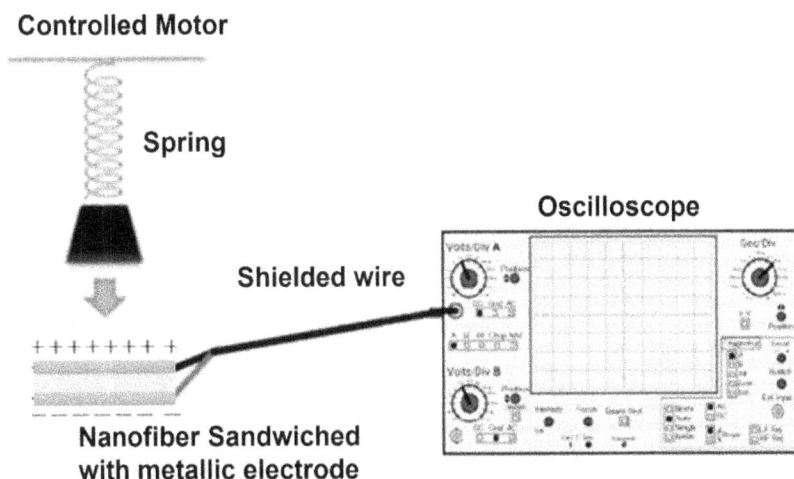

Figure 4.3. The cyclic piezoelectric schematic [67].

onto the sandwiched nanofibers and the output voltages have been measured using the oscilloscope.

In another famous and traditional analysis, the relation between applied force and corresponding generated voltage is analyzed for different piezoelectric nanofibers membranes. The applied force can be in the form of perpendicular force which causes pressure, shear/tension force, and bending impact. In the case of applied normal force, as shown in figure 4.3, the cyclic load has been applied on a nanofiber system using a lightweight spring that is built up in a vertical direction and controlled by a brushless DC motor that relates to an electronic speed controller.

The cyclic load is pressed on the sandwiched nanofiber with an applied range of forces, and then the corresponding peak-to-peak output voltage is measured by the oscilloscope. The forces can be controlled and calibrated by different methods, but the spring-based-force is one of the most convenient techniques to scale the applied force depending on both spring constant and displacement of the spring according to Hooke's law. In another correlated analysis, the applied force can be controlled to be of a constant value, but the cyclic frequency of oscillation can be variable. Hence, the relation between the generated voltage and the applied cyclic frequency can be measured. Mostly, the output voltage can have a saturation behavior with increment cyclic frequency above 10 Hz [21]. For piezo analysis under mechanical stretching impact or applied tension force, the corresponding output voltage can be measured depending on cyclic applied tension force. Here, we should take into consideration the mechanical stress–strain curve of the nanofibers membrane to detect the maximum elastic strength and strain that the synthesized mat can withstand without being cut. To control the stretching, the applied tension cyclic force can be controlled through hand-made set-ups using springs.

One recent technique is to apply the piezoelectric stretching test using a texture analyzer with a full and accurate control of the magnitude of the applied tension force along with the used cyclic frequency, as shown in figure 4.4. Therefore, the sample is sandwiched between two stretchable metallic electrodes, where electrodes are wired and connected to a high-impedance oscilloscope to measure the pulsed voltage corresponding to each stretching impact. Another voltage–force impact is related to applying a bending of the sandwiched nanofibers layer via a certain force at different angles to measure the output voltage for the various mechanical bending of the piezoelectric nanofiber. There are different controllers that have been used to make different bending angles, but some accessories of texture analyzer system are helpful to offer this bending control [40].

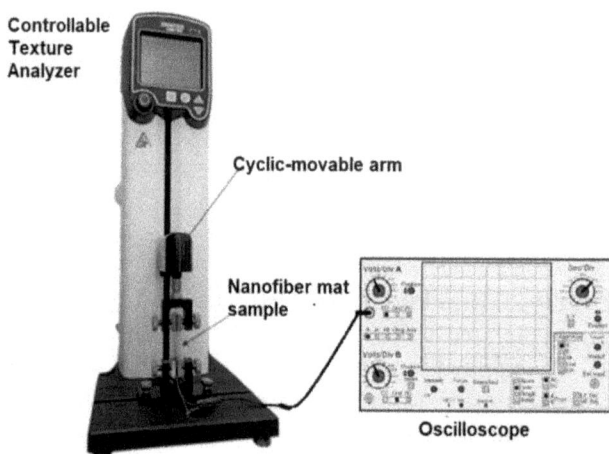

Figure 4.4. Tension-based-piezoelectric set-up using texture analyzer.

To measure the generated current, same set-ups of force–voltage analysis have been used but with replacing the oscilloscope with certain sensitive electrometers to detect the expected relatively low values of current within at least the nano/pico Amperes region, such as the Keithley electrometer of model 6517B. Within another technique, the piezoelectric output current amplitude can be calculated according to the frequency of the compressive applied force and the piezoelectric coefficient of the synthesized materials, as clarified in equation (4.7).

$$I = d_{33} \, dF/dt \qquad\qquad (4.7)$$

The piezoelectric performance of each sample based on different applied load resistances across the positive and negative electrode plates was studied. Different load resistances were applied across the electrodes as part of the characterization. The electroactivity of the sample can be analyzed along the load resistances also, as part of piezo characterization. Therefore, the piezoelectric voltage is measured across different load resistances connected through the electrode plate terminals. Mechanical stress in the form of impulse loading from a specific height is performed. Finally, the power value can be calculated by taking the ratio of voltage square to resistance load value. The generated power over the area of impact easily generates the power density values. The whole set-up is grounded by connecting it to the ground of oscilloscope, as shown in figure 4.3. During measurement, the static charges of the human body should be also grounded by connecting a grounded cable from the hands of the person conducting the experiment to the ground of the oscilloscope [41].

Another piezo characterization is to check the applicability of charging capacitors using the piezoelectric response of the synthesized piezoelectric nanofibers mat. The piezoelectric nanofibers mat can be added between two metallic electrodes tested for charging a commercial microfarad capacitor, based on the circuit clarified in figure 4.5. In this circuit, the energy harvesting piezoelectric unit is exposed to mechanical excitation through pressure/compression and then the pulsed generated output signal passes through a full wave diode rectifier which makes all pulses

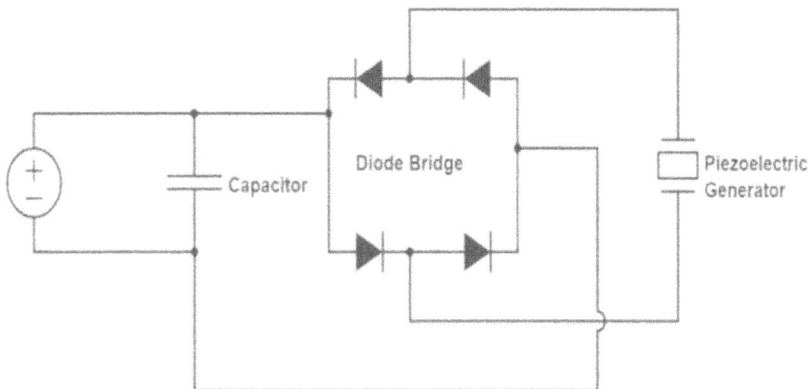

Figure 4.5. Charging circuit of the piezoelectric generator unit.

positively valued. Then, the capacitor can be charged by the signals generated beyond the full rectifier stage. Some boosting voltage circuits can be also used before the capacitor stage to enhance the voltage amplitude for a faster charging time of the capacitor [42].

In another important piezoelectric characterization, Sawyer–Tower circuit-based measurement has been used for the detection of the relationship between electric field and polarization, which is shown in figure 4.6. The Sawyer–Tower circuit was utilized at a certain frequency such as 10 Hz with a peak-to-peak voltage in the range of a few kilo Volts to record the ferroelectric hysteresis loop of the composite film. Then, as-prepared capacitors' ferroelectricity can be easily evaluated by Sawyer–Tower circuit, without any additional post treatment. The circuit mainly consists of a pair of capacitors connected in series where the total charge is shared between them. The capacitors included in the circuit are labelled as reference capacitor (C_r) and ferroelectric capacitor. The total charge is calculated after measuring the potential difference across the reference capacitor using an oscilloscope. Hence, the electric polarization value across the ferroelectric capacitor can be easily calculated by knowing the area of electrode 'A', based on the charge 'Q' value which is equivalent to $C_r V_r$ [43]. Materials that are basically insulators that do not allow the passage of electric charge directly are called dielectrics. Furthermore, an electric field applied through a dielectric leads to the aggregation of net positive charges on one side and the deposition of negative charges on the other side resulting in a potential difference between the surfaces, and this supports storage of the electric charges. There are two types of polymer dielectrics including linear and nonlinear, based on the nature of electric displacement developed from the applied electric field. The final hysteresis loop is influenced by another important factor called depolarization electric field due to the rapid oxidation of the aluminum bottom electrode. A linear type of dielectric behavior can be exhibited under a low electric field. In the presence of a low electric field, only amorphous dipoles in the piezoelectric films are polarized and the field strength was not enough for polarizing the dipoles with crystalline nature. As a result, the dipoles that are polarized can be easily reversed to its initial state at the discharging stage following its charge curve. Whenever the electric field is strong enough, the crystalline dipoles can easily be polarized along the electric field, at the same time the polarized crystals cannot be

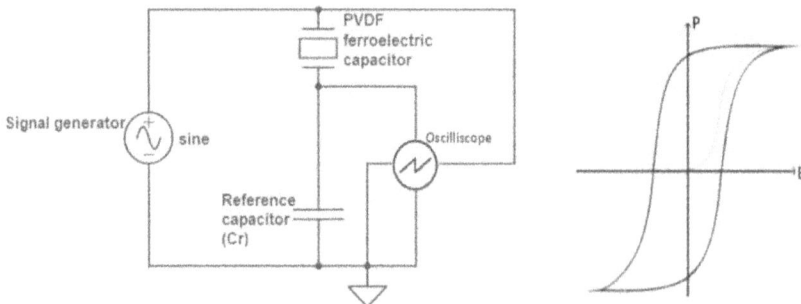

Figure 4.6. Sawyer–Tower circuit and corresponding generated P–E hysteresis loop.

reversed to their initial state during the discharge process. As a result, the $D–E$ or $P–E$ hysteresis loop may show a nonlinear nature and generates a polarization value at zero electric field, and then the $D–E$ curve can be generally used for analyzing the energy storage properties associated with the dielectrics [43]. The energy density stored (U_t) is expressed by the following equation:

$$U_l = \int E. \, dD = 0.5\varepsilon\varepsilon_0 E^2 \tag{4.8}$$

The stored energy and the energy loss is mentioned in both equations (4.9) and (4.10), respectively, as follows.

$$U_{nl} = \int E. \, dD \neq 0.5\varepsilon\varepsilon_0 E^2 \tag{4.9}$$

$$\eta = \frac{U_{nl}}{U_{nl} + U_{\text{loss}}} \tag{4.10}$$

In another important characterization, the atomic force microscope (AFM) with piezoresponse force microscopy (PFM) mode with a tungsten carbide coated tip and bias voltage up to 10 V has been used as characterization method for investigation of morphology and piezoelectric domains. In this analysis, the mechanical surface deformation is detected according to the applied electric voltage from the contact tip. The nanofibers mat is coated with a metallic sheet with relatively high surface conductivity, such as gold, as a contact for characterization of PVDF fibers by electrical modes. This reverse-piezoelectric measurement is another proof for the density of polarized dipoles inside the nanofibers.

4.4 Recent applications

In this section, we focus on the different applications of piezoelectric nanofibers related to self-powered units within PENGs and its correlated contribution in TENG, wearable electronics, acoustic energy harvesting, footstep generation, and vibrations sensors. Some applications are overlapped, but this chapter presents a deep focus on the details of each application's mechanism and its correlation to piezoelectric nanofibers mats. In addition, we focused on the recent used new materials and additives to reach the recently recorded maximum performance of each application.

4.4.1 Self-powered units

Self-powered piezoelectric units can be considered PENGs with expected high electric output response along mechanical flexibility features. In self-powered units, the applied mechanical excitation induces electric dipole moments that are known as polarized static charges, within aligned direction of such dipoles moments to generate electric field and consequently electric voltage. In general, the self-powered piezoelectric units can be used within wide variety of applications including wearable electronics for structural health monitoring, power generation/

transmission, and aerospace systems to estimate the system state for a better safety [44–46]. Recently, different nanocomposites have been extensively used as piezo-electric main materials inside self-power units such as lead PZT, zinc oxide (ZnO), and barium titanate (BTO). Recently, both ZnO nanoparticles and nanorods have gained increasing interest, compared to both PZT and BTO, due to their non-toxicity, increased piezoelectric coefficients, and higher sensitivity to mechanical disturbance [47]. Furthermore, other nanostructures and nanocomposites have been used for higher piezoelectric performance to sense impact location and reinforcing phases such as poly(3,4-ethylenedioxythiophene) (PEDOT)/CuSCN-coated ZnO nanorods on carbon fibers (CFs) that was fabricated by He et al [48]. This multi-structure nanocomposite was hosted by flexible PDMS [49]. The piezoelectric performance showed the incrementation by annealing of ZnO nanorods, along with relatively low-synthesis temperature which reduced the defects formation along with reduced free carriers. The aforementioned self-powered sensor has been used to detect the impact acceleration from 0.1 to 0.4 m s^{-2} with generated voltage up to 7.6 V [48]. In another work, Mokhtari et al developed a lightweight and mechanically-flexible nanogenerator of PVDF-doped-lithium chloride (LiCl) electrospun nano-fibers with an incremented output voltage from 1.3 to 5 V by adding LiCl within PVDF of a power density up to 0.3 μW cm^{-2}. In addition, the synthesized nanocomposite is found to detect different temperatures under static pressure [50]. Within another additive material, Liu et al presented a new nanocomposite including BZT-BCT/P(VDF-TrFE) electrospun nanofibers that were synthesized based on 40 wt.% content of BZT-BCT nanoparticles. The membrane showed an output voltage up to 13.01 V under an oscillatory relatively low applied mechanical force of 6 N with an output power of 1.44 μW [51].

Another additive of barium titanate (BaTiO$_3$) has been added within electrospun PVDF electrospun nanofibers with an enhanced electroactive phase of 91% β-phase at 10 wt.% BaTiO$_3$ nanoparticles. Due to the interfacial interaction between the the tetragonal structure of BaTiO$_3$ with the PVDF host including the ferroelectric nature. That leads the fabricated PENG nanocomposite to have a higher output voltage up to 50 V with a power density up to 4.07 mW m^{-2}. That power density is larger than most of the traditional similar generated powers with up to 10 times that of the case of a pristine PVDF nanofiber-based PENG device. The fabricated PENG device delivered an open-circuit voltage of ~50 V and short-circuit current density of ~0.312 mA m^{-2}. Also, the PVDF-BT nanofiber-based PENG device showed an output power density of ~4.07 mW m^{-2}, which is 10 times higher than that of a pristine PVDF nanofiber-based PENG device. Furthermore, the developed PENG has been newly demonstrated for self-powered real-time vibration sensing applica-tions such as for mapping of mechanical vibrations from faulty central processing unit (CPU) fans, hard disk drives, and electric sewing machines [52]. Yu et al used BaTi$_{0.88}$Sn$_{0.12}$O$_3$ (BTS) which is known for its relatively high piezoelectric coefficient and low T_c close to human body temperature is taken as an example for materials of this kind. Continuous piezoelectric BTS films were deposited on the flexible glass fiber fabrics (GFFs), self-powered sensors based on the ultrathin, super flexible, and polarization-free BTS-GFF/PVDF composite piezoelectric films are used for human motion sensing. In the low force region (1–9 N), the sensors have the outstanding performance with voltage sensitivity of 1.23 V N^{-1} and current sensitivity of 41.0 nA N^{-1}, with a stable performance up to 5000 cycles of usage [53].

4.4.2 Triboelectric generator

TENG converts mechanical disturbances into electric potential via triboelectricity generally called static electricity. Electric potential generated from the mechanical energy sources result from the separation of internal electric charges on the contact surfaces. The dynamic mechanical movements-based alternating potentials can be used for a variety of applications. The generated energy can be used for energizing a battery or powering up electrical energy-based devices such as wireless sensor or even a movement sensor for healthcare monitoring-based studies. There are four layers associated with a traditional triboelectric power generator. The first layer represents charge generation, the second layer deals with charge trapping, the third layer is called a charge collection layer and finally, the fourth layer is utilized for charge storage. Furthermore, four different modes are commonly available for triboelectric power generators. The first mode called vertical contact separation mode is based on two different stacked dielectric film structures that have electrodes on top and bottom. Opposite charged surfaces are formed due to the physical connection between the two dielectric films. Thus, oppositely charged surfaces separated by small gaps developed via vibrations results in the formation of an electric potential that can be utilized for powering up external loads. Whenever an opposite vibration closes the generated gap, the potential vanishes, and electrons flow back. Another mode is called plane contact sliding mode, which depends on a sliding movement for creating triboelectric charges on the two surfaces. The third mode, basically called single-electrode mode, is mainly utilized in mobile systems. In this case, the TENG material is not in electrical contact with the load and can be used in mobile system applications. The bottom part of the TENG is grounded for harnessing energy from some human motion-based activities such as a person walking across a floor. Here the local electric field distribution varies based on the movement between electrode and ground. The potential change in the electrode is balanced by the process of electron exchange between the ground and bottom electrode. Then, the fourth mode, called free standing triboelectric mode, is based on the principle of generation of static charges from dynamic objects because of their contact with air or other objects. This type of situation is mainly observed while a person walks across a carpet. A pair of symmetric electrodes beneath a dielectric layer is used for extracting the generated charges. The oscillatory motion of the object with respect to the electrodes results in an asymmetric distribution of charge in the media and thus an associated flow of electrons occurs between the two electrodes. It is not necessary for the moving objects to directly touch the top dielectric layer and this mode of performance reduces surface wear and friction, thus developing a highly durable TENG [54].

Nowadays, PENG-based nanofibers where the electric charge generation based on polarization changes via mechanical disturbances show an affinity towards

TENG. A graphene nanosheet-incorporated PVDF composite developed via electrospinning technology was successfully developed by Shi *et al* and the composite behaved as an excellent and promising TENG. A triboelectric output voltage of 1511 V with a short-circuit current of 189 mA and a maximum peak power density of 130.2 $W\,m^{-2}$ nearly eight times higher than that of PVDF-PA6 TENG is produced by this TENG. Here the studies mainly focus on the TENG comprising PVDF nanofibers and polyamide-6 (PA6) films. A 74.13 μJ energy/cycle is harnessed under impedance matching condition with a time-averaged output power density of 926.65 mW m^{-2}. Thus, triboelectric performance of PVDF-PA6 TENG is enhanced via incorporation of graphene nanosheets [55].

In another work, Kim *et al* developed Eu-doped PVDF nanofibers via electrospinning technique and the sample is treated as an active layer in TENG. The successful doping of europium ions in PVDF is confirmed by structural and optical investigations and revealed the induced discrete emissions of the electronic transitions. The β-phase content in PVDF nanofibers was found to be enhanced by the addition of Eu within the reduction of its diameter. An increase in the output power from 13 to 26 $\mu W\,cm^{-2}$ is observed by the addition of Eu up to 2.7 wt.% with some attractive optical features. The addition of Eu greatly enhanced the electrical power output, despite that the addition of Eu above a certain threshold level resulted in the formation of (NO_3) nitrate related complexes formation on the PVDF nanofiber surfaces thereby reducing the effective performance of TENG [56].

4.4.3 Wearable electronics

Wearable electronics have become popular in the last few years. Addition of functional values like sensing the environment around and reacting accordingly has converted the basic properties of conventional textiles into smart wearable textile materials. Such products are accompanied with wearable electronics which is often an indispensable part of intelligent textiles. Besides being a part of protection, intelligent textiles/smart fabric/electronic textiles have aesthetic appeal. Smart textile is a very broad field and it is an elegant structure capable of sensing stimuli from the surroundings to counter to that sensed stimuli and adapt to them by integration of suitable functionalities in the textile structure. These types of smart textile have generated an explosive growth in the field of wearable technology. An electronics/textiles wearable is a device worn on the human body that incorporates intelligence (sensor and electronics) into the clothes. A wearable device is essentially a tiny computer that has sensing, processing, storage and wireless communication capabilities. Smart textiles can react after the interpretation of data generated by conditions/stimuli due to their nature. Smart materials can sense and react to environmental conditions or stimuli according to thermal, mechanical, electrical magnetic or other bases [57]. Considering these situations where there is an increased demand for self-powered systems, sensing and mechanical energy harnessing is an important field where PENG and TENG are a necessity. TENG as wearable textile or fabric finds an important role and the field is developing rapidly. A pressure sensitive textile TENG sensor array was demonstrated by Wenjing Fan *et al* and the developed sensor array is used as a non-invasive method to monitor the signals created by sleep pane syndrome and cardiovascular disease [58].

An active sensor as a smart insole embedded with TENG that can be used for monitoring the walking pattern in real time has been developed by Lin *et al*, and the prototype was exceptionally durable with outstanding robustness mechanically. This was used for identifying abnormality in walking pattern or gait as part of rehabilitation assessment [59]. An ultralight single-electrode TENG based on textiles with nano/micro hybridised core–shell with a helical shape bundle was proposed by Ma *et al* for tracking human motion-based signals and biomechanical energy harnessing [60]. Besides fabrics, a variety of TENG-based systems are in the developing stage to meet the increased demand. A wearable wireless sweat sensing system based on TENG that is wireless and is self-powered was developed by Yu Song *et al* [61]. An underwater energy harvesting using stretchable bionic electro-cyte-based TENG using the principle of liquid electrification was demonstrated by Zou *et al* and the prototype mainly focused on rescue related applications in submarine and subaqueous movement detections [62]. Human–machine-based interface and intelligent systems based on wearable technology are arising at fast rate and, as part of it, haptic feedback-based smart gloves that can serve as a simple human–computer interaction method [63]. A highly flexible and scalable bio-inspired spider net coding interface was employed via a single-electrode TENG for multiple direction-based detection and control by Shiand Lee [64]. Also, a self-powered acoustic sensor having an ultrahigh sensitivity of 110 $mV\,dB^{-1}$ was developed via a TENG by Guo *et al* [65].

In a related application, a biomedical application based on wearable TENG for intracellular drug delivery has been discussed by different research publications. Liu *et al* designed a prototype that is used as a stable pulsed voltage source which can be used for triggering potential associated with a plasma membrane and its perme-ability [66]. This study revealed that the drug delivery system was efficient around 90% resulting in a survival rate of more than 94% for the associated cell. Apart from this, there are some studies based on an ionic TENG, fabricated from a gel that can be highly stretchable that is used in damaged tissues. In the damaged tissues, the biomechanical energy produced from TENG acts effectively and supports wound healing [67]. Thus, this detailed chapter clearly shows the scope of wearable TENG in the field of fitness, physiological signal detection and in the field of applications related to electric power sources. Theoretical model-based transformation of a variety of materials into nanogenerators promotes and realizes its usage for a variety of functions based on the advantages of each material. A simple thin film structure can be easily designed from the exceptional flexible nature of wearable nano-generators. Based on this concept, an extremely thin skinny TENG was developed by Jiang *et al*. Here, the conformal device is attached to the cuticle and has a single-electrode structure made of a translucent and extensible electrode [68]. Another stratified all nanofiber-based single electrode having a porous friction film was proposed by Peng *et al* [69]. This film was highly breathable, flexible, biodegradable and extensible. In addition, some wearable accessories such as shoes and other items are also prevalent. The efficiency of the TENG can greatly be improved by adding multiple layers and structures in three-dimensional mode. A three-dimensional intercalation electrode was designed by Gu *et al*, that can energize a capacitor of 1 microfarad from 0 to 8 V within 21 cycles [70], and a pre-strained monofilament-based composition was used by Ahn *et al* for generating a 3D textile framework PENG [71].

4.4.4 Acoustic sensors/harvesting

Piezoelectric materials can be used as a source for acoustic sensing or acoustic energy harvesting applications. Acoustic/sound energy detection has a wide range of applications in the field of environment protection, healthcare, industrial manufacture etc. Energy scavenging or detection, from acoustic signals is mainly associated with sound energy resources found in Nature such as noise, music or sound generated from instruments, industrial or machine generated sound and finally the sound generations from living beings. Noise or unwanted sound energy harnessing based on different nanostructures paves a way towards green energy technology. Our focus in this study is to concentrate on the polymer-based piezo active nanofiber-based materials that can sense acoustic signals. PVDF and its copolymer-based compositions are reported to be one of the best effective materials in nanofiber form for acoustic sensing applications. A PVDF-based electrospun nanofiber membrane transduced applied sound waves of different amplitude and frequency into electrical energy and this process is based on the piezoresponse activity of PVDF [72]. A mechanically elastic electrospun nanomembrane based on PVDF blended with thermoplastic polyurethane (PU) was fabricated and it was found to be very effective in acoustic sensing with higher mechanical strength and elasticity with a piezosensitivity of 667 ± 220 mV N^{-1}. An increase in the output generated voltage is found for applied sound waves of frequency up to 6 kHz with a maximum of 300 mV. A saturation behavior is observed beyond audible frequency (above 6–20 kHz) without any specific fluctuation [73]. Another study revealed the enhancement of acoustic energy conversion efficiency of PVDF by the addition of silver nanoparticles as dopants on its structure. Here, silver nanoparticles-incorporated PVDF electrospun nanofibrous membranes showed more effective response towards low frequency acoustic signals than mid and high frequency regions. Silver nanoparticles-based PVDF resulted in an increase of around 40% output power while comparing with pure PVDF nanomembrane [74]. An ultrasensitive acoustic nanogenerator was developed based on electrospun PVDF graphene composite nanomembrane doped with cerium (Ce^{3+}) ions. This novel acoustic nanogenerator produced an ac output potential of 3 V for an applied sound wave of 88 decibels [27]. Furthermore, both acoustic impedance and piezoactivity are the main feature that contributes to the effective response of PVDF and PVDF-based nanocomposites. Acoustic impedance represents the ability of a medium to ease the propagation of sound waves through it and this factor determines the acoustic response of the particular material. Acoustic impedance of PVDF is in the range of 3×10^6 kg m^{-2} s^{-1}. Some studies revealed the scope of nylon 6/6 in the field of energy harvesting and acoustic sensing due to its excellent acoustic impedance and piezo active properties [75].

In addition to the scope of piezoelectric materials in the field of acoustic sensing and energy harvesting, underwater acoustic sensing is another important scope of piezoelectric materials which can be used for several applications. This type of materials can be used for lowering unwanted sound or noise generated by underwater vehicles which may not be detected via a sonar system. Acoustic target strength of underwater vehicles can also be reduced to a great extent by using these types of materials. There are two important principles that are taken into account for a material to be sensitive to underwater acoustics, which are both its attenuation properties and impedance matching with water. The reflection of incident sound waves can be reduced to minimum if the characteristic impedance of the material matches with the water impedance. The penetrated sound waves can be absorbed to a great extent if the attenuation property of the materials is very high. The acoustic impedance of water is almost close to the impedance of polymer-based materials and thus a group of polymers is found to exhibit underwater acoustic sensing ability. The polymer structures used for the underwater sound absorption mainly include interpenetrating polymer networks (IPNs), polymer foams, gradient polymers etc with the inclusion of air voids and solids. In addition to voids and solid inclusions, nanofillers and photonic crystals can also be used to enhance the sensing ability. Some of the polymers having acoustic impedance matching to water impedance mainly include PU, PDMS and PVDF [76]. In nanoscale, the polymer chain undergoes compression and elongation under the effect of sound waves. Among all these polymers, PVDF is one of the best suited materials for hydroacoustic sensing, due to its matching acoustic impedance with water and piezoelectric properties [77]. One of the most important applications of PVDF in this field is its use as hydrophones to detect underwater sound pressure. PVDF-based hydrophones are highly suitable for underwater sound measurement due to their flat response, mechanical strength, flexible nature, and low acoustic impedance, and this greatly influences the sensitivity of the hydrophone. The matching acoustic impedance of PVDF with water makes it acoustically transparent [78]. Apart from the use of pure polymers, nanofillers inside pure polymers can greatly enhance the sound absorption under water. Some studies have clearly proved that carbon fibers-based fillers can greatly enhance the sound absorption beneath water surfaces [79]. Other studies based on PU showed that the addition of CNTs, up to 1 wt.% onto PU can greatly enhance the absorption of sound under water [80].

Furthermore, some of these piezo active materials play a major role in the field of flow sensing applications. Water flow sensing ability of PVDF membrane is proved in one of the studies showing the scope of a PVDF-based water flow meter. Here, the application of β-phase PVDF in the form of a tube-shaped sensor is demonstrated where the pressure changes based on water flow are sensed [81]. Another study presented an oscillatory flow sensor based on PVDF membrane which finds a major application in the field of marine science. These sensor-based membranes can be used to measure the flow velocity, pressure gradient, acceleration, and shear stress of water. This sensor generated a sensitivity of 0.75 V m^{-1} s^{-1} with a threshold detection limit of 3.4 mm s^{-1} [82].

4.4.5 Vibrational sensor

Smart devices and sensors have primary importance in the field of healthcare, robotics, mechanical equipment, sports, communication devices etc. These smart devices are based on the principle of sensing via detection of strain and vibrations occurring in objects. Most mechanical structures produce continuous and quick or slight shaking movements called vibrations associated with any type of disturbances occurred to them. This type of shaking movemens can generally be called vibrations, and sensors that can sense these types of vibrations are called vibrational sensors. In our daily lives, different kinds of vibration sources are available. Vibrations are generated by musical instruments, human body movements, tall buildings, cantilever, and machine parts. Sensing these vibrations is very important and there are several studies associated with this topic and most of the sensors developed are suitable for flat surfaces related vibration sources. Piezoelectric polymer-based nanosensors play a major role in the field of vibration sensing for detecting vibrations generated from flat and curved surfaces. The flexible nature of polymers enables them to sense vibrations from flat and curved surfaces. Out of the well-known piezoelectric polymers, PVDF plays a major role in sensing vibrations. A recent study based on electrospun PVDF nanofiber, developed a flexible thin self-operating sensor for measuring the vibrations in a string. The nanofibers are placed on a liquid crystal polymer (LCP) that acts as a flexible substrate for the successful fabrication of the sensor. The sensor was very successful in measuring each single smash and alternating periodic vibrations when they are attached to a racket string. The sensor showed an excellent performance towards different frequency range of vibrations and strain. The output potential generated by the sensor increased with an increase vibrational frequency from 2 up to 40 Hz and was then found to be decreasing for frequencies of 90 and 180 Hz. In addition to the generation of voltage, the frequency of vibration was sensed very accurately by the generated PVDF nanosample. A light hit onto the string with a hammer resulted in the generation of a repeatable potential output in the range 25–35 mV. These types of sensors can be used measuring the impact characteristics associated with the strings. Here the sensor is affected not only by strain, in addition strain rate is greatly influencing the output voltage [83].

Another study based on the copolymer of PVDF demonstrated a device that is proficient in sensing vibrations, orientation and acceleration. The P(VDF-TrFe)-based vibrational sensor was able to detect vibrations in the environment induced by the sound pressure in the range 60–80 dB. Around 6–14 μV alternating potential is generated by the vibrational sensor based on the sound intensity [84]. A recent study proved the effect and scope of PVDF/barium titanate nanocomposite for vibration-based energy harvesting. The composite was developed at various concentrations of barium titanate. Sample A represents 15 wt.% PVDF with 5 wt.% $BaTiO_3$, sample B contains 15 wt.% PVDF with 10 wt.% $BaTiO_3$ and sample C is a composition of 15 wt.% PVDF with 15 wt.% $BaTiO_3$. Excellent response was shown by 5 wt.% $BaTiO_3$ where a maximum of 0.02 μW cm^{-2} power density was generated under 10 MΩ. This sample was able to generate an open-circuit voltage of 2 V for a mechanical vibration of 15.17 Hz [85]. Another study where PVDF/BaTiO$_3$ electrospun composite nanofibers are developed based on the concept of replacing a microelectromechanical system (MEMS)-based hand tremor sensor. The new composite mixture was developed by adding 12 wt% BaTiO$_3$ on to 16 wt.% PVDF pellets polymer solution. The sample successfully sensed 6 hz frequency vibration which resembles the hand tremor frequency of a patient with Parkinson's disease [86]. A recent study based on PVDF-BT(10 wt%) nanogenerator showed an excellent performance with an open-circuit voltage of 50 V and output density of approximate 4.07 mW m^{-2}. This prototype successfully proved to be used for self-powered vibration sensing application. The mechanical vibrations associated with hard disk drives, electrical sewing machine, and faulty CPU fans can be sensed using this self-powered nanogenerator [52].

4.4.6 Footstep generation

Nowadays, human movement-based energy harvesting applications of piezoelectric nanofiber membrane are attracting much attention. Human movement mainly includes walking, jogging, skipping, running and jumping and the production of electrical energy based on the principle of piezoelectricity from these movements can be in general defined as footstep generation [87]. This innovative energy production system can be used in areas where a large number of pedestrians are found. Walking can be considered as one of the essential physical activities performed by any person in his life. The electric charge generation ability of a piezoelectric material upon application of a pressure or strain in the form of load can be utilized greatly here where the walking activity transfers load to the piezomaterials. Whenever a person walks, the foot falls on the ground through which the walking weight is exerted on the ground surface through which a person transfers his energy to the environment. Here the idea is to convert the mechanical energy from each footstep into electrical energy where an ac voltage is developed due to the building up of charges inside the piezomaterials. Piezoelectric materials mainly include crystals, ceramics and polymers and among these, direct piezoelectric effect is clearly evident in many crystals that are naturally occurring, such as quartz, Rochelle salt, human bone etc. In addition to crystals, engineered materials such as lithium niobate and PZT also exhibit piezoelectricity. Some of most common and abundant polycrystalline ceramics utilized for the piezoelectric applications include barium titanate, potassium niobate etc. Polymers are the third type of piezoelectric material that offers high flexibility, low cost, light weight, chemical resistance etc. Footstep generation-based energy production using polymers or polymer-based compositions has high demand due to many of these fascinating properties. Among polymers, PVDF and its copolymers, PLA and piezocomposites such as BaTiO$_3$, ZnO, graphene, MoS$_2$, KNN and PZT with PVDF and its copolymers for PENG find a wide scope associated with wearable nanogenerators for footstep generation. Our studies mainly focus on polymer and polymer composites-based nanogenerators for footstep generation systems [88].

The power generation from footsteps based on piezoelectricity in general can be classified into three processes. The first process involves the generation of ac voltage from piezo material by the impact of footsteps. The second process is associated with the rectification of these ac voltage into dc voltage, and finally the storage of direct voltage or utilizing the voltage for a specific requirement. The second and third parts in the footstep generation system related to rectification and storage are almost same regardless of the materials used for the voltage generation. PZT is the most common and well-known piezoceramic material utilized for footstep generation-based energy harvesting applications. We are interested in some of the common footstep generation systems based on ceramics where we can utilize the energy storage and rectification part while implementing a complete footstep generation system using polymer or polymer composites [89]. One of the recent studies used transducer in the system where pressure can be converted into electrical energy. They used piezo-electric materials similar to crystals and each piece is connected in parallel so that 40 piezoelectric pieces altogether contribute for generating maximum output current. Here an energy harvesting circuit containing LTC3588 IC is used for current generation. In addition, a microcontroller called Arduino is used for controlling energy harvesting process using an amplifier. This study showed that the model was able to charge two 12 V batteries by utilizing the energy generated from 80 steps [90].

Another proposed system showed water-cushioned-sole-based footwear where we apply pressure by exerting full bodyweight on to it. The applied pressure due to bodyweight forces the water to flow across the mini turbine. A small electric generator is dedicated for each shoe and thus an adequate amount of electric power is generated [91]. Another important study based on PZT was proposed for developing a piezoelectric tile. Here, the material used is PZT-SA and the generated voltage is controlled by an Arduino microcontroller along with a micro SD card. Each foot area consists of five piezosensors and the generated ac voltages are rectified to dc using a germanium diode (IN60). The number of piezosensors used for extracting the voltage is found to have a linear relationship with generated voltage. In this study, the voltage generated form running is found to be higher than walking [92]. Another research study demonstrated the use of a chain of PZT/PVDF with dc generator, battery and inverter as a single system and the system used around 20 piezoelectric sensors. Now, the footstep generation system can be designed from polymers and polymer-based composites. A recent study developed a shoe pad nanogenerator using PVDF nanofiber membrane with 110 m thickness and shaped into the form of a foot for a shoe pad. Aluminum foil sheet-based electrodes are attached with the sample to collect the extracted energy. Open-circuit voltage is found to be less while the person is walking, and the movement based on running generated more voltage. While running the frequency of applied force is higher than the case of walking and also the impact force is stronger. An optimal output power of 6.45 μW with a load resistance of 5.5 MΩ is generated by this shoe pad nanogenerator [93].

Another piezoelectric composite made of polymer PVDF with $CNF\text{-}BaTiO_3$ is fabricated and the composite membrane exhibited an excellent sensitivity to human footstep-based movement. This sample generated a voltage of around 9 V with a current of 739 nA from a single footstep [94]. Another study based on P(VDF:TrFE) sensors as a smart surface for pressure distribution in floor tiles is demonstrated. This work presented a smart floor idea where the floor tiles are constructed with piezoelectric sensor matrix printed on the surface. Around 130 μm thickness-based sensor matrices are printed on the tile and it offered a pressure sensitivity of 36 pC/N for an area of 1 cm^2 pixel size. This concept has a future application and scope in the field of energy harvesting based on human movements like walking, jumping, jogging, dancing and exercise movements [95].

A PENG based on a composite film fabricated from methyl ammonium lead iodide (MAPbI$_3$) perovskite and PVDF has attracted our attention. This composite generated a voltage in the range 16.7 to 17.8 V with a human motion such as finger tapping for a range of frequency from 0.5 to 5 Hz. Thus, this composite paves a new scope towards a footstep generation system [96]. Human movement such as finger pressing, releasing, wrist bending, foot stepping and finger tapping were applied to a PENG developed by incorporating barium titanate (BT) nanoparticles and graphene nanosheets onto PVDF. This PENG generated a maximum voltage of 7.8 and 2.8 V under stepping conditions by keeping the prototype under foot heal and toe, respectively [97].

Another flexible stretchable fiber nanogenerator was developed from silver nanowires and polytetrafluoroethylene (PTFE) coatings on a bare PU fiber and a sheath electrode of PDMS-silver nanowires (PDMS-AgNWs). This composite was able to distinguish between various knee related activities such as walking, jogging, knee extending and jumping and an average peak power density of 2.25 nW cm^{-2} was generated [98].

4.5 Summary

This chapter introduced an intensive review on the recent characterization and enhancements of piezoelectric nanofibers mats. In this review, we presented an overview of the most convenient materials used as piezoelectric nanocomposites along with a specific focus on PVDF nanofibers with the recently used dopants to enhance their piezoresponse. Furthermore, a quick overview of the fabrication techniques was presented. Moreover, a detailed survey of comprehensive literature was discussed related to the different techniques of piezo performance characterization set-ups to support the academic community who work in such piezoelectric nanocomposites concerning how to measure different characteristics related to piezoelectric analysis. Then, the chapter focused on the recent updated values of piezo performance of some piezo-nanofibers-based applications including self-powered units, TENG, vibration/acoustic transducers, and footstep generation, along with an explanation of the concept of each targeted application. The presented review in this chapter can offer an optimum venue for readers who would like to know the last few years progress of piezoelectric nanofibers in a comprehensive way, along with collecting most of the needed set-ups and techniques to measure piezoelectric parameters of nanofibers mats.

References

[1] Mishra R and Militky J 2019 *Nanoparticles and Textile Technology* (Elsevier)

[2] Zambrano-Zaragoza M L, Quintanar-Guerrero D, Mendoza-Muñoz N and Leyva-Gómez G 2020 Nanoemulsions and nanosized ingredients for food formulations *Handbook of Food Nanotechnology* (Academic) pp 207–56

[3] Nie G, Yao Y, Duan X, Xiao L and Wang S 2021 Advances of piezoelectric nanomaterials for applications in advanced oxidation technologies *Curr. Opin. Chem. Eng.* **33** 100693

[4] Li X, Sun M, Wei X, Shan C and Chen Q 2018 1D piezoelectric material based nanogenerators: methods, materials and property optimization *Nanomaterials* **8** 188

[5] Kapat K, Shubhra Q T H, Zhou M and Leeuwenburgh S 2020 Piezoelectric nano-biomaterials for biomedicine and tissue regeneration *Adv. Funct. Mater.* **30** 1909045

[6] Kalimuldina G, Turdakyn N, Abay I and Medeubayev A 2020 A review of piezoelectric PVDF film by electrospinning and its applications *Sensors (Basel)* **20** 5214

[7] Yang Z, Wang C and Lu X 2018 *Nanofibrous Materials* (Elsevier)

[8] Dahman Y 2017 Nanopolymers *Nanotechnology and Functional Materials for Engineers* (Elsevier) ch 6 pp 121–44

[9] Jacob J, More N, Kalia K and Kapusetti G 2018 Piezoelectric smart biomaterials for bone and cartilage tissue engineering *Inflamm. Regen.* **38** 2

[10] Sultana A, Alam M M, Pavlopoulou E, Berggren M, Crispin X and Zhao D 2021 Remarkable piezoelectric properties in thin films of cellulose nanobers after electrochemical poling https://doi.org/10.21203/rs.3.rs-1134434/v1

[11] Sencadas V, Garvey C, Mudie S, Kirkensgaard J J K, Gouadec G and Hauser S 2019 Electroactive properties of electrospun silk fibroin for energy harvesting applications *Nano Energy* **66** 104106

[12] Karan S K *et al* 2018 Nature driven spider silk as high energy conversion efficient bio-piezoelectric nanogenerator *Nano Energy* **49** 655–66

[13] Belbéoch C, Lejeune J, Vroman P and Salaün F 2021 Silkworm and spider silk electro-spinning: a review *Environ. Chem. Lett.* **19** 1737–63

[14] Persano L, Ghosh S K and Pisignano D 2022 Enhancement and function of the piezoelectric effect in polymer nanofibers *ACC. Mater. Res.* **3** 900–12

[15] Farahani A, Zarei-Hanzaki A, Abedi H R, Tayebi L and Mostafavi E 2021 Polylactic acid piezo-biopolymers: chemistry, structural evolution, fabrication methods, and tissue engineering applications *J. Funct. Biomater.* **12** 71

[16] Yu S *et al* 2022 Maximizing polyacrylonitrile nanofiber piezoelectric properties through the optimization of electrospinning and post-thermal treatment processes *ACS Appl. Polym. Mater.* **4** 635–44

[17] Jacob J, More N, Mounika C, Gondaliya P, Kalia K and Kapusetti G 2019 Smart piezoelectric nanohybrid of poly(3-hydroxybutyrate-co-3-hydroxyvalerate) and barium titanate for stimulated cartilage regeneration *ACS Appl. Bio. Mater.* **2** 4922–31

[18] Ramazanov M A, Maharramov A M, Shirinova H A and Palma L D 2020 Structure and electrophysical properties of polyvinylidene fluoride (PVDF)/magnetite nanocomposites *J. Thermoplast. Compos. Mater.* **33** 138–49

[19] Han G, Su Y, Feng Y and Lu N 2019 Approaches for Increasing the β-phase concentration of electrospun polyvinylidene fluoride (PVDF) nanofibers *ES Mater. Manuf.* **6** 75–80

[20] Zaarour B 2021 Enhanced piezoelectricity of PVDF nanofibers via a plasticizer treatment for energy harvesting *Mater. Res. Express* **8** 125001

[21] Shehata N *et al* 2022 Stretchable nanofibers of polyvinylidenefluoride (PVDF)/thermoplastic polyurethane (TPU) nanocomposite to support piezoelectric response via mechanical elasticity *Sci. Rep.* **12** 8335

[22] Ekbote G S, Khalifa M, Mahendran A and Anandhan S 2021 Cationic surfactant assisted enhancement of dielectric and piezoelectric properties of PVDF nanofibers for energy harvesting application *Soft Matter* **17** 2215–22

[23] Parangusan H, Ponnamma D and Al-Maadeed M A A 2018 Stretchable electrospun PVDF-HFP/Co-ZnO nanofibers as piezoelectric nanogenerators *Sci. Rep.* **8** 754

[24] Zhang W, Zaarour B, Zhu L, Huang C, Xu B and Jin X 2020 A comparative study of electrospun polyvinylidene fluoride and poly(vinylidenefluoride-co-trifluoroethylene) fiber webs: mechanical properties, crystallinity, and piezoelectric properties *J. Eng. Fiber Fabr.* **15** 1558925020939290

[25] Selvan R T, Jia C Y, Jayathilaka W A D M, Chinappan A, Alam H and Ramakrishna S 2020 Enhanced piezoelectric performance of electrospun PVDF-MWCNT-Cu nanocomposites for energy harvesting application *Nano* **15** 2050049

[26] Persano L, Ghosh S K and Pisignano D 2022 Enhancement and function of the piezoelectric effect in polymer nanofibers *Acc. Mater. Res.* **3** 900–12

[27] Garain S, Jana S, Sinha T K and Mandal D 2016 Design of *in situ* poled Ce^{3+}-doped electrospun PVDF/graphene composite nanofibers for fabrication of nanopressure sensor and ultrasensitive acoustic nanogenerator *ACS Appl. Mater. Interfaces* **8** 4532–40

[28] Chamankar N, Khajavi R, Yousefi A A, Rashidi A and Golestanifard F 2020 A flexible piezoelectric pressure sensor based on PVDF nanocomposite fibers doped with PZT particles for energy harvesting applications *Ceram. Int.* **46** 19669–81

[29] Du X, Zhou Z, Zhang Z, Yao L, Zhang Q and Yang H 2022 Porous, multi-layered piezoelectric composites based on highly oriented PZT/PVDF electrospinning fibers for high-performance piezoelectric nanogenerators *J. Adv. Ceram.* **11** 331–44

[30] He Q, Li X, Zhang H and Briscoe J 2023 Nano-engineered carbon fibre-based piezoelectric smart composites for energy harvesting and self-powered sensing *Adv. Funct. Mater.* **33** 2213918

[31] Shepelin N A *et al* 2021 Interfacial piezoelectric polarization locking in printable $Ti_3C_2T_x$ MXene-fluoropolymer composites *Nature Commun.* **12** 3171

[32] Capuana E, Lopresti F, Ceraulo M and La Carrubba V 2022 Poly-L-lactic acid (PLLA)-based biomaterials for regenerative medicine: a review on processing and applications *Polymers* **14** 1153

[33] Mutlu G, Calamak S, Ulubayram K and Guven E 2018 Curcumin-loaded electrospun PHBV nanofibers as potential wound-dressing material *J. Drug Deliv. Sci. Technol.* **43** 185–93

[34] Rabbani S, Jafari R and Momen G 2022 Superhydrophobic micro-nanofibers from PHBV-SiO2 biopolymer composites produced by electrospinning *Funct. Compos. Mater.* **3** 1

[35] Wang Q, Ma J, Chen S and Wu S 2023 Designing an innovative electrospinning strategy to generate PHBV nanofiber scaffolds with a radially oriented fibrous pattern *Nanomaterials* **13** 1150

[36] Alghoraibi I and Alomari S 2018 Different methods for nanofiber design and fabrication *Handbook of Nanofibers* (Springer International Publishing) pp 1–46

[37] Lee H, Kharaghani D and Kim I S 2018 Mechanical force for fabricating nanofiber *Novel Aspects of Nanofibers* (Rijeka: InTech)

[38] de K G, Monsores C, Oliveira da Silva A, de S, Oliveira S A, Weber R P and Dias M L 2022 Production of nanofibers from solution blow spinning (SBS) . *Mater. Res. Technol.* **16** 1824–31

[39] Elnabawy E *et al* 2021 Solution blow spinning of piezoelectric nanofiber mat for detecting mechanical and acoustic signals *J. Appl. Polym. Sci.* **138** 51322

[40] Omran N *et al* 2022 Solution blow spun piezoelectric nanofibers membrane for energy harvesting applications *React. Funct. Polym.* **179** 105365

[41] Shao H *et al* 2021 Single-layer piezoelectric nanofiber membrane with substantially enhanced noise-to-electricity conversion from endogenous triboelectricity *Nano Energy B* **89** 106427

[42] Li Y *et al* 2020 Multilayer assembly of electrospun/electrosprayed PVDF-based nanofibers and beads with enhanced piezoelectricity and high sensitivity *Chem. Eng. J.* **388** 124205

[43] Liu J *et al* 2020 Self-polarized poly(vinylidene fluoride) ultrathin film and its piezo/ferroelectric properties *ACS Appl. Mater. Interfaces* **12** 29818–25

[44] Rana M M *et al* 2020 Porosity modulated high-performance piezoelectric nanogenerator based on organic/inorganic nanomaterials for self-powered structural health monitoring *ACS Appl. Mater. Interfaces* **12** 47503–12

[45] Qing X, Li W, Wang Y and Sun H 2019 Piezoelectric transducer-based structural health monitoring for aircraft applications *Sensors (Switzerland)* **19** 545

[46] Güemes A, Fernandez-Lopez A, Pozo A R and Sierra-Pérez J 2020 Structural health monitoring for advanced composite structures: a review *J. Compos. Sci.* **4** 13

[47] Hu D, Yao M, Fan Y, Ma C, Fan M and Liu M 2019 Strategies to achieve high performance piezoelectric nanogenerators *Nano Energy* **55** 288–304

[48] He Q, Li X, Zhang H and Briscoe J 2023 Nano-engineered carbon fibre-based piezoelectric smart composites for energy harvesting and self-powered sensing *Adv. Funct. Mater.* **33** 2213918

[49] Sabry R S and Hussein A D 2019 PVDF: ZnO/BaTiO$_3$ as high out-put piezoelectric nanogenerator *Polym. Test.* **79** 106001

[50] Mokhtari F, Shamshirsaz M, Latifi M and Foroughi J 2020 Nanofibers-based piezoelectric energy harvester for self-powered wearable technologies *Polymers (Basel)* **12** 1–15

[51] Liu J *et al* 2020 Flexible and lead-free piezoelectric nanogenerator as self-powered sensor based on electrospinning BZT-BCT/P(VDF-TrFE) nanofibers *Sens. Actuators A Phys.* **303** 111796

[52] Athira B S, George A, Vaishna Priya K, Hareesh U S, Bhoje Gowd E, Surendran K P and Chandran A 2022 High-performance flexible piezoelectric nanogenerator based on electro-spun PVDF-BaTiO$_3$ nanofibers for self-powered vibration sensing applications *ACS Appl. Mater. Interfaces* **14** 44239–50

[53] Yu D, Zheng Z, Liu J, Xiao H, Huangfu G and Guo Y 2021 Superflexible and lead-free piezoelectric nanogenerator as a highly sensitive self-powered sensor for human motion monitoring *Nanomicro Lett.* **13**

[54] Kim D W, Lee J H, Kim J K and Jeong U 2020 Material aspects of triboelectric energy generation and sensors *NPG Asia Mater.* **12**

[55] Shi L *et al* 2021 High-performance triboelectric nanogenerator based on electrospun PVDF-graphene nanosheet composite nanofibers for energy harvesting *Nano Energy* **80** 105599

[56] Hong-Seok Kim K P 2018 Enhanced output power from triboelectric nanogenerators based on electrospun Eu-doped polyvinylidene fluoride nanofibers *J. Phys. Chem. Solids* **117** 188–93

[57] Ismar E, Kurşun Bahadir S, Kalaoglu F and Koncar V 2020 Futuristic clothes: electronic textiles and wearable technologies *Glob. Chall.* **4** 1900092

[58] Fan W *et al* 2020 Machine-knitted washable sensor array textile for precise epidermal physiological signal monitoring *Appl. Sci. Eng.* **6** eaay2840

[59] Lin Z *et al* 2019 A triboelectric nanogenerator-based smart insole for multifunctional gait monitoring *Adv. Mater. Technol.* **4** 1800360

[60] Ma L *et al* 2020 Continuous and scalable manufacture of hybridized nano-micro triboelectric yarns for energy harvesting and signal sensing *ACS Nano* **14** 4716–26

[61] Song Y, Min J H, Yu Y, Wang H B, Yang Y R, Zhang H X and Gao W 2020 Wireless battery-free wearable sweat sensor powered by human motion *Sci. Adv.* **6** eaay9842

[62] Zou Y *et al* 2019 A bionic stretchable nanogenerator for underwater sensing and energy harvesting *Nat. Commun.* **10** 2695

[63] Zhu M *et al* 2020 Haptic-feedback smart glove as a creative human-machine interface (HMI) for virtual/augmented reality applications *Appl. Sci. Eng.* **6** eaaz8693

[64] Shi Q and Lee C 2019 Self-powered bio-inspired spider-net-coding interface using single-electrode triboelectric nanogenerator *Adv. Sci.* **6** 1900617

[65] Guo H *et al* 2018 A highly sensitive, self-powered triboelectric auditory sensor for social robotics and hearing aids *Sci. Robot.* **3** eaat2516

[66] Liu Z *et al* 2019 Self-powered intracellular drug delivery by a biomechanical energy-driven triboelectric nanogenerator *Adv. Mater.* **31** 1807795

[67] Jeong S H, Lee Y, Lee M G, Song W J, Park J U and Sun J Y 2021 Accelerated wound healing with an ionic patch assisted by a triboelectric nanogenerator *Nano Energy* **79** 105463

[68] Jiang Y *et al* 2021 Stretchable, washable, and ultrathin triboelectric nanogenerators as skin-like highly sensitive self-powered haptic sensors *Adv. Funct. Mater.* **31** 2005584

[69] Peng X *et al* 2020 A breathable, biodegradable, antibacterial, and self-powered electronic skin based on all-nanofiber triboelectric nanogenerators *Sci. Adv.* **6** eaba9624

[70] Gu L *et al* 2020 Enhancing the current density of a piezoelectric nanogenerator using a three-dimensional intercalation electrode *Nat. Commun.* **11** 1030

[71] Ahn S *et al* 2020 Wearable multimode sensors with amplified piezoelectricity due to the multi local strain using 3D textile structure for detecting human body signals *Nano Energy* **74** 104932

[72] Shehata N, Hassanin A H, Elnabawy E, Nair R, Bhat S A and Kandas I 2020 Acoustic energy harvesting and sensing via electrospun PVDF nanofiber membrane *Sensors (Switzerland)* **20** 3111

[73] Shehata N, Nair R, Kandas I, Omran N and Hassanin A 2022 Elastic piezoelectric nanofibers mats for acoustic energy harvesting *Mater. Sci. Forum* **1075** 19–25

[74] Wu C M and Chou M H 2020 Acoustic–electric conversion and piezoelectric properties of electrospun polyvinylidene fluoride/silver nanofibrous membranes *Express Polym. Lett.* **14** 103–14

[75] Bloomfield P E, Lo W-J and Lewin P A 2000 Experimental study of the acoustical properties of polymers utilized to construct PVDF ultrasonic transducers and the acousto-electric properties of PVDF and P(VDF/TrFE) films *IEEE Trans. Ultrason. Ferroelec. Freq. Control* **47** 1397–405

[76] Fu Y, Kabir I I, Yeoh G H and Peng Z 2021 A review on polymer-based materials for underwater sound absorption *Polym. Test.* **96** 107115

[77] Eovino B T 2015 Design and analysis of a PVDF acoustic transducer towards an imager for mobile underwater sensor *Technical Report No. UCB/EECS-2015-154* UC Berkeley, CA http://eecs.berkeley.edu/Pubs/TechRpts/2015/EECS-2015-154.html

[78] Institute of Electrical and Electronics Engineers, Oceans 2019 *MTS/IEEE SEATTLE*

[79] Yuan B, Jiang W, Jiang H, Chen M and Liu Y 2018 Underwater acoustic properties of graphene nanoplatelet-modified rubber *J. Reinf. Plast. Compos.* **37** 609–16

[80] Gu B E, Huang C Y, Shen T H and Lee Y L 2018 Effects of multiwall carbon nanotube addition on the corrosion resistance and underwater acoustic absorption properties of polyurethane coatings *Prog. Org. Coat.* **121** 226–35

[81] Li Q, Xing J, Shang D and Wang Y 2019 A flow velocity measurement method based on a PVDF piezoelectric sensor *Sensors (Switzerland)* **19** 1657

[82] Shizhe T and Yijin W 2022 An artificial lateral line sensor using polyvinylidene fluoride (PVDF) membrane for oscillatory flow sensing *IEEE Access* **10** 15771–85

[83] Singh R K, Lye S W and Miao J 2019 PVDF nanofiber sensor for vibration measurement in a string *Sensors (Switzerland)* **19** 3739

[84] Persano L, Ghosh S K and Pisignano D 2022 Enhancement and function of the piezoelectric effect in polymer nanofibers *Acc Mater. Res.* **3** 900–12

[85] Güçlü H, Kasım H and Yazici M 2023 Investigation of the optimum vibration energy harvesting performance of electrospun PVDF/BaTiO$_3$ nanogenerator *J. Compos. Mater.* **57** 409–24

[86] Tamil Selvan R, Jayathilaka W A D M, Chinappan A, Alam H and Ramakrishna S 2020 Modelling and analysis of elliptical cantilever device using flexure method and fabrication of electrospun PVDF/BaTiO$_3$ nanocomposites *Nano* **15** 2050007

[87] Soni K, Jha N, Padamwar J, Bhonsle D and Rizvi T 2022 Footstep power generation using piezoelectric plate *Int. Res. J. Modern. Eng.* **04** 1736–40

[88] Liu Y *et al* 2021 Piezoelectric energy harvesting for self-powered wearable upper limb applications *Nano Select* **2** 1459–79

[89] Ali A, Khan U, Ahmad M, Aziz A and Neha N 2021 Footstep power generation using piezoelectric sensor *Proc. 2nd Int. Conf. on ICT for Digital, Smart, and Sustainable Development, ICIDSSD 2020 (New Delhi, 27-28 February)* https://eudl.eu/doi/10.4108/eai.27-2-2020.2303209

[90] Iswanto S, Suripto F, Mujahid K T, Putra N P, Apriyanto and Apriani Y 2018 Energy harvesting on footsteps using piezoelectric based on circuit LCT3588 and boost up converter *Int. J. Elect. Comp. Eng.* **8** 4104–10

[91] Mahmud I 2018 Electrical power generation using footsteps *Eur. Sci. J., ESJ* **14** 318

[92] Kamboj A, Haque A, Kumar A, Sharma V K and Kumar A 2017 Design of footstep power generator using piezoelectric sensors *2017 International Conference on Innovations in Information, Embedded and Communication Systems (ICIIECS) (Coimbatore)* pp 1–3

[93] Yu L, Zhou P, Wu D, Wang L, Lin L and Sun D 2019 Shoepad nanogenerator based on electrospun PVDF nanofibers *Microsyst. Technol.* **25** 3151–6

[94] Li M *et al* 2023 Flexible cellulose-based piezoelectric composite membrane involving PVDF and BaTiO 3 synthesized with the assistance of TEMPO-oxidized cellulose nanofibrils *RSC Adv.* **13** 10204–14

[95] Alvarez Rueda A *et al* 2023 Study of pressure distribution in floor tiles with printed P(VDF: TrFE) sensors for smart surface applications *Sensors* **23** 603

[96] Jella V, Ippili S, Eom J H, Choi J and Yoon S G 2018 Enhanced output performance of a flexible piezoelectric energy harvester based on stable MAPbI3-PVDF composite films *Nano Energy* **53** 46–56

[97] Shi K, Sun B, Huang X and Jiang P 2018 Synergistic effect of graphene nanosheet and BaTiO3 nanoparticles on performance enhancement of electrospun PVDF nanofiber mat for flexible piezoelectric nanogenerators *Nano Energy* **52** 153–62

[98] Cheng Y *et al* 2017 A stretchable fiber nanogenerator for versatile mechanical energy harvesting and self-powered full-range personal healthcare monitoring *Nano Energy* **41** 511–8

IOP Publishing

Energy Harvesting Properties of Electrospun Nanofibers
(Second Edition)

Jian Fang and Tong Lin

Chapter 5

Hi-performance piezoelectric nanofiber via advancing β-crystalline phase

Fatemeh Mokhtari, Pejman Heidarian, Russell J Varley and Joselito M Razal

Electrospinning has emerged as a versatile technique for the fabrication of nanofibers with unique properties and applications. Among various polymer materials, polyvinylidene fluoride (PVDF) has gained significant attention due to its exceptional mechanical, electrical, and chemical properties. The ability to control and enhance the β-crystalline structure in electrospun PVDF nanofibers holds great promise for various applications, including energy storage, water/gas filtration, and biomedical engineering. This chapter delves into the strategies employed to tailor the β-crystalline structure of PVDF fibers during the electrospinning process. These strategies comprehend the manipulation of solution properties, including solvent selection, polymer concentration, and additives, to enhance the nucleation and growth of the β-phase. Additionally, the influence of processing parameters, such as solution flow rate, applied voltage, and collector configuration, is explained to optimize the fiber alignment and crystallinity. Tailored post-treatment processes, such as annealing, mechanical stretching, and solvent vapor treatment, refine the β-crystalline structure by inducing molecular rearrangement and relaxation. This results in an increased β-phase content and improved crystallinity. Fine-tuned PVDF nanofibers have potential applications in high-performance membranes, piezoelectric devices, and scaffolds in tissue engineering. The characterization techniques utilized to assess the β-crystalline structure and morphology of electrospun PVDF fibers are comprehensively discussed, including x-ray diffraction (XRD), Fourier-transform infrared spectroscopy (FTIR), and differential scanning calorimetry (DSC). The effect of electrospinning parameters on β-phase formation was comprehensively explained in the previous version of this chapter. In the new edition, we extensively explore the techniques of fine-tuning the β-crystalline structure in PVDF nanofibers through precision electrospinning, with a special focus on the addition of fillers. Here we offer

doi:10.1088/978-0-7503-5487-5ch5

valuable insights into the optimization of electrospinning parameters, characterization techniques, and the resulting functional properties. This chapter serves as a valuable resource for researchers, scientists, and engineers involved in the design and fabrication of advanced materials for a wide range of applications, fostering further advancements in the field of electrospun PVDF fibers.

5.1 Introduction

The piezoelectric behavior of piezoelectric materials relies on the available orientations and alignment of dipoles, as well as their ability to endure significant strain under mechanical stress. Consequently, optimal piezoelectric performance is achieved when all the dipoles within the polymer are aligned in a unified direction or along the applied electric field [1]. Presently, a wide array of piezoelectric polymers finds applications in electronic devices, including nylons, polypropylene, polystyrene, PVDF, and PVDF copolymers. Among these, PVDF with predominantly polar phases holds utmost significance [2]. PVDF consists of monomer units (CH_2–CF_2) with fluorine atoms, which results in a significant dipole moment due to their electronegativity. Ferroelectricity and piezoelectricity in PVDF were first uncovered by Kaway in 1969. The electroactive characteristics of PVDF and its copolymers have rendered them highly valuable for various applications. The crystalline polymers exhibit five distinct phases: α, β, γ, δ, and ε [3]. While the α-phase is commonly encountered, it is the β-phase that takes center stage in terms of significance, primarily due to its exceptional piezoelectricity. The β-phase exclusively demonstrates a conformation of all-trans (TTTT). On the other hand, the δ- and α-phases display a conformation of TGTG', while the ε- and γ-phases exhibit a conformation of T3GT3G' [4]. Among the various phases, the β-phase exhibits a higher dipole moment (7×10^{-30} C·m) compared to the other phases [5]. Obtaining and maximizing the β-phase has always been of immense interest in this area of ongoing research. The summary of the different properties of the three phases in PVDF is presented in figure 5.1.

In recent years, there has been a significant amount of research and investigation into the utilization of piezoelectric polymers for the production of flexible piezoelectric nanogenerators (PENGs). These polymers, such as PVDF and its copolymers, have gained considerable attention. This can be attributed to their advantageous characteristics, including their lightweight nature, ease of processing, and exceptional flexibility [6]. The chemical structures of three PVDF copolymers are illustrated in figure 5.2. The random sequence copolymerization of PVDF with TrFE (–(CHF–CF_2)–) facilitates the enhanced development of the β-phase through copolymerization which eliminates the need for supplementary post-treatment methods. The inclusion of fluorine atoms introduces additional steric hindrance, effectively restraining the generation of the α-phase [7]. PVDF-hexafluoropropylene (HFP), the copolymer of PVDF-hexafluoropropylene, has attracted significant attention as a desirable material in recent years. This is primarily due to the incorporation of HFP groups, which endows PVDF-HFP with superior properties compared to PVDF. These properties include enhanced solubility, reduced

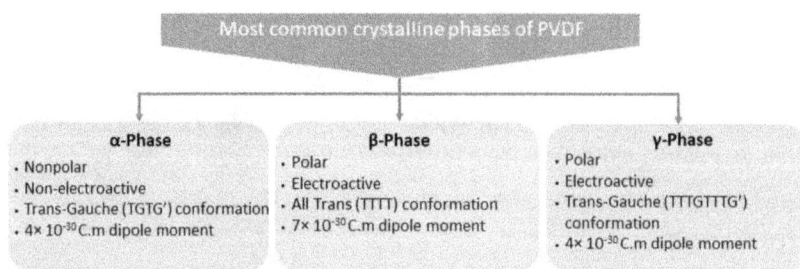

Most common crystalline phases of PVDF

α-Phase	β-Phase	γ-Phase
• Nonpolar	• Polar	• Polar
• Non-electroactive	• Electroactive	• Electroactive
• Trans-Gauche (TGTG') conformation	• All Trans (TTTT) conformation	• Trans-Gauche (TTTGTTTG') conformation
• 4× 10^{-30} C.m dipole moment	• 7× 10^{-30} C.m dipole moment	• 4× 10^{-30} C.m dipole moment

Figure 5.1. PVDF phases and their properties.

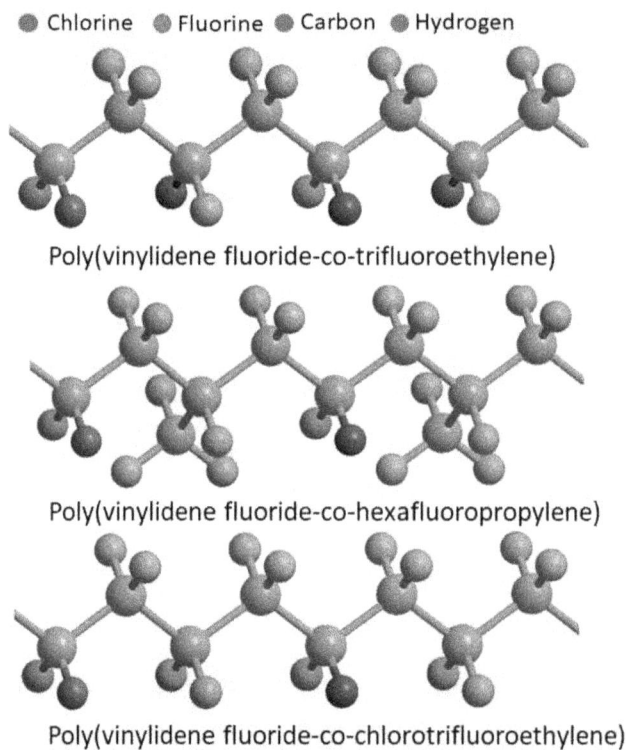

Chlorine   Fluorine   Carbon   Hydrogen

Poly(vinylidene fluoride-co-trifluoroethylene)

Poly(vinylidene fluoride-co-hexafluoropropylene)

Poly(vinylidene fluoride-co-chlorotrifluoroethylene)

Figure 5.2. Chemical structure of PVDF copolymer, PVDF-TrFe, PVDF-HFP and PVDF-CTFE. Reproduced from [12] with permission from the Royal Society of Chemistry.

crystallinity, increased free volume, and improved mechanical strength. Additionally, the inclusion of HFP groups elevates the fluorine content within PVDF-HFP, resulting in heightened hydrophobicity. However, the benefits of PVDF-HFP come with some disadvantages. One disadvantage is that PVDF-HFP films necessitate a high electric field during the poling process, making the poling procedure slightly challenging. Furthermore, PVDF-HFP possesses inherent limitations that restrict its usage. Despite exhibiting high dielectric breakdown values, the low piezoelectric voltage constant of PVDF-HFP renders it unsuitable for applications such as energy harvesting [8]. Polyvinylidene-co-chlorotrifluoroethylene (PVDF-CTFE) is an additional copolymer derived from vinylidene fluoride (VDF) and chlorotrifluoroethylene (CTFE). The characteristics and uses of PVDF-CTFE are contingent upon the relative mole percentages of VDF and CTFE present in the polymer composition. By adjusting these percentages, the properties of PVDF-CTFE can be tailored to suit specific requirements in various applications [9]. The α- to the β- and γ-phases are converted by pressing, stretching at high temperature, uniaxial tension, electrical poling, annealing, compression molding, the inclusion of fillers, etc. It is reported that the polar piezoelectric phases can co-exist together. Due to the highest piezoelectricity in its class, PVDF is widely adapted in different physical forms like films, membranes, and fibers to realize sensors, actuators, separators in supercapacitors, and mechanical energy harvesters [10]. The most prominent technique among various approaches to enhance the β-phase is the incorporation of filler materials. Researchers have extensively investigated different methods, such as surface modification of fillers and the formation of core–shell fillers, to specifically strengthen the interfacial interaction and enhance the polarity of composites based on PVDF. These techniques of surface or interface modification typically involve complex chemical processes. As a result, current research efforts are primarily focused on fillers with a high aspect ratio, as they possess a greater number of active sites that facilitate effective interfacial interaction [11].

When considering filler loading, composites can be categorized into two distinct groups: those with low or ultralow filler content (<1 wt.%) and those with medium-to-high filler content (>1 wt.%). The rationale behind employing medium-to-high filler content primarily revolves around incorporating the superior electrical properties of fillers into the polymer matrix. For instance, conductive fillers may be added up to the percolation threshold, or a significant proportion of ceramic nanoparticles may be introduced into the polymer matrix to enhance the dielectric performance. On the other hand, utilizing low and ultralow filler ratios is typically aimed at modifying the crystallinity, nanostructures, and facilitating the formation of new interfaces within the composite material.

This chapter will delve into various approaches employed to enhance the β-phase in PVDF, with particular emphasis on the utilization of different fillers. Our focus in this chapter is primarily on the significance of incorporating various types of fillers. Recognizing the crucial role fillers play in enhancing the properties of PVDF fibers, we delve into an in-depth examination of different filler materials and their effects when integrated into electrospun PVDF fibers. By highlighting the importance of fillers in this context, we aim to contribute to the understanding of how filler incorporation can significantly impact the overall performance and functionality of PVDF fibers, opening up new ways for innovative applications in diverse fields. The focus will be on investigating how the incorporation of fillers can positively impact the formation and stability of the desired β-phase in PVDF. Furthermore, we will explore the most prevalent characterization techniques used to assess the different phases of PVDF, namely FTIR spectroscopy, XRD, and DSC. These techniques enable the identification and quantification of different crystalline phases, as well as the determination of key thermal properties related to phase transitions.

5.2 Improvement of β-phase in PVDF

5.2.1 Effect of electrospinning parameters

Various methodologies are employed to induce the conversion of polymers into the desired polar β-phase. These techniques encompass a range of approaches, including but not limited to melt crystallization, unidirectional or bidirectional stretching, high-voltage poling, growth on specialized substrates utilizing inorganic fillers, spin coating, and electrospinning. Each of these techniques offers distinct advantages and mechanisms for promoting the transformation of polymers into the β-phase. Some of these methods come with their own set of drawbacks, such as intricate manufacturing processes, usage of massive solvents, time-consuming treatments, and limitations in small-scale production, among others [13]. Melt crystallization involves controlled heating and cooling processes to encourage the formation of the desired phase. Stretching, whether in a unidirectional or bidirectional manner, imparts mechanical stress that aids in aligning the polymer chains and promoting the β-phase. High-voltage poling utilizes the application of electrical potential to orient the molecular dipoles and induces the desired phase transition [14]. Growth on specialized substrates with the assistance of inorganic fillers provides a template for the desired phase formation [15]. Spin coating involves the deposition of a polymer solution or melt onto a substrate, resulting in the formation of a thin film with the desired phase [16]. Electrospinning, on the other hand, employs the electric force to generate fine fibers from a polymer solution or melt. This process facilitates the alignment and orientation of the polymer chains, leading to the desired β-phase conversion. Each technique offers its own set of advantages and considerations, and the choice of method depends on factors such as the specific polymer, desired phase transition, and intended application [17]. Figure 5.3 illustrates the significant factors influencing the electrospinning process, specifically targeting the enhancement of β-phase formation. Each of these parameters has been comprehensively discussed in our previous work [18]. In electrospinning process, the solvent and polymer influence surface tension and conductivity, while the viscosity primarily depends on the molecular weight and concentration of PVDF. Increasing viscosity leads to a decrease in solvent concentration, resulting in faster solvent evaporation and allowing sufficient time for the molecular chains to stretch before solidifying. This, in turn, leads to an increase in β-phase formation. However, excessively high viscosity poses challenges in stretching the polymer solution. Hence, it is important to select PVDF with an appropriate molecular weight and concentration for the electrospinning experiment [19, 20]. PVDF polymers with different molecular weights were utilized to produce nanofibers. The molecular weights of the PVDF polymers used were 180 000, 275 000, and 530 000 $g \cdot mol^{-1}$. These polymers exhibited varying degrees of fiber formation, as indicated by their respective $F(\beta)$ values [21]. As the concentration of PVDF increases, the polymer solution experiences a heightened stretching effect due to the electric field, resulting in an increase in $F(\beta)$. The improvement in $F(\beta)$ is observed to rise from 78% to 85.9% as the concentration reaches 20%. However, when the concentration is further increased to 26%, there is a subsequent decrease in $F(\beta)$ to 82.5% [22]. The findings

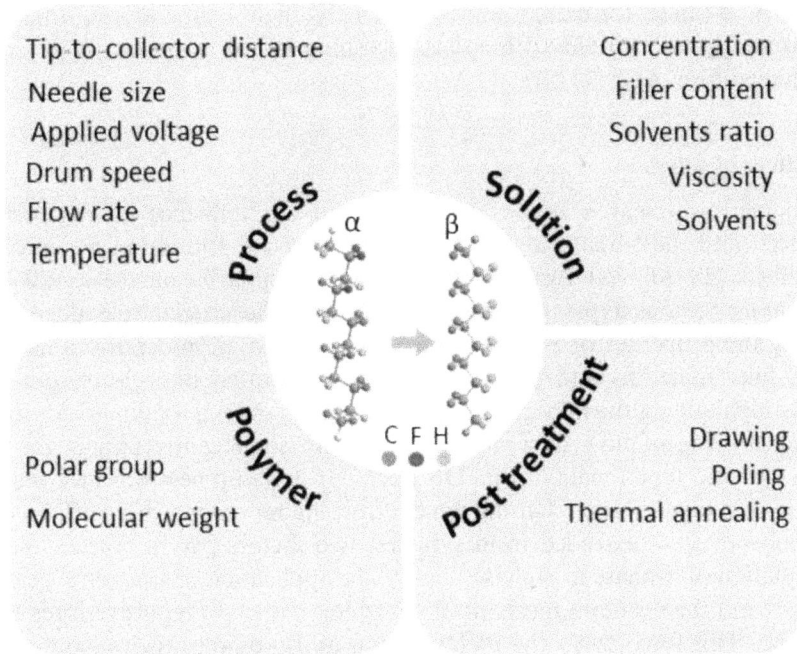

Figure 5.3. Effective parameters in enhancing β-phase formation in electrospun PVDF web.

of the study demonstrated that the nanofibers fabricated with dimethylformamide (DMF) as a solvent exhibited the highest $F(\beta)$ of 90.9%. Subsequently, dimethyl sulfoxide (DMSO) and N-methylpyrrolidon (NMP) showcased $F(\beta)$ values of 87.4% and 81.9% correspondingly, indicating their relatively lower performance compared to DMF [23]. Elevating the applied voltage from 14 to 20 kV exhibits an enhancement in the Xc value from 48% to 56.3%, resulting in a corresponding increase in $F(\beta)$ from 70% to 76%. However, as the applied voltage continues to rise up to 24 kV, both Xc and $F(\beta)$ experience a decline due to the diminished polarization of the PVDF solution during the stretching process [24]. Increasing the distance between the tip and collector from 9 to 15 cm results in an enhancement of the $F(\beta)$ parameter, which rises from around 77%–85.9%. This improvement can be attributed to the extended duration available for the nanofibers to elongate and the solvents to evaporate. However, as the tip-to-collector distance is further increased, the $F(\beta)$ value starts to decrease. This decline is primarily caused by a reduction in the intensity of the electric field and increasing in jet instability [25]. High relative humidity (RH) is commonly employed during electrospinning to achieve a substantial content of $F(\beta)$. Nevertheless, once the RH surpasses a critical threshold, the presence of water in the surroundings prevent the solvent's evaporation, rendering the electrospinning process difficult. Well-formed electrospun fibers were created in high relative humidity air (50%–70% RH) at ambient temperature [26]. Working temperature affect viscosity and surface tension of PVDF solution. Comparison of $F(\beta)$ at different temperature (5 °C, 15 °C, 25 °C, 35 °C, and 45 °C) shows that PVDF nanofibers having the highest $F(\beta)$ at 25 °C. Due to the lack of β-phase formation under extremely high temperatures, the ambient temperature typically employed for electrospinning PVDF nanofiber is the standard room temperature of 25 °C [27].

5.2.2 Effect of fillers

While numerous research studies have explored the influence of electrospinning parameters and post-treatment processes on β-phase formation in electrospun PVDF fibers [28–30], our focus in this chapter is primarily on the significance of incorporating various types of fillers. Recognizing the crucial role fillers play in enhancing the properties of PVDF fibers, we delve into an in-depth examination of different filler materials and their effects when integrated into electrospun PVDF fibers. By highlighting the importance of fillers in this context, we aim to contribute to the understanding of how filler incorporation can significantly impact the overall performance and functionality of PVDF fibers, opening up new ways for innovative applications in diverse fields. During the electrospinning process of a PVDF solution, as the polymer jet is extruded from syringes, two factors play a crucial role in the transformation of α-phase to β-phase. Firstly, the application of an electric field on the nanofibers and the resultant mechanical stretching induces a repulsive force between the charges. This force twists the PVDF chains and promotes their alignment along the fiber axis. Consequently, the electric dipoles in the PVDF solution align themselves. However, relaxation occurs after electrospinning, causing the PVDF polymeric chains to revert to their stable configuration. To address this issue, various particles are introduced as fillers into the structure of PVDF nanofibers during the electrospinning process. These particles serve to overcome the relaxation-induced defect, providing stability and enhancing the desired β-phase formation in the PVDF nanofiber structure [31]. Currently, the polarization coefficient of electrospun PVDF remains significantly low, thereby impeding the piezoelectric output of the device. Numerous experimental findings have suggested that augmenting the interfacial interaction can effectively heighten the β-phase transformation within the PVDF matrix when combined with nanofillers such as graphene, graphene oxide (GO), carbon nanotubes, MXene, and metal salts [32]. The incorporation of fillers not only improves the piezoelectric coefficient and sensitivity of the fibers but also enhances their mechanical properties, making them suitable for a wide range of applications including energy harvesting, sensors, actuators, and wearable electronics. Through the selection and incorporation of appropriate fillers, the piezoelectric performance of electrospun PVDF fibers can be substantially enhanced, unlocking new possibilities for advanced functional materials in various fields.

Carbon-based materials such as GO and reduced GO (rGO), are the most widely used as filler to enhance electrical properties of PVDF for their priorities in improving energy harvesting and storage. Fillers, such carbon nanotubes (CNTs), can significantly enhance the piezoelectric response of PVDF fibers by promoting the alignment of PVDF polymer chains and creating an efficient charge transfer pathway. These fillers serve as nucleating agents, facilitating the formation of a more ordered β-phase structure within the PVDF matrix. Results show that by increasing the mass ratio of GO and rGO up to 2 wt.%, the characteristic diffraction peak of α-phase almost disappeared, and the β-phase was the main crystallization of the electrospinning nanofiber mat (figure 5.4(a)) [33]. Carbon based materials and conductive fillers can significantly enhance the piezoelectric response of PVDF fibers by promoting the alignment of PVDF polymer chains and creating an efficient charge transfer pathway. These fillers serve as nucleating agents, facilitating the formation of a more ordered β-phase structure within the PVDF matrix. The interaction between MXene and PVDF-TrFE is typically attributed to electrostatic

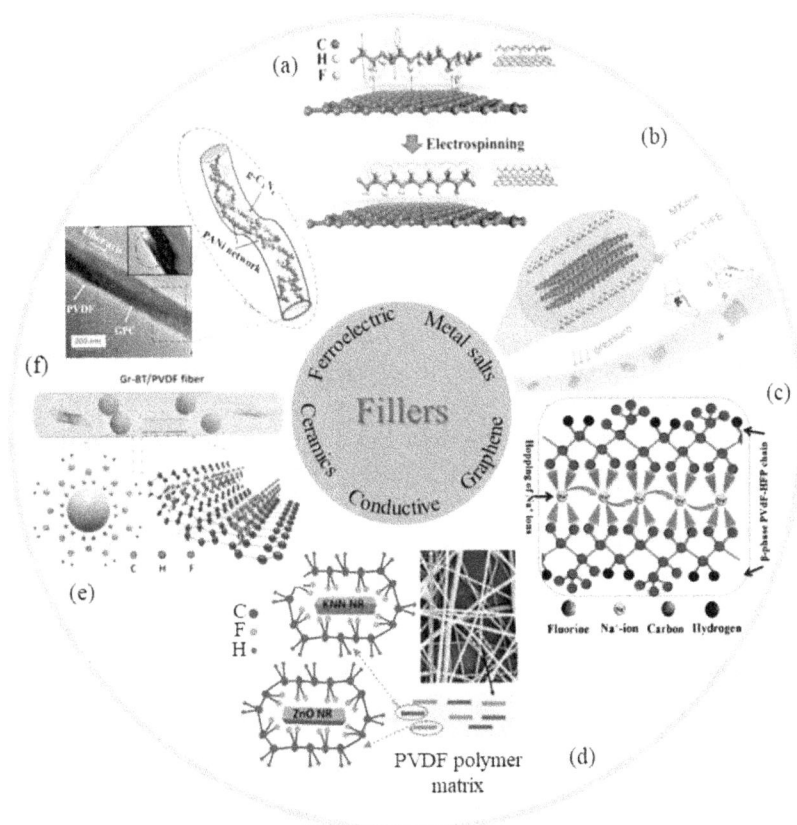

Figure 5.4. (a) interaction between the PVDF dipole and graphene increasing the β-phase fraction. Reprinted from [33], copyright (2021) with permission from Elsevier. (b) electrospun fiber of PVDF-TrFE/MXene. Reprinted from [34], copyright (2021) with permission from Elsevier. (c) Hopping of Na$^+$ ion via the electronegative channel of PVDF-HFP chains. Reprinted with permission from [35], copyright (2021) American Chemical Society. (d) a hybrid piezoelectric nanogenerator comprising of KNN/ZnO nanorods into PVDF electrospun web. [36], copyright (2020) John Wiley & Sons Ltd. (e) graphene nanosheet and BaTiO$_3$ nanoparticles interaction in electrospun PVDF nanofiber. Reprinted from [37], copyright 2018, with permission from Elsevier. (f) PANI nanorods network within an individual PVDF fiber. Reprinted with permission from [38], copyright (2019) American Chemical Society.

forces, hydrogen bonding, and Van der Waals interactions. The MXene nanosheets can serve as templates or sites for PVDF-TrFE crystal growth, guiding the arrangement of polymer chains and promoting the formation of the β-phase. The conductive MXene sheets generate induced charge in the electric field, and the adjacent sheets increase the local electric field intensity, making the dipoles of β-phase PVDF-TrFE crystals orient easily (figure 5.4(b)) [34]. Carbon-based materials like CNTs and graphene, as well as metallic nanoparticles such as silver and gold, are commonly used as conductive fillers. These materials possess excellent electrical conductivity and can be easily dispersed within the polymer matrix, ensuring good electrical contact throughout the material. The presence of conductive fillers facilitates the efficient transfer of electrical charges and helps to mitigate issues such as electrical noise and charge accumulation.

The β-phase PVDF-HFP exhibits an all-trans (TTTT) configuration where Na^+-ions interact with the polar $-CF_2$ group and hop along the chain. After adding the salt ionic liquid (SIL) solution (20 wt.%) in the polymer matrix, the α-phase crystalline peaks are reduced sharply and almost disappeared with the increase of SIL, and the new peak of the β-phase appearing at 21.07° corresponds to the (110/200) plane. The crystalline phase transformation from the α-phase into β-phase happened due to the interaction of the polar $-CF_2$ group of PVDF-HFP with cations of IL ($EMIM^+$) or salt (Na^+) presented in the SIL solution, resulting in chain conformation in the all-trans (TTTT) mode (figure 5.4(c)) [35]. In the hybrid nanogenerator of potassium sodium niobate (KNN)/zinc oxide (ZnO) and PVDF, the β-fraction values are increasing (92%) with increase in the ZnO nanorods up to a certain limit (3 wt.%) with the constant KNN fillers (3% KNN), after which β fraction values again decreases (figure 5.4(d)) [36]. Piezoelectric fillers are added to piezo polymers to enhance their piezoelectric properties, which involve the generation of an electric charge in response to applied mechanical stress or vice versa. These fillers increase the overall piezoelectric effect exhibited by the material, enabling higher sensitivity and improved performance in various applications. Commonly used piezoelectric fillers include piezoelectric ceramics such as lead zirconate titanate (PZT), barium titanate ($BaTiO_3$, BT), and zinc oxide (ZnO) particles. These fillers possess intrinsic piezoelectric properties, and when combined with the piezo polymer matrix, they enhance the overall piezoelectric response of the material. The addition of piezoelectric fillers helps to improve the sensitivity, signal-to-noise ratio, and energy conversion efficiency of the piezo polymer. BT nanoparticles have been used to enhance the piezoelectric property of polymers because of their high piezoelectric coefficient, and they are lead-free and low cost. β-phase content reaches a maximum value of 91.1% for nanocomposite fiber with 0.15 wt.% graphene and 15 wt.% BT (figure 5.4(e)) [37]. The utilization of conductive fillers like CNTs, graphene, and polyaniline (PANI) facilitates the creation of an interconnected network that enhances electrical conductivity. Consequently, this network leads to an enhanced efficiency in the conversion of mechanical energy in piezoelectric materials. Nevertheless, the practical application of CNTs and graphene is constrained due to their tendency to aggregate within the PVDF nanofibers, while

Table 5.1. Comparison of β-phase enhancement of PVDF-based nanofiber through adding fillers.

Filler type	Concentration	β-Phase enhancement	References
Silver nanoparticles (Ag-NPs)	0.4%–0.6%	8%	[39]
Graphene oxide (GO) + and halloysite (Hal)	0.05–3.2%	49%	[40]
Graphene-ZnO nanocomposite (G-ZnO)	1 wt.%	22%	[41]
rGO	2 wt.%	6%	[33]
2D nanoclay	15%	20%	[42]
MXene	0.15 wt.%	27%	[43]
MWCNTs	5 wt.%	290%	[44]
$Eu(TTA)_3(TPPO)_2$ and $FeCl_3 \cdot 6H_2O$	—	384%	[45]
$Mg_3Si_4O_{10}(OH)_2$ (Talc)	0.50 wt.%	48%	[46]
LiCl	0.001 33 wt.%	11%	[47]
$BiCl_3$	7 wt.%	13%	[48]
$BiCl_3$/ZnO	2 wt.%	21%	[49]
$ZnSnO_3$	3 wt.%	7%	[50]
Talc nanosheets	0.5 wt.%	48%	[51]
MoS_2 /TiO_2	—	126%	[52]
TiO_2–Fe_3O_4–GO	2 wt.%	80%	[53]
$CsPbBr3\%Ti_3C_2Tx$	0.4%/0.6%	20%	[54]

their use is also accompanied by high cost. Graphitic carbon nitride (g-C_3N_4) nanosheets are two-dimensional materials with excellent specific surface area, electron mobility, and low cost that have been explored in various applications. The dipole–dipole interaction between the oxygen atoms of graphitic carbon nitride nanosheets (g-C_3N_4) with CH_2 groups of PVDF further enhances the β-phase of PVDF which reached ~97%. Also, the –CF2 group of PVDF and (=N–H)$^+$ group of polyaniline (PANI) nanorods stimulate the dipole–dipole interaction (figure 5.4(f)) [38]. The choice of fillers depends on the specific requirements of the application and the desired performance characteristics of the piezo polymer. The concentration and dispersion of the fillers within the polymer matrix also play a crucial role in determining the overall performance of the composite material. Table 5.1 presents some fillers that are used to enhance the formation of the β-phase in PVDF nanofibers.

5.3 Quantification and determination of β-phase in PVDF

FTIR, XRD and DSC are the most common analysis techniques that prove the formation of the oriented β-phase within PVDF materials [55]. The XRD and FTIR can implement the qualitative identification and quantitative analysis of the β-phase. DSC is a thermoanalytical technique that has been complementary to the other identification techniques to identify the crystal phase of PVDF [56].

5.3.1 Fourier transform infrared (FTIR)

FTIR is crucial for analyzing PVDF's phase change, as it effectively detects and monitors crystalline structure variations [57]. FTIR can be used to characterize both the dipole orientation and crystallographic structure of nanofibers based on the sensitivity of CF_2 orientation changes [58]. FTIR analysis is an analysis technique that provides information about the chemical bonding or molecular structure of materials, whether organic or inorganic. It is used to identify unknown materials and material phases present in a specimen. Infrared (IR) spectroscopy is widely used to distinguish the principal PVDF phases, i.e. α, β and γ. Each crystalline phase exhibits distinct peaks, allowing for their differentiation. A closer examination reveals specific bands associated with the α polymorph, which can be observed at approximately 410, 489, 532, 614, 763, 795, 854, 975, 1149, 1209, 1383, and 1423 cm^{-1}. Conversely, the β-phase is characterized by exclusive peaks around 445, 473, and 1275 cm^{-1}. Lastly, the γ-phase manifests itself through bands located at approximately 431, 482, 811, 833, and 1234 cm^{-1} [59]. The F–C–F bending vibration is responsible for the absorption peaks observed at 480 and 507 cm^{-1}. The bending and wagging vibrations of CF_2 groups, as well as the rocking vibration in the PVDF chain, are indicated by the peaks at 614 and 764 cm^{-1}. The characteristic absorption peak at 838 cm^{-1} can be attributed to the H–C–H rocking mode and the symmetrical stretching F–C–F mode. The strong peak at 874 cm^{-1} is assigned to the CH_2 and CF_2 groups, which result from the CH_2 rocking and CF_2 stretching. The band observed at 1070 cm^{-1} is caused by the antisymmetric stretching of C–C, while the peaks at 1167 and 1230 cm^{-1} are associated with H–C–H wagging and rocking, respectively. Finally, the peak occurring at 1400 cm^{-1} corresponds to the C–F stretching vibration [60]. Besides these characteristic peaks, there are certain peaks that can arise from a combination of two out of the three mentioned phases. For instance, the presence of β and/or γ crystals can give rise to a band around 840 cm^{-1}. Therefore, this particular peak is not exclusive to a single phase but rather indicates the presence of either β or γ, or both. Moreover, infrared spectroscopy proves to be a valuable tool for quantifying the proportion of these electroactive polymorphs relative to the overall crystalline phase. By analyzing the infrared spectra, it becomes possible to determine the amount of α-, β-, and γ-phases present in a sample, providing quantitative insights into the distribution of these crystalline components. To determine the percentage of β-crystalline phase in each sample, the absorption peak of α- and β-phases, respectively, at wavelengths 840 and 763 cm^{-1} are evaluated. Percentage of β-crystalline phase is calculated using equation (5.1), where A_α and A_β are the absorption at 764 and 840 cm^{-1} and X_α and X_β are crystalline mass fraction of the α- and β-phases [61].

$$f(\beta) = \frac{X_\beta}{X_\alpha + X_\beta} \times 100 = \frac{A_\beta}{A_\beta + 1.26 A_\alpha} \times 100 \tag{5.1}$$

Figure 5.5 shows FTIR results for various content of ZnO in PVDF composite. The results shows that by increasing content of ZnO till 10 wt.%, the β-phase formation increased [62].

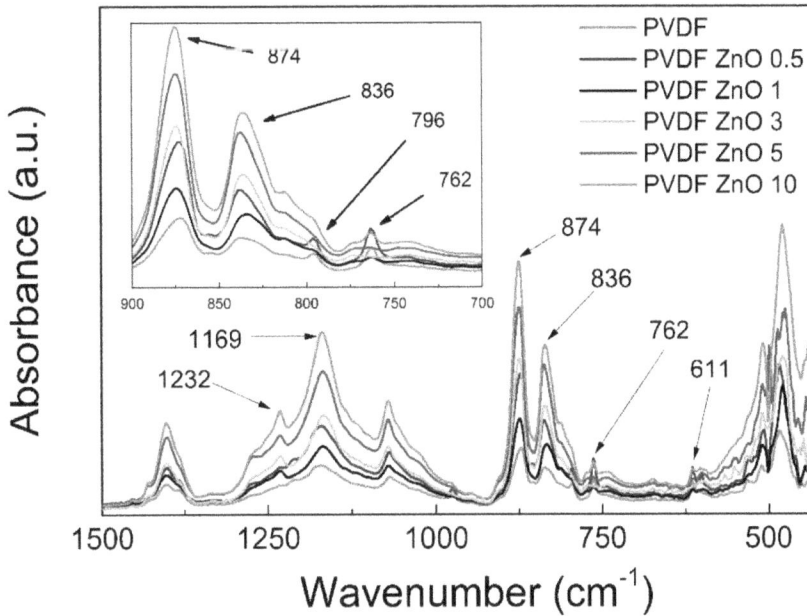

Figure 5.5. FTIR spectra for PVDF and the PVDF/ZnO composite in different concentrations. Reproduced from [62] CC BY 4.0.

Figure 5.6 depicts a clear correlation between the formation of crystallites and the applied voltage, highlighting the relationship between phase contents and voltage. Specifically, as the applied voltage is increased, there is an observable increase in the fraction of crystalline structure, while simultaneously decreasing the percentage of α-phase present within the crystallites [63].

5.3.2 X-ray diffraction (XRD)

X-ray diffraction (XRD) analysis is a common technique used to study the crystalline structure and phase identification of polymers. X-ray crystallography is a fundamental scientific technique used to investigate the intricate three-dimensional structures of crystals. By employing x-ray diffraction techniques, this discipline enables researchers to gain profound understanding of the material and molecular architecture of substances. When an intense beam of x-rays interacts with a crystalline lattice oriented in a particular manner, the x-rays undergo scattering phenomena that are distinctly influenced by the atomic arrangement within the crystal lattice. This scattering pattern provides crucial information about the positions, orientations, and interactions of the atoms, allowing for the determination of precise atomic structures and the elucidation of key properties and behaviours of the material under study [64]. In XRD analysis of electrospun PVDF web, distinct patterns of nonpolar α-phase peaks and reflections manifest at angles of 18.4° (020), 19.9° (110), and 26.6° (021). Meanwhile, the γ-phase exhibits peaks and reflections at 18.5° (020) and 20.2° (110). Notably, the crystalline β-phase displays a unique

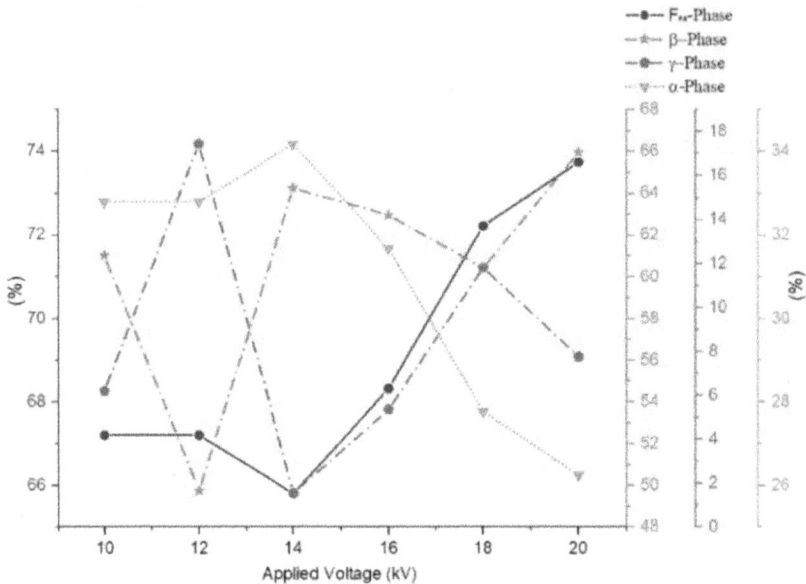

Figure 5.6. Relationship between the applied voltage during electrospinning and phase content in PVDF nanofiber [63] John Wiley & Sons. Copyright (2022) Society of Industrial Chemistry.

feature whereby peaks and reflections are observed at 20.6° (200) and (110), presenting as a composite peak composed of two individual peaks. This intriguing phenomenon suggests the presence of a zigzag conformation within the crystalline structure of the β-phase [65]. The total crystallinity (C_T) and β-phase crystallinity (C_β) of PVDF nanofiber mats produced by the electrospinning can be calculated by the peak integration method using the below equations, in which A_α, A_β, and A_γ indicate the total integral area from α-, β-, and γ-crystalline phases peaks, respectively [66]:

$$c_T(\%) = \left(\frac{A_{Cr}}{A_{Cr} + A_{amr}} \right) \times 100 \tag{5.2}$$

$$c_\beta(\%) = \left(\frac{A_\beta}{A_\alpha + A_\beta + A_\gamma} \right) \times 100 \tag{5.3}$$

A recent study elucidated the process by which the morphology of a PVDF hollow fiber affects its piezoelectric performance. The fabrication of PVDF hollow fibers involved coaxial electrospinning. In order to create the hollow structure, polyvinylpyrrolidone (PVP) with varying concentrations was utilized as the inner solution during the electrospinning process. Analysis using x-ray techniques revealed that the PVDF solid fiber displayed a weak peak corresponding to the nonpolar α-phase at an angle of $2\theta = 18.5°$, as well as a prominent peak corresponding to the polar β-phase at $2\theta = 20.6°$. These peaks were a result of mechanical stretching and polarization occurring during the electrospinning process. In contrast, the x-ray diffraction patterns of PVDF hollow fibers exclusively displayed the characteristic crystal structure of the β-phase. This observation can be primarily attributed to the more pronounced stretching of PVDF during coaxial electrospinning, which induces the formation of the β-phase. Comparisons between PVDF membranes prepared via spin coating and electrospinning, using the same solution, are depicted in figures 5.7(a) and (b). The spin-coated PVDF membrane exhibited an amorphous phase and diffraction peaks at angles of 17.5°, 18.5°, 19° (corresponding to the α-phase), and 20.5° (representing the β-phase). It is suggested that electrical poling influences the arrangement of phase components, with electrospinning playing a crucial role in facilitating inter-chain registration. Consequently, electrospinning not only significantly enhances β-phase formation but also improves the alignment of polarization within the resulting PVDF nanofibers. Figures 5.7(c) and (d) further corroborate the

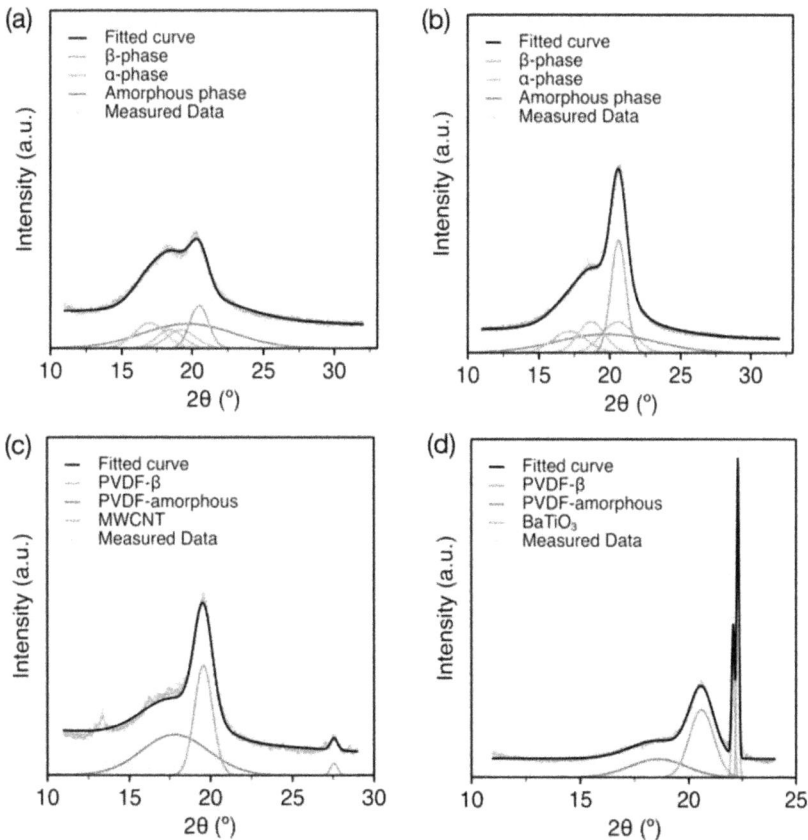

Figure 5.7. XRD patterns and corresponding characteristic peaks fitting showing improved β-phase crystalline content of nanofiber membranes prepared by electrospinning compared to spin-coated film from the same solution: (a) spin-coated PVDF, (b) PVDF nanofiber, Improved β-phase content ratio through electrospinning and additives in nanofibers: (c) CNT/PVDF and (d) BTO@PVDF. Reproduced from [67], with permission from Springer Nature.

positive effects of incorporating CNTs and barium titanate (BTO) particles on the formation of the β-phase in PVDF, consequently boosting the electrical properties of the device. CNTs possess semiconductive properties that depend on their tube diameter and chirality. During the electrospinning process of PVDF solutions containing dispersed CNTs, the CNTs act as nucleating agents, facilitating the elongation of the jet and promoting electrostatic interactions between the CF_2 groups of PVDF and the functional groups of CNTs. This interaction leads to a higher degree of crystallization. Similarly, BTO particles provide additional nucleation sites, facilitating the formation of a greater amount of β-phase in the PVDF structure [67].

5.3.3 Differential scanning calorimetry (DSC)

DSC is a valuable method employed for investigating the thermal behavior of polymers, including pivotal parameters like the melting point temperature (T_m) and the glass transition temperature (T_g). When exploring the thermal transitions in PVDF, DSC emerges as a significant tool. Its primary applications in PVDF studies encompass identifying the melting point, quantifying the sample's crystallinity percentage, monitoring solvent evaporation, and assessing the effects of annealing processes [6]. Depending on the crystalline phase of PVDF, different melting peaks appear on the DSC thermogram. The DSC characteristics not only depend on the crystalline phase, but also depend on the characteristics of the morphology such as defects and crystalline size, as well as the processing history of the sample, because of this reason the existing literature does not define a temperature range for the different phases [68]. The measurement of crystallinity degree (X_c) serves as an indicator of the alignment of nanofibers, playing a crucial role in determining the piezoelectric response. One of the studies had shown that a preferential poling in the electrospinning process can be achieved for the nanofibers formed without the need of any additional poling step needed for PVDF films with the use of a drum collector. The rotational motion of the drum introduces an additional mechanical stretch, resulting in the preferential alignment of CF_2 dipoles during the electro-spinning process [69]. The degree of crystallinity (X_c) of the samples was calculated from the DSC thermograms according to:

$$X_c(\%) = \frac{\Delta H_m}{\Delta H_0} \times 100 \qquad (5.4)$$

where ΔH_m is the enthalpy of melting for the nanofiber sample and ΔH_0 is the melting enthalpy for 100% crystalline PVDF which is considered as 104.7 J g^{-1} [70]. Distinguishing between the α- and β-phases using DSC studies is not recommended due to the proximity of their respective peaks within the melting endotherm [71]. The DSC heating scans analysis showed that the higher melting temperature is related to high content of α-phase of PVDF Pellet or fibers spun at low humidity. Also, for these samples, the melting peak is much narrower than for fibers with higher β-phase content, that is, electrospun at high humidity. The higher β-phase content provides higher melting heat (ΔH_m) and lower melting temperature (T_m). These results seem very reasonable as the β-phase is considered as the low-temperature phase of PVDF.

Moreover, it is generally accepted that the β-phase, because of strong polar interactions between parallel dipoles existing in the TTT conformation, should have much higher melting enthalpy than the α-phase [72]. The rise in melting temperature observed with an increasing content of multi-walled carbon nanotubes (MWCNTs) can be attributed to a corresponding increase in the β-phase content. A comparison between the pristine PVDF nanofiber and those supplemented with MWCNTs reveals a DSC curve where the peak melting temperature shifts towards higher temperatures (figure 5.8). Additionally, the presence of MWCNT leads to a decrease in the melting enthalpy of the melting point features. These observations suggest that the crystalline structure of PVDF fibers with added MWCNTs exhibits a greater prevalence of the β-phase. It is widely recognized that the melting temperature of the β-phase surpasses that of the α-phase [73].

Research shows that adding 0.50 wt.% talc increases polar β-phase in the PVDF matrix up to 89.6% which augmenting its piezoelectric response. The DSC plots presented in figures 5.9(a) and (b) illustrate the characteristics of electrospun PVDF and talc/PVDF nanocomposite fabrics. When talc was incorporated into the PVDF matrix, both the crystallization temperature (T_c) and melting temperature (T_m) of the nanofabrics increased. This increase in T_c and T_m can be attributed to the presence of talc layered crystals, which provide nucleation sites that facilitate the crystallization process. Moreover, the presence of talc layered crystals hindered the mobility of PVDF chains, resulting in a downward shift of T_c and T_m values when the nanofibers contained talc concentrations above 0.5 wt.% [46].

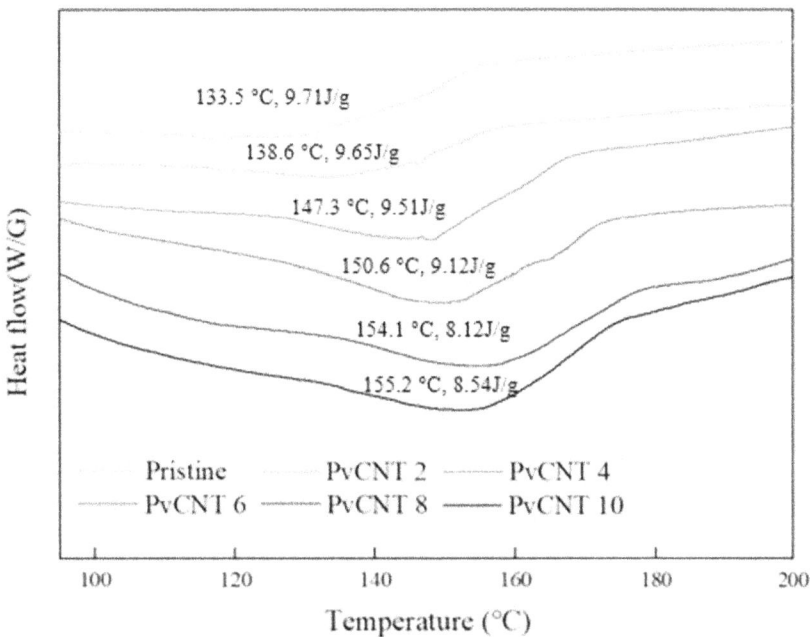

Figure 5.8. DSC curves of the change in the PVDF crystal phase with the MWCNT content 0, 0.002, 0.004, 0.006, 0.008 and 0.01 wt.%. Reproduced from [73] CC By 4.0.

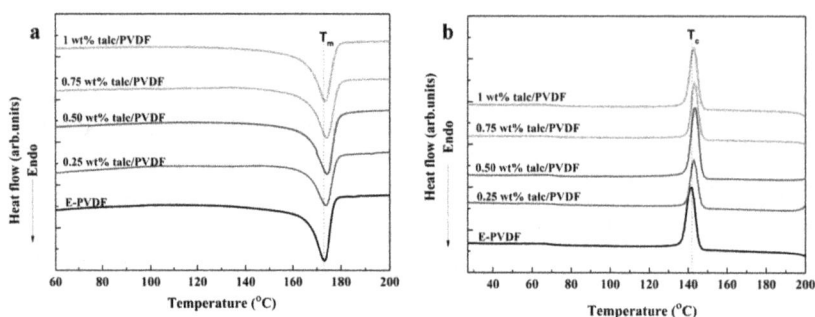

Figure 5.9. DSC traces of E-PVDF and talc/PVDF composite webs: (a) first cooling cycle and (b) second heating cycle. Reproduced from [46] with permission from the Royal Society of Chemistry.

5.4 Conclusion

In conclusion, the addition of fillers to PVDF nanofibers has been shown to enhance the formation of the β-phase in the polymer. The β-phase is desirable in PVDF due to its high piezoelectric and pyroelectric properties, making it suitable for various applications such as sensors, actuators, and energy harvesting devices. By incorporating fillers into the PVDF matrix, the crystallinity and alignment of the beta phase can be improved, leading to enhanced performance. The choice of filler material plays a crucial role in determining the effectiveness of β-phase formation. Fillers with high aspect ratios, such as CNTs or graphene, have been found to be particularly effective in promoting the β-phase formation. These fillers not only act as nucleating agents but also provide structural templates for the alignment of PVDF chains, resulting in enhanced crystallization and β-phase content. Although significant progress has been made in understanding the effects of fillers on β-phase formation in PVDF nanofibers, there are still several areas that warrant further research. Some potential directions for future work include:

1. Optimization of filler concentration: The optimal concentration of fillers may vary depending on the specific application and the type of filler used. Further studies are needed to determine the ideal filler loading that maximizes β-phase formation while maintaining the mechanical and electrical properties of the composite.

2. Characterization techniques: Advanced characterization techniques, such as x-ray diffraction (XRD), Fourier-transform infrared spectroscopy (FTIR), and scanning electron microscopy (SEM), can provide detailed insights into the crystalline structure, alignment, and distribution of the β-phase. More in-depth characterization studies are required to better understand the mechanisms behind the enhancement of beta phase formation by fillers.

3. Influence of processing parameters: The processing parameters, such as electrospinning conditions and post-treatment methods, can significantly affect the morphology and crystallinity of PVDF nanofibers. Investigating the influence of these parameters on β-phase formation in the presence of

fillers would help optimize the fabrication process and improve the overall performance of the composite.

4. Performance evaluation: Further research is needed to evaluate the functional properties of PVDF nanofibers with fillers, including their piezoelectric, pyroelectric, mechanical, and thermal properties. Performance testing and comparison with existing materials will help assess the effectiveness of filler-enhanced PVDF nanofibers for specific applications.

5. Scale-up and industrial applicability: Scaling up the fabrication process of filler-enhanced PVDF nanofibers and assessing their feasibility for large-scale production is an important step towards practical applications. Investigations into the scalability, cost-effectiveness, and potential challenges of industrial implementation will be valuable for commercialization.

By addressing these areas of future work, researchers can continue to advance the understanding and development of filler-enhanced PVDF nanofibers with enhanced β-phase formation, paving the way for their widespread use in various technological applications.

References

[1] Aydemir D *et al* 2023 Electrospinning of PVDF nanofibers incorporated cellulose nanocrystals with improved properties *Cellulose* **30** 885–98

[2] Gebrekrstos A, Muzata T S and Ray S S 2022 Nanoparticle-enhanced β-phase formation in electroactive pvdf composites: a review of systems for applications in energy harvesting, emi shielding, and membrane technology *ACS Appl. Nano Mater.* **5** 7632–51

[3] Porwal C *et al* 2023 Bismuth vanadate-reduced graphene oxide-polyvinylidene fluoride electrospun composite membrane for piezo-photocatalysis *Nano-Struct. Nano-Obj.* **34** 100969

[4] Panicker S S, Rajeev S P and Thomas V 2023 Impact of PVDF and its copolymer-based nanocomposites for flexible and wearable energy harvesters *Nano-Struct. Nano-Obj.* **34** 100949

[5] Mokhtari F 2022 *Self-Powered Smart Fabrics for Wearable Technologies* (Springer Nature)

[6] Mokhtari F, Latifi M and Shamshirsaz M 2016 Electrospinning/electrospray of polyvinylidene fluoride (PVDF): piezoelectric nanofibers *J. Text. Inst.* **107** 1037–55

[7] He Z *et al* 2021 Electrospun PVDF nanofibers for piezoelectric applications: a review of the influence of electrospinning parameters on the β phase and crystallinity enhancement *Polymers* **13** 174

[8] Chakhchaoui N *et al* 2021 Improvement of the electroactive β-phase nucleation and piezoelectric properties of PVDF-HFP thin films influenced by TiO_2 nanoparticles *Mater. Today Proc.* **39** 1148–52

[9] Bicy K *et al* 2022 Lithium-ion battery separators based on electrospun PVDF: a review *Surf. Interfaces* **31** 101977

[10] Pusty M and Shirage P M 2022 Insights and perspectives on graphene-PVDF based nanocomposite materials for harvesting mechanical energy *J. Alloys Compd.* **904** 164060

[11] Sasmal A *et al* 2023 Two-dimensional metal-organic framework incorporated highly polar PVDF for dielectric energy storage and mechanical energy harvesting *Nanomaterials* **13** 1098

[12] Shepelin N A *et al* 2019 New developments in composites, copolymer technologies and processing techniques for flexible fluoropolymer piezoelectric generators for efficient energy harvesting *Energy Environ. Sci.* **12** 1143–76

[13] Liu Z *et al* 2022 Fabrication of β-phase-enriched PVDF sheets for self-powered piezoelectric sensing *ACS Appl. Mater. Interfaces* **14** 11854–63

[14] Li H and Lim S 2023 Self-poled and transparent polyvinylidene fluoride-co-hexafluoropropylene-based piezoelectric devices for printable and flexible electronics *Nanoscale* **15** 4581–90

[15] Yu R *et al* 2023 Flexible and ultrathin waterproof conductive cellular membranes based on conformally gold-coated PVDF nanofibers and their potential as gas diffusion electrode *Mater. Des.* **225** 111441

[16] Ravisankar M, Pramod K and Gangineni R 2023 Effect of the top electrode on local piezoelectric and the ferroelectric response of PVDF thin films in PVDF/Au/Si and Ag/PVDF/Au/Si multilayers *Appl. Phys. A* **129** 146

[17] Prakash O, Tiwari S and Maiti P 2022 Fluoropolymers and their nanohybrids as energy materials: application to fuel cells and energy harvesting *ACS Omega* **7** 34718–40

[18] Mokhtari F, Foroughi J and Latifi M 2019 *Energy Harvesting Properties of Electrospun Nanofibers* (Bristol: IOP Publishing) pp 5-1–5-28

[19] Shao H *et al* 2015 Effect of electrospinning parameters and polymer concentrations on mechanical-to-electrical energy conversion of randomly-oriented electrospun poly (vinylidene fluoride) nanofiber mats *RSC Adv.* **5** 14345–50

[20] Griffin A *et al* 2022 Scalable methods for directional assembly of fillers in polymer composites: creating pathways for improving material properties *Polym. Compos.* **43** 5747–66

[21] Zaarour B, Zhu L and Jin X 2019 Controlling the surface structure, mechanical properties, crystallinity, and piezoelectric properties of electrospun PVDF nanofibers by maneuvering molecular weight *Soft Mater.* **17** 181–9

[22] Zhong G *et al* 2011 Understanding polymorphism formation in electrospun fibers of immiscible Poly(vinylidene fluoride) blends *Polymer* **52** 2228–37

[23] Gee S, Johnson B and Smith A 2018 Optimizing electrospinning parameters for piezoelectric PVDF nanofiber membranes *J. Membr. Sci.* **563** 804–12

[24] Jiyong H *et al* 2017 Mixed effect of main electrospinning parameters on the β-phase crystallinity of electrospun PVDF nanofibers *Smart Mater. Struct.* **26** 085019

[25] Zaarour B *et al* 2018 Controlling the secondary surface morphology of electrospun pvdf nanofibers by regulating the solvent and relative humidity *Nanoscale Res. Lett.* **13** 285

[26] Slack J J *et al* 2020 Impact of polyvinylidene fluoride on nanofiber cathode structure and durability in proton exchange membrane fuel cells *J. Electrochem. Soc.* **167** 054517

[27] Lee S H *et al* 2020 Fabrication and characterization of piezoelectric composite nanofibers based on poly (vinylidene fluoride-co-hexafluoropropylene) and barium titanate nanoparticle *Fibers Polym.* **21** 473–9

[28] Mokhtari F, Shamshirsaz M and Latifi M 2016 Investigation of β phase formation in piezoelectric response of electrospun polyvinylidene fluoride nanofibers: LiCl additive and increasing fibers tension *Polym. Eng. Sci.* **56** 61–70

[29] Mokhtari F *et al* 2016 Advances in electrospinning: the production and application of nanofibres and nanofibrous structures *Text. Prog.* **48** 119–219

[30] Mokhtari F *et al* 2017 Modeling of electrospun PVDF/LiCl nanogenerator by the energy approach method: determining piezoelectric constant *J. Text. Inst.* **108** 1917–25

[31] Feng Z *et al* 2023 Piezoelectric effect polyvinylidene fluoride (PVDF): from energy harvester to smart skin and electronic textiles *Adv. Mater. Technol.* **8** 2300021

[32] Chen G *et al* 2022 Electrospun flexible PVDF/GO piezoelectric pressure sensor for human joint monitoring *Diam. Relat. Mater.* **129** 109358

[33] Yang J *et al* 2021 Piezoelectric nanogenerators based on graphene oxide/PVDF electrospun nanofiber with enhanced performances by in-situ reduction *Mater. Today Commun.* **26** 101629

[34] Wang S *et al* 2021 Boosting piezoelectric response of PVDF-TrFE via MXene for self-powered linear pressure sensor *Compos. Sci. Technol.* **202** 108600

[35] Mishra R *et al* 2021 Polar β-phase PVdF-HFP-based freestanding and flexible gel polymer electrolyte for better cycling stability in a Na battery *Energy Fuels* **35** 15153–65

[36] Bairagi S and Ali S W 2020 A hybrid piezoelectric nanogenerator comprising of KNN/ZnO nanorods incorporated PVDF electrospun nanocomposite webs *Int. J. Energy Res.* **44** 5545–63

[37] Shi K *et al* 2018 Synergistic effect of graphene nanosheet and $BaTiO_3$ nanoparticles on performance enhancement of electrospun PVDF nanofiber mat for flexible piezoelectric nanogenerators *Nano Energy* **52** 153–62

[38] Khalifa M and Anandhan S 2019 PVDF nanofibers with embedded polyaniline–graphitic carbon nitride nanosheet composites for piezoelectric energy conversion *ACS Appl. Nano Mater.* **2** 7328–39

[39] Issa A A *et al* 2017 Physico-mechanical, dielectric, and piezoelectric properties of pvdf electrospun mats containing silver nanoparticles *C* **3** 30

[40] Abbasipour M *et al* 2017 The piezoelectric response of electrospun PVDF nanofibers with graphene oxide, graphene, and halloysite nanofillers: a comparative study *J. Mater. Sci., Mater. Electron.* **28** 15942–52

[41] Hasanzadeh M, Ghahhari M R and Bidoki S M 2021 Enhanced piezoelectric performance of PVDF-based electrospun nanofibers by utilizing *in situ* synthesized graphene-ZnO nano-composites *J. Mater. Sci., Mater. Electron.* **32** 15789–800

[42] Tiwari S *et al* 2019 Enhanced piezoelectric response in nanoclay induced electrospun PVDF nanofibers for energy harvesting *Energy* **171** 485–92

[43] Liu X *et al* 2023 $BaTiO_3$/MXene/PVDF-TrFE composite films via an electrospinning method for flexible piezoelectric pressure sensors *J. Mater. Chem. C* **11** 4614–22

[44] Lin B *et al* 2023 Preparation of MWCNTs/PVDF composites with high-content β form crystalline of PVDF and enhanced dielectric constant by electrospinning-hot pressing method *Diam. Relat. Mater.* **131** 109556

[45] Fu G *et al* 2022 Enhanced piezoelectric performance of rare earth complex-doped sandwich-structured electrospun P(VDF-HFP) multifunctional composite nanofiber membranes *J. Mater. Sci., Mater. Electron.* **33** 22183–95

[46] Shetty S, Mahendran A and Anandhan S 2020 Development of a new flexible nanogenerator from electrospun nanofabric based on PVDF/talc nanosheet composites *Soft Matter* **16** 5679–88

[47] Mokhtari F *et al* 2017 Comparative evaluation of piezoelectric response of electrospun PVDF (polyvinilydine fluoride) nanofiber with various additives for energy scavenging application *J. Textile Inst.* **108** 906–14

[48] Chen C *et al* 2020 Enhanced piezoelectric performance of $BiCl_3$/PVDF nanofibers-based nanogenerators *Compos. Sci. Technol.* **192** 108100

[49] Zhang D *et al* 2022 Enhanced piezoelectric performance of PVDF/BiCl₃/ZnO nanofiber-based piezoelectric nanogenerator *Eur. Polym. J.* **166** 110956

[50] Kavarthapu V S *et al* 2023 Electrospun ZnSnO₃/PVDF-HFP nanofibrous triboelectric films for efficient mechanical energy harvesting *Adv. Fiber Mater.* **5** 1685–98

[51] Shetty S *et al* 2022 Evaluation of piezoelectric behavior and biocompatibility of poly (vinylidene fluoride) ultrafine fibers with incorporated talc nanosheets *J. Appl. Polym. Sci.* **139** e52631

[52] Bhatt A *et al* 2023 Enhanced piezoelectric response using TiO₂/MoS₂ heterostructure nanofillers in PVDF based nanogenerators *J. Alloys Compd.* **960** 170664

[53] Haji Abdolrasouli M, Abdollahi H and Samadi A 2022 PVDF nanofibers containing GO-supported TiO₂–Fe₃O₄ nanoparticle-nanosheets: piezoelectric and electromagnetic sensitivity *J. Mater. Sci., Mater. Electron.* **33** 5970–82

[54] Xue Y *et al* 2023 Heterojunction engineering enhanced self-polarization of PVDF/CsPbBr₃/ Ti₃C2Tₓ composite fiber for ultra-high voltage piezoelectric nanogenerator *Adv. Sci.* **10** 2300650

[55] Andrew J S and Clarke D R 2008 Effect of electrospinning on the ferroelectric phase content of polyvinylidene difluoride fibers *Langmuir* **24** 670–2

[56] Ruan L *et al* 2018 Properties and applications of the β phase poly(vinylidene fluoride) *Polymers* **10** 228

[57] Huang Z-X *et al* 2020 β-Phase formation of polyvinylidene fluoride via hot pressing under cyclic pulsating pressure *Macromolecules* **53** 8494–501

[58] Sánchez J A G *et al* 2014 *29th Symp. on Microelectronics Technology and Devices (SBMicro)* pp 1–3

[59] Brunengo E *et al* 2020 Double-step moulding: an effective method to induce the formation of β-phase in PVDF *Polymer* **193** 122345

[60] Li J *et al* 2022 Highly sensitive, flexible and wearable piezoelectric motion sensor based on PT promoted β-phase PVDF *Sens. Actuators, A* **337** 113415

[61] Meng X *et al* 2022 Microfluidic fabrication of β-phase enriched poly(vinylidene fluoride) microfibers toward flexible piezoelectric sensor *J. Polym. Sci.* **60** 1718–26

[62] Chamakh M M *et al* 2020 Vibration sensing systems based on poly(vinylidene fluoride) and microwave-assisted synthesized ZnO star-like particles with controllable structural and physical properties *Nanomaterials* **10** 2345

[63] Nugraha A S *et al* 2022 Effects of applied voltage on the morphology and phases of electrospun poly(vinylidene difluoride) nanofibers *Polym. Int.* **71** 1176–83

[64] Ge X *et al* 2022 Preparation of polyvinylidene fluoride–gold nanoparticles electrospinning nanofiber membranes *Bioengineering* **9** 130

[65] Oflaz K and Özaytekin İ 2022 Analysis of electrospinning and additive effect on β phase content of electrospun PVDF nanofiber mats for piezoelectric energy harvester nano-generators *Smart Mater. Struct.* **31** 105022

[66] Jin L *et al* 2020 Enhancement of β-phase crystal content of poly(vinylidene fluoride) nanofiber web by graphene and electrospinning parameters *Chin. J. Polym. Sci.* **38** 1239–47

[67] Li J *et al* 2023 A self-powered piezoelectric nanofibrous membrane as wearable tactile sensor for human body motion monitoring and recognition *Adv. Fiber Mater.* **5** 1417–30

[68] Martins P, Lopes A C and Lanceros-Mendez S 2014 Electroactive phases of poly(vinylidene fluoride): determination, processing and applications *Prog. Polym. Sci.* **39** 683–706

[69] Singh R K, Lye S W and Miao J 2021 Holistic investigation of the electrospinning parameters for high percentage of β-phase in PVDF nanofibers *Polymer* **214** 123366

[70] Mokhtari F *et al* 2021 Dynamic mechanical and creep behaviour of meltspun PVDF nanocomposite fibers *Nanomaterials* **11** 2153

[71] Mohamadi S and Sharifi-Sanjani N 2016 Crystallization of PVDF in graphene-filled electrospun PVDF/PMMA nanofibers processed at three different conditions *Fibers Polym.* **17** 582–92

[72] Szewczyk P K *et al* 2020 Enhanced piezoelectricity of electrospun polyvinylidene fluoride fibers for energy harvesting *ACS Appl. Mater. Interfaces* **12** 13575–83

[73] Eun J H *et al* 2021 Effect of MWCNT content on the mechanical and piezoelectric properties of PVDF nanofibers *Mater. Des.* **206** 109785

IOP Publishing

Energy Harvesting Properties of Electrospun Nanofibers
(Second Edition)

Jian Fang and Tong Lin

Chapter 6

Acoustoelectric energy conversion of nanofibrous materials

Chenhong Lang, Jingye Jin, Jian Fang and Tong Lin

6.1 Introduction

Sound is a wave that is formed and propagated due to the vibration of an object in a medium, and it exists everywhere around us. In general, sound is specifically referred to those acoustic vibrations in air. It is well known that our human ears can only hear a limited section of sounds, in the frequency range of 20–20 000 Hz. Beyond this range, sound is inaudible to us, and sounds below 20 Hz are termed as infrasound while those above 20 kHz are ultrasound. Within the hearable range, sounds are further classified according to the frequency, into low-frequency (<200 Hz), mi-frequency (200–6000 Hz) and high-frequency (6000–20 000 Hz) sound.

Sound pressure level (SPL), the amplitude of acoustic vibrations, is an important parameter to indicate sound intensity and loudness. Human ears are sensitive to sound intensity, and when SPL is greater than 90 dB, the sounds are usually too loud to be tolerable, and therefore considered as noise. It has been widely accepted now that noise is a type of environmental pollution causing health issues, and considerable attention has been devoted to monitor and control harmful noise in working environments and living spaces.

Traditional acoustic technologies mainly involved sound recording and reproduction. They have been gradually broadened to sound detection, cancellation, energy conversion, and harvesting, including various applications in information communication, entertainment, environment protection [1], industrial manufacturing [2], healthcare [3], medical settings [4], and defense [5]. Due to the considerable amount of energy carried by acoustic vibration of sounds, there have been growing efforts in recent years to convert acoustic energy into electric signals through energy harvesting and conversion techniques. The main applications of acoustoelectric energy conversion can be divided into two main categories: acoustic sensors functioning to detect

doi:10.1088/978-0-7503-5487-5ch6

sound sources, and acoustic energy generators that convert sounds into electrical power high enough to drive other microelectronic devices. For all these proposed applications, the most critical component in achieving efficient acoustoelectric energy conversion is acoustic transducer. Acoustic transducers can convert acoustic vibrations into electric signals with different energy conversion mechanisms, such as electromagnetic induction, the piezoelectric effect, and the triboelectric effect. Electromagnetic induction-based transducers have been conventionally used in many applications, such as microphones and speakers. With the growing needs of miniatured device dimensions and flexible electronics in recent years, more research efforts have been devoted to the development of piezoelectric and triboelectric acoustic transducers and energy conversion systems. These recent developments have benefited from promising advances in functional piezoelectric and triboelectric materials, such as nanofibers. Nanofibrous materials, including electrospun polymeric nanofibers and many other inorganic nanowires, can be effectively prepared with different sizes, morphologies and fibrous structures. They are lighter than their microsized counterparts, and much more sensitive to acoustic vibrations, and therefore can achieve much higher acoustoelectric energy conversion efficiency.

In this chapter, after introducing basic principles of acoustic transducing, common materials, and key evaluation parameters, we will summarize recent developments in using nanofibrous materials for acoustic energy conversion, particularly for sensing and energy harvesting applications.

6.2 Conventonal acoustic transducers

6.2.1 Basic principles

The basic function of an acoustic transducer is to convert sound vibrations to testable electric signals. This can be achieved by a few different principles (table 6.1), such as electromagnetic induction [6–10], piezo-resistance [11], piezo-capacitance [12], piezo-optics [13], the piezoelectric effect [14, 15], and the triboelectric effect [16–18]. For example, a diaphragm vibrates when it receives air vibrations caused by sound, inducing changes in resistance, capacitance, or light intensity, which are further converted into voltage changes or digital signals. Consequently, some acoustoelectric conversion processes need an external power to drive the conversion of diagram vibration into electric signals, as illustrated in figure 6.1(a). In some other cases, electric signals can be directly generated from sound vibrations through electromagnetic induction, piezoelectric effects, or triboelectric effects (figure 6.1(b)); therefore, no external power supply is required.

In an electromagnetic acoustoelectric energy conversion device, a magnet is normally attached to a flexible membrane fixed with an electrical coil. The membrane movement caused by acoustic wave induces electrical current and voltage within the coil. For piezoelectric effect, electric charges with opposite polarities are generated on piezoelectric material surfaces under a mechanical strain of the material. Piezoelectric membranes are typically used for acoustoelectric conversion, and the vibration of piezoelectric membrane leads to an alternating voltage signal. The triboelectric effect generates opposite electric charges on material surfaces when two materials with

Table 6.1. Comparison of conventional acoustic transducers.

Methods	Principle	Sensing element materials	Devices	External power	Sensor	Energy harvester	Application example
Piezoresistive [11]	Change of electrical resistance under acoustic vibration	Metallic plates, carbon granules	A vibration diaphragm, two metallic plates, separated by carbon granules, connected with external power source	√	√	×	Button microphone, repeater, etc
Piezo-capacitive [12]	Change of capacitance under acoustic vibration	Metallic plates, a dielectric medium	A vibration diaphragm, two metallic plates, separated by a dielectric medium, connected with external power source	√	√	×	DC-biased microphones, radio frequency (RF) or high frequency (HF) condenser microphones, electret microphone, etc
Piezo-optical [13]	Change of light intensity under acoustic vibration	Optical fiber	A laser source, optical fibers, a vibration and reflecting diaphragm, a photo detector, a transmission or recording	√	√	×	Industrial and surveillance acoustic monitoring, audio calibration and measurement, high-fidelity recording and law enforcement, etc

(Continued)

Table 6.1. (*Continued*)

Methods	Principle	Sensing element materials	Devices	Applications External power	Sensor	Energy harvester	Application example
Electromagnetic [6–10]	Electromagnetic induction under acoustic vibration	Diamagnetic, ferromagnetic and paramagnetic materials	A vibration diaphragm, a magnet, a fixed coil placed in a coil holder, a resonator (acoustic energy harvester)	×	√	√	Stage microphone; acoustic energy harvester, etc
Piezoelectric [14, 15]	Piezoelectric charge generation under acoustic vibration	Piezoelectric materials	A vibration diaphragm, two electrodes, piezoelectric materials, a resonator (acoustic energy harvester)	×	√	√	Contact microphone, amplify sound from acoustic musical instruments, and to record sound in challenging environments, such as underwater under high pressure; acoustic energy harvester, etc
Triboelectric [16–18]	Triboelectric charge generation under acoustic vibration	Positively charged materials: metal, such as Cu; negatively charged materials: fluorinated polymers, such as polytetrafluoro ethylene (PTFE) and PVDF.	A vibration diaphragm, two electrodes, triboelectric materials, a resonator (acoustic energy harvester)	×	√	√	Wearable and flexible electronics, military surveillance, jet engine noise reduction, low-cost implantable human ear, and wireless technology applications, etc

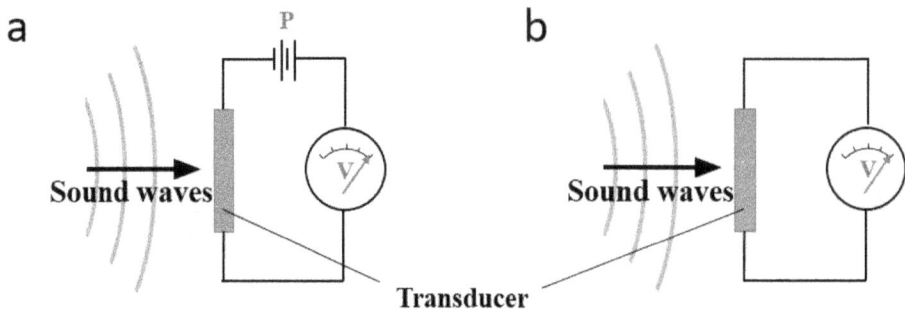

Figure 6.1. Acoustoelectric conversion (a) with and (b) without external power source.

different electric properties are brought together and then separated; an electric potential is therefore formed across these two surfaces. In comparison with electromagnetic systems that are normally with a bulky device structure, piezoelectric and triboelectric energy converters have a simpler device structure and are more dependent on material performance.

6.2.2 Materials

In this section, the material selection for the above-mentioned three types of acoustic transducers without an external power supply are briefly introduced.

6.2.2.1 Electromagnetic transducer
A permanent magnet and a fixed coil are the key components in an electromagnetic acoustoelectric conversion device. A neodymium iron boron magnet is usually used for its high residual magnetic flux density. Diamagnetic, paramagnetic, and ferromagnetic materials can be used for making the coil, and diamagnetic materials are preferred over others because they result in a very weak repletion with the magnet [19].

6.2.2.2 Piezoelectric transducer
Piezoelectric materials can convert acoustic energy to electrical signals by the piezoelectric effect. Common materials include piezoelectric ceramics (e.g. ZnO, PZT, AIN) and organic piezoelectric materials (e.g. PVDF, PVDF-HFP and P(VDF-TrFE)). Under the same strain level, piezoelectric ceramics can normally generate higher electric potentials than organic piezoelectric materials. However, much higher acoustic energy is required to activate ceramic materials. By contrast, piezoelectric polymers have the advantages of being lightweight, flexible, stretchable, and biocompatible, which make them much more sensitive to low-intensity acoustic vibrations for a much wider range of applications.

6.2.2.3 Triboelectric transducer
The most familiar example of triboelectric charge generation to us is by rubbing a glass rod with a piece of silk fabric. In theory, the contact and separation of any two

different insulating materials can generate triboelectric charges on material surfaces. For evaluating the charge generation capacity of a triboelectric material, the triboelectric series has been constructed. On this series, the materials that are placed further apart can generate more triboelectric charges than those that are placed closer together. The materials on the top end of the series are more likely to lose electrons and carry positive charges, while the materials at the other end are more easily negatively charged [20].

6.2.3 Evaluation of acoustoelectric conversion

Acoustic electrical energy conversion can be evaluated by various parameters, as summarized below [21].

6.2.3.1 Sound waves

- **Sound pressure** is the local pressure deviation from the ambient (average or equilibrium) atmospheric pressure, caused by a sound wave. In air, sound pressure can be measured using a microphone. The SI unit of sound pressure is the Pascal (Pa).
- **SPL** is a logarithmic measure of the effective pressure of a sound relative to a reference value. SPL(L_p) is measured in dB and defined by

$$L_p = 20 \lg (P_1 / P_0)$$

where P_1 is the sound pressure, and P_0 is the reference sound pressure of 0.000 02 Pa.

6.2.3.2 Acoustic sensors

- The **signal-to-noise ratio** (SNR) measures the signal quality of an acoustic sensor. This ratio can be obtained using the following equation:

$$\text{SNR} = 20 \log_{10} \frac{\frac{1}{m}\sum_{i=1}^{1}(V_{p-p})_i}{\frac{1}{n}\sum_{i=1}^{1}(V_{\text{noise}})_i}$$

Here, $V_{p\text{-}p}$ is the amplitude of the voltage signal, and V_{noise} is the amplitude of the noise.
- **Sensitivity** is another key characteristic of sensor device. The sensitivity (S) of the sensor device can be calculated using the following equation:

$$S = \frac{P(\text{Pa})}{V} = \frac{P_o \cdot 10^{\frac{L_p(\text{dB})}{20}}}{V}$$

where $P(\text{Pa})$ is the acoustic pressure in Pascal, V is the voltage output of the sensor, P_o is the reference sound pressure of 0.000 02 Pa, and $L_p(\text{dB})$ is the SPL in decibels.

- **Frequency bandwidth** is the range of sound frequency that the device can respond to, which decides the usage range of an acoustic sensor.
- **Sound pressure response** is the range of sound pressure that the device can respond to with electric signal outputs.

6.2.3.3 Acoustic energy harvesters

- **Voltage** and **current** outputs are the basic performance measurements of an acoustic conversion device. They are often characterized by the maximum open-circuit voltage, and maximum output short-circuit current.
- **The instantaneous power** of an acoustic energy harvester can be calculated by the equation

$$P = UI$$

where U and I represent the voltage and current outputs on the optimum load, respectively.

- **Instantaneous area power density** can be calculated by the equation:

$$P_A = P/A$$

where P is the instantaneous power, and A is the area of the energy conversion element.

- **Instantaneous volume power density** can be calculated by the equation

$$P_V = P/V$$

where V is the volume of the energy conversion element.

  Average power is defined as the average of instantaneous power over one cycle. The average power over a certain time period is equal to the total amount of energy within that time divided by time. The average power can be calculated by the equation:

$$P_{\text{avg}} = E/T$$

where E is the total energy inside this time $(t_1, \ldots, t_2)$, which can be calculated as follows:

$$E = \int_{t_1}^{t_2} p(t)\,dt$$

If this division by the time is inserted, the average power can be calculated as follows:

$$P_{\text{avg}} = \frac{1}{t_2 - t_1} \int_{t_1}^{t_2} p(t)\,dt$$

- **The sound power** P_a of a sound source can be calculated according to the equation

$$P_a = \frac{Ap^2}{\rho c} \cos\theta$$

where A is the surface area of the device, p is sound pressure, ρ is the air density of $1.292\ \mathrm{kg\,m^{-3}}$, c is sound speed in air ($343.2\ \mathrm{m\,s^{-1}}$), and θ is the angle between the direction of the sound propagation and the normal to the surface (zero degree in this case).

- **Energy conversion efficiency** is calculated as the ratio of output electric power and input acoustic power.

$$\eta = P_{\mathrm{avg}}/P_a \times 100\%$$

6.3 Acoustic sensors

6.3.1 Electrospun piezoelectric polymer nanofibers

Acoustic sensors made of piezoelectric polymer films such as PVDF, PVDF-HFP, and PVDF-TrFE can now be found on commercial market. However, highly controlled multistep processes are often involved in the production of these piezo-electric films, leading to high prices [22]. The electrospinning process applies simultaneous mechanical stretching and electric poling effects during fiber forma-tion, and electrospun nanofibers have been reported to have strong piezoelectricity without extra stretching and poling treatments [23–33], opening up promising potential for applications in sensing, energy conversion, and harvesting.

In 2016, Lang *et al* [34] from Tong Lin's research group reported that PVDF nanofiber membrane produced by electrospinnig showed incredible acoustic-to-electric conversion ability. They prepared a sensor device by sandwiching a piece of electrospun PVDF nanofiber web with two transparent polyethylene terephthalate (PET) films which were gold sputter-coated on the inner side. The gold-coated surfaces were contacted with the nanofiber layer (shown in figure 6.2(a)), and they functioned as electrodes to collect electrical signals. A through-hole was cut on each plastic film to allow the nanofibers directly receive sound waves, as schematically

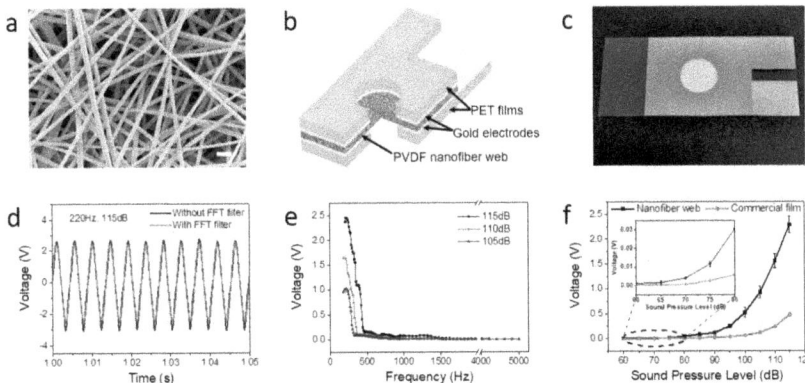

Figure 6.2. (a) SEM image of the electrospun PVDF nanofibers (scale bar, 1 µm), (b) schematic illustration of the acoustic sensor structure, (c) digital photo of the device, (d) voltage outputs of the device under sound with and without FFT treatment, and the effects of (e) sound frequency and (f) SPL (sound frequency = 220 Hz) on devie voltage outputs. Reproduced from [34] CC BY 4.0.

shown in figure 6.2(b). A photo of the actual device is shown in figure 6.2(c). The sandwich nanofiber sensor devices can detect low-frequency sound with a sensitivity of 266 mV Pa^{-1}, more than five times higher than that of a commercial piezoelectric PVDF dense film. The highest voltage output measured under 115 dB, 220 Hz sound waves was 3.10 V.

Cui *et al* [35] developed a piezoelectric nanofiber membrane/polymer composite membrane sound sensor with a helical silver electrode structure (figure 6.3). The piezoelectric sound sensor can detect low-frequency (from 50 to 400 Hz) sound wave. The maximum open-circuit voltage of the piezoelectric sound sensor reach 1.8 V. By controlling the tension of the device, the resonance frequency of the device can be changed. When the nanofiber membrane's tension was increased in the device, the resonant frequency increased from about 120 to 200 Hz.

Viola *et al* [36] designed a series of drum-shaped clamping structures of different sizes based on the frequency response mechanism of cochlear hair cells. PVDF-TrFE nanofiber membranes were clamped in a circular PLA framework, and copper tape was attached to the inner surface of the PLA as electrodes. The author compared the test results of non-oriented and oriented nanofiber membranes, and the experimental results showed that the resonance frequencies of devices with different circular diameters were different. For non-oriented nanofiber membranes, the resonance frequency increases with the decrease of circular diameter. As the diameter of the oriented nanofiber membrane decreases, the resonance frequency does not change much.

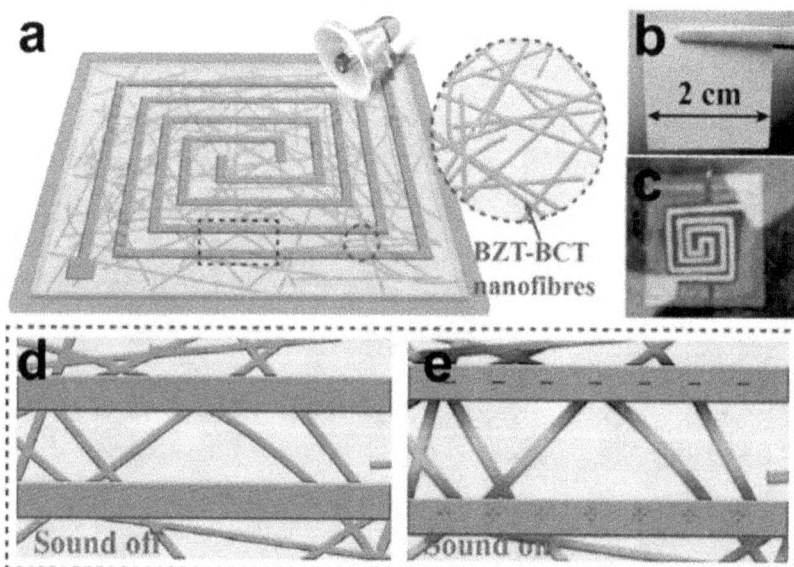

Figure 6.3. (a) Schematic illustration of the piezoelectric nanogenerator, (b) digital photo of the BZT–BCT nanofiber membrane, (c) digital photo of the piezoelectric nanogenerator, (d) and (e) working mechanism of the piezoelectric nanogenerator. Reproduced from [35] CC BY 3.0.

In terms of frequency response, piezoelectric nanofiber membranes mainly have good response to medium and low frequencies, while their response effect at high frequencies is relatively lower. However, frequency response is an important parameter that limits the application of sensors, so research on frequency response has always been a focus of attention for researchers. Wang *et al* [37] prepared a transparent and oriented PVDF-TrFE nanofiber film using near-field electrospinning technology (figure 6.4). Due to the low packing density and sparse structure, the nanofiber mesh is transparency and air permissive, which is beneficial for air vibration and forced vibration of the nanofibers. Therefore, compared with the dense film, it has a wider frequency response (200–5000 Hz), showing relatively good frequency response even at frequencies more than 1000 Hz.

More recently, electrospun polyacrylonitrile (PAN) nanofiber membranes were reported by Tong Lin's research group to have strong piezoelectricity [38]. PAN is an amorphous vinyl-type polymer composed of a cyano (–CN) group in each repeating unit. It has a larger unit dipole moment, nearly 3.5 Debye, than PVDF. PAN nanofibers prepared by electrospinning have high zigzag conformation content and orientated structure, responsible for the high piezoelectricity. Since discovering the high piezoelectric properties of PAN, Tong Lin's research group has also applied it to the study of acoustoelectric conversion.

Shao *et al* [39] from Tong Lin's research group reported a sandwich acoustic sensor based on electrospun polyacrylonitrile (PAN) nanofibers for human sound detection. The performance of the acoustic sensor at low decibels was tested in this work, for the reason that the sound of human speech is generally between 40–70 decibels. The nanofiber sensor shows a sensitivity as high as 23 401 mV Pa^{-1} to 70 dB sound. The nanofiber sensors can also distinguish sounds from different instruments, which are used as standard sound sources to display the high-precision of nanofiber devices. The experimental results show that the device can recognize people's voices with high resolution, and the impact of background noise on speech recognition is minimal.

Peng *et al* [40] from Tong Lin's research group demonstrated the excellent sounding sensing ability of electrospun PAN nanofiber membranes. In comparison to electrospun PVDF nanofiber membranes, PAN nanofiber membranes show larger acoustoelectric conversion capability, broader response bandwidth, larger

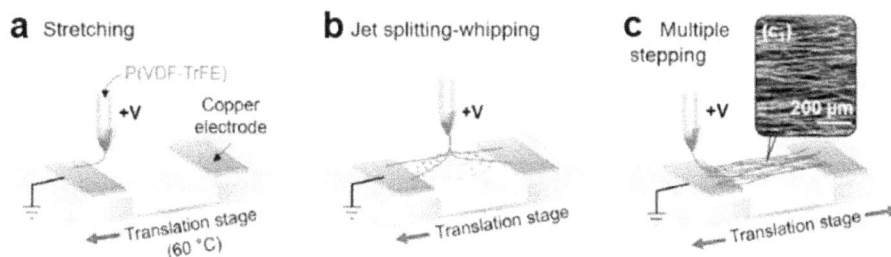

Figure 6.4. The process of dynamic near-field electrospinning to make a transparent P(VDF-TrFE) nanofiber acoustic sensor. Reproduced from [37] CC BY 4.0.

sensitivity, and higher fidelity. The brilliant detection accuracy originates from the high piezoelectricity and rigidity of PAN nanofibers, facilitating deformation recovery and reduction of the interference from harmonics. The nanofiber sensor device can detect low-to-middle frequency sound (100–600 Hz) at the middle sound pressure level (60–95 dB) precisely, with a signal-to-noise ratio as high as 57.2 dB and fidelity as high as 0.995.

6.3.2 Electrospun nonpiezoelectric polymer nanofibers

Inspired by the flow-based auditory system of mosquitoes, Zhou *et al* [41] prepared a directional sensor based on electromagnetic induction using a 2D nanofiber mesh as the sensing material. By depositing Au on a freestanding electrospun polymethyl methacrylate (PMMA) nanofiber mesh, they fabricated an Au–PMMA nanomesh. Within the 2D mesh, most of the fibers align between the two electrodes. Output signals are induced by the motion of the nanomesh in a magnetic field. Compared to the noisy signal recorded by the reference microphone, the nanomesh flow sensor can exclusively capture the signals at 400 Hz and reject the unwanted noise signals.

6.4 Acoustoelectric harvesters

Piezoelectric acoustic energy harvesters, also known as acoustoelectric generators, produce a higher level of electric outputs than acoustic sensors, and are targeted by researchers so that the generated electricity can be indirectly or directly used for powering other microelectronic devices or systems. A diaphragm and energy conversion element are also required for an acoustoelectric harvester. Besides these, in some cases, a resonator is used to increase the vibrations caused by sound and an energy storage device is used to store the electric energy.

6.4.1 Piezoelectric harvesters

6.4.1.1 Electrospun piezoelectric polymer nanofibers
Lang *et al* [42] from Tong Lin's group designed a multihole electrode structure on the basis of the previous sound sensor, and used PVDF-TrFE with better piezo-electric performance as the core material to make a nanofiber sound generator (figures 6.5(a)–(d)). The multihole design makes the device vibrate more violently under high sound pressure levels, thus enabling more sound energy to be converted into device vibration mechanical energy, and thus into electrical energy. The open-circuit voltage, short-circuit current, and the power output with the load resistance were tested. Under a sound (SPL 115 dB, frequency 210 Hz) activation, it generated periodic voltage and current outputs with peak values as high as 14.5 V and 28.5 μA (figures 6.5(e) and (f)). The device showed the maximum power output of 141.3 μW at the external resistance of 470 kΩ (figures 6.5(g) and (h)). The voltage output performance is five times that of the commercial piezoelectric film with the same structure (figure 6.5(i)).

The holes in the electrodes were found to have significant effects on the output signals of the device. The effect of hole number on device outputs is shown in figure 6.6(a). By maintaining the overall hole area, the voltage output increases with

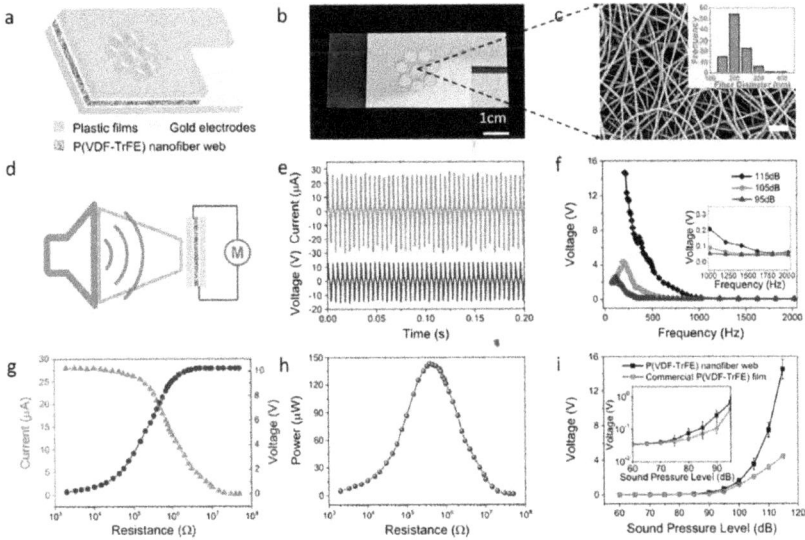

Figure 6.5. (a) Schematic illustration of the acoustic energy harvester device, (b) digital photo of the device, (c) SEM image of the electrospun PVDF-TrFE nanofibers (scale bar, 2 μm), (d) schematic illustration of the acoustoelectric conversion testing, (e) voltage and current outputs of the device under sound, (f) effect of sound frequency on the voltage output of the device, (g) dependence of the peak voltage and current output on the resistance of the external load, (h) dependence of the peak power output on the external load resistance, (i) effect of SPL on devices made of nanofiber web and commercial dense film. (eight-hole device, hole diameter = 4.9 mm; web thickness = 20 μm; film area = 12 cm^2; (a)–(e) and (g)–(i): SPL = 115 dB and sound frequency = 210 Hz). Reproduced from [42], copyright (2017), with permission from Elsevier.

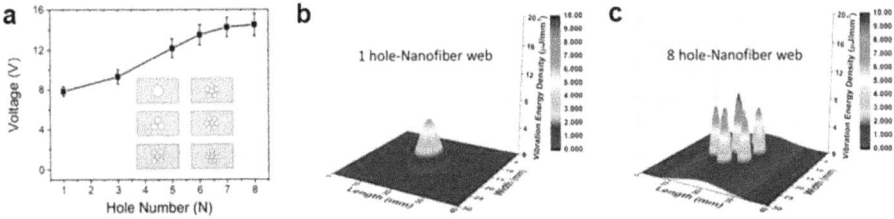

Figure 6.6. (a) Voltage output as a function of the number of holes (sound frequency = 210 Hz; SPL = 115 dB; web area = 12 cm^2; web thickness = 20 μm), Modelled vibration energy density of nanofiber webs in (b) eight-hole and (c) one-hole device (sound frequency = 210 Hz; sound pressure level = 115 dB). Reproduced from [42], copyright (2017), with permission from Elsevier.

the increasing number of holes until there are six. When the number of holes continues to increase, the voltage output remains at around 14 V. Figure 6.6(b) and (c) illustrates the vibration energy profiles of the nanofiber web in the one-hole and eight-hole devices. The central part of the hole shows a higher vibration energy density than the periphery of the hole. The eight-hole nanofiber web and PET film both have much higher vibration energy than the one-hole nanofiber web and PET film. This is because nanofibers in the hole vibrate more strongly than those covered with electrodes. Therefore, a multihole structure should include multiple vibration spots in the fibrous web, which would facilitate transport of vibration energy from the exposed areas to the entire device.

With the same multihole sandwich structure, Shao *et al* [43] from Tong Lin's research group reported that electrospun PAN membranes can convert high decibel noise (sound pressure level above 115 dB) into a large voltage output (up to 58 V) with an energy conversion efficiency as high as 85.9%.

Zheng *et al* [44] from Tong Lin's group found that thermally stable electrospun PAN nanofiber membranes prepared through a heat treatment program still exhibit strong piezoelectric properties. Then their group for the first time applied the parallel arranged PAN nanofiber membranes after heat treatment to thermoacoustic conversion (figures 6.7(a)–(e)). Under 118 dB, 230 Hz sound, the nanofiber device can generate an open-circuit voltage of up to 118 V and a short-circuit current of 12 μA (power density 392 mW m^{-2}) within a wide temperature range (room temperature −450 °C). While the temperature has little effect on the acoustic and electrical output. Then a micro thermal acoustic engine was used to generate thermal sound at temperatures exceeding 480 °C, with a frequency of 358 Hz. Under the thermal sound wave, the nanofiber device was heated to 280 °C and the device operates stably for at least 2 h. A single device can generate an open-circuit voltage of 102 V and short-circuit current of 10 μA (figures 6.7(f)–(h)).

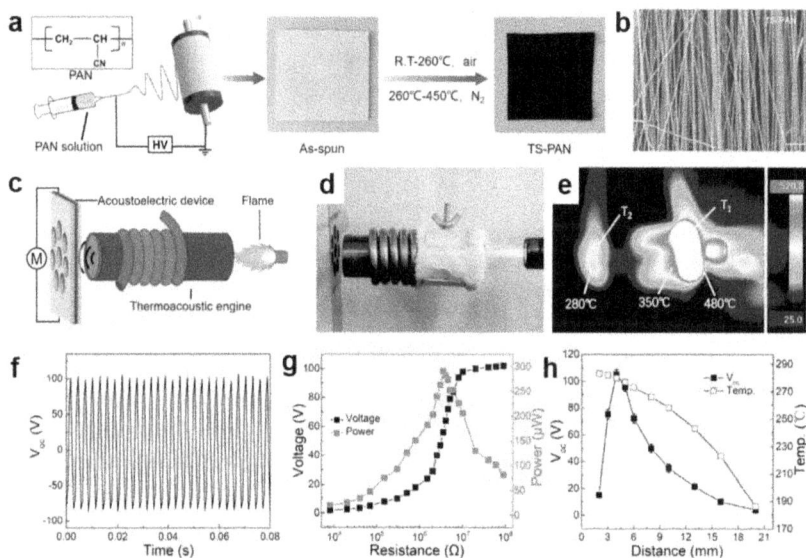

Figure 6.7. (a) Preparation process of thermally stabilized PAN nanofiber membranes. (b) SEM pictures of thermally stabilized PAN nanofiber membranes (scale bar: 2 μm). (c) Schematic picture of a thermoacoustic engine combined with acoustoelectric power generator. (d) Digital photos and (e) infrared thermal image of thermoacoustic engine/acoustoelectric power generator during working. (f) The open-circuit voltage output of the generator under thermoacoustic waves. (g) Dependence of the peak voltage and instantaneous peak power outputs on the resistance of the external load. (h) Effect of working distance on the open-circuit voltage output and surface temperature of the acoustoelectric power generator. Reprinted from [44], copyright (2022), with permission from Elsevier.

Peng *et al* [45] from Tong Lin's group designed a slit electrode, which is combined with oriented PAN nanofibers to achieve enhanced acoustic and electrical bandwidth and total power output (figures 6.8(a) and (b)). When PAN nanofibers are oriented perpendicular to the slit ($\perp$), the bandwidth of this device is much wider than parallel devices ($\parallel$) and devices using randomly oriented nanofiber membranes (RDM) (figures 6.8(c)–(h)). Compared to PAN nanofibers made of randomly nanofiber membrane, PAN nanofibers with parallel slits exhibit similar bandwidth but larger electrical output. The size of the slit affects bandwidth and output. However, the number of slits has almost no impact on bandwidth, but it has an impact on electrical output. Under the action of sound, electrode vibration causes misalignment of the slits on both sides. The anisotropic tensile properties of oriented nanofiber membranes allow fibers to stretch in different ways based on their arrangement angle with the slit. The parts perpendicular to the slit are subjected to stronger stretching, which helps to obtain a wider bandwidth. Under 115 dB noise, a five-slit device (size 4×3 cm^2) with PAN nanofibers perpendicular to the slit

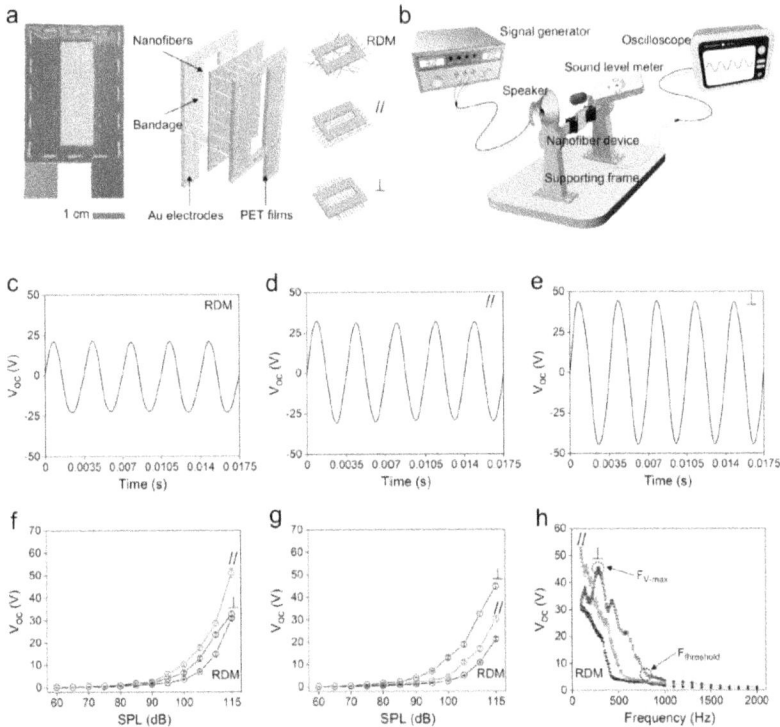

Figure 6.8. (a) A digital photo (scale bar, 1 cm) and a schematic diagram of the acoustoelectric device. (b) The acoustoelectric test setup. Voltage signal waveform for (c) RDM, (d) $\parallel$, and (e) $\perp$ (sound condition: 286 Hz and 115 dB). Effect of SPL on Voc for RDM, $\parallel$, and $\perp$ at (f) 100 Hz and (g) 286 Hz. (h) Dependency of Voc on sound frequency for RDM, $\parallel$, and $\perp$ (SPL: 115 dB) (slit length 30 mm and width 10 mm; nanofiber membrane thickness 33 μm). Reprinted with permission from [45], copyright (2023) American Chemical Society.

produced a peak open-circuit voltage of 39.85 ± 1.34 V and a peak short-circuit current of 6.25 ± 0.18 µA with a bandwidth of 100–900 Hz. The wide frequency response and large electrical output allow this type of acoustoelectric device to be used for power supply.

Another commonly used method to improve conversion efficiency and output energy is to modify piezoelectric nanofibers to improve the performance of PVDF β-phase content, thereby improving piezoelectric output or sensitivity. For enhancing the piezoelectric effect, different additives have been introduced into PVDF or its copolymer nanofibers.

Adhikary et al [46] fabricated composite electrospun nanofibers of Eu^{3+}-doped PVDF-HFP/graphene for an ultrasensitive wearable piezoelectric acoustic transducer. The addition of Eu^{3+} and graphene also increases the β-phase content and crystallinity of the nanofibers. They demonstrated a sensitivity of 11 $V\,Pa^{-1}$ to musical sound at 82 dB.

Maity et al [47] reported a sensitive and efficient piezoelectric acoustic transducer made of few layers MoS_2 nanosheets incorporated into electrospun PVDF nanofiber webs. The FTIR, XRD, and XPS results show that the PVDF-MoS_2 composite nanofibers present a higher β-phase content and crystallinity than neat PVDF nanofibers. As a result, the acoustic transducer shows an ultrahigh sensitivity of 19 $V\,Pa^{-1}$ to musical sound at 90 dB, which is superior to most of the recently reported piezoelectric acoustic sensors.

Ramasamy et al [48] incorporated phenylisocyanate functionalized graphene oxide (IGO) into electrospun PVDF nanofibers, resulting in rough surface morphology, enhanced crystallinity, and electroactive β-phase of PVDF nanofibers. The ultimate tensile strength and modulus and the acoustic sensitivity of PVDF nanofibers doped with IGO were increased by 303%, 332% and 63.09%, respectively, compared to pristine PVDF nanofibers.

The reported acoustoelectric transducers based on doped nanofibers are summarized as shown in table 6.2.

6.4.1.2 Nanowires

Other than continuous electrospun nanofibers, other forms of 1D nanomaterials have also been used in acoustic energy harvesting. Wang et al [58] developed a nanowire energy harvester that can be driven by an ultrasonic wave to generate continuous DC output. The energy harvester was based on vertically aligned ZnO nanowire arrays with a zigzag metal electrode on the top and a conductive electrode on the bottom. The zigzag electrode is used to create, collect and output electricity generated from the nanowires. The output power of the energy harvester is about 1 pW per 2 mm^2 area substrate.

Cha et al [59] reported an energy harvesting device based on piezoelectric ZnO nanowires driven by 100 Hz frequency sound. The energy harvester was made on a polyethersulfone (PES) substrate with a PdAu coating as the electrode. Well-aligned ZnO nanowires were grown on the top of the PES substrate, and covered by a vibration plate. When a sound wave irradiates the device, it causes vibration of the top contact electrode and compression of the ZnO nanowires, which results in the

Table 6.2. Comparison of the reported acoustoelectric transducers based on doped nanofibrous materials.

Material	Doping material	Doping method	β-Phase contents			Reference
			Before doping	After doping	Increase ratio	
PVDF nanofibers	Ce^{3+} and graphene	Mixed in the solution with ((NH_4)$_4$Ce-(SO_4)$_4$·$2H_2O$ and grapheme.	96%	99%	3.1%	[49]
PVDF-HFP nanofibers	Eu^{3+} and graphene	Mixed in the solution with Eu(NO_3)$_3$·$6H_2O$ and grapheme.	96%	99%	3.1%	[46]
PVDF nanofiber	MoS_2	Mixed in the solution with MoS_2.	92%	95%	3.2%	[47]
PVDF nanofiber	MAPbBr	Mixed in the solution with synthesized methylammonium lead bromide ($CH_3NH_3PbBr_3$) (MAPbBr).	76%	91%	19.7%	[50]
PVDF nanofiber	TiO_2	Mixed in the solution with TiO_2.	80%	93%	16.3%	[51]
PVDF nanofibers	ZnO	Immobilize ZnO on PVDF fibers by a dip-coating technique.	70.8%	73.2%	3.4%	[52]
PVDF nanofiber	ZnS nanorods	Mixed in the solution with synthesized ZnS nanorods.	82%	97%	18.3%	[53]
PVDF nanofiber	Ag nanoparticles	Mixed in the solution with Ag nanoparticles.	83%	94%	13.3%	[54]
PVDF nanofiber	Phenylisocyanate functionalized graphene oxide	Mixed in the solution with phenylisocyanate functionalized graphene oxide (IGO).	68.7%	97.1%	41.3%	[48]
PVDF nanofiber	[Cd(II)-µ-I4] two-dimensional metal–organic framework	Mixed in the solution with synthesized CdI_7-NAP, [CdI_2(NAP)]n.	Not mentioned	98%	—	[55]
PAN and PVDF nanofiber	MWCNTs	Mixed in the solution with multi-walled carbon nanotubes MWCNTs.	52.6%	70.0%	33.1%	[56]
PAN nanofiber	Cellulose nanocrystals	Mixed in the solution with cellulose nanocrystals.	55.59% (Zigzag conformation content)	89.9% (Zigzag conformation content)	61.7%	[57]

generation of an electric potential. The device can generate 50 mV of output voltage under sound with an intensity of 100 dB.

Lee *et al* [60] fabricated a hybrid cell that harvests both acoustic and solar energy. Sound wave-driven energy harvesting was demonstrated using both laterally bonded single wires and vertically aligned nanowire arrays, in the frequency range of 35–1000 Hz. Figure 6.9(a) schemetically shows the experimental setup for single nanowires. An alumina rod was firmly attached to the sound wave generator at one end, and its other end was mounted to the energy harvesting device. The sound wave generated by the transducer was directly transmitted to the single nanowires. The vibration of the end surface of the alumina rod was measured with an accelerometer. As shown in figure 6.9(b), the voltage output is relatively uniform over the entire frequency range under investigation. The voltage output does not significantly depend on the sound frequency.

The aligned nanowires were combined with CdS (n-type)/CdTe (p-type) nano-particles to make a hybrid device to harvest both acoustic and solar energy. Figure 6.9(c) presents a diagram of the hybrid device. Simulated sunlight illumina-tion and acoustic vibration are supplied through the top PET/ITO layer and the bottom Au electrode, respectively. Figure 6.9(d) shows the energy band diagram of the hybrid device. The upper and lower diagrams correspond to the solar harvester and nanowire energy harvester, respectively. The solar energy harvester is supposed to generate DC output (figure 6.9(e)), while the sound harvester generates AC voltage. A capacitor was added to allow the DC output to be stored (figure 6.9(f)).

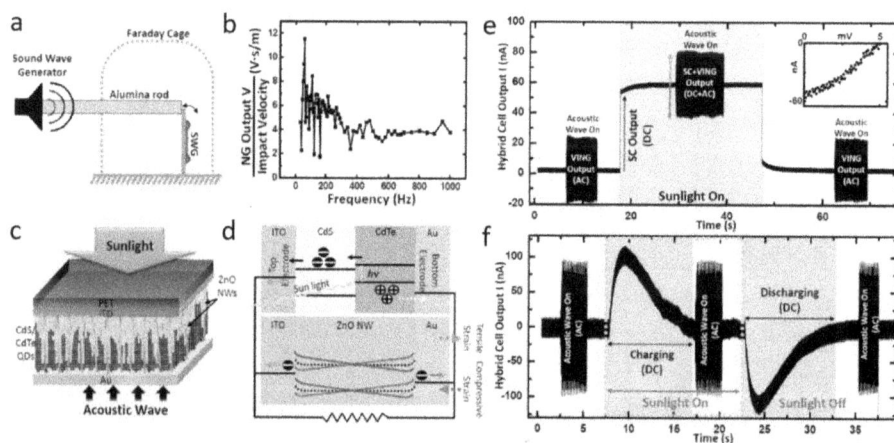

Figure 6.9. (a) Schematic of a single nanowire energy harvesting setup, and (b) effect of sound frequency on the ratio of single nanowire energy harvester voltage output to the applied mechanical impact velocity. (c) Schematic of a hybrid device with two different energy sources, (d) energy band diagram of the hybrid device showing electron flow under sunlight illumination and acoustic vibration, (e) current output of the hybrid device, and (f) current output of the hybrid device after connecting a capacitor. Reprinted with permission from [60], copyright (2010) American Chemical Society.

6.4.2 Triboelectric acoustic generators

6.4.2.1 Electrospun PVDF nanofibers

In recent years, acoustic triboelectric energy harvesters have also been investigated. Due to the wave characteristics of sound propagation, a resulting acoustic pressure separates and pushes back the two films at a certain frequency. On account of the triboelectric effect, an electrical signal with the same frequency is generated between the two contact surfaces of the two different materials, which have different electronegativities. The surface of the material with a higher electronegativity is negatively charged after it is separated, and the other material's surface has an equal amount of positive charges. Generally, polyamide and metal are used as positively charged materials in a triboelectric energy harvester, while negatively charged materials are mostly fluorinated polymers, such as PTFE, fluorinated ethylene propylene (FEP) and PVDF [20].

Cui *et al* [16] reported a mesh-web-based triboelectric energy harvester with a sandwich structure of two substrates and one vibrating web. As shown in figure 6.10(a), a piece of aluminum (Al)-covered stainless steel mesh acts as the contact surface and bottom electrode, i.e. substrate B. Another piece of stainless steel mesh with a layer of electrospun PVDF nanofibers acted as the top electrode, i.e. substrate A (figure 6.10(b)). A piece of polyethylene film with Al coated on one surface and PVDF nanofibers on the other surface was used as the vibrating membrane of the device. The PVDF nanofiber and Al act as negatively and positively charged materials, respectively. The mesh structure allows the air circulation and benefits the penetration of the sound wave. The thin and lightweight vibrating film facilitates absorbtion of low-frequency sound wave energy, and displays a sensitive response to small sound wave pressure. Furthermore, the porous mesh substrate facilitates the spread of sound waves, and reduces the sound energy loss in the substrate. The triboelectric energy harvester device can be driven by sound waves in a broad bandwidth ranging from 50 to 425 Hz, and it obtains a short-circuit current density as high as 45 mA m^{-2} and peak power density of 202 mW m^{-2}.

Liu *et al* [61] demonstrated an integrated triboelectric nanogenerator with a three-dimensional structure. From top to bottom, layer A consists of a spacer, a porous polyvinyl chloride (PVC) shelf coated with Al, and a layer of PVDF nanofibers, buliding the basic structure. Layer B consists of a spacer, an Al layer, a polyethylene

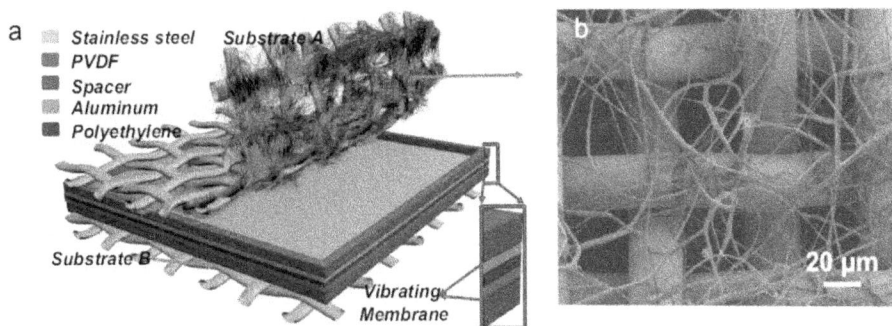

Figure 6.10. (a) Schematic of a sound-driven triboelectric energy harvester, and (b) SEM image of the PVDF nanofibers on the stainless steel mesh. Reprinted from [16], copyright (2015), with permission from Elsevier.

(PE) substrate, and a layer of PVDF nanofibers, serving as a vibrating film. The device is manufactured by alternating the stacking of layers A and B. The collaborative work of different multifunctional integration layers is beneficial for the propagation and adsorption of sound, improving output performance. This device can generate an open-circuit voltage of up to 232 V and a short-circuit current of up to 2.1 mA.

Chen *et al* [62] constructed an integrated triboelectric nanogenerator with polymer tubes, as shown in figure 6.11(a). The nanogenerator based on electrospun PVDF nanofibers is attached to one end of a polyvinyl chloride (PVC) polymer tube, which is nested by another larger diameter PVC tube. This unique tube structure design can increase sound pressure, thereby enhancing the force applied to the vibrating film and achieving high output performance. Figure 6.11(b) shows the nanogenerator with a sandwich structure, mainly composed of PVDF nanofiber triboelectric layer (figure 6.11(c)), conductive fabric (figure 6.11(d)), and Kapton spacer. Under a sound frequency of 170 Hz and sound pressure of 115 dB, the nanogenerator can deliver an open-circuit voltage and short-circuit current of 400 V and 175 μA, respectively. The generator can operate stably in the frequency range of 20–1000 Hz.

Qiu *et al* [63] designed a sound-driven triboelectric nanogenerator based on porous foam copper and sandwich-shaped friction pairs, mainly composed of Cu foam, PVDF nanofibers, and nylon fabric (figure 6.12). Cu foam was used as a sound collector, conductor, and amplifier, as well as the friction layer and conductive layer of the triboelectric nanogenerator. PVDF nanofiber membrane and nylon fabric are used as vibration membranes. Under a sound frequency of 170 Hz and sound pressure of 115 dB, this nanogenerator has a maximum current

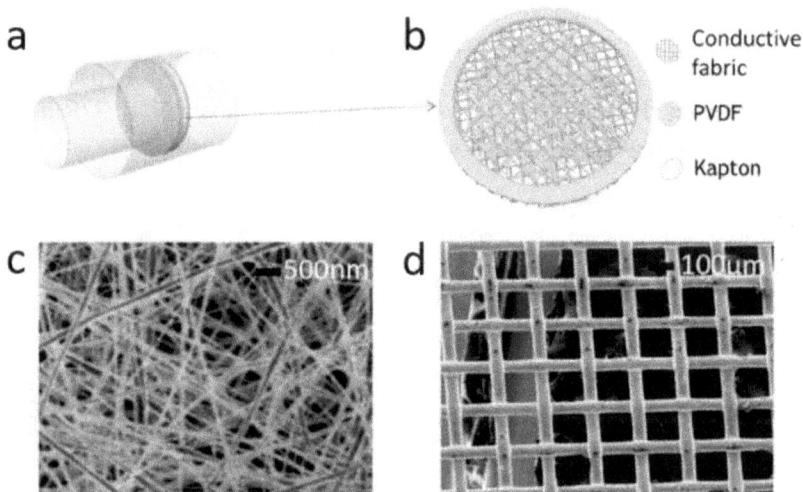

Figure 6.11. (a) Schematic structure of the triboelectric nanogenerator integrated with the polymer tubes, (b) schematic picture of the sandwich-structure triboelectric nanogenerator, SEM images for the (c) PVDF nanofibers and (d) the conductive fabric. Reprinted from [62], copyright (2019), with permission from Elsevier.

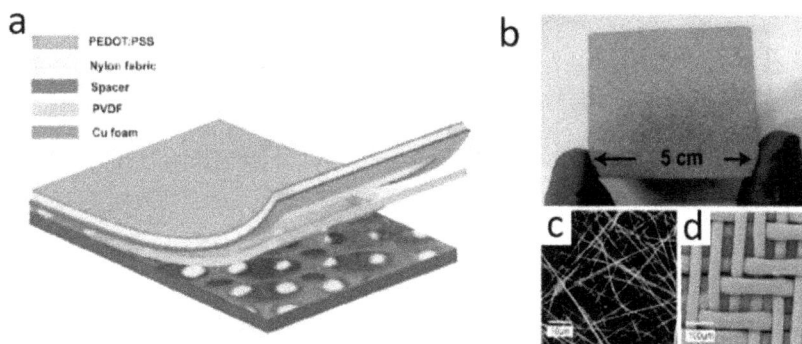

Figure 6.12. (a) The schematic structure of the triboelectric nanogenerators, (b) digital photo of the copper foam, SEM image of (c) the electrospun PVDF nanofiber and (d) the nylon fabric. Reprinted from [63], copyright (2020), with permission from Elsevier.

density of 25.01 mA m^{-2} and a charging rate of 20.91 μC s^{-1}, with a charging conversion efficiency of up to 59.85%. Moreover, it has good working stability and can work continuously for 7 days without attenuation under speaker drive.

Wang *et al* [64] proposed a hybrid triboelectric nanogenerator that can simultaneously capture wind and sound energy, inspired by windmills (figure 6.13). The sound-driven nanogenerator was based on two conductive fabric layers sandwiched with electrospun PVDF-TrFE nanofibers, with one conductive fabric electrode shared by a wind-driven and sound-driven nanogenerator. The unique structural design enables the electrical output of the sound-driven nanogenerator to be enhanced by the wind-driven nanogenerator. The wind-driven nanogenerator can provide a maximum open-circuit voltage of 530 V and short-circuit current of 10.5 μA at wind speeds of 10 m s^{-1}. with a peak power of approximately 1.5 mW. Under a frequency of 180 Hz and a sound pressure of 115 dB, the corresponding electrical outputs of the sound-driven TENG are 80 V, 19 μA, and 0.5 mW, respectively.

Shao *et al* [65] from Tong Lin's group reported a composite electrospinning nanofiber membrane by blending PAN and PVDF polymer (figure 6.14). Under 117 dB 250 Hz sound, devices made of PAN-PVDF nanofiber membranes can generate open-circuit voltage of 94.1 V and short-circuit current of 17.4 μA. with a power density of 250.1 mW m^{-2} and energy conversion efficiency of 25.6%. Under the same conditions, the voltage output was 3.9 times that of a single component PAN nanofiber and 4.5 times that of a single component PVDF nanofiber. The group proposed that the conversion of noise to electricity originates from the endogenous frictional electric effect within a single-layer fiber membrane, which is caused by the microphase separation of PAN and PVDF within the nanofibers, exposing both components to the fiber surface. In addition, the ratio of two polymer components in nanofibers affects the microstructure, molecular conformation, and energy conversion performance of the polymers in the fibers.

Niu *et al* [66] from Tong Lin's group added a small amount of nylon-6 nanofibers to the PVDF fiber matrix to improve the electrical output of a single-layer PVDF nanofiber membrane (figure 6.15). Nylon-6 solution and PVDF solution were

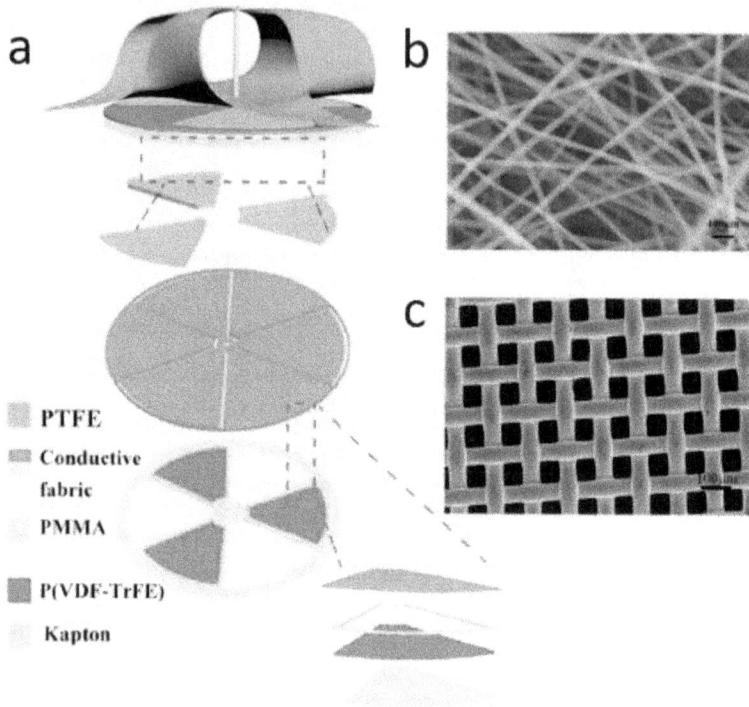

Figure 6.13. (a) Schematic illustration of the hybridized triboelectric nanogenerator, SEM images of (b) P (VDF-TrFE) nanofiber membrane and (c) conductive fabrics. Reprinted from [64], copyright (2020), with permission from Elsevier.

electrospun separately using different electrospinning nozzles, one charged by a positive power supply and the other by a negative power supply. Adding only 4.3 wt% of nylon-6 nanofibers greatly improved the acoustic electric conversion. At 230 Hz 118 dB SPL, the device can generate up to 201.4 V and 17.6 µA (power density 1.30 W m^{-2}). Compared with pure PVDF nanofiber counterparts, the voltage and current output are 2.7 and 2.6 times higher, respectively. Nylon-6/PVDF devices have also expanded their frequency response band, covering the frequency range of 230–800 Hz. The interface polarization between PVDF and nylon-6 nanofibers forms a frictional electric nanogenerator (TENG), which makes nylon-6/PVDF nanofiber devices like many small PENG and TENG components. When nanofibers deform, the electrical output of the device comes from the collective effect of TENG and PENG.

Xu *et al* [67] designed a laminated electrospun nanofiber triboelectric sound energy generator. The structural design diagram is shown in the figure 6.16(a), basically composed of four layers, including conductive porous PU foam (figure 6.16 (b1)) and Cu mesh as bottom and top electrodes, electrospun PVDF nanofiber membrane and nylon nanofiber membrane (figure 6.16(b2)) as negative and positive triboelectric layers, respectively. When sound waves penetrate the interior of the device, a large number of interconnected cavities and voids in the porous material will facilitate air vibration (figures 6.16(c)–(e)). Therefore, porous PU foam and

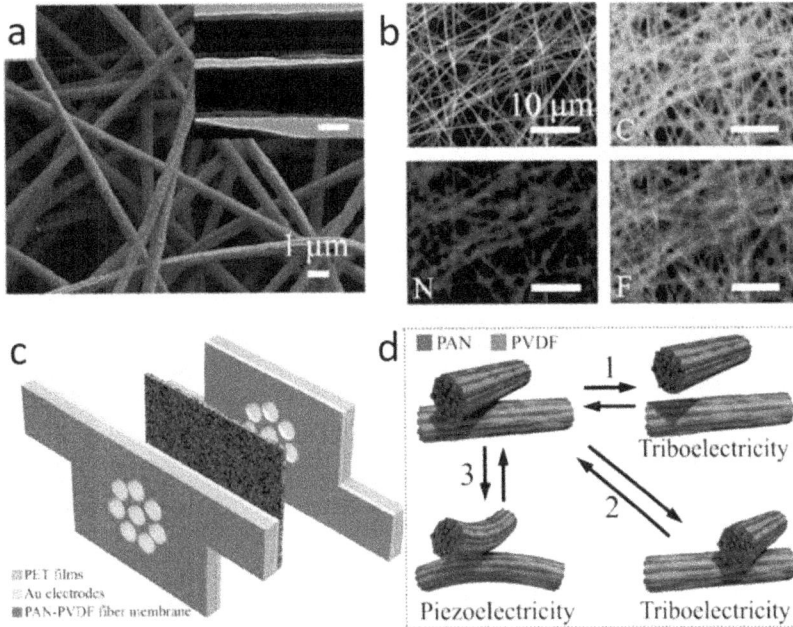

Figure 6.14. (a) SEM image of the PAN-PVDF nanofibers; Inset: TEM image (scale bar: 200 nm). (b) The uniform distribution of carbon (C), nitrogen (N), and fluorine (F) in the PAN-PVDF nanofibers shown in SEM-EDX mapping images, schematic picture of (c) the PAN-PVDF noise generator and (d) the proposed charge generation mechanism within the PAN-PVDF nanofiber membrane. Reprinted from [65], copyright (2021), with permission from Elsevier.

nanofiber membrane are conducive to the transmission and capture of acoustic vibration energy.

Inspired by the eardrum, Jiang *et al* [68] reported an ultrathin eardrum-like triboelectric acoustic sensor composed of fully flexible nanofibers. The sensor consists of electrospun PAN nanofiber and polyamide 6 (PA6) nanofiber, coated with silver (Ag) on the nanofibers through magnetron sputtering. The entire device thickness is only 40 μm and has a sensitivity of up to 228.5 mV Pa^{-1} at 95 dB. In addition, the device has a wide frequency response (20–5000 Hz), and the frequency response can be adjusted by changing the size and shape of the sensor. When the device is combined with artificial intelligence algorithms, the recognition accuracy of speech conversion can reach as high as 92.64%.

6.4.2.2 Nanowires

Yang *et al* [17] used a Helmholtz resonator with a size-tunable narrow neck to enhance the vibration and triboelectrification effect of an energy harvester. They also optimized the parameters of the resonator to increase the output and broaden the frequency response. Figure 6.17(a) and (b) schematically show the structure of the device. The circular multilayer membrane on the surface of the cavity is the core part of the device. An Al thin film with a nanoporous surface plays the roles of an

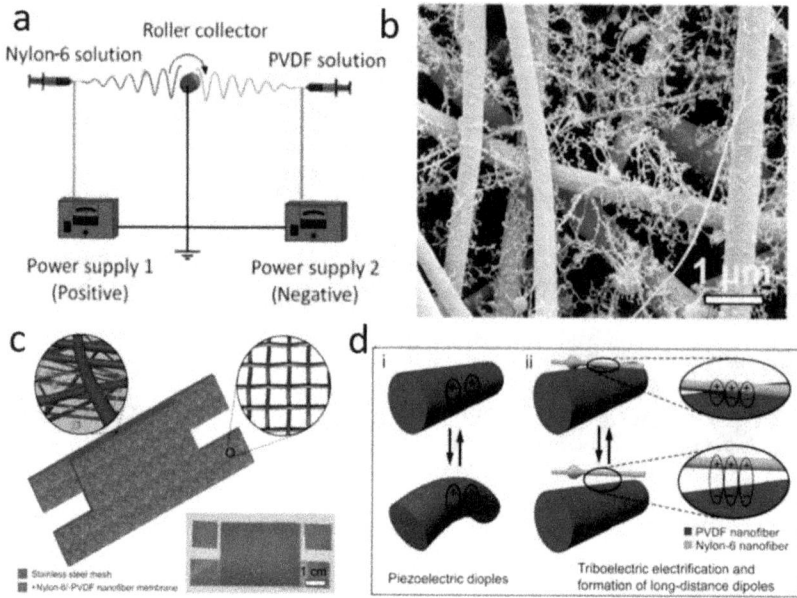

Figure 6.15. (a) Schematic illustration of the electrospinning setup for producing nanofiber membranes. (b) SEM image of + nylon-6/-PVDF nanofibers. Schematic illustration of (c) the generator structure (Inset is a digital photo of the device), and (d) acoustoelectric charge generation within nylon-6/PVDF nanofiber membrane. Reprinted from [66], copyright (2021), with permission from Elsevier.

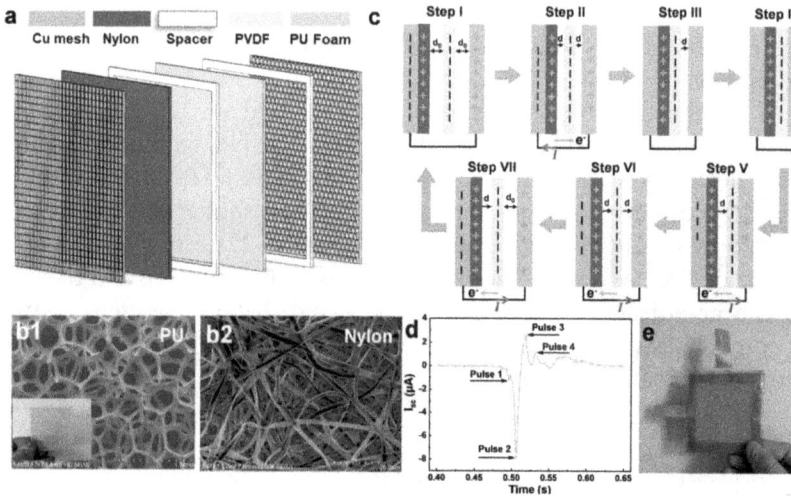

Figure 6.16. (a) Schematic illustration of the laminated structure of tiboelectric acoustic energy harvester. (b1–b2) SEM images of the PU foam and the electrospun nylon fiber membrane. (c) Acoustoelectric energy transfer mechanism of the device. (d) Magnified short-circuit current output of the device during one working cycle. (e) Digital photo of the device. Reprinted from [67], copyright (2022), with permission from Elsevier.

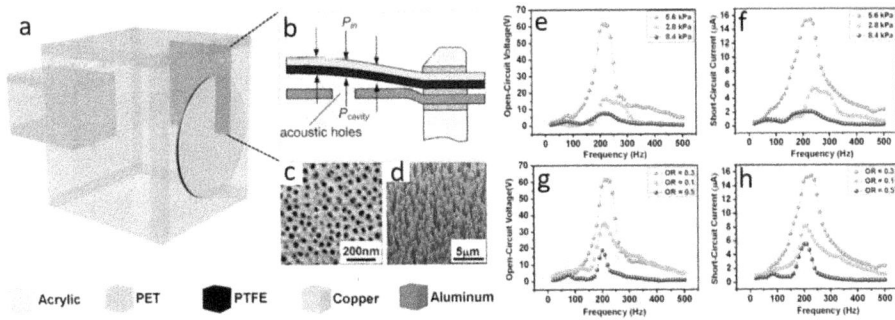

Figure 6.17. (a) Schematic and (b) cross-sectional view of the triboelectric energy harvester. (c) SEM image of nanopores on the Al electrode, and (d) SEM image of PTFE nanowires fabricated on the film surface by plasma etching. Effect of sound frequency on (e) voltage outputs and (f) current outputs with different prestresses. Effect of sound frequency on (g) voltage outputs and (h) current outputs with different open ratios. Reprinted with permission from [17], copyright (2014) American Chemical Society.

electrode and a contact surface. Figure 6.17(c) shows a SEM image of nanopores on the Al. A PTFE film with copper acts as another electrode. Inductively coupled plasma reactive ion etching was used to fabricate the aligned nanowires on the PTFE surface, which largely enhances the triboelectrification. Figure 6.17(d) shows a SEM image of the PTFE nanowires. A suitable initial prestress of 5.6 kPa and an optimum open ratio of 0.3, which can optimize the oscillation coupling between the air trapped in the cavity and PTFE thin film, are used to maximum the electrical output (figures 6.17(e)–(h)). Moreover, the natural frequencies of the devices can be designed by changing the parameters of the Helmholtz resonator. Four differently sized devices were used to broaden the working bandwidth from 10 to 1700 Hz.

Fan *et al* [18] reported a 125 μm thickness, rollable paper-based triboelectric energy harvester for sound wave energy harvesting. The device, which contains no Helmholtz resonator, can achieve a maximum power density of 121 mW m^{-2} under an SPL of 117 dB. The paper substrate was coated with copper, which acts as a positively charged material that generates triboelectric charges upon contact with a thin PTFE nanowires membrane, as schematically shown in figure 6.18(a). Figure 6.18(b) shows a photograph of the multihole paper electrode. Figure 6.18(c) shows a SEM image of PTFE nanowires, which acts as negatively charged materials. A digital photo of a paper-based acoustic energy harvester is presented in figure 6.18(d). ANSYS software was used to simulate the sound wave-induced PTFE membrane vibration under various frequencies, as shown in figure 6.18(e). From the simulation results, it can be seen that under sound with different frequencies, the PTFE membrane vibrated with different deformation regions and magnitudes. A cycle of electricity generation process induced by vibration under sound is schematically depicted in figure 6.18(f). Under sound, the PTFE will be separated from the copper. Then the altered inner dipole moment drives free electrons to flow from the electrode with PTFE nanowires to the multihole paper copper electrode until the maximum separation state is reached. The free electrons will flow in a reverse direction from maximum separation state toward the maximum contact state. With that, a full cycle of electricity generation

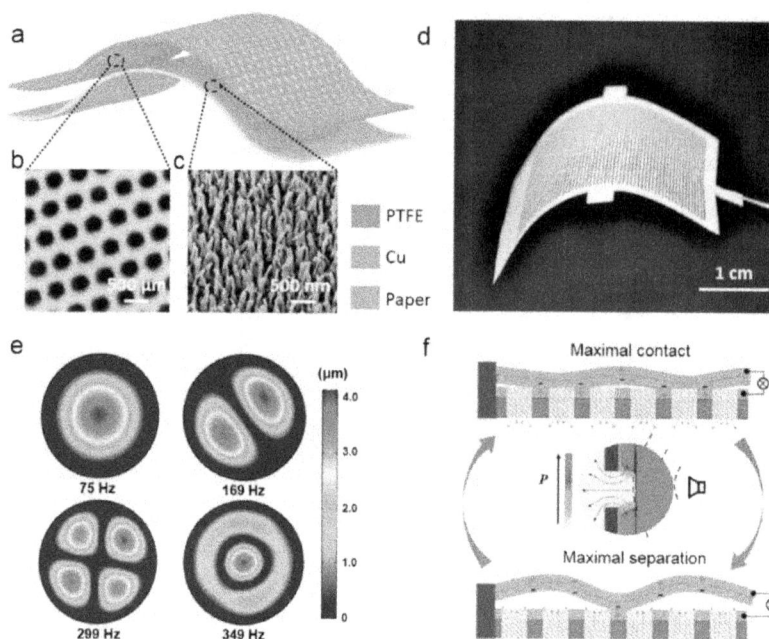

Figure 6.18. (a) Schematic illustration of the paper-based triboelectric energy harvester, (b) digital photo of the multihole paper electrode, (c) SEM image of the PTFE polymer nanowires, and (d) digital photo of the paper-based triboelectric energy harvester. (e) ANSYS analysis of PTFE membrane vibration under various sound frequencies. (f) Illustration to interpret the PTFE membrane vibration under sound and electricity generation. Reprinted with permission from [18], copyright (2015) American Chemical Society.

is completed. The authors studied the influence of the hole shape, central hole distribution, hole diameter, void-to-surface ratio, electrode thickness, and electrode substrate material on the electrical output and device frequency response.

Yu *et al* [69] designed a triboelectric nanogenerator for a sound-driven self-powered communication unit. The nanogenerator is composed of two frictional layers made of different materials. One contact surface is polytetrafluoroethylene (PTFE) film, which is surface modified by inductively coupled plasma (ICP) reactive ion etching. A copper (Cu) film was deposited as the electrode on the back of the PTFE film, which adheres to an acrylic plate with circular holes. The other contact surface is a copper film deposited on top of the Kapton film, which adheres to a circular acrylic plate. The resonant frequency of the nanogenerator is approximately 1.5 kHz, effectively driving electronic optical devices within a bandwidth of 1.3–1.65 kHz. Under 1.5 kHz frequency sound wave, the open-circuit voltage and short- circuit current are approximately 8.13 V and 10.31 μA, respectively. The maximum effective electrical power of the device is 18.38 μW when the load resistance is 0.29 MΩ.

6.4.2.3 Nanopillars
Kanik *et al* [70] constructed core–shell nanostructures, with PES as the core and As_2Se_3 as in the shell, for building a 3D-printed triboelectric energy harvester device.

In addition to an increase in the total surface area by using nanoscale materials, the application of multilayer device structure is also important to improve the overall energy output. Measurements at sound frequencies between 10 and 100 Hz revealed that the device has a better response at 10 Hz with a peak-to-peak open-circuit voltage of 107.3 V and a peak-to-peak short-circuit current of 0.49 μA. The output power density (max. 5.47 mW m^{-2}) of the 3D triboelectric energy harvester decreases with increasing sound frequency.

6.5 Potential applications

6.5.1 Sound sensor

With fantastic acoustoelectric energy conversion property, sound sensors based on nanofibrous materials can be applied in many scenarios, such as sound detection, sound recording and recognition.

6.5.1.1 Sound detection

The nanowire sound sensor with a Helmholtz resonator developed by Yang *et al* [17] can be used to localize acoustic sources with an error less than 7 cm in a plane (2 × 1.8 m^2). In their experiment, three nanowire sound sensors were arranged in an L shape and anchored in a two-dimensional plane (figure 6.19(a) and b)). When a balloon was burst on the two-dimensional plane, the output voltages were detected by the nanowire sound sensor and the localization of sound source can be calculated by functions. Two correlation functions of the three output voltages were derived from the cross-correlation function (in figure 6.19(c) and (d)). The time lag at the highest peaks of their cross-correlation functions is used to estimate the time difference between the two output voltages. The time lags of the two cross-correlation functions are zero (AS$_1$ and AS$_2$, AS$_2$ and AS$_3$), which indicates the same distance from the acoustic source to the ASs. This calculation result is well consistent with the experiments.

A laminated electrospun nanofibers triboelectric sound sensor reported by Xu *et al* [67] can be used as a warning sensor for real-time noise decibel monitoring, as shown in figure 6.20(a). The sound sensor first captures the noise energy, and the converted electrical output signal will be transmitted to the cloud platform for further processing by the charge amplification module. The real-time collected decibel signal will be displayed on a computer or mobile phone, as shown in figure 6.20(b). Once the collected decibel value exceeds the set threshold, an alarm signal will flash on the interface, and the decibel value will be stored in the background.

6.5.1.2 Sound recording and recognition

In addition to sensing the SPL and frequency of noise, nanofibrous sound sensors also have the ability to record human voices, for example, the pronunciation of 'one, two, three, four, five' [34], 'A, B, C, D, E' [47], and 'hello, bye' [46]. Figures 6.21(a) and (b) show voice-activated voltage outputs of a sound sensor made of electrospun PVDF nanofiber web and a commercial microphone for comparison. The black,

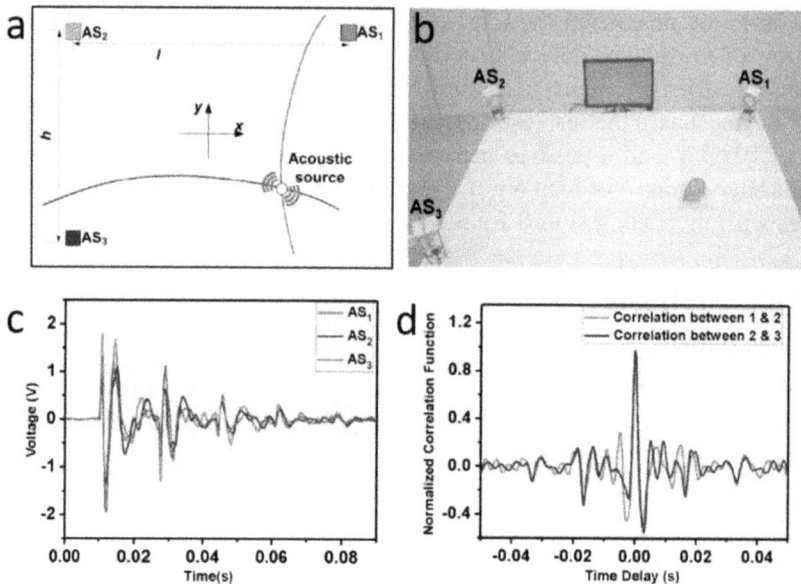

Figure 6.19. (a) Schematic illustration and (b) digital photo showing the working mechanism of sound sensors for sound localization, (c) output signals from three sound sensors when a balloon burst. (d) Correlation functions of the output signals from different sound sensors. Reprinted with permission from [17], copyright (2014) American Chemical Society.

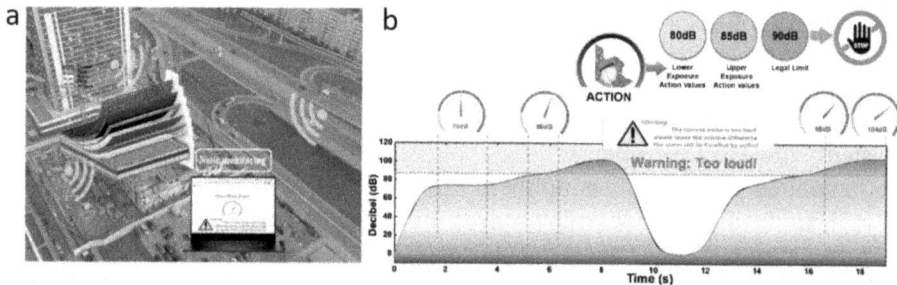

Figure 6.20. (a) and (b) Schematic illustration and the real-time decibel monitoring by the triboelectric device as a warning sensor for noise decibel monitoring. Reprinted from [67], copyright (2022), with permission from Elsevier.

orange, blue, pink, and green lines represent voltage outputs and FFT profiles of the words 'one', 'two', 'three', 'four', and 'five', respectively. From the voltage outputs and FFT-processed frequency spectra, the nanofibrous sound sensor exhibits excellent ability to record human voices. This sensing ability effectively enables the sensors to distinguish between male and female voices [46].

With a broad working bandwidth, the multihole paper-based sound sensor produced by Fan *et al* [18] was capable of acting as an active self-powered microphone for human voice recording (figure 6.22(a)). Compared to other existing

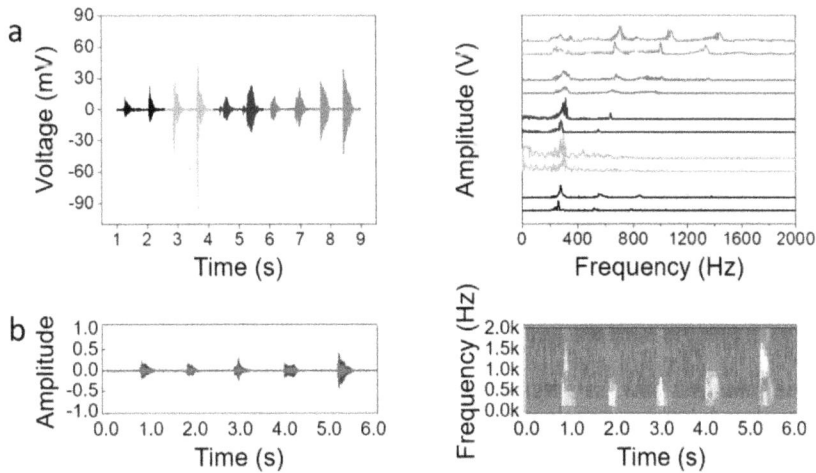

Figure 6.21. Voltage outputs of human voices saying 'one', 'two', 'three', 'four', and 'five', and the FFT-processed frequency spectrum recorded by (a) a sound sensor made of an electrospun PVDF nanofiber web and (b) a commercial microphone, respectively (SPL, 70–80 dB). Reprinted from [34] CC BY 4.0.

acoustic sensors, this paper-based sensor is rollable (figures 6.22(b) and (c)). The output of the rollable sensor with a symmetric circle pattern is independent of the sound wave's incident direction, as shown in figure 6.22(d). This feature makes the rollable sensor suitable for various sensing circumstances, such as theatre stage live recording, military surveillance, and omnibearing sensors.

Due to the fluctuation of people's voices with changes in environment, time, and emotions, the sensor is not suitable as a standard sound for scientific characterization. Shao *et al* [39] used the sound generated by standard instruments as a model to test the sound distinguishing performance of sensors. They selected eight instruments for testing, namely piano, cello, bass coil, organ, flute, guitar, trumpet, and violin. When these instruments play 'do re mi fa sol la ti' at C4 pitch, the waveform and frequency response characteristics of different notes and instruments are unique, which are ideal standard sound sources for evaluating acoustic sensors.

To demonstrate the excellent sensing ability, Peng *et al* [40] used nanofiber devices to record a piece of music 'High mountains and flowing water' and then played the electrical signals generated using music playing software on a computer. The waveform recorded by the PAN device had almost the same wave feature as the original sound wave. When the electrical signals were played on a computer, the one from the PAN device showed similarity to the original music.

6.5.2 Power supply

If the converted energy is large enough or harvested in a storage device, the acoustoelectric energy conversion based on nanofibrous materials can be applied in scenarios of power generation and supply, such as powering electronic devices or electrochemical reactions.

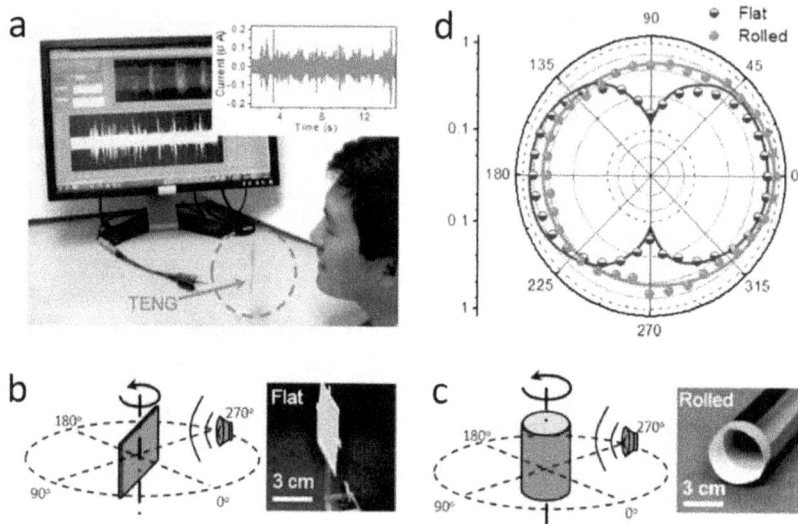

Figure 6.22. (a) Digital photo of a paper-thin triboelectric sound sensor as a self-powered microphone for sound recording. Inset is the output signals. Schematics showing the measurement of the directional patterns of the (b) flat and (c) rolled paper-thin triboelectric sound sensor. To the right are digital photos of the devices. (d) Directional patterns of the flat and rolled sound sensors. Reprinted with permission from [18], copyright (2015) American Chemical Society.

These acoustoelectric energy conversion devices that can generate a high level of electric signals can be directly used as power supplies for many microelectronic devices and electrochemical processes. Lang *et al* [42] reported a power generation device from a small piece of electrospun PVDF-TrFE nanofibers sandwiched with two pieces of PET electrode films, each of which had eight holes. Without pre-storage in a capacitor, the voltage generated from this device can light up 27 commercial LEDs (figure 6.23). Such an acoustoelectric generator may offer an effective solution to harvesting noise pollution into usable electricity for self-powered electronics.

Cui *et al* [16] reported a triboelectric sound energy harvester made of an electrospun PVDF nanofiber layer, two Al layers as triboelectric materials, and a copper mesh layer as the electrode. The energy harvester could instantaneously light up 138 LEDs under sound with an SPL 114 dB and a frequency 160 Hz. After continuously working for 7 days and vibrating 100 million cycles, there was no decay in current outputs.

The electricity generated by nanofiber devices after rectification can not only be directly used to power LEDs, but also run other electronic devices. The PAN-PVDF nanofiber device proposed by Shao *et al* [65] can directly light up 43 commercial LEDs simultaneously, as shown in figure 6.24(a). The nanofiber device can also directly power commercial liquid crystal displays (LCDs) electronic devices, such as timers and calculators. When noise is applied, the digital timer can operate (figure 6.24(b)). Calculators can also perform calculation processes under sound, i.e. '8 × 6 = 48' (figure 6.24(c)).

Figure 6.23. Electrospun nanofibrous acoustoelectric generators can light up commercial LEDs. Reprinted from [42], copyright (2017), with permission from Elsevier.

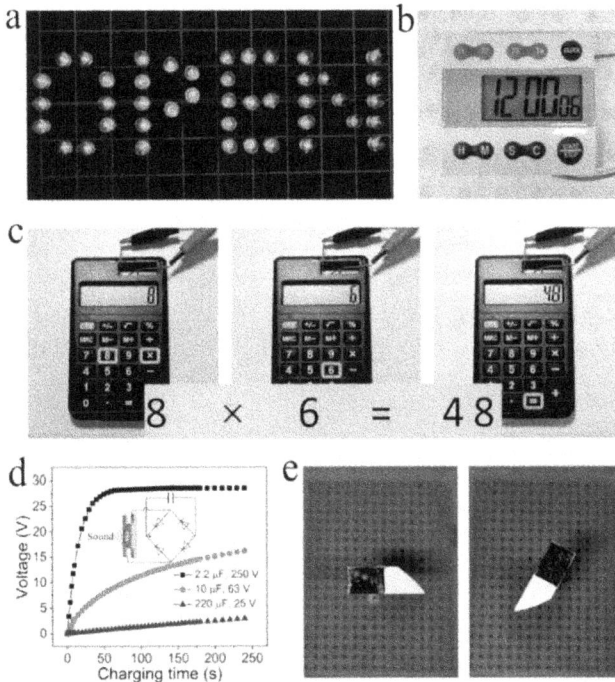

Figure 6.24. The electric power generated by nanofiber sound harvester can run (a) 43 commercial LEDs, (b) a timer, and (c) a digital calculator, separately. (d) The relationship of voltage and charging time for various capacitors powered by the nanofiber sound harvester. (e) Digital photos showing the rotation of a motor powered by the electric power generated by nanofiber sound harvester and stored in the capacitor. (The nanofibers were prepared from 7 wt% PAN + 7 wt% PVDF solution; nanofiber membrane thickness: 40 µm; sound source SPL 117 dB, 250 Hz, the electric energy was used after rectification.) Reprinted from [65], copyright (2021), with permission from Elsevier.

The electricity generated by the nanofiber device after rectification can also be stored in the energy bank for further use. Figure 6.24(d) shows the voltage changes of various capacitors charged by nanofiber acoustic devices. Under sound conditions of 117 dB 250 Hz, it took around 60 s to charge a 2.2 µF capacitor from 0 to 28.0 V.

In addition, charging a 10 μF or a 220 μF capacitor for 240 s can reach 16.2 or 2.9 V, respectively. After charging a 220 μF capacitor for 180 s, the power in the capacitor can drive a small electronic motor (figure 6.24(e)).

The combination of thermoacoustic engine/heat-treated PAN acoustoelectric equipment reported by Zheng *et al* [44] can convert different heat sources into electrical energy. When solar energy was used as a heat source to run the thermoacoustic generator, the heat-treated PAN device generates an output of over 100 V, which can directly charge the lithium battery (7.4 V, 300 mAh), and the voltage rises to 1.05 V within two hours. In addition, the recharged lithium battery can be used to run GPS, mobile phones, and smartwatches to assist in outdoor exploration (figure 6.25).

Peng *et al* [45] designed a self-powered wireless detector consisting of a five-slit acoustoelectric device and a wireless transmitter. The system can wirelessly report detection results without consuming external power. As shown in figures 6.26(a)–(c), the wireless transmitter consists of a five-slit unit (i.e. power supply, POW) and a commercial capacitor (C, 4.7 μF), a rectifier (LTC-3588), a wireless transmitter (FS1000A), a regulating resistor (R, 0–2 MΩ), an adjustment transistor (Q), and related circuits that connect them together. It can detect noise in various scenarios, such as high-speed trains, airports, highway traffic, and industry (figures 6.26(d)–(h)).

The electricity generated by nanofiber devices after rectification can also be used for electrochemical reactions. Lang *et al* [42] reported a power generation device from a small piece of electrospun P(VDF-TrFE) nanofibers sandwiched with two

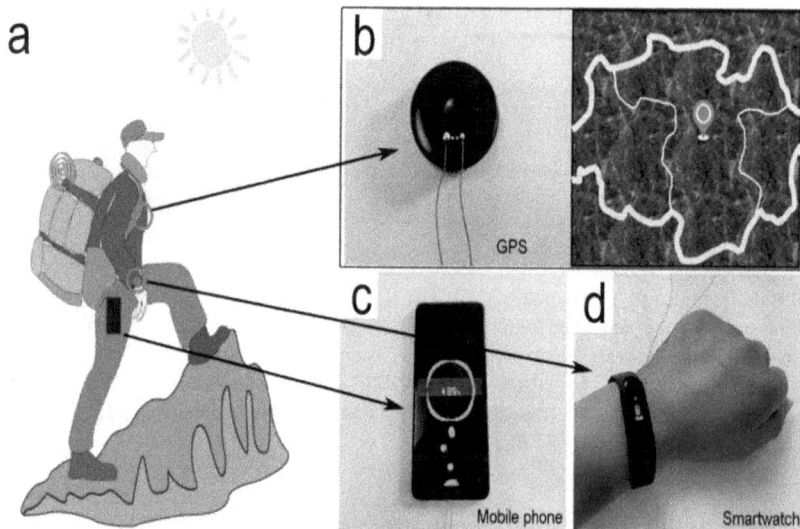

Figure 6.25. (a) Diagram to illustrate the idea of using thermoacoustic engine/heat-treated PAN energy harvester for an outdoor emergency power supply, (b) GPS locating, (c) mobile phones and (d) smartwatches (heat source for a–d: alcohol burner). Reprinted from [44], copyright (2022), with permission from Elsevier.

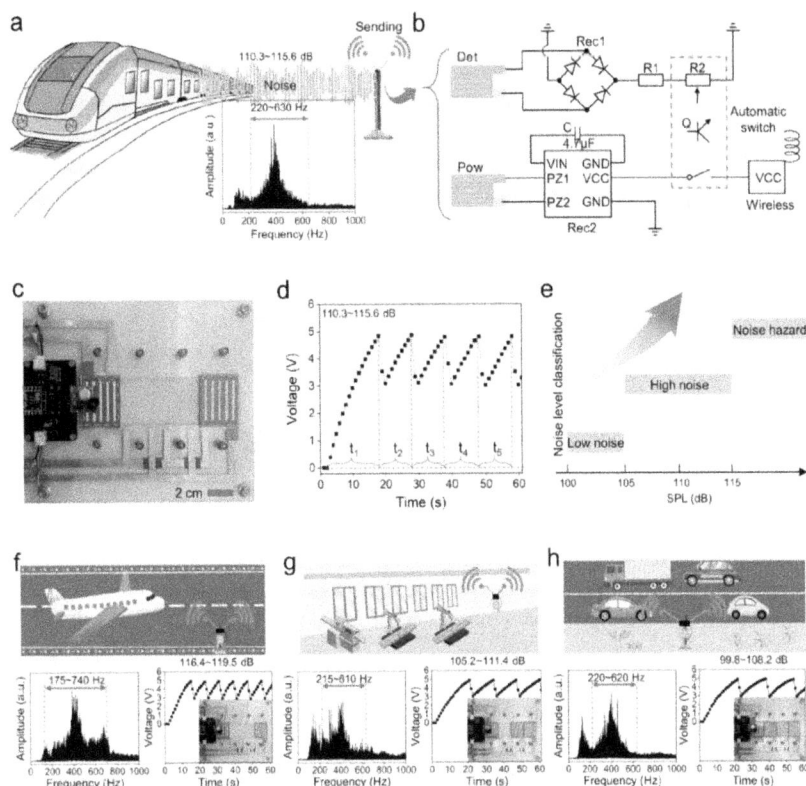

Figure 6.26. (a) Self-powered wireless noise sensor applied to high-speed train noise monitoring. (b) Circuit schematic diagram and (c) digital photo of self-powered wireless noise sensor. (d) Voltage across commercial capacitors (SPL: 110.3–115.6 dB). (e) Noise level classification. Self-powered wireless noise sensor applied to the (f) airport, (g) factory, and (h) highway noise monitoring. (a–e) One nanofiber device for the power supply unit. (f) One nanofiber device for power supply unit, (g) two and (h) three nanofiber devices for power supply unit; five-slit ⊥ device. Reprinted with permission from [45], copyright (2023) American Chemical Society.

pieces of PET electrode films, each of which had eight through holes. Without pre-storage in a capacitor, the voltage generated from this device can run electrochemical polymerization reactions and perform corrosion protection of metals (figure 6.27(a) and (b)).

Liu *et al* [61] demonstrated Fe(OH)₃ sol–gel electrophoresis using the electrical energy converted by the three-dimensional integrated triboelectric nanogenerator with five units under the sound wave. As electrophoresis progresses, the color near the anode changes from red to colorless, and the color near the cathode changes to dark red.

The single-electrode triboelectric nanogenerator for harvesting the acoustic energy reported by Qiu *et al* [63] can be used to carry out a self-powered electrochromin device (figures 6.28(a)–(c)). Under sound, the electrochromic device gradually changes from transparent white to dark blue (figure 6.28(d)). When the

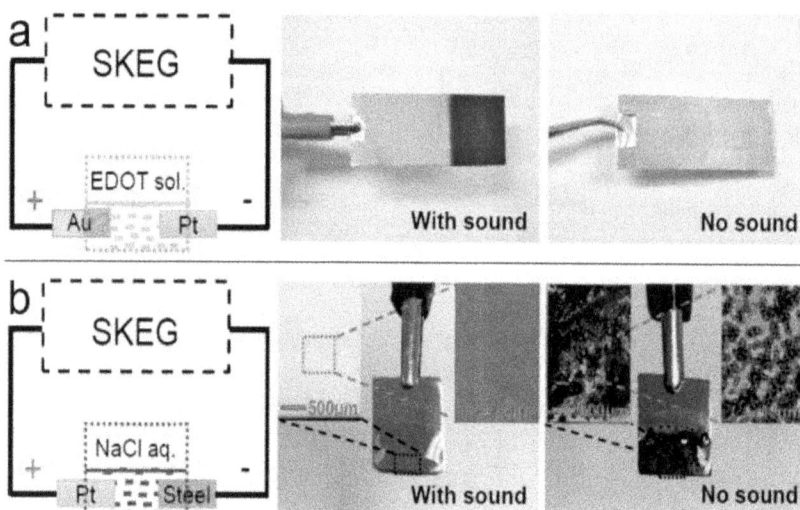

Figure 6.27. Electrospun nanofibrous acoustoelectric generators can (a) run polymerization of EDOT, (b) perform metal corrosion protection (no energy storage device was used in two demonstrations). Reprinted from [42], copyright (2017), with permission from Elsevier.

switch dote is reversed, the color of the electrochromic film will become transparent white. This indicates that triboelectric nanogenerator can be widely applied in many electronic fields, such as switchable mirrors, flexible electronic paper, etc.

Table 6.3 lists most of the current reports on acoustoelectric transducers based on nanofibrous materials. Acoustoelectric transducers made of electrospun piezoelectric polymer nanofibers show high output voltage both based on piezoelectricity and triboelectrification principles. However, acoustoelectric transducers made of nanowires based on triboelectrification are better energy conversion than those based on piezoelectricity.

6.6 Conclusions

Owing to their high aspect ratio, high specific surface area, and surface charge density, acoustoelectric sensors and energy harvesters based on piezoelectric nanofibrous materials are not only produced by simpler fabrication processes, but also show higher sensitivity and energy conversion efficiency than those based on conventional techniques and materials. Nanofibrous materials, especially for electrospun nanofibers, are lightweight and have great flexibility, which makes them particularly suitable for the emerging applications of wearable electronics.

Electrospinning is a one-step process to produce piezoelectric nanofibers with strong piezoelectricity. It is not a tedious multistep method involving mechanical stretching and high-voltage poling at an elevated temperature. At the same time, there is a wide range of materials on the triboelectric series that can be made into nanofibers for triboelectric energy harvesting. The research on acoustoelectric energy conversion is still at its early stage, with ongoing efforts in improving materials fabrication, transducer device structure, electrical outputs, and responsive frequency range, and it can be expected that many new applications will be explored with nanofibrous materials.

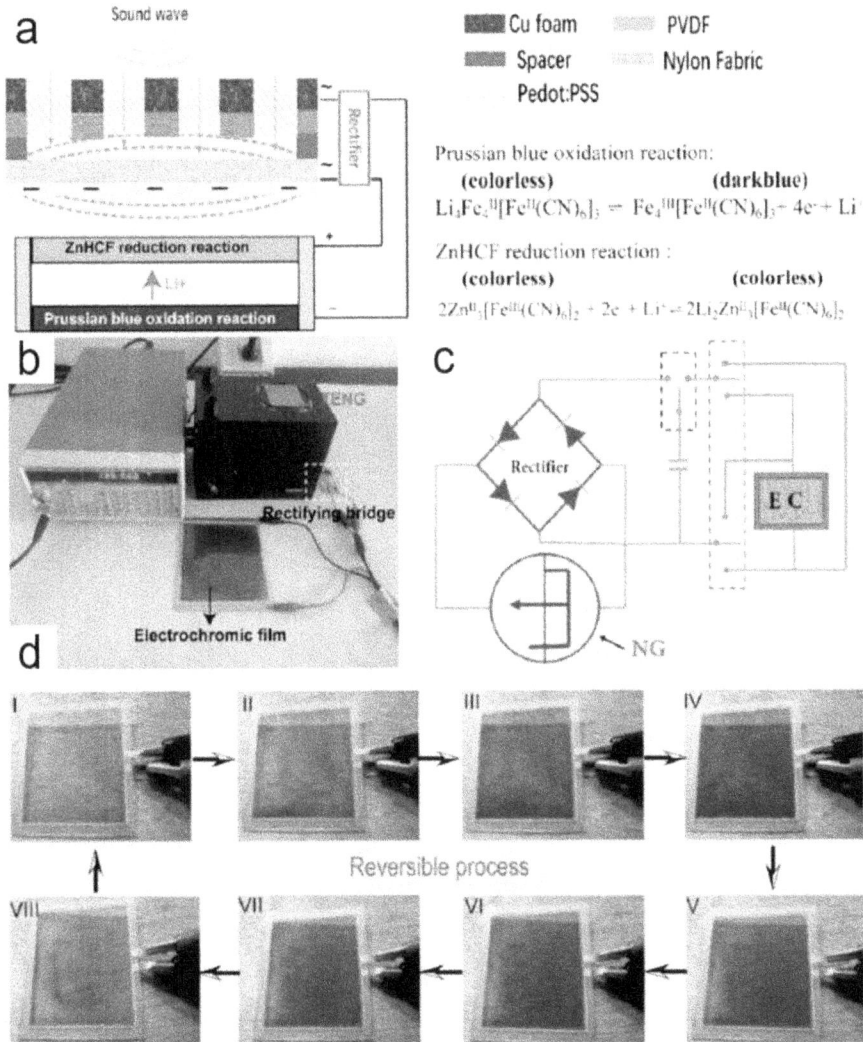

Figure 6.28. (a) Mechanism illustration of electrochromic film driven by triboelectric nanogenerator. (b) The optical photo of the experiment setup. (c) The schematic illustration of a self-powered EC system. (d) The color change process of the electrochromic film after charging by triboelectric nanogenerator (I) Initial moment (II) after 5 s (III) after 10 s (IV) after 15 s (V–VIII) The color change process of the electrochromic film after reduction. Reprinted from [63], copyright (2017), with permission from Elsevier.

Table 6.3. Comparison of the reported acoustoelectric transducers based on nanofibrous materials.

Material	Design	Principle	Device parameters		Acoustic wave			Energy output				References
			Working unit size	R_{int} (kΩ)	SPL (dB)	f_r (Hz)	V_m (V)	I_m (μA)	Sensitivity (V Pa^{-1})	P (μW)	P_V/P_A	
					Acoustic sensors							
PVDF nanofibers	Electrospun PVDF nanofiber web clamped by two gold-coated PET films with one hole.	Piezoelectric	$4 \times 3 \times 0.004$ cm^3		115	220	3.1		0.27			[34]
PVDF-HFP composite nanofibers	Electrospun Eu^{3+}-doped PVDF-HFP/graphene composite nanofibers.	Piezoelectric			82				11			[46]
BZT–BCT nanofibers	Polydimethylsiloxane (PDMS) and BZT–BCT nanofibers are assembled into a piece of membrane. The spiral silver electrode was deposited on the BZT–BCT membrane through magnetron sputtering using a shadow mask.	Piezoelectric	$2 \times 2 \times 0.0005$ cm^3			300	1.8	0.67				[35]
PVDF-TrFE nanofibers	Electrospun PVDF-TrFE nanofiber membrane clamped by two copper electrodes on PLA with circular holes.	Piezoelectric	Circular hole diameter: 10, 16, 20, and 30 mm; thickness, 60 μm		78	100–400	1.7×10^{-2}					[36]
PVDF-TrFE nanofibers	The PVDF-TrFE nanofiber mesh acoustic sensor fabricated by dynamic	Piezoelectric	Nanofiber mesh area:1.8 cm^2;		95	250	0.1		8.9×10^{-2}			[37]

(Continued)

Table 6.3. (*Continued*)

Material	Design	Principle	Device parameters		Acoustic wave				Energy output			References
			Working unit size	R_{int} (kΩ)	SPL (dB)	f_r (Hz)	Vm (V)	Im (μA)	Sensitivity (V Pa^{-1})	P (μW)	Pv/P$_A$	
PVDF nanofibers	near-field electrospinning. Electrospun PVDF nanofibers membrane.	Piezoelectric	thickness, 307 nm			4000	0.33		2.3×10^{-4}			[71]
PVDF composite nanofibers	Electrospun PVDF-IGO composite nanofibers.	Piezoelectric	Fiber diameter, 110 nm		94	15 000	0.0125		3.115×10^{-3}			[48]
PAN nanofibers	Electrospun PAN nanofiber membrane clamped by two gold-coated PET films with eight holes.	Piezoelectric	4×4 cm^2 $3 \times 4 \times 0.003$ cm^3		115	230	40.7		23.401			[39]
PAN nanofibers	Electrospun PAN nanofiber membrane clamped by two gold-coated PET films with one hole.	Piezoelectric	$3 \times 4 \times 0.003$ cm^3		115	230	4.46		0.3519			[40]
PVDF nanofibers	Electrospun PVDF nanofiber membrane clamped by two gold-coated electrospun PU nanofiber membrane electrode with a supporting frame with open windows. A parylene layer was deposited to improve the mechanical durability at the fiber-to-fiber joints.	Piezoelectric and Triboelectric	$2.5 \times 2.5 \times 0.000\,26$ cm^3		115	250	70.9		10.0506			[72]
PVDF	A dispersion of single-walled carbon nanotubes was sprayed onto the both sides of	Piezoelectric	$5 \times 5 \times 0.02$ cm^3			150						[73]

Material	Description	Mechanism	Dimensions									Ref.
	PVDF webs using a coaxial nozzle.											
PVDF nanofiber	Electrospun wave-shaped PVDF nanofiber membrane with Al electrodes on both sides.	Piezoelectric			110	100	2.8×10^{-3}					[74]
PMMA nanofibers	Au-coated aligned electrospun PMMA nanofiber mesh placed in a magnetic field.	Electromagnetic	$0.8 \times 0.8 \times 0.0003$ cm^3									[41]
Acoustic energy harvesters												
PVDF composite nanofibers	Electrospun Ce^{3+}-doped PVDF/graphene composite nanofibers.	Piezoelectric	3×4 cm^2		88	110	2.6					[49]
W-ZnS nanorods and PVA nanofibers	Electrospun W-ZnS nanorods and PVA nanofiber membrane clamped with the aluminum foil and a conductive adhesive carbon tape.	Piezoelectric	Three-layer stacking membrane: $3.5 \times 2.5 \times 0.028$ cm^3	40 000	100		4	2	0.4165		$1.7\ \mu$W cm^{-3}	[75]
PVDF-TrFE nanofibers	Electrospun PVDF-TrFE nanofiber membrane clamped by two gold-coated PET films with eight holes.	Piezoelectric	$4 \times 3 \times 0.002$ cm^3	470	115	210	14.5	28.5	1.3	141.3	$306.5\ \mu$W cm^{-3}	[42]
PVDF composite nanofibers	Electrospun PVDF-MoS$_2$ composite nanofibers.	Piezoelectric	80×0.015 cm^3		90		12	2×10^{-3}	19	2×10^{-4}	$1.67 \times 10^{-4}\ \mu$W cm^{-3}	[47]
PVDF nanofibers	Electrospun MAPbBr doped PVDF nanofiber membrane clamped by nickel and copper coated polyester conducting fabric.	Piezoelectric	9.5×8.5 cm^2	850	85		5		13.8	4		[50]
PVDF nanofibers	Electrospun TiO$_2$ doped PVDF nanofiber	Piezoelectric	$9 \times 7.5 \times 0.0250$ cm^3		90		8.75	26	4		$2.37\ \mu$W cm^{-3}	[51]

(Continued)

Table 6.3. (*Continued*)

Material	Design	Principle	Device parameters		Acoustic wave				Energy output			References
			Working unit size	R_{int} (kΩ)	SPL (dB)	f_r (Hz)	Vm (V)	Im (μA)	Sensitivity (V Pa^{-1})	P (μW)	Pv/P$_A$	
	membrane clamped by two conducting fabrics.											
PVDF nanofiber	Electrospun ZnS-NRs doped PVDF nanofiber membrane clamped by two conducting fabrics.	Piezoelectric	Conducting fabric: 8 × 8 cm^2	1000	100	86	6		3		0.15 μW cm^{-2}	[53]
PVDF nanofiber	Electrospun PVDF–ZnO nanofiber membrane is clamped by PET films with copper tapes stuck on. Nine through holes were equally distributed on each electrode.	Piezoelectric	PVDF–ZnO membrane:3 × 3 × 0.004 cm^3 PET films:4 × 4 × 0.01 cm^3	700	116	140	1.7			1.792	50 μW cm^{-3} (0.2 μW cm^{-2})	[76]
PVDF nanofiber	Electrospun PVDF membrane as the active layer and SWCNT-coated PVDF membrane as durable electrodes.	Piezoelectric	PVDF web:4 × 4 × 0.0331 cm^3	1750		220	5.24			4.8		[77]
PAN nanofibers	Electrospun PAN nanofiber membrane clamped by two gold-coated PET films with eight holes.	Piezoelectric	4 × 3 × 0.003 cm^3	4000	117	230	58	12		210.3		[43]
PVDF nanofber	Electrospun MOF/PVDF composite nanofiber membrane sandwitched between two ITO coated PET sheet.	Piezoelectric	4 × 4 × 0.006 cm^3	1000	120	110	6		0.95	6.25		[55]

PAN nanofiber	Electrospun PAN nanofiber membrane sandwiched between two aluminum foils and PI films with eight through holes.	Piezoelectric	$3 \times 3 \times 0.004$ cm^3	3500	118	230	118	12	352.8 μW	39.2 μW cm^{-2}	[44]
PVDF nanofibers	Electrospun PVDF nanofibers and Al layer with stainless steel mesh electrodes.	Triboelectric	Vibrating membrane $10 \times 10 \times 0.0005$ cm^3	0.1	100	175	90	180		202 mW m^{-2}	[16]
PVDF nanofiber	A multihole PVC plate covered with a layer of Al. A PE film with Ag electrode on one side and PVDF nanofibers on the other side (Part A), a PE film with Ag electrode on one side (Part B), and two spacers are glued along the edges of the PVC plate on both sides.	Triboelectric	10×10 cm^2 Thickness: PVC: 0.5 mm PE: 5 μm Spacer:30 μm × 2		110	200	72	6.6×10^2			[78]
PVDF nanofiber	Layer A is composed of one spacer, one Al-coated porous PVC shelf and a layer of PVDF nanofibers. Layer B consists of one spacer, one Al layer, one polyethylene (PE) substrate and a layer of PVDF nanofibers.	Triboelectric	9.5×9.5 cm^2 Thickness: PVC: 0.5 mm Al: 100 nm × 2 PE: 5 μm Spacer:30 μm × 2	0.1	100.5	200	232	2.1×10^3	48 869.47	5414.9 mW m^{-2}	[61]
PVDF nanofiber	Electrospun PVDF nanofiber membrane clamped by two conductive fabrics.	Triboelectric	38 cm^2	2000	115	170	400	175	28 000	7 W m^{-2}	[62]

(Continued)

Table 6.3. (*Continued*)

Material	Design	Principle	Device parameters		Acoustic wave				Energy output			References
			Working unit size	R_{int} (kΩ)	SPL (dB)	f_r (Hz)	Vm (V)	Im (μA)	Sensitivity (V Pa^{-1})	P (μW)	Pv/P$_A$	
PVDF nanofiber	PVDF nanofibers were deposited on the Cu foam by electrospinning. PEDOT: PSS was spin-coated on the surface of the nylon fabric to make a conductive film. A spacer layer was adhered to the four sides of the nylon fabric, and PVDF nanofibers were attached onto it.	Triboelectric	Area: 25 cm² Thickness PVDF:20 μm Cu foam: 1.5 mm nanowire: 21.4μm Nylon fabric: 96.6 μm Spacer: 20μm	5000	125	110	546.3	60.9		1640		[63]
P(VDF-TrFE) nanofibers,	The conductive fabric, Kapton and P(VDF-TrFE) constructed sandwiched TENG glued to one side of the acrylic sheet framework.	Triboelectric	Thickness Kapton: 65 μm Conductive fabric: 80μm	1000	115	180	80	19		500		[64]
PVDF nanofiber and nylon nanofiber	Electrospun PVDF and nylon nanofiber membranes are utilized as the negative and positive triboelectric layers, and attached to the PU foam and Cu mesh, respectively.	Triboelectric		10 000	104	200	170	33	53.6		1.28×10^2 μW cm^{-2}	[67]
PAN nanofiber and PA6 nanofiber	Electrospun PAN and PA6 nanofiber membranes coated	Triboelectric	Diameters of the circular film: 30,		104	400	0.612		0.715			[68]

Material	Description	Type	Dimensions							Output	Ref.
PVDF nanofibers	with Ag are assembled together.										
	Coaxial electrospun PVDF hollow fiber membrane clamped by PDMS valve and Iron mesh.	Piezoelectric and triboelectric	40, 50 mm Thickness: 40 μm; 4.5 × 3 × 0.0026 cm³; two polyimide films:10 × 10 cm², with a hole of 3.5 × 2 cm²; iron meshes: 10 × 10 cm²; wire diameter, 0.2 mm; density, 400 mesh	10 000	117.6	150	105.5	16.7	650	0.92 W m^{-2}	[79]
PAN-PVDF nanofiber	Electrospun PAN-PVDF nanofiber membrane clamped by two gold-coated PET films with eight holes.	Piezoelectric and triboelectric	3 × 4 × 0.004 cm³	6600	117	250	94.10	17.40	300.1	6252.5 μW cm^{-3} 250.1 mW m^{-2}	[65]
PVDF nanofiber	Electrospun PVDF nanofiber and a small number of nylon-6 nanofiber mebrane clamped by two stainless steel meshes.	Piezoelectric and triboelectric	3 × 4 × 0.004 cm³	4700	118	230	201.4	17.6	1.57 × 10³	130.8 μW cm^{-2}	[66]
PAN-PVDF nanofiber	Electrospun PAN/PVDF/ MWCNTs nanofiber membrane and a frame/spacer clamped by two copper meshes and two polyethylene PET outside frames.	Piezoelectric and triboelectric	4 × 4 × 0.192 cm³ Effect area:2.5 × 2.5 cm²	10 000	116	200	126.5	30.2	1.41 mW	2.25 W m^{-2}	[56]

(Continued)

Table 6.3. (*Continued*)

Material	Design	Principle	Device parameters		Acoustic wave		V_m (V)	I_m (μA)	Energy output			References
			Working unit size	R_{int} (kΩ)	SPL (dB)	f_r (Hz)			Sensitivity (V Pa^{-1})	P (μW)	Pv/P$_A$	
PAN nanofiber	The PS/PA66 and PAN/CNC bilayer membranes clamped by two pieces of aluminum foil and then sealed on both sides with PET film.	Piezoelectric		100	110	230	10.92	12.9		7.8		[57]
PAN nanofiber	Electrospun PAN oriented nanofiber membrane clamped by two gold-coated PET films with five slits.	Piezoelectric	$3 \times 4 \times 0.0033$ cm^3	6150	115	382	52.72	7.97		152.41		[45]
ZnO nanowires	Aligned ZnO nanowires covered by a zigzag silicon electrode coated with platinum.	Piezoelectric	Area: 0.02 cm^2, nanowires: density, 10^9 cm^{-2}, height, 0.0001 cm; diameter, 0.000 004 cm	3.56		Ultrasonic wave 41 000	7×10^{-4}	1.50×10^{-4}		1.00×10^{-6}		[58]
ZnO nanowires	Flexible PdAu-coated PES substrate acting as a top electrode and ZnO nanowire arrays on a GaN/sapphire substrate.	Piezoelectric	Nanowires: length, 0.001 cm; diameter, 0.000 015 cm		100	100	5×10^{-2}					[59]
ZnO nanowires	Laterally bonded single wires or vertically aligned nanowire arrays.	Piezoelectric	Nanowires: length, 0.01 cm; diameter, 0.000 08 cm;	100		100	5×10^{-4}	1.2×10^{-2}			5 pW cm^{-2}	[60]

Material	Type	Description	Dimensions								Ref.
PTFE nanowires	Triboelectric	Nanowire-based PTFE thin film triboelectric energy harvester with a Helmholtz resonator and nanoporous Al electrode and copper electrode.	Radius, 6.5 cm; nanowires: diameter, 0.000 0054 cm; length, 0.000 11 cm	6000	110	240	60.5	15.1			[17]
PTFE nanowires	Triboelectric	Nanowire-based PTFE thin film triboelectric energy harvester with multihole copper electrode.	Thickness, 0.0025 cm	800	117	250	22	17	194	968.0 μW cm^{-3}	[18]
PTFE nanowires	Triboelectric	One contact surface is PTFE nanowires film with a Cu thin film, another is a copper thin film deposited on top of a Kapton film. Both of two are adhered onto an acrylic plate with a circular hole, respectively.	Circular hole diameter was 55/60 mm Thickness: Acrylic plate: 20 mm, Cu film: 40 nm	290		1500	8.125	10.31		18.38	[69]
PES nanopillars	Triboelectric	Core-shell nanostructures (PES in the core and As$_2$Se$_3$ in the shell) for building a 3D-printed multilayered device.	$17.5 \times 5 \times 4.9$ cm^3		87.98	10	53.65	0.245	52.5	5.47 mW m^{-2}	[70]

Acknowledgments

Funding support from the National Natural Science Foundation of China (Grant No. 51803108), Fujian Provincial Natural Science Foundation of China (Grant No. 2021J01977), and Science Foundation of Zhejiang Sci-Tech University (ZSTU) (Grant No. 22202251-Y, No. 22202253-Y) is acknowledged.

References

[1] Blumstein D T *et al* 2011 *J. Appl. Ecol.* **48** 758
[2] Liang S Y, Hecker R L and Landers R G 2004 *J. Manuf. Sci. Eng.-Trans. ASME* **126** 297
[3] Stansfeld S, Haines M and Brown B 2000 *Rev. Environ. Health* **15** 43
[4] Barnett G H, Kormos D W, Steiner C P and Weisenberger J 1993 *J. Neurosurg.* **78** 510
[5] Ghiurcau M V, Rusu C, Bilcu R C and Astola J 2012 *Signal Process.* **92** 829
[6] Lai T, Huang C and Tsou C 2008 Presented at *Symp. on Design, Test, Integration and Packaging of MEMS/MOEMS, 2008. MEMS/MOEMS 2008*
[7] Khan F U 2013 Presented at *16th Int. Multi Topic Conf. (INMIC), 2013*
[8] Bhat R 2014 *Int. J. Sci. Res.* **3** 6
[9] Chen C-C, Yan W-Y, Wu Y-Y and Ting C-C 2014 *Int. J. Eng. Technol.* **4** 86
[10] Khan F U 2016 *Sādhanā* **41** 397
[11] Scheeper P R, van der Donk A G H, Olthuis W and Bergveld P 1994 *Sens. Actuators* A **44** 1
[12] Torkkeli A, Rusanen O, Saarilahti J, Seppa H, Sipola H and Hietanen J 2000 *Sens. Actuators* A **85** 116
[13] Consales M, Ricciardi A, Crescitelli A, Esposito E, Cutolo A and Cusano A 2012 *ACS Nano* **6** 3163
[14] Toda M and Thompson M L 2006 *IEEE Sens. J.* **6** 1170
[15] Xu J, Dapino M J, Gallego-Perez D and Hansford D 2009 *Sens. Actuators* A **153** 24
[16] Cui N, Gu L, Liu J, Bai S, Qiu J, Fu J, Kou X, Liu H, Qin Y and Wang Z L 2015 *Nano Energy* **15** 321
[17] Yang J, Chen J, Liu Y, Yang W, Su Y and Wang Z L 2014 *ACS Nano* **8** 2649
[18] Fan X, Chen J, Yang J, Bai P, Li Z and Wang Z L 2015 *ACS Nano* **9** 4236
[19] Izhar Khan F U 2015 *J. Micromech. Microeng.* **25** 023001
[20] Zhu G, Peng B, Chen J, Jing Q and Lin Wang Z 2015 *Nano Energy* **14** 126
[21] Lang C, Wang H, Fang J, Jin J, Peng L and Lin T 2024 *Nano Energy* **130** 110117
[22] Lovinger A J 1983 *Science* **220** 1115
[23] Fang J, Wang X and Lin T 2011 *J. Mater. Chem.* **21** 11088
[24] Mandal D, Yoon S and Kim K J 2011 *Macromol. Rapid. Commun.* **32** 831
[25] Persano L, Dagdeviren C, Su Y, Zhang Y, Girardo S, Pisignano D, Huang Y and Rogers J A 2013 *Nat. Commun.* **4** 1633
[26] Wang Y R, Zheng J M, Ren G Y, Zhang P H and Xu C 2011 *Smart Mater. Struct.* **20** 045009
[27] Ren G, Cai F, Li B, Zheng J and Xu C 2013 *Macromol. Mater. Eng.* **298** 541
[28] Fang J, Niu H, Wang H, Wang X and Lin T 2013 *Energy Environ. Sci.* **6** 2196
[29] Li B, Xu C, Zheng J and Xu C 2014 *Sensors* **14** 9889
[30] Mandal D, Henkel K and Schmeisser D 2014 *Phys. Chem. Chem. Phys.* **16** 10403
[31] Adhikary P, Garain S and Mandal D 2015 *Phys. Chem. Chem. Phys.* **17** 7275
[32] Chang C, Tran V H, Wang J, Fuh Y K and Lin L 2010 *Nano Lett.* **10** 726

[33] Hansen B J, Liu Y, Yang R and Wang Z L 2010 *ACS Nano* **4** 3647

[34] Lang C, Fang J, Shao H, Ding X and Lin T 2016 *Nat. Commun.* **7** 11108

[35] Cui N, Jia X, Lin A, Liu J, Bai S, Zhang L, Qin Y, Yang R, Zhou F and Li Y 2019 *Nanoscale Adv.* **1** 4909

[36] Viola G *et al* 2020 *ACS Appl. Mater. Interfaces* **12** 34643

[37] Wang W, Stipp P N, Ouaras K, Fathi S and Huang Y Y S 2020 *Small* **16** 2000581

[38] Wang W, Zheng Y, Jin X, Sun Y, Lu B, Wang H, Fang J, Shao H and Lin T 2019 *Nano Energy* **56** 588

[39] Shao H, Wang H, Cao Y, Ding X, Fang J, Wang W, Jin X, Peng L, Zhang D and Lin T 2021 *Adv. Electron. Mater.* **7** 2100206

[40] Peng L, Jin X, Niu J, Wang W, Wang H, Shao H, Lang C and Lin T 2021 *J. Mater. Chem.* C **9** 3477

[41] Zhou J, Li B, Liu J, Jones W E and Miles R N 2018 *J. Micromech. Microeng.* **28** 095003

[42] Lang C, Fang J, Shao H, Wang H, Yan G, Ding X and Lin T 2017 *Nano Energy* **35** 146

[43] Shao H, Wang H, Cao Y, Ding X, Fang J, Niu H, Wang W, Lang C and Lin T 2020 *Nano Energy* **75** 104956

[44] Zheng Y, Wang W, Niu J, Jin X, Sun Y, Peng L, Li W, Wang H and Lin T 2022 *Nano Energy* **95** 106995

[45] Peng L, Niu J, Jiang P, Han X, Jin X, Liu X, Wang W, Lang C, Wang H and Lin T 2023 *ACS Appl. Mater. Interfaces* **15** 29127

[46] Adhikary P, Biswas A and Mandal D 2016 *Nanotechnology* **27** 495501

[47] Maity K, Mahanty B, Sinha T K, Garain S, Biswas A, Ghosh S K, Manna S, Ray S K and Mandal D 2017 *Energy Technol.* **5** 234

[48] Sekkarapatti Ramasamy M, Rahaman A and Kim B 2021 *Ceram. Int.* **47** 11010

[49] Garain S, Jana S, Sinha T K and Mandal D 2016 *ACS Appl. Mater. Interfaces* **8** 4532

[50] Sultana A, Alam M M, Sadhukhan P, Ghorai U K, Das S, Middya T R and Mandal D 2018 *Nano Energy* **49** 380

[51] Alam M M, Sultana A and Mandal D 2018 *ACS Appl. Energy Mater.* **1** 3103

[52] Cheong O J, Lee J S, Kim J H and Jang J 2016 *Small* **12** 2567

[53] Sultana A, Alam M M, Ghosh S K, Middya T R and Mandal D 2019 *Energy* **166** 963

[54] Wu C M and Chou M H 2020 *Express Polym. Lett.* **14** 103

[55] Roy K, Jana S, Mallick Z, Ghosh S K, Dutta B, Sarkar S, Sinha C and Mandal D 2021 *Langmuir* **37** 7107

[56] Sun W, Ji G, Chen J, Sui D, Zhou J and Huber J 2023 *Nano Energy* **108** 108248

[57] Fan Z, Wu S, Fang K, Tang F, Zhang L and Huang F 2023 *J. Mater. Chem.* A **11** 13378

[58] Wang X, Song J, Liu J and Wang Z L 2007 *Science* **316** 102

[59] Cha S N, Seo J S, Kim S M, Kim H J, Park Y J, Kim S W and Kim J M 2010 *Adv. Mater.* **22** 4726

[60] Lee M, Yang R, Li C and Wang Z L 2010 *J. Phys. Chem. Lett.* **1** 2929

[61] Liu J, Cui N, Gu L, Chen X, Bai S, Zheng Y, Hu C and Qin Y 2016 *Nanoscale* **8** 4938

[62] Chen F, Wu Y, Ding Z, Xia X, Li S, Zheng H, Diao C, Yue G and Zi Y 2019 *Nano Energy* **56** 241

[63] Qiu W, Feng Y, Luo N, Chen S and Wang D 2020 *Nano Energy* **70** 104543

[64] Wang F, Wang Z, Zhou Y, Fu C, Chen F, Zhang Y, Lu H, Wu Y, Chen L and Zheng H 2020 *Nano Energy* **78** 105244

[65] Shao H, Wang H, Cao Y, Ding X, Bai R, Chang H, Fang J, Jin X, Wang W and Lin T 2021 *Nano Energy* **89** 106427

[66] Niu H, Zhou H, Shao H, Wang H, Ding X, Bai R and Lin T 2021 *Nano Energy* **90** 106618

[67] Xu W *et al* 2022 *Nano Energy* **99** 107348

[68] Jiang Y, Zhang Y, Ning C, Ji Q, Peng X, Dong K and Wang Z L 2022 *Small* **18** 2106960

[69] Yu A *et al* 2016 *ACS Nano* **10** 3944

[70] Kanik M, Say M G, Daglar B, Yavuz A F, Dolas M H, El-Ashry M M and Bayindir M 2015 *Adv. Mater.* **27** 2367

[71] Shehata N, Hassanin A H, Elnabawy E, Nair R, Bhat S A and Kandas I 2020 *Sensors* **20** 3111

[72] Nayeem M O G, Lee S, Jin H, Matsuhisa N, Jinno H, Miyamoto A, Yokota T and Someya T 2020 *Proc. Natl Acad. Sci.* **117** 7063

[73] Lim J and Kim H S 2021 *Sens. Actuators* A **330** 112840

[74] Xu F, Yang J, Dong R, Jiang H, Wang C, Liu W, Jiang Z, Zhang X and Zhu G 2021 *Adv. Fiber Mater.* **3** 368

[75] Sultana A, Alam M M, Biswas A, Middya T R and Mandal D 2016 *Transl. Mater. Res.* **3** 045001

[76] Sun B, Li X, Zhao R, Ji H, Qiu J, Zhang N, He D and Wang C 2019 *J. Mater. Sci.* **54** 2754

[77] Hwang Y J, Choi S and Kim H S 2019 *Sens. Actuators* A **300** 111672

[78] Gu L, Cui N, Liu J, Zheng Y, Bai S and Qin Y 2015 *Nanoscale* **7** 18049

[79] Yu Z, Chen M, Wang Y, Zheng J, Zhang Y, Zhou H and Li D 2021 *ACS Appl. Mater. Interfaces* **13** 26981

IOP Publishing

Energy Harvesting Properties of Electrospun Nanofibers
(Second Edition)

Jian Fang and Tong Lin

Chapter 7

Polyacrylonitrile as piezoelectric materials working at high-temperature

Wenyu Wang, Xin Jin, Junzhu Tao and Xuekai Zheng

7.1 Introduction

Compared with inorganic piezoelectric materials, piezoelectric polymer materials have been extensively studied as key materials for small kinetic energy harvesting because of their high flexibility, good processability, and the possibility of a large-area and curved surface preparation. They have been used widely in energy harvesting applications such as piezoelectric nanogenerators (PENG), electronic information, artificial intelligence and sensors. Polymer piezoelectric materials such as polyvinylidene fluoride (PVDF) and its copolymers, poly (L-lactic acid) (PLLA), and polyamide-11 (PA-11) have been extensively investigated as key materials for small kinetic energy harvesting in the past decades. Among them, PVDF and PVDF-based copolymer have subsequently become the most studied piezoelectric polymer in the past half century. However, these polymer piezoelectric materials tend to exhibit low piezoelectric properties at room temperature, and especially lose piezoelectric properties at high temperatures [1–3].

In recent years, the rapid growth in the automotive, petroleum, nuclear and aerospace industries have stimulated demand for pressure sensors for testing related to this sector, sustainable self-powered self-storage devices, and smart clothing for health testing of operators in high-temperature environments [1–3]. The urgent demand for high-temperature, high-stability piezoelectric materials in aerospace, deep oil exploration and other fields has gradually expanded, and people have begun to pay attention to piezoelectric materials that can work stably under high-temperature conditions. So far, efforts have been made to explore the design of high-performance polymer piezoelectric materials. The mainstream high-temperature piezoelectric materials are usually inorganic materials such as lead zirconate titanate (PZT), $BaTiO_3$, and potassium sodium niobate (KNN). However, they exhibit

doi:10.1088/978-0-7503-5487-5ch7

deficiencies in high-temperature environments mainly in terms of brittleness, toxicity, low strain and high manufacturing cost. Inorganic materials usually have high stiffness and hardness, making it difficult to achieve flexibility and bendability, and poor piezoelectric sensitivity to detect small vibrations.

In high-temperature environments, automotive (100 °C–350 °C), petroleum (150 °C–300 °C), nuclear power plants (30 °C–250 °C), and aerospace (−60 °C–900 °C), non-destructive inspection and continuous monitoring of tiny components or curved structures, or the physical protection and health inspection of operators are required [4, 5]. All these segments require piezoelectric devices with high-temperature stability, flexibility, high sensitivity, and curved surface properties. Compared with ceramic materials, polymeric piezoelectric materials that can work at high temperature have unique advantages, such as flexibility, lightweight, and curved surface property [6]. Particularly in the *in situ*, inspection and continuous monitoring of tiny components or surfaces in high-temperature environments need to be high-temperature stable, flexible, and piezoelectrically sensitive for operator protection and health detection [7–9].

However, the organic piezoelectric materials that currently exist are not ideal for operation in high-temperature environments. Due to the molecular structure and thermal stability limitations of organic materials, they have low Curie temperatures, leading to a decrease in piezoelectric performance under high-temperature conditions. Therefore, there is an urgent need to develop high-temperature polymer piezoelectric materials, while there are significant challenges. Efforts have been made to increase the operating temperatures of piezoelectric polymers, such as molecular modification to increase polymer chain rigidity and crystallinity, cross-linking to immobilize chain movement, and compositing with inorganic filter, but with limit success.

For a long time, there has been a big challenge for designing a polymer piezoelectric material working at high-temperature conditions (e.g. >200 °C) because it is difficult for diploe groups to maintain orientation at high temperatures and there is also a conflict between the high-temperature resistant molecular structure and processability. For crystalline polymers, such as PVDF, the piezoelectric property would start to decrease at 80 °C due to the degree of crystalline orientation starting to weaken and lose its original structure at the melting point of 170 °C [10]. For a non-crystalline polymer, such as polyacrylonitrile (PAN), the piezoelectric performance would become weak due to the order of diploe groups decreasing due to the conformational transformations at temperature of T_g. In addition, trapezoidal or cyclic structures are expected for heat-resistant polymers, which give it poor solubility and non-melting characteristics. It is essential to solve the above challenges to develop a novel high-temperature resistant piezoelectric polymer material possessing flexibility, high piezoelectric performance both at room temperature and high temperature, and excellent cycling stability at high temperature.

In summary, the flexible, curved surface and large-area fabrication advantages of organic piezoelectric materials make them a viable solution to the limitations of inorganic piezoelectric materials. However, efforts to develop high-temperature resistant organic piezoelectric materials are needed to achieve piezoelectric properties for operation in high-temperature environments. In recent years, PAN has entered the public eye as a highly promising piezoelectric polymer material, which has excellent performance not only at both room temperature, but also at extremely high temperature of 500 °C. Through continuous research and innovation, it is believed that more advanced and diverse applications of high-temperature piezoelectric materials can be realised in the future.

7.2 Brief history of PAN as piezoelectric materials

PAN has shown promise as a piezoelectric material, with initial research conducted in the 1970s. During this period, researchers recognized PAN's potential due to the large dipole moment of the cyano group on the PAN side chain. This suggested that PAN could become a highly reactive polymer electret and exhibit piezoelectric properties. As early as 1978, Stupp and Carr [11] found a very large dipole moment for the cyano group on the PAN side chain and indicated that PAN had the potential to become a highly reactive polymer electret. Then, Ueda and Carr [12] prepared NH_4NO_3 and $NaNO_3$-doped PAN cast films and found that the piezoelectric constants of the films varied with temperature.

Until then, two key concepts of piezoelectricity have been discovered for PAN, but it has failed to exhibit good piezoelectric properties. The first key advantage of PAN as a piezoelectric material is the permanent molecular dipole. A strongly polar cyano (–CN) group exists on each repeating unit of PAN macromolecules, which has high dipole moment as high as 3.5 Debye and also has high density of dipole groups [13]. When an external force is applied, the distance or direction between atoms or molecules changes, and the electric dipole moment also changes, thus changing the polarization strength and polarization direction of the crystal. This polarization due to deformation is the positive piezoelectric effect. Typically, a piezoelectric constant, such as $|d_{33}|$, is used to describe the strength of the piezoelectric effect that occurs when a force or electric field is applied to a piezoelectric material along the direction of its polarization. Thus, in theory, this large dipole moment grants PAN excellent piezoelectric properties, making it a promising piezoelectric material.

The second key concept for PAN as a piezoelectric material is the orientation of dipoles. Semi-crystalline polymers can be piezoelectric if they have permanent molecular dipoles and crystalline regions, as established. PAN has two typical conformations in solid-state, planar zigzag conformation and 3^1-helical conformation, and the latter endows PAN with piezoelectricity [14–16]. Through thermal poling, mechanical stretching and electrospinning, the dipoles will be efficiently oriented and aligned and the change in dipole corresponds to a conformational shift [17]. Henrici-Olivé and Olivé [13] reported that the zigzag confirmation has an all-transform (TTTT) structure with the dipole moment of –CN on the PAN side chain as high as 3.5 Debye. Thus, the piezoelectricity of PAN mainly stems from two main aspects. On one hand, sufficiently large dipole moments in the repeating structural units, are mainly determined by the molecular structure. On the other hand, oriented dipole arrangement in the solid state, usually involves a multi-step process of large-scale stretching and electrolysis at high temperatures. That is, the molecular conformation of a piezoelectric polymer material determines the size of its dipole, and the size of the dipole determines the piezoelectricity of the polymer [18].

However, research on PAN's piezoelectric properties waned in the following years, with more emphasis placed on its mechanical properties as a precursor for carbon fiber production [19]. PAN's strength characteristics made it a popular choice in the composite materials industry. It was not until 2019 that Wang *et al* [20] increased the content of the planar sawtooth conformation in PAN by electrospinning, allowing PAN to be rediscovered as a new piezoelectric polymer material for energy harvesting. Their unexpectedly high voltage results have rekindled researchers' enthusiasm for studying the piezoelectric properties of PAN.

It was at this point that the third piezoelectric key concept of PAN was discovered, retention of the aligned diploes, what made it the most important in ensuring high-voltage electrical characteristics. For a polymer to be piezoelectric, the dipoles in its macromolecular chains must maintain a certain orientation under certain working conditions. In general, most piezoelectric polymers have the following characteristics, permanent molecular dipoles that can be aligned and oriented, and the dipoles can maintain their state after the oriented alignment [21]. When the polarization is dominated by dipole orientation mostly, the polymers with a larger dipole moment typically have larger piezoelectricity [22]. Various processing and proper adjustment techniques could be applied, such as poling at a slightly higher temperature than the glass transition temperature, to ensure better alignment and locking of dipoles, or removing the electric field when the temperature drops to room temperature, which helps maintain tightly aligned polarized structures of dipoles for long-lasting locking [23]. These different processes contribute to the preparation and improved properties of PAN piezoelectric polymers. Since then, researchers have focused on investigating PAN as a piezoelectric polymer material. Efforts are underway to better understand PAN's piezoelectric behavior and optimize its performance in various technological applications.

The last key concept of high-temperature-resistant piezoelectric properties of polymers is the tendency to form a high-temperature-resistant backbone structure, like a ring structure. The higher melting point of 317 °C and the reaction of ring formation during heat treatment gives PAN the potential to form high-temperature-resistant main chains [24, 25]. As a result, the thermally stable PAN no longer melts and exhibits high thermal stability, which provides PAN fibers with more opportunities for high-temperature piezoelectric applications. More research on the piezoelectric properties of PAN is also gradually focusing on their high-temperature-resistant piezoelectric properties, which has become one of the most promising research prospects about PAN at present.

For the piezoelectric property of PAN, its molecular conformation and aggregation structure is important. Studies have been conducted to increase and maintain the content of planar zigzag conformation in PAN macromolecular chains mainly based on suitable fabrication technique and combining with other substances. The performance of polymer piezoelectric materials is strongly dependent on the fabrication techniques. Various processing methods, including mechanical poling, thermal poling, and the addition of substances, have been shown to increase the content of planar zigzag conformation in PAN macromolecule chains, which improves their piezoelectric properties. This chapter summarizes the fabrication techniques of PAN piezoelectric materials and the forms of PAN piezoelectric materials can be divided into three categories, i.e., in the order of 1D, 2D and then 3D.

7.2.1 1D fibrous piezoelectric material

The 1D fibrous piezoelectric material is an optimal candidate for wearable flexile energy supply. Wet spinning combined with mechanical stretching is a sample preparation technique to obtain PAN piezoelectric fiber, in which a mechanically oriented stress field can induce PAN to transform from disordered to ordered. If the electromechanical coupling properties of those samples prepared by mechanically directional stress file are comparable to those of conventional high-voltage polarization, the post-processing process becomes easier. Mechanical stretching is a method of using external forces to alter the fiber orientation structure, describing the process of fiber being stretched uniaxially or biaxially to multiple times its original length [26]. Mechanical stretching specifically orients the dipole by applying an external force in the axial direction of the fiber and, as a result of the force, the intermolecular spacing is reduced and the molecular chains elongated. The orientation of the molecular chains is reorganized as the direction of stretching changes, thus enabling the molecules to be oriented.

As shown in figure 7.1(a), during wet spinning, which involves repeated mechanical stretching, generally divided into three successive stages: jet stretching, wet stretching (first stretching) and hot stretching (second stretching). The polymer

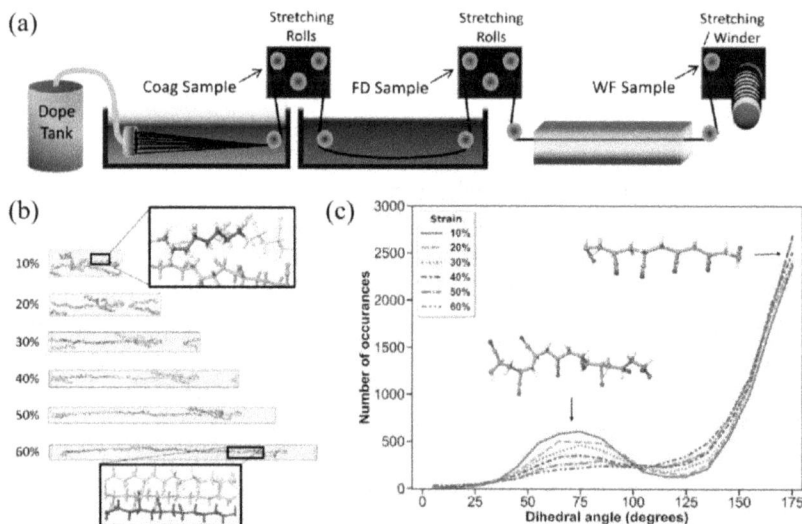

Figure 7.1. (a) Preparation of 1D PAN fibers by combining wet spinning and mechanical stretching; (b) molecular dynamics visualization of the increasing strain for atactic polyacrylonitrile showing the transformation from a two-phase to a single phase system and (c) the dihedral angle. Reproduced with permission from reference [27] John Wiley & Sons. Copyright (2021) Wiley Periodicals LLC.

dope is heated to 65 °C, fed by a metering pump through a pressure regulator and heated, filtered prior to wet spinning. At a constant rate of 40 ml min^{-1}, the dope flow was extruded through a 1000-hole spinneret with a 50 μm hole diameter into a heated coagulation bath, containing a mixture of dimethylsulfoxide (DMSO) and water [27]. Zhang *et al* [28, 29] report that for the I_{1230}/I_{1250} for PAN fibers at different stages of the fiber formation process, when external forces are applied to PAN fibers, the molecules in the fibers undergo a number of changes, such as twisting, elongation and flipping, and are aligned and arranged in a certain way. For the samples with different isotacticity, Fourier transform infrared (FTIR) and I_{1230}/I_{1250} analyses were carried out, and the results showed that as the isotacticity increased, the conformation content of the 3^1-helical chain in the macromolecules increased, the I_{1230}/I_{1250} values increased accordingly. It is reasonable to assume this process can cause changes in the orientation structure of PAN fibers, the 3^1 helical conformation statistic of the PAN macromolecules in the fibers decreases and the planar sawtooth conformation statistic increases with the gradual increase of the drafting force field [28, 29]. Figures 7.1(b) and (c) show that the action of the nitrile side groups of PAN molecular chains becomes stronger when the draft multiplicity increases, and the molecular chains of PAN are made alternatively into merit-based distribution along the fiber axis under the action of the strong dipole moment so the ordered phase is formed by syndiotactic chain sequences [30, 31]. Therefore, the most important thing in the wet spinning process is the mechanical force pulling, which makes the bond angle inside the molecule change, the overall orientation of the molecular chain is enhanced, and the strong polar group –CN of PAN itself will be subject to greater pulling force, and thus be more likely to undergo orientation. Some studies have demonstrated that mechanical stretching can increase the content of the planar sawtooth conformation in PAN molecular chains, but that is where it stops, without ever getting into the study of the intrinsic connections between conformational and piezoelectric properties [27, 32, 33].

Therefore, subsequent studies could delve into the effects of mechanical stretching on the internal structure of the molecule and explore the intrinsic link between molecular conformation and piezoelectric properties.

7.2.2 2D piezoelectric polymer film

The preparation of 2D piezoelectric polymer films is divided into two types: thermal polarization and electrostatic spinning. Earlier scholars orientated the molecules of polar polymers by high-temperature conditions. Later on, the macromolecular orientation of electrospun PAN nanofibers was preserved by electrospinning, which combines polarization and stretching.

The 2D piezoelectric polymer films prepared by casting combined thermal poling have unique properties that make them idea as a dynamic strain sensor for life signal monitoring application on the surface of human skin or implanted inside the body. Thermal poling is a method of achieving molecular orientation of polar polymers by increasing temperature and applying external conditions such as enhanced electric fields or high voltages. Specifically, the molecular orientation of polar polymers can be altered under high-temperature conditions due to increased molecular thermal motion [18]. In general, the process of thermal poling will include dipole orientation, surface charge deposition and charge injection [34].

In figure 7.2(a), the PAN film to be polarized is sandwiched between a pair of planar electrodes and immersed below the liquid surface of silicone oil. The silicone oil is extremely well insulated to avoid flying arcs at the edges of the electrodes, and a high voltage can be applied. After heating to a certain temperature, a high DC voltage is applied to the two electrodes so that the electric field inside the film is polarized to tens or even hundreds of megavolts per meter. There are two main transitions of PAN close to 100 °C and 150 °C, called Minami α_2 and α_1 relaxations, respectively [35–37]. The high-temperature relaxation α_1 is located in the amorphous region while the low-temperature absorption α_2 is located in the paracrystalline region [38]. When highly oriented PAN is heated, glass transition can be observed at 100 °C, and the melting/degradation still occurs at 340 °C [39, 40]. The poling temperature is usually chosen to be between the glass transition temperature T_g and

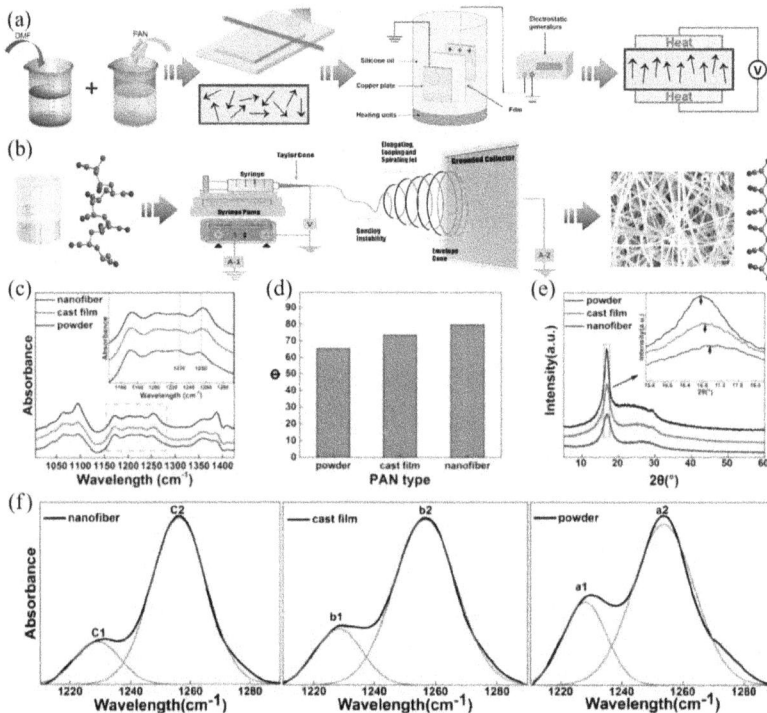

Figure 7.2. (a) Preparation of 2D PAN cast film combined with thermal poling. Reproduced with permission from [141]. Copyright 2023, Elsevier. (b) Preparation of 2D PAN fibers by electrospinning. Reproduced with permission from [19]. Copyright 2012, Elsevier. (c) FTIR spectra of PAN powder, PAN cast film and electrospun PAN nanofiber membrane (roller speed $100 \, \text{mm s}^{-1}$). (d) FTIR-based zigzag conformation content estimation; (e) x-ray diffraction (XRD) patterns of PAN powder, PAN cast film and electrospun PAN nanofiber membrane; (f) FTIR characteristics of PAN nanofiber, PAN cast film and PAN powder peak-to-peak fitting diagram. Reproduced with permission from [20]. Copyright 2019, Elsevier.

the melting point T_m, so that the electret dipole can be formed with good charge stability. The polarization is carried out for a period of time, depending on the situation, keeping the electric field in place, and the sample is cooled to room temperature before the field is withdrawn in order to 'freeze' the dipole in its direction.

The heating process and the electric field are two important factors of the thermal polarization process. Usually, the heating temperature chosen is higher than the glass transition temperature (T_g), the molecular mobility increases, and the molecular dipoles are more easily rearranged after being polarized. Secondly, the electric field is chosen to induce the alignment of the molecular dipoles, and the proper duration allows more dipoles to be aligned, resulting in a greater increase in conductivity. Several studies have demonstrated that the polarization resulting from the orientation of the dipole charge in the polymer along the electric field during thermal polarization and maintaining the orientation after cooling can even increase the residual polarization intensity of ferroelectric thin films [41–43]. This also demonstrates that the combination of high temperature and electric field can serve to align the dipoles within the polymer, but related studies have not advanced further and there is still a gap regarding the effect of the thermal polarization process on the piezoelectric properties of PAN.

As shown in figure 7.2(b), electrospinning is the other method that can orient the molecules of polar polymers, integrated polarization and stretching, with the possibility of retaining the macromolecular orientation in the electrospun PAN nanofibers, which is the advantage of preparing planar sawtooth conformation PAN piezoelectric films [44]. Isotactic PAN uneven charge distribution within the molecule, with a high dipole moment and the stretching during electrospinning will contribute to the alignment of the semi-crystalline regions, both of which will improve the piezoelectricity of electrospun PAN. Importantly, breaking macroscopic polar domains into nanoscale polar regions through decreasing the diameter of PAN fiber has a strong promotion on improving the piezoelectric property. And with the decrease of nanofiber diameter after mechanical stretching, exponential increasing of piezoelectric voltage response was exhibited [45, 46].

Figures 7.2(c) and (e) show the FTIR spectra and the XRD patterns of PAN nanofiber film, PAN cast film and PAN powder. The vibration bands at 1250 and 1230 cm^{-1} in the figure correspond to the planar sawtooth conformation and 3^1-helix conformation of PAN, respectively (figure 7.2(f)). The planar sawtooth conformation content can be calculated by fractionally fitting the characteristic peaks at 1250 and 1230 cm^{-1}. Let Φ be the peak area of 1250 cm^{-1} accounted for the sum of the areas of the two peaks, and the Φ of PAN nanofiber film is 79.7%, which is higher than that of cast film ($\Phi = 73.6\%$), while the raw material PAN powder shows a lower Φ (65.4%) (figure 7.3(d)) [47].

The preparation of high-performance piezoelectric fibers by electrospinning has also become increasingly popular in recent years. In 2010, Chang et al [48] demonstrated the effect of fiber size on piezoelectric properties, while demonstrating that the polarity of electrospun nanofibers was determined by naturally aligning the dipoles in the nanofiber crystals through a strong electric field (greater than 107 V m^{-1}) and stretching forces in electrostatic spinning, causing the nonpolar phase

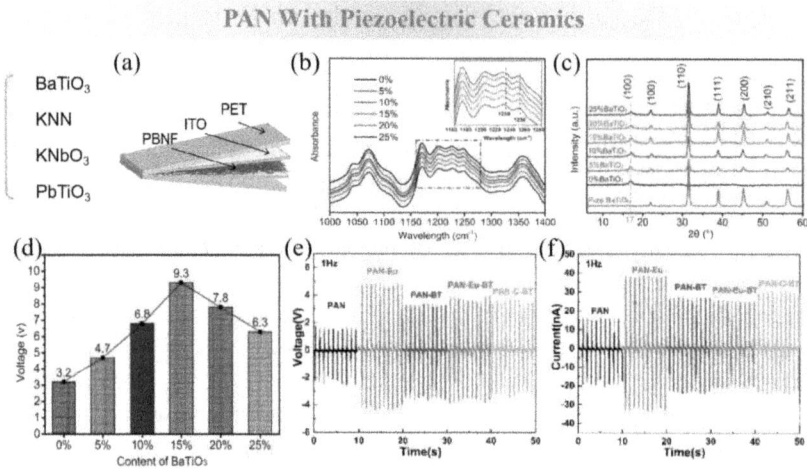

Figure 7.3. PAN with piezoelectric ceramics. (a) Typical device structure of PBNF (PAN with BaTiO$_3$ nanofiber) nanogenerator; (b) FTIR spectra of the PAN with BaTiO$_3$ nanofiber; (c) XRD patterns of BaTiO$_3$ powder, PAN and different types of PAN/BaTiO$_3$ nanofiber membranes; (d) the voltage outputs of PAN nanofiber membranes with different contents of BaTiO$_3$. Reproduced with permission from [64]. Copyright 2021, Elsevier. (e) Output voltage and (f) current of the five samples of PAN, PAN-Eu, PAN-BT, PAN-Eu-BT and PAN-C-BT fibers. Reproduced with permission from [63]. Copyright 2022, Elsevier.

(random orientation of the dipoles) to transform into a polar phase. After, in 2019, Street *et al* [49] electrospun PAN with different isotacticities by concentration adjustment and made some preliminary and simple explorations of the piezoelectric properties of PAN. They find the 25% and 52% isotactic electrospun PAN samples showed piezoelectricity, with $d_{33,\ \text{eff}}$ in the range of 1–2 pC/N at a given electro-spinning condition of 1 ml h^{-1} pump speed and 15 kV applied voltage. Conversely, the commercially available PAN sample manifested minimal-to-nonexistent signal, rendering it undetectable. The inherent piezoelectricity of electrospun anisotropic PAN was demonstrated, and compared to free-radical polymerized PAN, electrospun PAN always maintained higher crystallization. During electrospinning, *in situ* polarization and stretching occur due to the application of a high electric field, which induces a herringbone conformation and isotacticity [50]. Moreover, Wang *et al* [20] found that by changing the rotation speed of the drum collector during the electrospinning process, the fiber orientation within the fiber membrane is also adjusted. At an applied voltage of 23 kV, a flow rate of 0.5 ml h^{-1} of polymer solution, and a surface speed of the spinning drum collector of 1200 mm s^{-1}, the average diameter of the fibers was reduced to 350 ± 50 nm and the average orientation angle was reduced to 12.88°. Also, they compressively shocked a 5 cm^2 PAN nanofiber nonwoven membrane, which can generate up to 2.0 V of voltage with an electrical output of 1.0 μA. These different experimental conditions can produce different results, e.g., solution pumping speed, voltage magnitude in the case of electrostatic spinning, and impact conditions and film dimensions (e.g., working area and thickness) as in the case of piezoelectric output testing, thus demonstrating the critical role of processing technology and testing conditions in the development of piezoelectric materials.

In conclusion, 2D nanofibers fabricated by electrospinning techniques typically have a larger specific surface area, controlled porosity, extremely long lengths, oriented alignment at the molecular level, and light weight. It is apparent that the electrospinning combines well the advantages of thermal polarization and mechanical stretching, both in inducing changes in the orientation direction of the dipole and in regulating the fiber diameter. By adjusting the conditions during electrospinning, the herringbone conformation content within the PAN fiber is increased to ensure large dipole moments, demonstrating that electrospinning can improve the piezoelectric properties of PAN fabrics.

7.2.3 3D Organogel method

In addition, during the process of the 3D organogel method, using PAN as a raw material and its good piezoelectric properties, the mechanical energy can be converted into electrical energy to achieve the self-powered function of the sensor. The organogel method refers to a semi-solid system of organic solvents as liquid fillers in a cross-linked three-dimensional network [51–53]. The continuous organic solvent liquid phase usually has a strong affinity for networks, which are made up of low molecular-weight organogelators, polymeric organic gelators, or covalently bound polymer networks [54].

By one-pot organogel method, Li et al [55] used PAN to prepare ionic conductive organogels with high flexibility, strong adhesion, recyclable and anti-freezing properties. The specific reason that gives the organogel its piezoelectric properties is the nucleophilic addition reaction of the macromolecular chain of the polymer PAN with $HONH_3Cl$ to produce amidoxime groups, followed by cross-linking of the metal Zn^{2+} with –OH and $–NH_2$ on the amidoxime groups through complexation [56–58]. Although there is no electric or force field, the force of metal coordination causes a change in the spatial arrangement of the PAN molecular chains, resulting in a more ordered molecular chain arrangement, and generates a non-zero vector sum of dipole moments, allowing the PAN 3D organogel film to show certain piezoelectric properties under pressure [20, 59]. In addition, the mechanical interaction of metal coordination bonds, as well as the reaction of carbon and nitrogen triple bonds in the nitrile group to form carbon and nitrogen double bonds, leads to an increase in the conformational content of the unidirectional arrangement of PAN macromolecular chains. The self-powered ion skin device made from this gel is able to continuously output a stable voltage of 60 mV and a current of 1.7 mA at a cyclic impact force of 2 Hz and 13 N [55].

Deng et al [60] prepared a PVA-PAN self-generating hydrogel, showing that dipole interactions between PAN chains and hydrogen bonding between PVA and PAN chain segments promoted the transfer of PAN to a planar zigzag conformation, and the maximum d_{33} of the hydrogel was 32p/CN. The transition of the PVA-PAN-0 to PVA-PAN-5, the area of the characteristic absorption peak at $1250\,cm^{-1}$ gradually increased, and the 3^1 helical conformation within the hydrogel shifts to a planar zigzag conformation as the content of AN increases, improving the piezo-electricity of the hydrogel.

Although the voltage output is not as high as that of the PAN nanofiber membrane obtained by electrostatic spinning, it is sufficient to power the sensor. This can be attributed to two main factors. Firstly, the –CN groups in PAN undergo reactions to form amidoxime groups, potentially altering the piezoelectric properties. Additionally, the inherent conductivity of the ionic organogel may limit the generation of high electrical potential. In the future, we can try to improve the piezoelectric properties of PAN ionic skin organogels from the above two points.

7.3 Enhancement of PAN piezoelectricity at room temperature

In previous studies, due to two bottlenecks, a low power output and poor sensing capability, making it hard for the outputs from pure PAN-based nanogenerators to meet the requirements for fabricating high-efficiency power-generating devices. To address these issues, some studies have effectively improved the piezoelectric properties of PAN by adding single or double layers of fillers to PAN that have interaction forces with PAN, resulting in an increased content of planar herringbone conformations within the PAN composite piezoelectric nanofiber membrane. Most of these inorganic piezoelectric materials are added in the form of microparticles and nanoparticles, which can form piezoelectric composites with polymeric piezoelectric materials, not only to improve the conformation of the piezoelectric polymer but also to give it the excellent properties of inorganic piezoelectric materials. Commonly used additives are usually classified as: (a) piezoelectric ceramics such as $BaTiO_3$ nanoparticles [61–64]; (b) semiconductors such as copper oxide (CuO) nanorods [65], zinc oxide (ZnO) nanorods [66, 67], and TiO_2 nanoparticles [68]; (c) inorganic salt $NaNO_3$, $CaCl_2$, and NaCl nanoparticles to name a few [69].

7.3.1 Composite with piezoelectric ceramic materials

Piezoelectric ceramics have been the most common piezoelectric materials over the past century and are widely used in sensors and brakes. Various ceramic materials with good piezoelectricity, such as barium titanate ($BaTiO_3$) [62, 64], PZT [70], KNN [71], potassium niobate ($KNbO_3$) [72], lead titanate ($PbTiO_3$) [73] and lithium niobate ($LiNbO_3$) [74], etc. Among them, most of the piezoelectric materials, led by PZT, are lead-based materials, which are hazardous to the environment and human health. Therefore, lead-free piezoelectric materials such as $BaTiO_3$ and KNN, which have high dielectric constants and low dielectric losses, are being selected for blending with PAN and are receiving increasing attention.

Figure 7.3(a) shows a simple energy harvesting device fabricated by Cai *et al* [62] by mixing $BaTiO_3$ nanoparticles with PAN. Based on the electrospinning technique, $BaTiO_3$ nanoparticles were rapidly precipitated and uniformly distributed in the PAN nanofiber membrane, which improved the tensile strength and the ratio of planar herringbone conformation. The amount of added $BaTiO_3$ nanoparticles significantly affected the output voltage and current, with the voltage reaching its peak at 15 wt% of $BaTiO_3$ at 1.5 V, and the output current underwent a similar change. From the PAN/BaTiO$_3$ (figures 7.3(b) and (c)), it can be seen that after the addition of filler, the vibration bands of the planar zigzag conformation and 3^1-helical conformation at 1250 and 1230 cm^{-1} change significantly, and the calculations indicate that the introducing of BaTiO$_3$ results in the improving of the content of zigzag conformation, which is the reason of enhancing the piezo-electric properties after adding BaTiO$_3$. Moreover, Yuan *et al* [64] developed a PAN/BaTiO$_3$ piezoelectric nanofiber membrane sensor with 15 wt% BaTiO$_3$ added for real-time damage detection of composites. A sensor fabricated using pure PAN nanofibers membrane (with a thickness of 150 μm and an area of 1 × 5 cm^2) generated a voltage of 3.2 V when subjected to repetitive bending loads (at a deflection rate of 0.1 mm s^{-1}). In figure 7.3(d), under the same test conditions, the PAN/BaTiO$_3$ composite membrane sensor exhibited a good response to electrical signals during mechanical loading, with a significantly improved mechanical-to-electrical output of 9.3 V peak, enabling real-time monitoring of delamination faults. He *et al* [63] noted the effects of single filler and double fillers on the conformation of the PAN polymer molecular chain. With the Eu^{3+} doped PAN/BaTiO$_3$ piezoelectric composite fibers prepared by electrospinning (spinning time of 2 h, spinning voltage of 19 kV, and syringe pump extruding at a rate of 1.0 ml h^{-1}), the ratio of transformation of PAN from 3^1-helical conformation to planar zigzag is the highest. PAN-EU-BT fiber membranes can reach a dielectric constant of 3.5, which corresponds to a twofold increase, while the dielectric loss is as low as 0.01, which nicely improves their piezoelectric properties. In figures 7.3(e) and (f), the addition of filler greatly increases the output voltage and output current of the material, and the output voltage (∼4 V) and output current (∼25 nA) of PAN-Eu-BT are relatively large. With the same 4 × 4 cm^2 area and the same pressure of 12.5 kPa, gradually increasing the frequency (1–3 Hz), the output voltage of the PAN-Eu-BT increases from ∼4 to ∼6 V and the output current from ∼25 to ∼45 nA. It suggests that the impact frequency of the higher force can stimulate the flow of external electrons in a short period of time, whereas at the higher frequency, the surface of the fiber layer charge will not be completely neutralized, which may lead to an increase in output.

The aforementioned example illustrates that the incorporation of lead-free piezo-electric ceramics, such as BaTiO$_3$ and KNN nanoparticles, during the preparation of PAN nanofiber, using electrospinning technique can increase the proportion of planar zigzag conformation, thus improving the piezoelectric performance and enhancing the tensile strength of the resulting films. It can be inferred that the addition of other piezoelectric ceramics may also have a positive effect on the piezoelectric performance, which warrants further exploration.

7.3.2 Composite with semiconductor materials

A semiconductor is a material with conductive properties between those of a conductor, and an insulator at room temperature can be used to control the flow of electrical currents. The addition of semiconductors to polymeric piezoelectric materials usually enhances the piezoelectric effect and increases the electrical conductivity. This is because the structure and material properties of semiconductors and polymers are very different, resulting in a change in the molecular structure of the composite material. Furthermore, the doping of semiconductor materials enhances the conductivity of the polymer, thus making it easier to transfer charges and enhancing the piezoelectric properties of such composites. The doping of PAN with such semiconductor materials as CuO nanorods [65], ZnO nanorods [66], and TiO$_2$ nanoparticles [68] also has an optimizing effect on the piezoelectric properties of the polymer.

As shown in figure 7.4(a), Sun *et al* [66] fabricated PAN/ZnO nanofibrous membranes by electrostatic spinning. In figure 7.4(b), PAN powder, PAN nanofibers and PAN/ZnO nanofabrics all have similar FTIR spectral profiles, whereas PAN/ZnO nanofabrics have the highest peak area at 1250 cm^{-1}. Similarly, in figure 7.4(c), it is shown that PAN/ZnO nanofabrics have the highest content of planar zigzag conformation, followed by PAN nanofiber film and PAN powder. In figures 7.4(e) and (f), Bairagi *et al* [65] fabricated PAN/CuO nanorods-based electrospun nanocomposite made a flexible lightweight lead-free piezoelectric nanogenerator, the PAN/0.5% CuO-based nanogenerator has shown a maximum output voltage of 1.25 V, a maximum short-circuit current of 118 nA and the power density is 0.75 μW cm^{-2} when pressure is applied by a standard testing method. The addition of CuO provides a conductive pathway into the PAN matrix, significantly reducing the resistance of the PAN and thus allowing more efficient energy harvesting in PAN polymer-based piezoelectric nanogenerators.

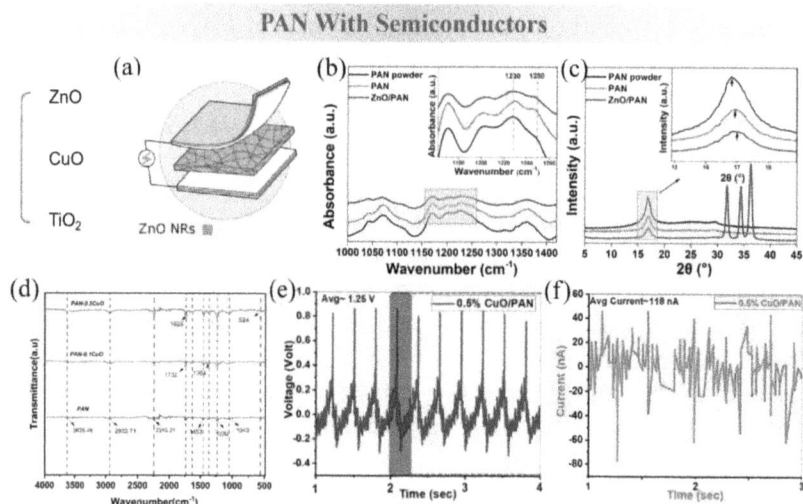

Figure 7.4. PAN with semiconductors. (a) Fabrication of PAN/ZnO NR-based electrospun nanocomposite; (b) FTIR spectra of the PAN powder, PAN nanofiber membrane, and ZnO/PAN nanofabric; (c) XRD patterns of the PAN powder, PAN nanofiber membrane, and ZnO/PAN nanofabric. Reproduced with permission from [66]. Copyright 2020, American Chemical Society. (d) FTIR spectra of PAN, PAN/0.1% CuO and PAN/0.5% CuO; (e) short-circuit current when pressure is applied by standard testing instrument a pure PAN electrospun nanocomposite and (f) a PAN/0.5% CuO electrospun nanocomposite. Reproduced with permission from [65]. Copyright 2022, Springer Nature.

Semiconductors are tunable materials that respond significantly to external stimuli such as light, heat and electricity, which the polymer can flexibly handle. Thus, by adding semiconductors to PAN piezoelectric polymers the material can be endowed with a variety of functions and there is great scope for development.

7.3.3 Composite with inorganic salt materials

Moreover, the addition of some inorganic salts to pure PAN is also beneficial in improving the piezoelectric properties and piezoelectric stability of PAN nanofiber membranes. Inorganic salts have both positive and negative ions in their molecular structure and they are strongly polarized internally. When inorganic salts are added to polymer solutions they affect the self-assembly process, resulting in many new polarized layers on the surface of the polymer particles. Qin *et al* [75] conducted a study to examine how various salts affect the electrospinning of PAN polymer solution. They found that LiCl was the most conductive when added to different concentrations of PAN solution, followed by $NaNO_3$, $CaCl_2$, NaCl, and no salt added, in that order.

As shown in figure 7.5(a), Wu *et al* [69] doped PAN with $NaNO_3$ and prepared $PAN/NaNO_3$ fibrous membranes by electrospinning (voltage of 22 kV, flow rate of 0.5 ml h^{-1}, spinning time of 5 h) and made a piezoelectric device. In figure 7.5(b), comparison of the data measured with different content of $NaNO_3$ doping in PAN reveals that when $NaNO_3$ doping is 0.9 wt%, the crystal spacing is the smallest, the content of planar sawtooth conformation is the largest. The piezoelectric voltage and current of $NaNO_3$-doped PAN fiber membranes showed an increasing and then decreasing trend with the increase of $NaNO_3$ dosage, and the maximum values were 7.1 V and 7.160 µA. Overall, it appears that the voltage has increased by 31.48% and the current by 111.33%. Moreover, Shi *et al* [76] successfully prepared composite PAN nanofibers with a small amount of 1-allyl-3butylimidazolium tetrafluoroborate ionic liquids (ILs) and ferric chloride hexahydrate ($FeCl_3 \cdot 6H_2O$) by electrospinning technique. The saturation polarisation is 0.4 µC cm^{-2}, which is twice the pure PAN film and is expected to be used in applications such as dielectric energy storage devices, flexible piezoelectric sensors and electronic skin.

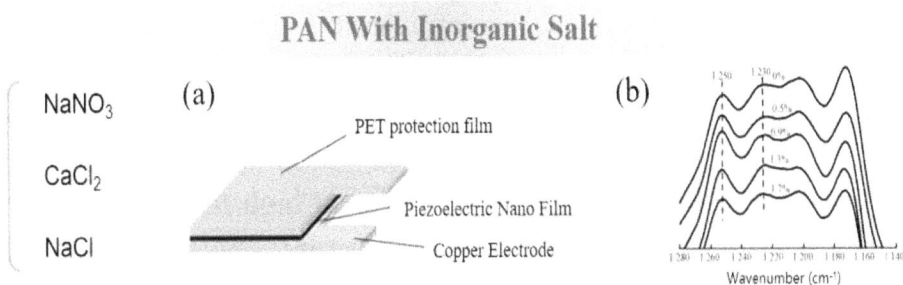

Figure 7.5. PAN with inorganic salt. (a) Schematic diagram of the preparation of $PAN/NaNO_3$ composite fiber membrane; (b) FTIR spectra of PAN fiber membrane doped with different mass fraction of $NaNO_3$. Reproduced from [70]. Copyright 2019, with permission from Elsevier.

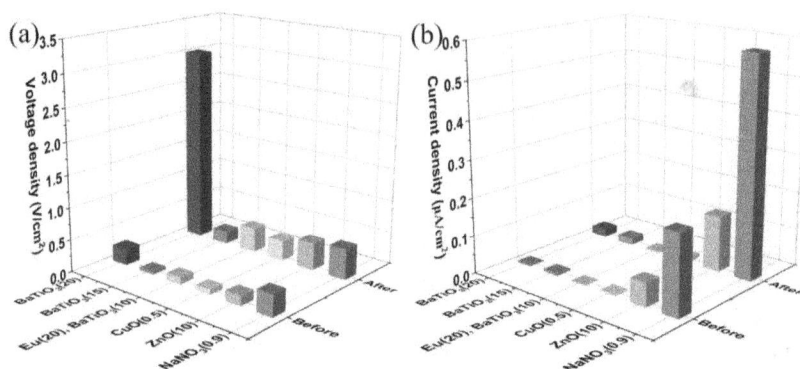

Figure 7.6. (a) Comparison before and after adding different filler materials to PAN-based piezoelectric nanogenerators to enhance voltage and (b) current density. Reproduced with permission from [77], copyright 2023, Elsevier.

However, the specific reasons for changing the inorganic salt content to enhance the piezoelectric properties of PAN composite nanofibers have not been mentioned in detail and need to be explored.

Figures 7.6(a) and (b) outline the comparison plots before and after adding different filler materials to the PAN-based piezoelectric nanogenerator to increase the voltage and current density. With the introduction of suitable levels of substances, we can initially see that the addition of inorganic salts increases the current significantly, while piezoelectric ceramic materials and semiconductors mainly increase the output voltage substantially. In the graph, PAN/NaNO$_3$ fibrous membranes exhibit the highest current, and we review their experimental content and find that it may be due to the fact that the electrostatic spinning duration is up to 5 h, suggesting that the film thickness also has an effect on the piezoelectric properties. Not only the three common additives of piezoelectric ceramics, semiconductors, and inorganic salts, but also other additives such as metal–organic framework (MOF) nanoparticles have been reported in the literature to be added to PVDF, which improves its dimensional homogeneity and prevents bead formation, and which one can try to add to PAN as well [78].

These previous works are summarized to provide the basis and ideas for the subsequent design of PAN-based nanogenerators. The future research direction can focus on exploring the principle of the preparation method to improve the content of zigzag conformation of PAN, the fundamental principle of the microscopic force between the added nanoparticles and the additives, so as to improve the piezo-electricity performance of PAN in a more in-depth investigation (table 7.1).

7.4 Improvement of PAN piezoelectricity at high temperature

For a long time, a key issue with polymer-based mechanical–electrical energy conversion devices has been their inability to operate at high temperatures (e.g., 300 °C–500 °C). This problem is particularly evident for devices made of piezoelectric polymers, since most existing piezoelectric polymers have low Curie temperatures. They usually operate only under mild conditions, e.g. below 140 °C. Above this critical temperature,

Table 7.1. Ways to improve the piezoelectric properties of PAN and PVDF piezoelectric material, the preparation, and output for different test conditions

Ways to improve the piezoelectric properties	Piezoelectric material Filler materials and contents (wt%)	Electrode material	Preparation methods	Sample size area/thickness (cm² μm⁻¹)	Test condition force/frequency (N Hz⁻¹)	Outputs Before enhancement Voltage density (V cm⁻²)	Current density (μA cm⁻²)	After enhancement Voltage density (V cm⁻²)	Current density (μA cm⁻²)	References
Blended with piezoelectric ceramics	PAN-C/BaTiO$_3$(20)	Silver paste	NPs solution cast film	4/1000	0.15/–	0.0625	0.0015	3	0.02	[61]
	PAN/BaTiO$_3$(15)	Conductive PET	Electrospinning	8/60	2/–	0.0375	0.005	0.19	0.0175	[62]
	PAN/BaTiO$_3$(15)	Silver fabrics	Electrospinning	5/150	—	0.64	—	1.86	—	[64]
	PAN/Eu(20)/BaTiO$_3$(10)	—	Electrospinning	16/–	–/3	0.106	0.001	0.38	0.0028	[63]
Doped semiconductor	PAN/CuO(0.5)	Al	Electrospinning	9/–	20/7	0.14	0.004	0.56	0.011	[65]
	PAN/ZnO(10)	Copper tapes	Electrospinning and the low-temperature hydrothermal synthesis	16/–	8/2	0.15	0.063	0.42	0.148	[66]
Added inorganic salts	PAN/NaNO$_3$(0.9)	Cu	Electrospinning	16/–	0.25/110	0.34	0.2119	0.47	0.58	[69]
Solution blending and adding inorganic fillers	PAN/ILs(6)-FeCl$_3$·6H$_2$O(8)	Silver foil	Electrospinning	16/–	—	0.2	0.047	0.33	0.094	[76]
The incorporation of metal–organic framework	PVD/MOFs(5)	Cu	Electrospinning	3/–	5/5	0.136	—	0.18	—	[78]
Doped semiconductor	PVDF/ZnO(15)	Aluminum tape	Electrospinning	4/120	—	0.079	—	0.28	—	[67]
	PVDF/ZnO(7)	Aluminum foils	Electrospinning	4/120	1.56/6	0.079	—	0.16	—	[79]
	PVDF/ZnO(150)	Aluminum tape	Electrospinning and electrospraying	4/41	–/4	0.06	—	0.75	0.064	[80]

the performance deteriorates rapidly with time. At higher temperatures, violent chain movements can interfere with dipole orientation.

Different piezoelectric polymers have different operating temperature limits, which is strongly dependent on their characteristic temperature, for example, their glass transition temperature (T_g), Curie temperature (T_c), melting point (T_m), and decomposition temperature (T_d). When the temperature is higher than T_c, polarization fading will cause the piezoelectricity start to decrease or disappear. Most polymers have a melting temperature in the range of 160 °C–260 °C and a decomposition temperature below 300 °C.

We propose for the first time to classify piezoelectric materials according to their service temperature, i.e., general purpose (<100 °C), engineering (100 °C–150 °C), high-temperature engineering (150 °C–200 °C), and ultra-high-temperature engineering piezoelectric polymer (>200 °C). The maximum working temperature of polymer piezoelectric materials, such as PVDF and its copolymers [48, 81, 82], polyurea-9(PUA-9) [83], polyamide-11(PA-11) [84], polylactic acid(PLA), and polyvinyl chloride(PVC) [85] are summarized in figure 7.7(a). The most common

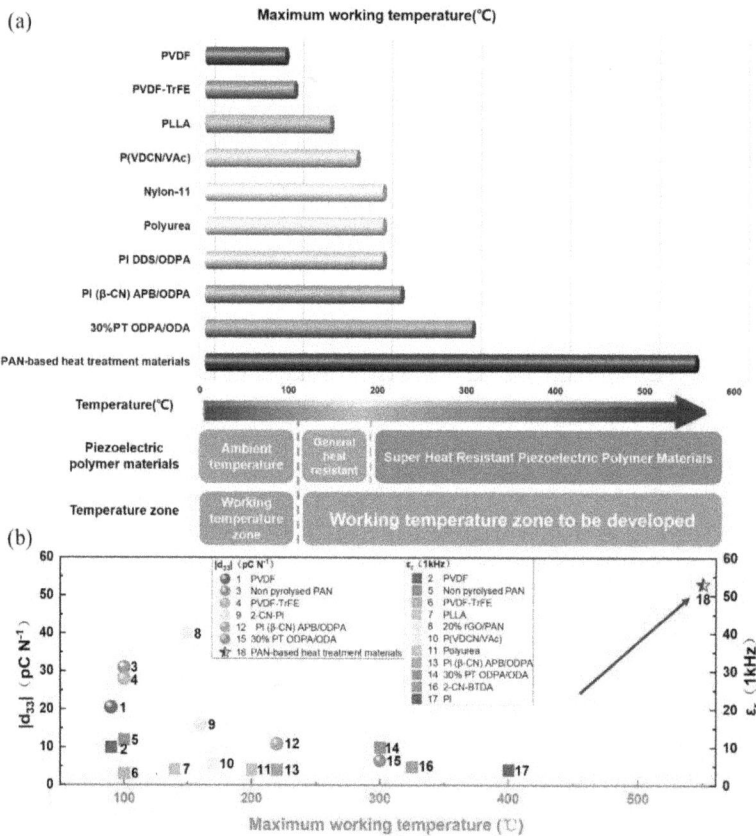

Figure 7.7. (a) Maximum working temperature of different polymer piezoelectric materials. (b) Piezoelectric/dielectric properties of polymer piezoelectric materials at different temperatures [23, 56, 87–98]. Reproduced with permission from [77]. Copyright 2023, Elsevier.

piezoelectric polymer material, PVDF, has a Curie temperature of 120 °C and a melting point of 159 °C. However, the operating temperature range of PVDF (−20 °C– 70 °C) is much lower than its Curie temperature and can be used for general purpose. The operating temperature of PVDF-copolymers does not exceed 120 °C and the maximum working temperatures of PUA-9, PA-11, and PLLA are not higher than 200 °C [86]. In figure 7.7(b), a collation of the piezoelectric or dielectric properties of currently common piezoelectric polymers at their highest operating temperatures is shown, and it is clear that many efforts have been made by researchers to address the issue of high-temperature resistance of piezoelectric polymers. The ultra-high engineering piezoelectric polymer that can service over 200 °C is a challenge to design. Based on the above premise, this section will expand on the potential advantages and challenges of PAN's high-temperature piezoelectric performance.

7.4.1 Potential advantages

PAN is commonly used as a carbon fiber precursor because of its good mechanical properties and because it is heat stabilized to provide excellent resistance to high temperatures. It has been established that thermally stabilized PAN can form a ladder-like conjugated macrocycle structure through the chemical bonding of the cyclized chain structures at an elevated temperature. The ladder-like structures have considerable steric hindrances, and the ladder chains may interlink each other through the cyclized structure. The higher melting point of 317 °C and the reaction of ring formation during heat treatment gives PAN the potential to form high-temperature-resistant main chains [24, 25].

Ge *et al* [99] investigated the structure and properties of PAN fibers at the thermal oxidative stabilization stage and found that the cyclization reaction dominated in the early stage and the oxidation reaction dominated in the later stage, especially after 250 °C when the cross-linking structure increased significantly. Figures 7.8(a) and (b) demonstrate that the rate of growth of the dehydrogenation reaction with increasing temperature is similar to that of the cyclization reaction, since the formation of the ring structure is a prerequisite for the oxidation reaction. Thus, the ring structure seriously prevents chain rotation and motion at high temperatures, responsible for the excellent high-temperature resistance and fire retardancy. Figures 7.8(c) and (d) show the XRD spectra FTIR spectra of thermal oxidative stabilization fibers with different temperature under air, the crystal size decreases gradually with increasing temperature, and a clear cyclization reaction occurs during the synthesis. Such high-temperature characteristics have been widely used to prepare carbon fibers and develop flame-retardant fabrics. Therefore, an in-depth study of the advantages and challenges of high-temperature piezoelectric properties of PAN is important for the development of high-performance piezoelectric materials.

7.4.2 Challenge

Analysis in terms of structural design, the high-temperature piezoelectric polymer should have polar groups in high-temperature-resistant backbone, like conjugated ring structure, which is not easily polarized by the external field. Intuitively, in order

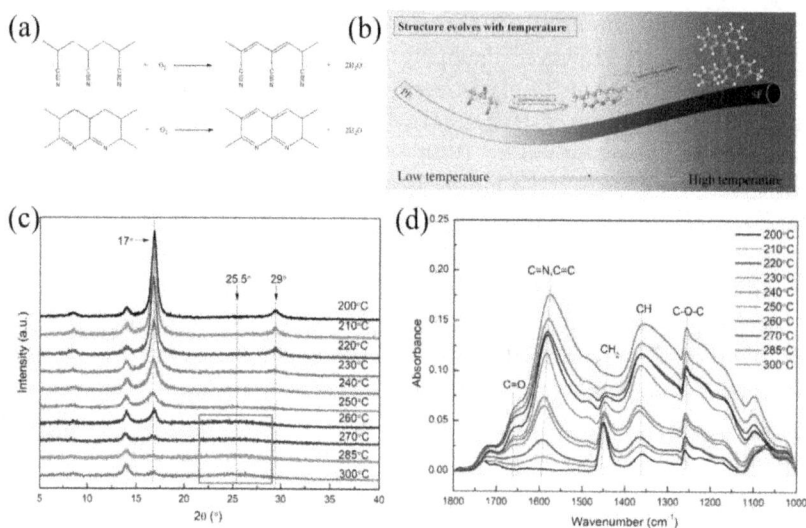

Figure 7.8. (a) Dehydrogenation reaction during thermal oxidative stabilization process. (b) Schematic diagram of structural evolution during microwave pre-oxidation. (c) XRD pattern and (d) FTIR spectra of thermal oxidative stabilization fibers with different temperature under air. (a), (c), (d) Reproduced from [99] John Wiley & Sons © 2023 Wiley Periodicals LLC. (b) Reproduced from [142], copyright 2023, with permission from Elsevier

to obtain a high piezoelectric effect at both room temperature and high temperature, piezoelectric materials need to satisfy two contradictory conditions: (1) polarization is easily perturbed by the external field, (2) the piezoelectric activity is stable at higher temperature.

However, the higher the Curie temperature, the higher the stability of the polarization state of the material, and thus its stability under temperature, electric field, and pressure conditions is increased, which results paradoxically in obtaining piezoelectric polymer with higher Curie temperature and also high room temperature piezoelectric property. Therefore, it is an insurmountable gap to designing a polymeric piezoelectric material that can operate and remain stable over a wide temperature range from room temperature to 500 °C.

The challenge also comes from the tension between the high-temperature-resistant structure design and processability of such a polymer. In terms of structural characteristics, to design a polymer material that has the ability of working at high temperature, macromolecular backbone with strong conjugation, trapezoidal, semi-trapezoidal, or aromatic heterocycles structures should be formed, which directly manifests itself in excellent high-temperature resistance (high T_g, high T_d, excellent high-temperature dimensional stability, etc), excellent dielectric properties, excellent environmental stability (resistance to chemical, atmospheric and space environments, etc), and good mechanical properties. However, such structural characteristics also tend to cause poor processing and process performance of piezoelectric polymer compared to the polymers with linear structure, which needs to be solved.

7.4.3 Solution

Scientists have been working on how to design high-temperature and ultra-high engineering piezoelectric polymers. Examples include increasing the serviceable temperature of piezoelectric polymers by introducing rigid groups to increase the rigidity of the molecular chains [100], as well as increasing the T_g of polymers by orienting or crystallizing the molecular chains at high temperatures through heat treatment and heat setting. Several researchers have used cross-linking [101, 102], or the addition of inorganic materials prepared as composites to increase the serviceability temperature of polymers [4, 103]. This is a far cry from existing piezoelectric materials with single-crystal structures (working temperature range 500 °C–1700 °C, complex preparation process, small scale and mostly at laboratory stage at present) and a far cry from piezoelectric ceramics (working temperature range 200 °C–500 °C) [104]. It has been shown that PAN with imide structure has potential application as piezoelectric materials in the temperature range of 200 °C–400 °C and thermal stabilized PAN with trapezoidal structure has piezoelectric performance at 550 °C, which has potential application as an ultra-high engineering piezoelectric polymer.

Thus, the techniques that fabricate high-temperature resistance of PAN piezoelectric nanofiber membranes can be fundamentally improved through three main points: (a) modification of the molecular structure; (b) blend with inorganic materials; and (c) blend of heat-resistant polymer.

7.4.3.1 Modification of the molecular structure

It is difficulty for a polymer with rigid chains and conjecture ring structure to be fabricated by a one-step method, i.e., producing high-temperature piezoelectric materials directly from solution/melt, because a polymer with these structures is difficult to solve and melt. One possible way to fabricate high-temperature-resistant piezoelectric materials is to design the molecular structure of PAN with highly oriented, thermally stable aromatic ring structure in the main chain [105–107]. To resolve the conflict between high-temperature-resistant structure and processibility, a two-step method for PAN is a new and alternative approach, which involves the solution contained polar groups, under processing, and subsequently cyclizing through thermal treatment (figures 7.9(a)–(f)).

In terms of thermally stabilized PAN piezoelectric material, PAN nanofiber was electrospun and then molecular structure was modified by thermal stabilization. The overall chemical reaction target is to form a high-temperature-resistant structure, in which high-density dipole groups are oriented. The chemical reactions during the heat treatment process were complicated and carefully controlled at three different stages, as shown in figure 7.9(a). In the first stage of dehydrogenation reaction, the reaction was in the temperature of 50 °C–260 °C in the O_2 atmosphere, in which the oxygen atoms incorporating into the polymer chains help the –C≡N– groups converted to –C=N, and then dehydrogenation reaction leads to the formation of stable –C=N in the PAN macromolecule chains. The double bonds formed by the dehydrogenation reaction improve the thermal stability of the PAN and play a positive role in reducing chain breakage during the following further heat treatment

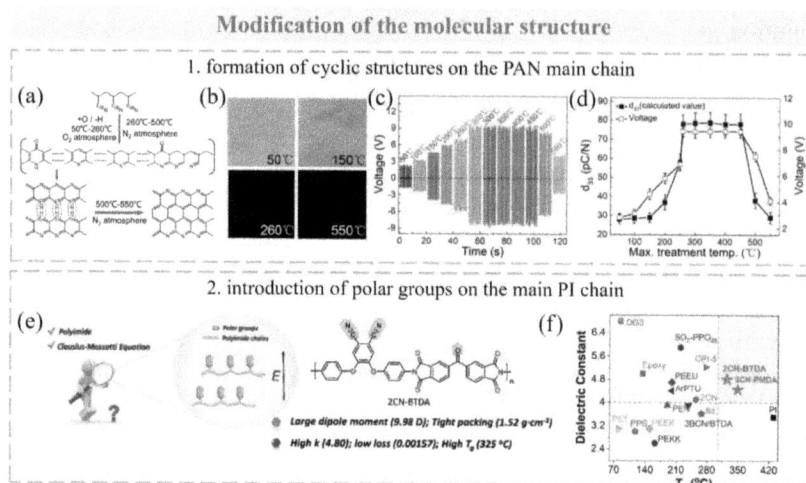

Figure 7.9. Modification of the molecular structure. (a) Chemical reactions involved in the heat treatment of the PAN membranes; (b) digital photos showing the appearance of the PAN membranes prepared in different temperature steps; (c) voltage outputs for the PAN membranes treated under different temperature conditions; (d) output and d_{33} data of the treated PAN membranes. Reproduced from [90] with permission from the Royal Society of Chemistry. (e) Schematic diagram for designing a high-performance polyimide dielectric; (f) dielectric constants of different polymers measured at 1 kHz. Reproduced with permission from [97] John Wiley & Sons, copyright 2023 Wiley Periodicals LLC.

[108]. Then in the second stage of cyclization reaction controlled at 260 °C–500°C in the N_2 atmosphere, the PAN molecular chain forms a ring structure, in which the nitrile group reacts with neighbouring groups to form a stable trapezoidal ring structure and keep it in order [109]. Finally, in the third stage reaction at 500 °C–550 °C, as we can see in figure 7.9(b), in the N_2, larger conjugate structure will be established. During the heat treatment, PAN fibers undergo a change from white to yellow and finally to black-brown heat-stabilized fibers. The change in fiber color is also suggestive of a change in PAN's molecular structure, the appearance of the black-brown color being thought to be due to the formation of an aromatic ring structure [110, 111]. After such smart post-treatment, the chain-like structure of the PAN molecule is transformed into a conjugated trapezoidal ring structure combined with alignment of dipole groups of –CN, which solved the bottleneck problem and exhibits outstanding piezoelectricity from 50 °C–550 °C in figures 7.9(c) and (d) [110, 111]. Surprisingly, a $2.5 \times 2.5 \, cm^2$ device made of PAN thermally stabilized at 500 °C can produce 9.7 V and a current of 4 μA with a maximum output power density of 26.4 Mw m^{-2} at a working temperature of 450 °C. And the 550 °C heat-treated PAN nanofiber membrane did exhibit excellent stability, as it could operate at 500 °C in air under repeated compression and release (shock frequency 1 Hz) for at least 2 h 46 min (10 000 cycles) without changing the electromechanical conversion capability. This is the first example to date of stable operation of a polymeric piezoelectric material at up to 500 °C [110, 111].

In addition, some other studies did some tentative research on designing high-temperature piezoelectric polymer materials by introduction of strong dipole groups of –CN into the main chain of polyimide (PI) [112]. In terms of introducing –CN into rigid PI backbones and thermal imidization shown in figures 7.9(e) and (f). Their research mainly focused on the molecular design and the properties characterization, such as polarization strength (P_r), dielectric constant (ε_r), and piezoelectric constant (d_{33}), while there was a lack of actual measurement of piezoelectric properties.

As shown in figure 7.9(e), Zhu et al [97] introduced two ortho-positioned aromatic nitrile groups into a rigid PI backbone, which greatly improved its polarization properties. The designed PI (2CN-BTDA) can be seen in figure 7.9(f) showing a high ε_r of 4.80 at 1 kHz (25 °C) and T_g values up to 325 °C. The spatial location of –CN as a small size polar group favours directional polarization, which in turn combines with the rigid polymer backbone PI to some extent to maximize polarization rates and the tightest possible chain stacking. Subsequently, Gonzalo et al [113] synthesized three PIs, and they differ in the position of the –CN dipoles in the aromatic rings (poly2–4 and poly2–6) and the number of –CN dipoles in the repeat units (poly2–6 and poly2CN). When two CNs were present in the main chain and adjacent positions, their glass transition temperature was lower ($T_g = 176$ °C) despite the introduction of most of the ether bonds, and PI(2CN) has the highest residual polarization strength ($P_r = 8.4$ mC m^{-2}). This leads to the highest dipole concentration in the monomer unit, which has some high-temperature piezoelectric potential, providing some guidance for further introduction of –CN into the PI backbone and thus improving its high-temperature piezoelectric properties. By a two-step procedure, Maceiras et al [98] synthesized a series of PIs containing –CN in the main chain. The first step is a polycondensation reaction of 4,4'oxidiphthalic anhydride (ODPA) with one or two aromatic diamines. Secondly, each poly(amic acid) or copoly(amic acid) was converted into its corresponding polyimide film by thermal cyclodehydration. In these copolymers, the thermal stability of the amorphous polyimides, 2,6-bis(3-aminophenoxy)benzonitrile/4,4'oxidiphthalic anhydride ((β-CN) APB/ODPA), had a d_{33} of up to 16 pC/N. When the temperature was increased to 160 °C, the d_{33} remained at 8 pC/N, proving that it can be used in operating environments above 100 °C, solving the problem that conventional PVDF piezoelectric polymer materials cannot be used at high temperatures. It has become an ideal candidate for high-temperature piezoelectric sensors.

Therefore, the two-step method which combines materials forming and post-thermal treatment is an effective way to resolve the conflict between high-temperature-resistant structure and processibility, thus allowing modification of molecular structure and fabrication of high-temperature-resistant piezoelectric materials by using PAN or polymers containing aromatic nitrile groups.

7.4.3.2 Blend with inorganic materials

Researchers still take approaches that blending PAN with inorganic materials or heat-resistant polymer enhances high-temperature resistance and improved piezoelectric behaviour, which is easier to control while the performance of designed

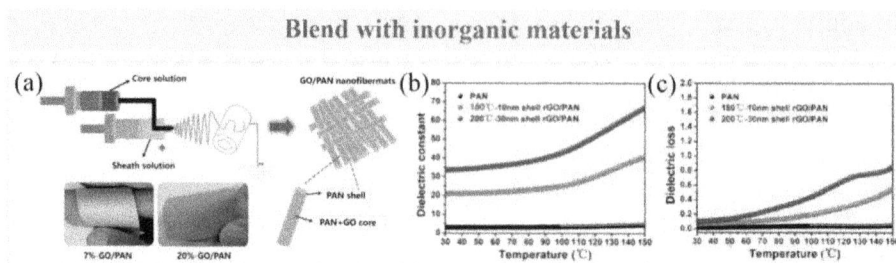

Figure 7.10. Blend with inorganic materials. (a) Mechanism of coaxial electrostatic spinning for the preparation of GO/PAN core–shell structured nanofiber mats; (b) dielectric constant and (c) dielectric loss of 200 °C–30 nm shell rGO/PAN and 180 °C–10 nm shell rGO/PAN as a function of the temperature at 1 kHz. Reproduced with permission from [91]. Copyright 2020, American Chemical Society.

PAN-based piezoelectric is poorer compared with the above modification of molecular structure [114–117]. The main research work of blending with inorganic materials or heat-resistant polymer are illustrated in figure 7.10.

Kambale *et al* [71] first proposed filled lead-free ferroelectric ceramics in 2017 to improve the dielectric properties of PAN composites. By selecting potassium sodium niobate, the thermal stability of PAN was improved by the addition of the lead-free ferroelectric ceramic KNN, in which KNN particles have a ferroelectric Curie temperature of 420 °C as the second phase, and PAN as the matrix material. The prepared the PAN/KNN composites were prepared by hot-pressing at or above the melting temperature and the maximum decomposition temperature increased from 353 °C to 394 °C. In addition, by coaxial electrostatic spinning, as shown in figure 7.10(a), Su *et al* [91] prepared PAN nanofiber mats containing reduced graphene oxide (rGO) with a core–shell structure, which were then hot-pressed into dense composite films. The rGO/PAN composites exhibit a significantly elevated dielectric constant ($\varepsilon' = 40$, 150 °C) and low dielectric loss (tan$\delta = 0.55$, 150°C), concurrently maintaining stable dielectric properties over a broad temperature spectrum (figures 7.10(b) and (c)). These attributes indicate prospects for the application of this material in high-temperature electronic systems. Madakbas *et al* [118] produced PAN/huntite composites by adding different proportions of huntite to PAN, the glass transition temperature of which increased with the content of huntite. Also, the dielectric constant of the samples was increased, so the authors suggest that it may be the interaction of the huntite particles with the PAN that increases the polarization rate.

7.4.3.3 Blend with heat-resistant polymer
In addition to blending with high-temperature-resistant piezo fillers, PAN was blended with other high-temperature-resistant polymers and the combination of mass fraction and mass ratio of the two organic solutions was adjusted to prepare piezoelectric nanofiber membranes with improved high-temperature resistance as well. The advantage of blending is that the composite formed by mixing different polymers can combine the advantages and reduce the disadvantages of each

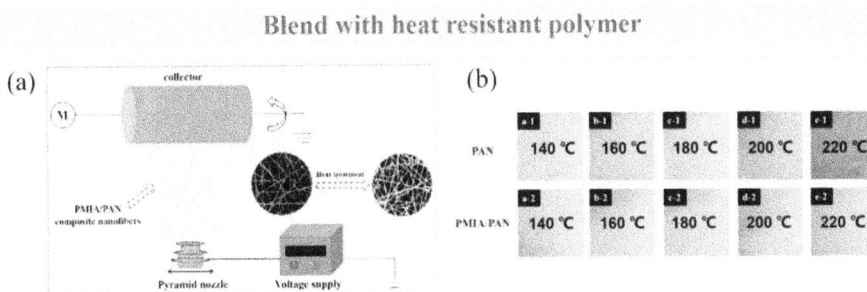

Figure 7.11. Blend with heat-resistant polymer. (a) Schematic diagram of the electrostatic spinning device for the preparation of composite nanofibers of PMIA/PAN; (b) physical comparison of PAN and PMIA/PAN composite nanofiber membranes treated at different temperatures. Reproduced with permission from [119]. Copyright 2021, Elsevier.

respective polymer, thus improving the mechanical properties, thermal stability and other application properties of the composite.

As is shown in figure 7.11(a), Zhang *et al* [119] used a mixture of PAN and high-performance poly(m-phenylene dicarboxamide) (PMIA) as raw materials, and then prepared PMIA/PAN composite nanofiber membranes with excellent mechanical properties, stable filtration performance and good high-temperature resistance. The advantages of the two components were combined to enhance the mechanical properties, which increased the tensile strength of the composite nanofiber membrane from 3.8 to 6.6 MPa. After treatment at 220 °C, in figure 7.11(b), we can see the PAN nanofiber membrane gradually carbonized in color, but the PMIA/PAN composite nanofiber membrane remained almost unchanged, with significantly better high-temperature resistance than the PAN nanofiber membrane. Producing composites by blending PAN with other high-temperature-resistant polymers, its excellent thermal stability and rigidity can provide the nanofibers with more support and cross-linking power, thus enhancing the high-temperature resistance and mechanical properties, which provide a potential possibility to design PMIA/PAN composite high-temperature-resistant piezoelectric material.

In summary, polymer structure modification has been recognized as the most effective strategy for preparing high-temperature piezoelectric polymer materials in the range of 200 °C–550 °C and mixing PAN with other polymers with high T_g values or with piezoelectric ceramic fillers has the potential to achieve improved heat resistance and piezoelectric properties in the temperature range of 120 °C–200 °C.

7.5 Application of PAN piezoelectric materials

The unique structure gives PAN outstanding overall performance, like excellent piezoelectric properties, mechanical properties, hydrophilicity, and ultra-high-temperature resistance after thermal stabilization compared with traditional PVDF, which endow PAN piezoelectric materials with potential applications in a variety of fields and even in extreme conditions. Importantly, PAN also has greater flexibility, the ability of mass production, qualifying it to be prepared in a wide range of applications compared with piezoelectric ceramic materials, facilitating the realization of flexible power generation and microforce detection in a variety of outdoor scenarios [120, 121]. The application of PAN piezoelectric materials, such as power generation, acoustic and electrical conversion, sensors, thermoacoustic conversion, and potential application in the future are summarized in this chapter, in which the relationship between the structure of PAN and advantages in applications are discussed below.

7.5.1 Nanogenerator

A piezoelectric nanogenerator (PENG), relies on the positive piezoelectric effect in piezoelectric materials to convert mechanical energy into electrical energy. Under the action of an external force, the body and surface charges generated by the piezoelectric material create a voltage drop across the PENG electrodes, which drives the electron motion of the external circuit to generate current when the PENG is connected to an external load. In recent years, PENGs based on polymer piezoelectric materials have attracted worldwide attention because of their light weight, low cost, high efficiency in low-frequency energy harvesting, and stability in a humid environment compared to conventional power generation technologies. Recently, PAN piezoelectric materials have demonstrated advantages of high piezoelectricity, enabling them to be studied as a new type of piezoelectric material for nanogenerators (shown in figure 7.12).

Generally, the preparation of PAN-based PENG is similar to other polymer-based PENG [63–65, 122]. For example, as shown in figures 7.12(a)–(c), Wang et al [20] fabricated an energy device by sandwiching a nanofiber membrane between two electrodes. To ensure a secure enclosure, the device was sealed on both sides using a polyethylene terephthalate (PET) membrane. In figure 7.12(b), the device was then mounted on a stable stand during testing, and a compressive force was applied to it, and real-time measurements of voltage and current output from the device were recorded using an electrochemical workstation. Moreover, the stability of the nanofiber device, shown in figure 7.12(c), demonstrated that the device could work for a long time without losing its energy conversion capability through experiments lasting up to 5000 s.

In figures 7.12(d)–(f), Cai et al [62] found that the piezoelectricity of PAN electrospun nanofiber films was significantly enhanced by the addition of $BaTiO_3$, where $BaTiO_3$/PAN fibers were fabricated based on a one-step electrostatic spinning technique, as shown in figure 7.12(d), and when applied to a simple rectifier circuit, a commercial LED could be lit by repeated tapping (figure 7.12(e)) and the capacitor could be charged to 2 V within 2 min, when PAN with $BaTiO_3$ nanofibers was placed on different body parts, the response was also regular (figure 7.12(f)). The sensitive response signal and high energy output are sufficient to demonstrate the promise of PAN combined with inorganic materials for nanogenerators in the field of energy harvesting. In figures 7.12(g)–(i), Bairagi et al [65] used an electrostatic

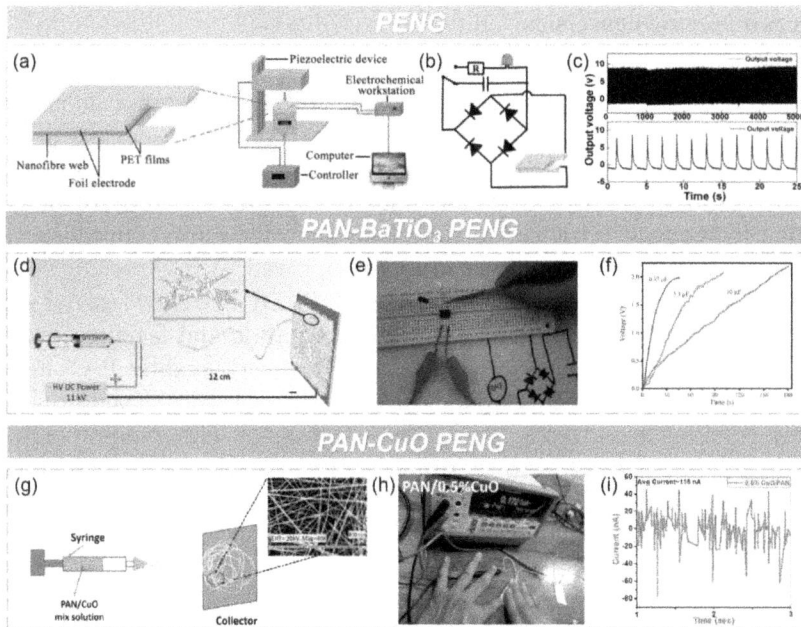

Figure 7.12. PAN-based materials for piezoelectric nanogenerator applications; (a) schematic illustration of device structure and energy conversion tests; (b) circuit for charging a capacitor and lighting a LED; (c) long-term working stability of nanofiber device. Reproduced with permission from [20]. Copyright 2019, Elsevier. (d) A schematic representation of the preparation of nanofibers by a one-step electrospinning technique; (e) lights up commercial LEDs in the dark with repetitive taps (3.3 μF, multimedia view); (f) charging voltage and time of the three capacitors with different capacitance values. Reproduced with permission from [62]. Copyright 2020, IOP Publishing. (g) fabrication of PAN/CuO NR-based electrospun nanocomposite; (h) output current of the piezoelectric nanogenerator in finger tapping mode; (i) output short-circuit current of the nanogenerators. Reproduced with permission from [65]. Copyright 2022, Springer Nature.

spinning method to prepare a flexible and lightweight lead-free PENG based on PAN and CuO nanorods (figure 7.12(g)). When the device consisting of CuO/PAN nanocomposites is tapped with finger pressure, the PENG shows a certain output voltage and current (figure 7.12(h)). This is due to the fact that the cyano group in PAN is a strong electron-absorbing group, and the site resistance and electron repulsion effects result in increased rigidity of the main chain, which leads to an easier orderly arrangement of the main chain and better crystallization properties. This gives PAN fibers excellent mechanical properties, including high strength, stiffness, toughness and wear resistance, all of which are highly beneficial for their use as nanogenerators. Under external forces, the high stiffness of PAN enables it to produce large charge separation effects with small deformations, thereby using the piezoelectric effect to output electrical energy (figure 7.12(i)). Therefore, PAN is used to create efficient nanogenerators that not only have efficient power generation performance, but also ensure the stability and lifetime of the nanogenerator. Such nanogenerators are an effective power generation device and are expected to replace conventional battery-powered devices.

7.5.2 Acoustoelectric conversion

Electrospun piezoelectric nanofiber membranes can be used to make not only piezoelectric nanogenerators but also successful high-performance acoustoelectric generators. Studies have found that acoustoelectric conversion devices made from PVDF nanofiber membranes can produce at least five times the electrical output of cast films [123, 124]. Although PVDF and its copolymer PVDF-TrFE have an application in acoustoelectric conversion, their acoustoelectric conversion capability, sensitivity, and fidelity are expected to be improved.

Whereas PAN fiber membranes are stiffer than PVDF and show greater strain for the same stress, PAN membranes show less deformation and faster recovery from deformation [125, 126]. The better high elastic recovery enables PAN to receive sound signals and convert them into the corresponding electrical signals quickly and accurately, resulting in a fast response to the sound signal. The common preparation of PAN membrane devices is to sandwich a piece of electrostatically spun PAN membrane between two electrodes-coated PET membranes of the same size [127, 128]. In figure 7.15(g), eight through holes were cut in each PET membrane to receive sound signals, and the sound pressure was measured with a sound level meter, the current outputs were measured by an electrochemistry workstation with a low-noise current preamplifier.

In figures 7.13(a)–(c), Shao *et al* [127], for the first time, reported that electrospun PAN membranes have converted high decibel noise (figure 7.13(a)), such as 115 dB, into a large voltage output. It also shows the remarkable property of electrospun PAN nanofiber membranes to convert audio noise into electric energy, with an energy conversion efficiency as high as 85.9% (figure 7.13(b)). Without the

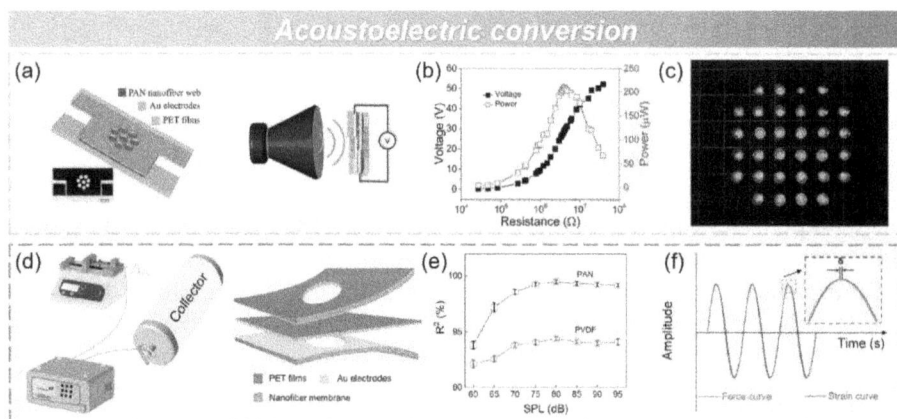

Figure 7.13. PAN-based materials for acoustoelectric conversion applications. (a) Schematic diagram of PAN acoustoelectric device; (b) effect of external load on device voltage outputs and instantaneous peak power outputs; (c) an image to show the lighting of commercial LEDs powered by the acoustoelectric device directly. Reproduced with permission from [127]. Copyright 2020, Elsevier. (d) Schematic illustration of the electrospinning process and schematic diagram of the acoustoelectric device, effect of (e) SPL (sound frequency 200 Hz) on the *F* value, dynamic mechanical analysis for (f) PAN nanofiber membrane. Reproduced from [129] with permission from Royal Society of Chemistry.

requirement of pre-accumulation in an energy storage device, the acoustoelectric conversion device made of PAN can be stored in a power bank and generate enough electricity to run commercial electronic equipment (shown in figure 7.13(c)). In figures 7.13(d)–(f), Peng *et al* [129] studied the piezoelectric properties and the capability to harvest sound energy for power generation purposes of PAN nanofiber membranes (figure 7.13(d)), which showed larger acoustoelectric conversion capability. The nanofiber sensor device shows broader response bandwidth, larger sensitivity, and higher fidelity, with a signal-to-noise ratio of up to 57.2 dB and fidelity of up to 0.995 (shown in figure 7.13(e)). Electrical output stability is an important indicator of the practicality of acoustoelectric conversion devices, and the better acoustoelectric conversion capability is attributed to the strong vibration response properties of PAN nanofibers (figure 7.13(f)). The excellent piezoelectricity and high rigidity of PAN nanofiber membranes under the same forces allow them to exhibit incredible sound detection properties compared with PVDF and its copolymer.

In addition, acoustoelectric conversion devices made from PAN show higher voltage output at low and medium frequency sounds, especially below 500 Hz, filling the gap of inaccurate and incomplete collection of low and medium frequency sound signals by acoustoelectric conversion devices made from PVDF [123]. This gives it a very large potential application in the collection of noise energy in airports, industry and public places [130]. Thus, electrospun PAN nanofiber devices are expected to be newer and more advanced acoustoelectric sensors for accurate sound detection and even other high-tech applications.

7.5.3 Sensors

In today's rapidly evolving technology, the development of renewable, environmentally friendly, cost-effective energy or self-powered sensors is becoming increasingly important in order to meet the energy needs of the future. The main piezoelectric polymer materials used for sensors are PVDF [131–133] and PAN [134, 135], usually prepared as electrostatically spun films, and they are promising candidates for wearable pressure sensors due to their good flexibility, light weight and ease of manufacture. Compared to many electronic devices, these ionomer devices are more flexible and resilient and can even sense a wide range of external stimuli [136]. In the case of flexible wearable piezoelectric sensors made of PAN, the key technology for energy conversion and signal capture in this device is the efficient conversion of external stimuli into electrical signals.

In figures 7.14(a)–(c), Li *et al* [55] prepared ionic conductive organogels with high flexibility, high viscosity, recyclability, and frost resistance by a one-pot method using PAN as raw material. This microstructured organogel with a layered surface produces a stable electrical signal output in response to cyclic impact forces, giving the sensor a highly sensitive response over a wide detection range. Monitoring movement signals, converting them into electrical energy, and obtaining integrated devices with self-powered performance make ionic skins made of PAN a favourable candidate in the field of soft machinery. This is mainly because, in the field of

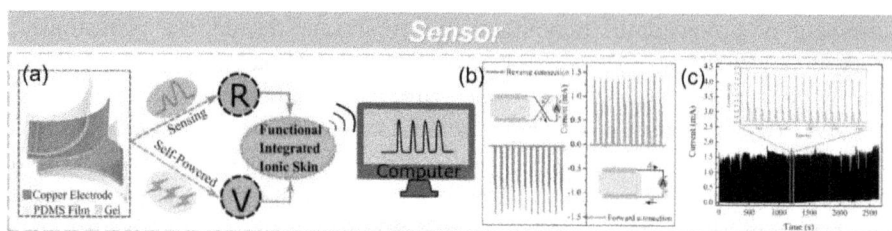

Figure 7.14. (a) A schematic diagram of functionalized integrated ionic skin; (b) test of self-powered ionic skin device based on H-ICOG-Zn-20% current output polarity by its reversing connection to the instrument electrode; (c) cycle stability of current outputs of the self-powered ionic skin device under about 5000 cycles. Reproduced from [55], copyright 2022, with permission from Elsevier.

sensors, the ion transport rate is the key to achieving high sensitivity and fast response [137]. The faster the ion transport rate, the shorter the response time of the sensor and the faster it can detect and respond to the environment or a substance. For a specific PAN-based piezoresistive sensor and self-powered sensor, in figure 7.14(a), the preparation method uses copper foil as the electrodes, and the gel was sandwiched between the two electrodes. When PAN is used as a macro-molecular polymer for complexation with other substances, the ionic conductivity of the gel film is effectively improved. In figures 7.14(b) and (c), the current output polarity of the H-ICOG-Zn-20%-based PAN sensor is demonstrated, as well as the cycling stability under the current output of the self-powered ionic skin device for 5000 cycles at 13 N of force.

In the future, the development of self-powered chemical sensors will face significant challenges concerning a continuous, stable, and long-lasting energy supply [138]. The high ion transport rate of PAN materials, therefore, allows for excellent performance in the field of sensors, as well as sensitivity response, and also good cycling stability and durability. It can be used to develop sensors for the detection of chemical substances, biomolecules, and contaminants in the environment, and even has a wide range of applications in the fields of environmental monitoring, biomedicine, and food safety.

7.5.4 Working at high-temperature

It has been difficult to apply most piezoelectric polymers to high-temperature environments due to their poor thermal stability, but by processing and changing the structure of PAN, breakthroughs have been made that other piezoelectric polymer cannot reach at operating temperatures [90]. The thermally stable electrospun PAN nanofiber films offer greater elasticity and strain, and also show excellent piezoelectric properties at temperatures up to 550 °C, with a higher acoustic and electrical output than piezoelectric ceramics.

As shown in figures 7.15(a)–(d), with a special thermal stabilization process, Wang *et al* [90] prepared a flexible high-temperature piezoelectric membrane from electrospun PAN nanofibrous membranes. During the preparation process, figure 7.15(b), the membrane showed different colors from light yellow to dark

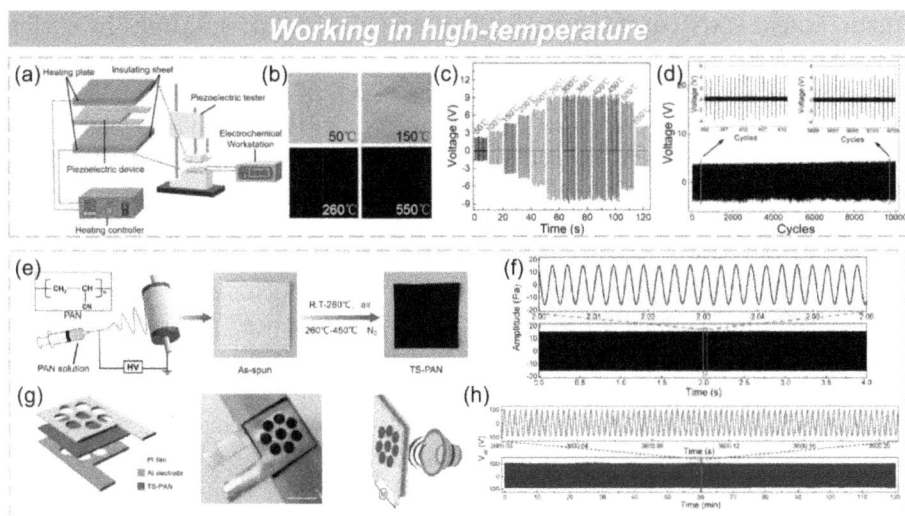

Figure 7.15. (a) The apparatus for testing the electrical outputs of PAN membranes; (b) digital photos showing the appearance of the PAN membranes prepared in different temperature steps; (c) the voltage outputs of the PAN nanofiber mat before and after charge removal treatment; (d) working stability of the 550 °C treated PAN membrane device at 500 °C. Reproduced from [90] with permission from Royal Society of Chemistry. (e) Preparation process of thermally stabilized PAN nanofiber membranes, (f) sound spectrum produced by the thermoacoustic engine; (g) schematic diagram of TS-PAN acoustoelectric device; (h) working stability of the TS-PAN device during acoustoelectric conversion. Reproduced from [139], copyright 2022, with permisison from Elsevier.

brown by heat treatment at different temperatures. In figures 7.15(a) and (c), a system of instruments and heating devices for testing the electrical output of PAN membranes is demonstrated, and a device made with this with an area of 2.5 × 2.5 cm^2 can produce 9.7 volts at 450 °C. The energy device can operate stably in high-temperature environments, powering commercial LEDs and charging capacitors. Figure 7.15(d) shows that membranes treated at 550 °C show the best stability and can be operated in air at 500 °C for more than 10 000 cycles without changing the mechanical–electrical conversion capability. Zheng *et al* [139] converted thermoacoustic waves into electric power by using an acoustoelectric device prepared from PAN nanofiber membranes. With the preparation process shown in figure 7.15(e), PAN nanofiber membranes were prepared using the electrostatic spinning method, then heat stabilized using a multi-step heat treatment method. Under the sound spectrum curve produced by the thermoacoustic engine (figure 7.15(f)), the PAN-based acoustoelectric device was heated to 280 °C, worked stably for at least 2 h, as in figure 7.15(h). A single unit (3 × 3 cm^2) generates electrical energy by converting thermoacoustic waves with a peak output of 102 V and 10 μA and is also capable of running commercial electronic devices such as 30 LEDs. The device can also store electrical energy in lithium batteries or capacitors for further use. This high-temperature-resistant acoustoelectric conversion material also enables the development of acoustoelectric energy harvesters for different fields and applications. This is because it has been established that thermally stable PAN can form ladder-like conjugated cyclic structures by cyclizing chain structures at high temperatures [140]. The ladder structure has a considerable spatial potential resistance and the ladder chains can be interconnected by the cyclized structure. This severely prevents the chain from rotating and moving at high temperatures and has excellent high-temperature resistance and flame retardancy. This high-temperature property has been used extensively in the preparation of carbon fibers and in the development of flame-retardant fabrics. It explains the high-temperature-resistance piezoelectric properties of heat-treated PAN films.

As a result, PAN piezoelectric materials have outstanding comprehensive performance, for example high piezoelectric performance in wide temperature range, high mechanical property, hydrophilic property, and piezoelectric stability. PAN and thermally stabilized PAN, have excellent mechanical-to-electrical conversion capabilities at both room temperature and both extreme high temperatures of 550 °C, which can be developed for flexible devices for energy harvesters, sensors and other applications in harsh, extreme high-temperature conditions.

7.6 Conclusion

In summary, the main field of application regarding PAN are in PENGs, acoustic-electric conversion, sensor, high-temperature resistant piezoelectric, and thermoacoustic wave conversion to electrical energy. In PENG, PAN-based materials efficiently harvest energy from mechanical vibrations due to their excellent piezoelectric response. The field of acoustoelectric conversion benefits from the excellent piezoelectric coefficient and elastic recovery of PAN to convert acoustic waves into electrical signals with high fidelity. PAN-based sensors have higher sensitivity due to their high ionic transport rate and can be used for precise detection and monitoring. In addition, PAN is treated to withstand high temperatures and exhibits excellent piezoelectric properties, making it ideal for applications in extreme conditions. PAN can also convert thermoacoustic waves into electrical energy, enabling efficient energy harvesting from temperature gradients. Overall, PAN's superior piezoelectric properties make it a versatile material for energy harvesting and conversion and sensing applications, providing increased performance and reliability in a variety of fields.

This chapter was reprinted from [77], copyright (2023), with permission from Elsevier

References

[1] Chandrasekaran S, Bowen C, Roscow J et al 2019 Micro-scale to nano-scale generators for energy harvesting: self powered piezoelectric, triboelectric and hybrid devices Phys. Rep. **792** 1–33

[2] Karan S K, Maiti S, Lee J H et al 2020 Recent advances in self-powered tribo-/piezoelectric energy harvesters: all-in-one package for future smart technologies Adv. Funct. Mater. **30** 2004446

[3] Zhang D, Wang D, Xu Z et al 2021 Diversiform sensors and sensing systems driven by triboelectric and piezoelectric nanogenerators Coord. Chem. Rev. **427** 213597

[4] Sun Y, Chen J, Li X *et al* 2019 Flexible piezoelectric energy harvester/sensor with high voltage output over wide temperature range *Nano Energy* **61** 337–45

[5] Kim N-I, Chang Y-L, Chen J *et al* 2020 Piezoelectric pressure sensor based on flexible gallium nitride thin film for harsh-environment and high-temperature applications *Sens. Actuators, A* **305** 111940

[6] Bakhtar L J, Abdoos H and Rashidi S 2023 A review on fabrication and *in vivo* applications of piezoelectric nanocomposites for energy harvesting *J. Taiwan Inst. Chem. Eng.* **148** 104651

[7] Rahaman M S A, Ismail A F and Mustafa A 2007 A review of heat treatment on polyacrylonitrile fiber *Polym. Degrad. Stab.* **92** 1421–32

[8] Rezakazemi M, Sadrzadeh M and Matsuura T 2018 Thermally stable polymers for advanced high-performance gas separation membranes *Prog. Energy Combust. Sci.* **66** 1–41

[9] Kuo Y-H, Luo J, Steier W *et al* 2005 Enhanced thermal stability of electrooptic polymer modulators using the Diels–Alder crosslinkable polymer *IEEE Photonics Technol. Lett.* **18** 175–7

[10] Zheng T, Wu J, Xiao D *et al* 2018 Recent development in lead-free perovskite piezoelectric bulk materials *Prog. Mater Sci.* **98** 552–624

[11] Stupp S and Carr S 1978 Spectroscopic analysis of electrically polarized polyacrylonitrile *J. Polym. Sci.: Polym. Phys. Ed.* **16** 13–28

[12] Ueda H and Carr S 1984 Piezoelectricity in polyacrylonitrile *Polym. J.* **16** 661–7

[13] Henrici-Olivé G and Olivé S 1979 Molecular interactions and macroscopic properties of polyacrylonitrile and model substances *Chemistry. Advances in Polymer Science* **vol 32** (Springer)

[14] Grobelny J, Sokól M and Turska E 1984 A study of conformation, configuration and phase structure of polyacrylonitrile and their mutual dependence by means of WAXS and 1H BL-n.m.r *Polymer* **25** 1415–8

[15] Hobson R J and Windle A H 1993 Crystalline structure of atactic polyacrylonitrile *Macromolecules* **26** 6903–7

[16] Rizzo P, Auriemma F, Guerra G *et al* 1996 Conformational disorder in the pseudohexagonal form of atactic polyacrylonitrile *Macromolecules* **29** 8852–61

[17] Cooper A, Oldinski R, Ma H *et al* 2013 Chitosan-based nanofibrous membranes for antibacterial filter applications *Carbohydr. Polym.* **92** 254–9

[18] Furukawa 1989 Piezoelectricity and pyroelectricity in polymers *IEEE Trans. Electr. Insul.* **24** 375–94

[19] Nataraj S, Yang K and Aminabhavi T 2012 Polyacrylonitrile-based nanofibers—a state-of-the-art review *Prog. Polym. Sci.* **37** 487–513

[20] Wang W, Zheng Y, Jin X *et al* 2019 Unexpectedly high piezoelectricity of electrospun polyacrylonitrile nanofiber membranes *Nano Energy* **56** 588–94

[21] Kepler R G and Anderson R 1992 Ferroelectric polymers *Adv. Phys.* **41** 1–57

[22] Kawai H 1969 The piezoelectricity of poly (vinylidene fluoride) *Jpn. J. Appl. Phys.* **8** 975

[23] Datta A, Choi Y S, Chalmers E *et al* 2017 Piezoelectric nylon-11 nanowire arrays grown by template wetting for vibrational energy harvesting applications *Adv. Funct. Mater.* **27** 1604262.1–10

[24] Bahl O P and Mathur R B 1979 Effect of load on the mechanical properties of carbon fibres from pan precursor *Fibre Sci. Technol.* **12** 31–9

[25] Cato A D and Dan D E 2003 Flow behavior of mesophase pitch *Carbon* **41** 1411–7

[26] Kim G H, Hong S M and Seo Y 2009 Piezoelectric properties of poly (vinylidene fluoride) and carbon nanotube blends: β-phase development *Phys. Chem. Chem. Phys.* **11** 10506–12

[27] Moskowitz J D, Jackson M B, Tucker A *et al* 2021 Evolution of polyacrylonitrile precursor fibers and the effect of stretch profile in wet spinning *J. Appl. Polym. Sci.* **138** 50967

[28] Zhang J, Shen Z and Gao A 2019 Conformation transformation of PAN fiber molecular chain during fiber formation *High-tech fibers and applications* **44** 21–4+50

[29] Hinrichsen G 1972 Structural changes of drawn polyacrylonitrile during annealing *J. Polym. Sci.* C **38** 303–14

[30] Sokół M, Grobelny J and Turska E 1987 Investigation of structural changes of poly-acrylonitrile on swelling. Wide-angle X-ray scattering study *Polymer* **28** 843–6

[31] Fleck J N 1972 *Bibliography on Fibers and Composite Materials—1969–1972* (Battelle-Columbus, Metals and Ceramics Information Center)

[32] Sawai D, Kanamoto T, Yamazaki H *et al* 2004 Dynamic mechanical relaxations in poly (acrylonitrile) with different stereoregularities *Macromolecules* **37** 2839–46

[33] Gribanov A V and Sazanov Y N 2008 Polyacrylonitrile: carbonization problems *Russ. J. Appl. Chem.* **81** 919–32

[34] Farmer B, Hopfinger A and Lando J 1972 Polymorphism of poly (vinylidene fluoride): potential energy calculations of the effects of head-to-head units on the chain conformation and packing of poly (vinylidene fluoride) *J. Appl. Phys.* **43** 4293–303

[35] Okajima S, Ikeda M and Takeuchi A 1968 A new transition point of polyacrylonitrile *J. Polym. Sci. Part A-1 Polym. Chem.* **6** 1925–33

[36] Imai Y, Minami S, Yoshihara T *et al* 1970 Preparation and characterization of amorphous polyacrylonitrile *J. Polym. Sci.* B **8** 281–8

[37] Minami S 1974 Morphology and mechanical properties of polyacrylonitrile fibres *ACS Applied Polymer Symp.* **25** 145–57

[38] Rizzo P, Guerra G and Auriemma F 1996 Thermal transitions of polyacrylonitrile fibers *Macromolecules* **29** 1830–2

[39] Bashir Z 2001 The hexagonal mesophase in atactic polyacrylonitrile: a new interpretation of the phase transitions in the polymer *J. Macromol. Sci.* B **40** 41–67

[40] Hori T, Zhang H S, Shimizu T *et al* 1988 Change of water states in acrylic fibers and their glass transition temperatures by DSC measurements *Textile Res. J.* **58** 227–32

[41] Choi A C, Pramanick A, Misture S T *et al* 2018 Polarization mechanisms in P(VDF-TrFE) ferroelectric thin films (Phys. Status Solidi RRL 10/2018) *Phys. Status Solidi (RRL)—Rapid Res. Lett.* **12** 1870331

[42] Xiao M, Zhang W, Zhang Z *et al* 2017 The influence of preferred orientation and poling temperature on the polarization switching current in PZT thin films *Appl. Phys.* A **123** 487

[43] Ye Y, Guo T-L, Jiang Y-D *et al* 2012 Investigation on poling and electric properties of PVDF films *J. Univ. Electron. Sci. Technol. China* **41** 463–6

[44] Zhang L F, Aboagye A, Kelkar A *et al* 2014 A review: carbon nanofibers from electrospun polyacrylonitrile and their applications *J. Mater. Sci.* **49** 463–80

[45] Comstock R, Stupp S and Carr S 1977 Thermally stimulated discharge currents from polyacrylonitrile *J. Macromol. Sci., Part B: Phys.* **13** 101–15

[46] Azmi S, Hosseini Varkiani S-M, Latifi M *et al* 2022 Tuning energy harvesting devices with different layout angles to robust the mechanical-to-electrical energy conversion performance *J. Ind. Text.* **51** 9000S–16S

[47] Zheng Y 2022 Study of piezoelectric and acoustic-electric conversion properties of polyacrylonitrile nanofiber membranes *PhD Thesis* Tiangong University

[48] Chang C, Tran V H, Wang J *et al* 2010 Direct-write piezoelectric polymeric nanogenerator with high energy conversion efficiency *Nano Lett.* **10** 726–31

[49] Street R M, Minagawa M, Vengrenyuk A *et al* 2019 Piezoelectric electrospun polyacrylonitrile with various tacticities *J. Appl. Polym. Sci.* **136** 47530

[50] Yu S, Milam-Guerrero J, Tai Y Y *et al* 2022 Maximizing polyacrylonitrile nanofiber piezoelectric properties through the optimization of electrospinning and post-thermal treatment processes *ACS Appl. Polym. Mater.* **4** 635–44

[51] Sagiri S, Behera B, Rafanan R *et al* 2014 Organogels as matrices for controlled drug delivery: a review on the current state *Soft Mater.* **12** 47–72

[52] Bastiat G, Plourde F, Motulsky A *et al* 2010 Tyrosine-based rivastigmine-loaded organogels in the treatment of Alzheimer's disease *Biomaterials* **31** 6031–8

[53] Yang Y, Huang Q, Niu L *et al* 2017 Waterproof, ultrahigh areal-capacitance, wearable supercapacitor fabrics *Adv. Mater.* **29** 1606679

[54] Zeng L, Lin X, Li P *et al* 2021 Recent advances of organogels: from fabrications and functions to applications *Prog. Org. Coat.* **159** 106417

[55] Li W, Zhang J, Niu J *et al* 2022 Self-powered and high sensitivity ionic skins by using versatile organogel *Nano Energy* **99** 107359

[56] Wang L and Daoud W A 2019 Hybrid conductive hydrogels for washable human motion energy harvester and self-powered temperature-stress dual sensor *Nano Energy* **66** 104080

[57] Guo X, Zhang C, Shi L *et al* 2020 Highly stretchable, recyclable, notch-insensitive, and conductive polyacrylonitrile-derived organogel *J. Mater. Chem.* A **8** 20346–53

[58] Huang J, Peng S, Gu J *et al* 2020 Self-powered integrated system of a strain sensor and flexible all-solid-state supercapacitor by using a high performance ionic organohydrogel *Mater. Horizons* **7** 2085–96

[59] Chun K-Y, Seo S and Han C-S 2021 Self-powered, stretchable, and wearable ion gel mechanoreceptor sensors *ACS Sens.* **6** 1940–8

[60] Deng W, Tu L, Fu R *et al* 2022 Polyacrylonitrile-based self-powered hydrogel for flexible sensing *Sci. Sin. Chim.* **52** 494–503

[61] Zhao G R, Zhang X D, Cuo X *et al* 2018 Piezoelectric polyacrylonitrile nanofiber film-based dual-function self-powered flexible sensor *ACS Appl. Mater. Interfaces* **10** 15855–63

[62] Cai T T, Yang Y, Bi T *et al* 2020 $BaTiO_3$ assisted PAN fiber preparation of high performance flexible nanogenerator *Nanotechnology* **31** 24LT01

[63] He Y Q, Fu G M, He D Y *et al* 2022 Fabrication and characterizations of Eu^{3+} doped PAN/$BaTiO_3$ electrospun piezoelectric composite fibers *Mater. Lett.* **314** 131888

[64] Yuan L, Fan W, Yang X *et al* 2021 Piezoelectric PAN/$BaTiO_3$ nanofiber membranes sensor for structural health monitoring of real-time damage detection in composite *Compos. Commun.* **25** 100680

[65] Bairagi S, Chowdhury A, Banerjee S *et al* 2022 Investigating the role of copper oxide (CuO) nanorods in designing flexible piezoelectric nanogenerator composed of polyacrylonitrile (PAN) electrospun web-based fibrous material *J. Mater. Sci.-Mater. Electron.* **33** 13152–65

[66] Sun Y, Liu Y, Zheng Y D *et al* 2020 Enhanced energy harvesting ability of ZnO/PAN hybrid piezoelectric nanogenerators *ACS Appl. Mater. Interfaces* **12** 54936–45

[67] Sorayani Bafqi M S, Bagherzadeh R and Latifi M 2015 Fabrication of composite PVDF-ZnO nanofiber mats by electrospinning for energy scavenging application with enhanced efficiency *J. Polym. Res.* **22** 130

[68] Ding D, Li Z W, Yu S *et al* 2022 Piezo-photocatalytic flexible PAN/TiO$_2$ composite nanofibers for environmental remediation *Sci. Total Environ.* **824** 153790

[69] Wu H, Jin X, Wang W Y *et al* 2019 Preparation and piezoelectric properties of polyacrylonitrile/sodium nitrate nanofiber membrane *J. Textile Res.* **41** 26–32

[70] Lian Y, He F, Wang H *et al* 2015 A new aptamer/graphene interdigitated gold electrode piezoelectric sensor for rapid and specific detection of *Staphylococcus aureus Biosens. Bioelectron.* **65** 314–9

[71] Kambale K R, Goyal R, Butee S P *et al* 2017 Novel polyacrylonitrile/potassium sodium niobate composites with superior dielectric and thermal properties *Compos. Commun.* **5** 8–12

[72] Egerton L and Dillon D M 1959 Piezoelectric and dielectric properties of ceramics in the system potassium—sodium niobate *J. Am. Ceram. Soc.* **42** 438–42

[73] Shabbir G and Kojima S 2014 Anomalous variations in elastic properties of lead zirconate niobate-lead titanate single crystals in the vicinity of its ferroelectric phase transition *EPL (Europhys. Lett.)* **105** 57001

[74] Guo Y, Kakimoto K-I and Ohsato H 2004 Phase transitional behavior and piezoelectric properties of (Na$_{0.5}$K$_{0.5}$)NbO$_3$–LiNbO$_3$ ceramics *Appl. Phys. Lett.* **85** 4121–3

[75] Qin X H, Yang E L, Li N *et al* 2007 Effect of different salts on electrospinning of polyacrylonitrile (PAN) polymer solution *J. Appl. Polym. Sci.* **103** 3865–70

[76] Shi Q, He S, He Y *et al* 2023 Enhanced the dielectric and piezoelectric properties of polyacrylonitrile piezoelectric composite fibers filled with ionic liquids *J. Appl. Polym. Sci.* **140** e53824

[77] Tao J Z, Wang Y F, Zheng X K *et al* 2023 A review: polyacrylonitrile as high-performance piezoelectric materials *Nano Energy* **118** 108987

[78] Moghadam B H, Hasanzadeh M and Simchi A 2020 Self-powered wearable piezoelectric sensors based on polymer nanofiber–metal–organic framework nanoparticle composites for arterial pulse monitoring *ACS Appl. Nano Mater.* **3** 8742–52

[79] Sorayani Bafqi M S, Sadeghi A-H, Latifi M *et al* 2021 Design and fabrication of a piezoelectric out-put evaluation system for sensitivity measurements of fibrous sensors and actuators *J. Ind. Text.* **50** 1643–59

[80] Mirjalali S, Bagherzadeh R, Abrishami S *et al* 2023 Multilayered electrospun/electro-sprayed polyvinylidene fluoride + zinc oxide nanofiber mats with enhanced piezoelectricity *Macromol. Mater. Eng.* **308** 2300009

[81] Lang S B and Muensit S 2006 Review of some lesser-known applications of piezoelectric and pyroelectric polymers *Appl. Phys.* A **85** 125–34

[82] Fang J, Wang X and Lin T 2011 Electrical power generator from randomly oriented electrospun poly(vinylidene fluoride) nanofibre membranes *J. Mater. Chem.* **21** 11088–92

[83] Hattori T, Takahashi Y, Iijima M *et al* 1996 Piezoelectric and ferroelectric properties of polyurea thin films copolymerized by vapor deposition polymerization of aliphatic and aromatic monomers *Jpn. J. Appl. Phys.* **35** 2199–204

[84] Ma B, Liu F, Li Z *et al* 2019 Piezoelectric Nylon-11 nanoparticles with ultrasound assistance for high-efficiency promotion of stem cells osteogenic differentiation *J. Mater. Chem.* B **11** 1847–54

[85] Paria S, Karan S K, Bera R et al 2016 A facile approach to develop a highly stretchable PVC/ZnSnO₃ piezoelectric nanogenerator with high output power generation for powering portable electronic devices *Ind. Eng. Chem. Res.* **55** 10671–80

[86] Furukawa T and Fukada E 2010 Piezoelectric relaxation in poly(γ-benzyl-glutamate) *J. Polym. Sci.: Polym. Phys. Ed.* **14** 1979–2010

[87] Vinogradov A M, Hugo Schmidt V, Tuthill G F et al 2004 Damping and electromechanical energy losses in the piezoelectric polymer PVDF *Mech. Mater.* **36** 1007–16

[88] Fukada E 1998 New piezoelectric polymers *Jpn. J. Appl. Phys.* **37** 2775

[89] Zhao C, Zhang J, Wang Z L et al 2017 A poly(L-lactic acid) polymer-based thermally stable cantilever for vibration energy harvesting applications *Adv. Sustain.Syst.* **1** 1700068

[90] Wang W, Zheng Y, Sun Y et al 2021 High-temperature piezoelectric conversion using thermally stabilized electrospun polyacrylonitrile membranes *J. Mater. Chem.* A **9** 20395–404

[91] Su Y, Zhang W, Lan J et al 2020 Flexible reduced graphene oxide/polyacrylonitrile dielectric nanocomposite films for high-temperature electronics applications *ACS Appl. Nano Mater.* **3** 7005–15

[92] Wang D H, Kurish B A, Treufeld I et al 2015 Synthesis and characterization of high nitrile content polyimides as dielectric films for electrical energy storage *J. Polym. Sci., Part A: Polym. Chem.* **53** 422–36

[93] Jacquemin J L, Ardalan A and Bordure G 1978 Electrical conductivity and dielectric constant of heat-treated polyacrylonitrile in an AC regime *J. Non-Cryst. Solids* **28** 249–57

[94] Jian G, Jiao Y, Meng Q Z et al 2021 Excellent high-temperature piezoelectric energy harvesting properties in flexible polyimide/3D PbTiO₃ flower composites *Nano Energy* **82** 105778

[95] Ramadan K S, Sameoto D and Evoy S 2014 A review of piezoelectric polymers as functional materials for electromechanical transducers *Smart Mater. Struct.* **23** 033001

[96] Khanbareh H, Hegde M, Bijleveld J C et al 2017 Functionally graded ferroelectric polyetherimide composites for high temperature sensing *J. Mater. Chem.* C **5** 9389–97

[97] Zhu T W, Yu Q X, Zheng W W et al 2021 Intrinsic high-k-low-loss dielectric polyimides containing ortho-position aromatic nitrile moieties: reconsideration on Clausius–Mossotti equation *Polym. Chem.* **12** 2481–9

[98] Maceiras A, Martins P, San Sebastián M et al 2014 Synthesis and characterization of novel piezoelectric nitrile copolyimide films for high temperature sensor applications *Smart Mater. Struct.* **23** 105015

[99] Ge Y, Zhang H, Yu H et al 2023 Effect of structural evolution in high-temperature thermal oxidative stabilized polyacrylonitrile fibers on carbonized fibers *J. Appl. Polym. Sci.* **140** e54692

[100] Zhai Y, Wang N, Mao X et al 2014 Sandwich-structured PVdF/PMIA/PVdF nanofibrous separators with robust mechanical strength and thermal stability for lithium ion batteries *J. Mater. Chem.* A **2** 14511–8

[101] Tang Y Y, Li P F, Zhang W Y et al 2017 A multiaxial molecular ferroelectric with highest curie temperature and fastest polarization switching *J. Am. Chem. Soc.* **139** 13903–8

[102] Akiyama M, Morofuji Y, Kamohara T et al 2007 Preparation of oriented aluminum nitride thin films on polyimide films and piezoelectric response with high thermal stability and flexibility *Adv. Funct. Mater.* **17** 458–62

[103] Li Q, Chen L, Gadinski M R et al 2015 Flexible high-temperature dielectric materials from polymer nanocomposites *Nature* **523** 576–9

[104] Kobayashi M, Jen C K, Bussiere J F *et al* 2009 High-temperature integrated and flexible ultrasonic transducers for nondestructive testing *NDT E Int.* **42** 157–61

[105] Wiles K B 2001 Determination of reactivity ratios for acrylonitrile/methyl acrylate radical copolymerization via nonlinear methodologies using real time FTIR *Master's Thesis* Virginia Tech

[106] Ji M, Wang C, Bai Y *et al* 2007 Structural evolution of polyacrylonitrile precursor fibers during preoxidation and carbonization *Polym. Bull.* **59** 527–36

[107] Zhang C, Liu J, Guo S *et al* 2018 Comparison of microwave and conventional heating methods for oxidative stabilization of polyacrylonitrile fibers at different holding time and heating rate *Ceram. Int.* **44** 14377–85

[108] Fitzer E and Müller D 1975 The influence of oxygen on the chemical reactions during stabilization of pan as carbon fiber precursor *Carbon* **13** 63–9

[109] Peebles J R L, Peyser P, Snow A *et al* 1990 On the exotherm of polyacrylonitrile: pyrolysis of the homopolymer under inert conditions *Carbon* **28** 707–15

[110] Friedlander H, Peebles J R L, Brandrup J *et al* 1968 On the chromophore of polyacrylonitrile. VI. Mechanism of color formation in polyacrylonitrile *Macromolecules* **1** 79–86

[111] Burlant W and Parsons J 1956 Pyrolysis of polyacrylonitrile *J. Polym. Sci.* **22** 249–56

[112] Maceiras A, Costa C, Lopes A *et al* 2015 Dielectric relaxation dynamics of high-temperature piezoelectric polyimide copolymers *Appl. Phys.* A **120** 731–43

[113] Gonzalo B, Vilas J, Breczewski T *et al* 2009 Synthesis, characterization, and thermal properties of piezoelectric polyimides *J. Polym. Sci., Part A: Polym. Chem.* **47** 722–30

[114] Goyal R K, Katkade S S and Mule D M 2013 Dielectric, mechanical and thermal properties of polymer/BaTiO$_3$ composites for embedded capacitor *Compos. Part B: Eng.* **44** 128–32

[115] Liu S, Xue S, Zhang W *et al* 2014 Enhanced dielectric and energy storage density induced by surface-modified BaTiO$_3$ nanofibers in poly(vinylidene fluoride) nanocomposites *Ceram. Int.* **40** 15633–40

[116] Parali L, Şabikoğlu İ and Kurbanov M A 2014 Piezoelectric properties of the new generation active matrix hybrid (micro-nano) composites *Appl. Surf. Sci.* **318** 6–9

[117] Goyal R K, Madav V V, Pakankar P R *et al* 2011 Fabrication and properties of novel polyetheretherketone/barium titanate composites with low dielectric loss *J. Electron. Mater.* **40** 2240–7

[118] Madakbas S, Celik Z, Dumludag F *et al* 2014 Preparation, characterization and electrical properties of polyacrylonitrile/huntite composites *Polym. Bull.* **71** 1471–81

[119] Zhang H, Xie Y, Song Y *et al* 2021 Preparation of high-temperature resistant poly (m-phenylene isophthalamide)/polyacrylonitrile composite nanofibers membrane for air filtration *Colloids Surf., A* **624** 126831

[120] Wekin A, Richards C, Matveev K *et al* 2008 Characterization of piezoelectric materials for thermoacoustic power transduction *Proc. of the ASME Int. Mechanical Engineering Congress and Exposition*

[121] Dewi A, Mufti N, Fibriyanti A *et al* 2020 The improvement of triboelectric effect of ZnO Nanorods/PAN in flexible Nanogenerator by adding TiO$_2$ nanoparticle *J. Polym. Res.* **27** 1–10

[122] Kalimuldina G, Turdakyn N, Abay I *et al* 2020 A review of piezoelectric PVDF film by electrospinning and its applications *Sensors* **20** 5214

[123] Lang C, Fang J, Shao H *et al* 2016 High-sensitivity acoustic sensors from nanofibre webs *Nat. Commun.* **7** 11108

[124] Lang C, Fang J, Shao H *et al* 2017 High-output acoustoelectric power generators from poly (vinylidenefluoride-co-trifluoroethylene) electrospun nano-nonwovens *Nano Energy* **35** 146–53

[125] Farsi A H, Pullen A D, Latham J *et al* 2017 Full deflection profile calculation and Young's modulus optimisation for engineered high performance materials *Sci. Rep.* 7

[126] Mirjalali S, Mahdavi Varposhti A, Abrishami S *et al* 2022 A review on wearable electrospun polymeric piezoelectric sensors and energy harvesters *Macromol. Mater. Eng.* **308** 2200442

[127] Shao H, Wang H X, Cao Y Y *et al* 2020 Efficient conversion of sound noise into electric energy using electrospun polyacrylonitrile membranes *Nano Energy* **75** 104956

[128] Lang C, Fang J, Shao H *et al* 2017 High-output acoustoelectric power generators from poly (vinylidenefluoride-co-trifluoroethylene) electrospun nano-nonwovens *Nano Energy* **35** 146–53

[129] Peng L, Jin X, Niu J R *et al* 2021 High-precision detection of ordinary sound by electrospun polyacrylonitrile nanofibers *J. Mater. Chem.* C **9** 3477–85

[130] Leventhall H G 2004 Low frequency noise and annoyance *Noise Health* **6** 59–72

[131] Kim J Y, Jang M, Jeong G-H *et al* 2021 MXene-enhanced β-phase crystallization in ferroelectric porous composites for highly-sensitive dynamic force sensors *Nano Energy* **89** 106409

[132] Lin M-F, Cheng C, Yang C-C *et al* 2021 A wearable and highly sensitive capacitive pressure sensor integrated a dual-layer dielectric layer of PDMS microcylinder array and PVDF electrospun fiber *Org. Electron.* **98** 106290

[133] Haghayegh M, Cao R, Zabihi F *et al* 2022 Recent advances in stretchable, wearable and bio-compatible triboelectric nanogenerators *J. Mater. Chem.* C **10** 11439–71

[134] Qureshi Y, Tarfaoui M and Lafdi K 2021 Electro-thermal–mechanical performance of a sensor based on PAN carbon fibers and real-time detection of change under thermal and mechanical stimuli *Mater. Sci. Eng.: B* **263** 114806

[135] Fu X, Li J, Li D *et al* 2022 MXene/ZIF-67/PAN nanofiber film for ultra-sensitive pressure sensors *ACS Appl. Mater. Interfaces* **14** 12367–74

[136] Jayaraman P, Sengottaiyan C and Krishnan K 2020 Self-assembled polymer thin films towards nanoarchitectonics for respiration monitoring *J. Nanosci. Nanotechnol.* **20** 2893–901

[137] Chang Y, Wang L, Li R *et al* 2021 First decade of interfacial iontronic sensing: from droplet sensors to artificial skins *Adv. Mater.* **33** 2003464

[138] Aaryashree, Sahoo S, Walke P *et al* 2021 Recent developments in self-powered smart chemical sensors for wearable electronics *Nano Res.* **14** 3669–89

[139] Zheng Y, Wang W, Niu J *et al* 2022 Thermoacoustic energy harvesting using thermally-stabilized polyacrylonitrile nanofibers *Nano Energy* **95** 106995

[140] Zhu D, Xu C, Nakura N *et al* 2002 Study of carbon films from PAN/VGCF composites by gelation/crystallization from solution *Carbon* **40** 363–73

[141] Cho B-G, Lee J-E, Jeon S-Y and Chae H G 2023 A study on miscibility properties of polyacrylonitrile blending films with biodegradable polymer, shellac *Polymer Testing* **121** 107983

[142] Zhao G, Chen J, Zhang C, Zeng J, Zhou Z, Liu J and Guo S 2023 Study on the relationship between ring formation and properties of pre-oxidized polyacrylonitrile-based fibres *J. Mol. Struct.* **1284** 135412

IOP Publishing

Energy Harvesting Properties of Electrospun Nanofibers
(Second Edition)

Jian Fang and Tong Lin

Chapter 8

Self-powered electronic skins constructed of electrospun nanofibers

Faqiang Wang, Jianyong Yu, Bin Ding and Zhaoling Li

With the growing demand for flexible electronics, exploring novel wearable electronic skins is of significance. The self-powered electronic skins constructed of electrospun nanofibers are highly sought after for their distinctive features and numerous potential applications. This chapter aims to provide a comprehensive overview of the typical self-power modes, including capturing thermal, solar, and mechanical energy from the surrounding environment and then converting them into electrical energy to achieve the driving of the electronic skins. This is followed by a systematic discussion of electronic skins integrated with individual and multifunctional sensing systems, and the properties that should be available in today's electronic skins according to practical needs. Meanwhile, the main application scenarios of electronic skins are described with a view to promoting the development of this field.

8.1 Introduction

As one of the key components of the human body, biological skin is not only a natural physical barrier, but also the primary medium through which people interact with complex environments. The skin can provide an advanced perceptual network that converts external mechanical stimuli, temperature as well as humidity changes, etc into physiological signals, which are received and interpreted by the brain and further guide the body's responses [1]. This compelling phenomenon has prompted a great deal of research effort devoted to the field of skin simulation. Subsequently, with the emergence of artificial intelligence, automation, micro–nano manufacturing and other emerging technologies, electronic skin has developed rapidly and shown good application prospects in many fields such as wearable electronics, personal health monitoring, artificial neural networks, and so on [2–5].

doi:10.1088/978-0-7503-5487-5ch8

In general, electronic skin as a sensing platform was composed of multiple components including substrate layer, electrode layer, sensing layer and so on [6–8]. Among them, wearable power supply unit is an indispensable cornerstone for electronic skin [9, 10]. However, the shortcomings of traditional rigid batteries, in terms of portability, sustainability and safety greatly limit the practical applications of electronic skins in harsh environments [11–15]. Therefore, based on the consideration of practical demands, it turns out that developing wearable self-powered systems is an urgent challenge that needs to be solved. And the utilization of the human body or the environment as an energy source to drive electronic skin without an external power source is considered as a promising method that has attracted a great deal of interest [16–19].

It is worth mentioning that electronic skins based on films, rubbers, hydrogels, and aerogels often suffer from the disadvantages of poor permeability, complex preparation process, and high wastage. This is undoubtedly unable to meet the current demands for flexible and wearable electronics. In contrast, electrospun nanofibers with inherent specific surface area, outstanding flexibility, and superior breathability are ideal candidates for the preparation of highly sensitive electronic skins due to their characteristics of controllable structures and properties [20–22]. Overall, nanofibers-based self-powered electronic skin has great potential and high research value.

8.2 Self-powered systems

8.2.1 Thermal harvester integrated self-powered systems

The strategies of recycling waste heat for power generation are gaining immense popularity. A typical example of this is thermoelectric generator (TEG), which is defined as equipment that converts thermal energy into electrical energy by means of a thermal gradient [23, 24]. Seebeck effect (figure 8.1(a)) and Peltier effect (figure 8.1(b)) are the basis for the operation of thermoelectric devices [24, 25]. Generally, TEGs are made up of P-type and N-type semiconductor pairs, the former with excess holes and the latter with excess electrons to carry current. They are connected thermally in parallel and electrically in series. As the temperature gradient rises through the thermoelectric material from the hot surface to the cold surface, the free charge moves, allowing for the conversion of heat to electrical energy. Furthermore, the pyroelectric nanogenerators are based on the pyroelectric effect, i.e., when a crystal is heated or cooled, the change in temperature leads to a spontaneous change in spontaneous polarization intensity, resulting in the surface polarization charge in one direction of the crystal [26–28]. Thus, by heating or cooling the pyroelectric material in intervals, the electric current can be generated.

8.2.2 Photovoltaic cell powered sensing systems

Integrated photovoltaic systems provide an efficient way of establishing self-powered sensing systems by harvesting solar energy and converting it efficiently into electricity. Based on the photovoltaic effect, when exposed to light, the state of charge distribution within the solar cells changes, resulting in the generation of

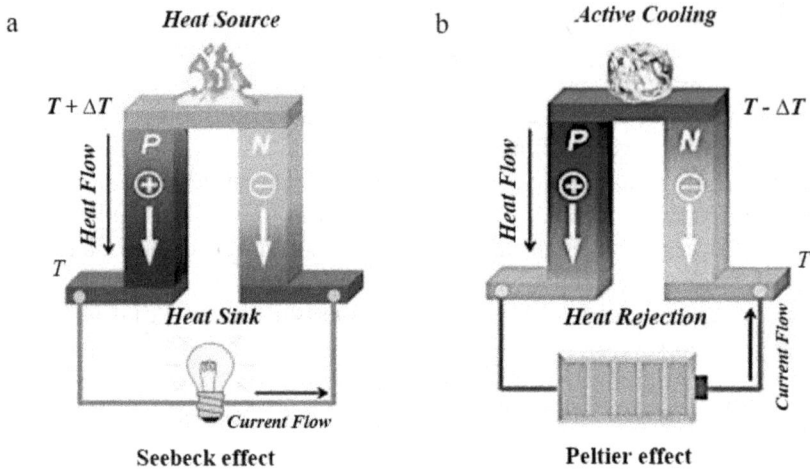

Figure 8.1. Schematic of the working principle of thermoelectric devices: (a) Seebeck effect and (b) Peltier effect. Reprinted from [24], copyright 2015, with permission from Elsevier.

electric potential and current [29–31]. In other words, it is the case where PN junction of a semiconductor is illuminated by a light source, a photogenerated voltage will appear on both sides of a PN junction, and when the PN junction is shorted, a current will be generated.

8.2.3 Mechanical harvester integrated self-powered systems

The mechanical energy generated by elongation, distortion, rotation and other movements can be found everywhere in life. As one kind of green energy, it has advantageous characteristics including being sustainable, renewable and clean. Among them, the self-powered concept, that mechanical energy collected through piezoelectric nanogenerators (PENGs) or triboelectric nanogenerators (TENGs) and subsequently converted into electrical energy to drive electronic skin, is considered as an attractive strategy for the development of wearable electronic devices [32, 33].

Piezoelectricity can be broadly divided into two categories, called inverse piezo-electric effect and positive piezoelectric effect. Compared with the former, the latter has gained more favor in the field of self-powered electronic skins [33, 34]. With the piezoelectric material in its initial relaxation mode (figure 8.2) [5, 35], that is, without any additional external forces, the dielectric crystals inside the material will not be polarized, therefore no charge is generated. However, when the piezoelectric material is deformed under pressure, based on the relative displacement of spatially separated polarization charge centers, the electric dipole changes, causing two opposite surfaces to accumulate charges of opposite sign. At this point, for the purpose of balancing the potential, the electrons are moved through an external circuit, thus generating an identifiable electrical pulse. Relatively, when the pressure is released, the piezoelectric material behaves as electrically neutral, and the external circuitry outputs the opposite current. With its high sensitivity, fast response and

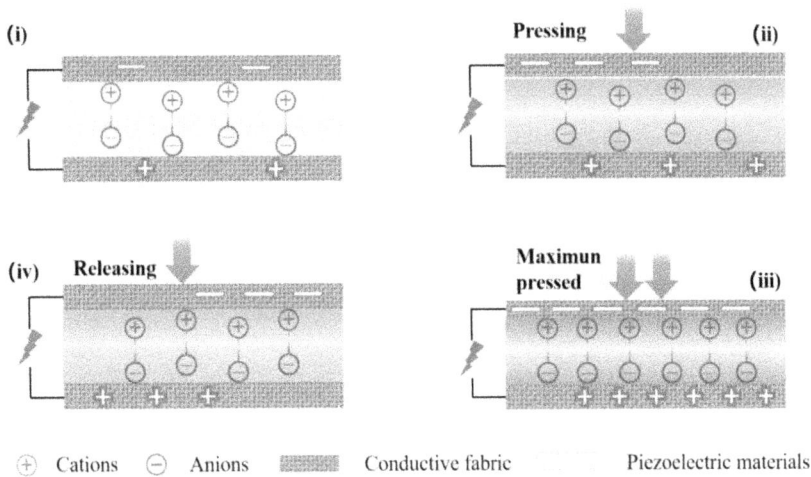

Figure 8.2. Mechanism of piezoelectric sensing. Reproduced from [5] with permission from Springer Nature.

outstanding output performance, the piezoelectric effect has been extensively employed in self-powered systems. It is worth mentioning that PENG has material limitation in terms of preparation, i.e., only materials showing piezoelectric property can be used, commonly inorganic (ZnO, MoS_2, $BaTiO_3$, etc) and polymer materials (PVDF and its copolymer).

According to the coupling effect of contact electrification and electrostatic induction, a TENG can convert mechanical energy generated by external movements into electrical energy and thus generate electrical signals [36–38]. Theoretically, any two different kinds of materials, even the same material, can be charged by frequent friction. Taking into account the different structural designs and configurations, TENG can be classified into four typical operating modes comprising vertical contact-separation (CS) mode, lateral-sliding (LS) mode, single-electrode (SE) mode, and freestanding triboelectric-layer (FT) mode (figure 8.3) [35].

The CS mode is composed of two different materials with dissimilar triboelectric polarity and two electrodes. Under the application of pressure, the two materials are in contact with each other, the equal electrical charge is generated on the surface, and manifested as positive potential and negative potential, respectively. When the force is released, the two materials gradually separate and a potential difference is created, prompting free electrons to flow between the two connected electrodes and generating a transient current. Subsequently, as external pressure is applied again, the induced charge flows back to its original electrode and a reverse current can be detected. If the CS process is repeated, correspondingly, the TENG produces a periodic electrical output. LS mode has similar structure and working principle to CS mode. In place of the original vertical orientation, the relative displacement of this mode changes to a direction parallel to the interface and exhibits a higher energy harvesting rate.

Unlike the two modes mentioned above, the SE mode shows a change in its own construction, which has only one electrode and a frictional electric layer. For the

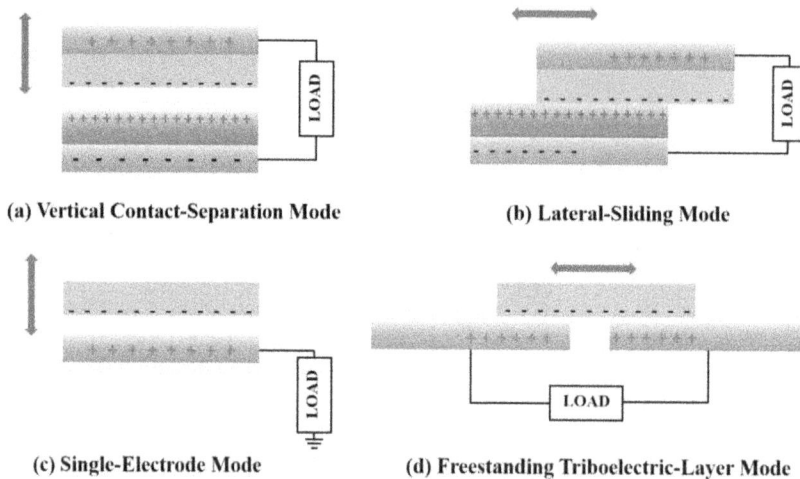

(a) **Vertical Contact-Separation Mode**

(b) **Lateral-Sliding Mode**

(c) **Single-Electrode Mode**

(d) **Freestanding Triboelectric-Layer Mode**

Figure 8.3. Four typical operating modes of TENG. Reproduced from [35] John Wiley & Sons, copyright 2018 WILEY-VCH Verlag GmbH & Co. KGaA, Weinheim.

other frictional electric layer, it is often necessary to use other external objects, such as human skins. With the ground as the reference electrode, when the external object moves, the current will flow between the electrode and the ground. This mode, although simple in construction, has a low output due to the existence of a single electrode, and is more inclined to be used in areas such as fingertip drive energy collection. Finally, the FS mode consists of two stationary electrodes and a friction layer. The movement of the friction layer backwards and forwards along the underlying electrode (it may or may not be in direct contact with the electrode) produces an unbalanced charge distribution. For the purpose of balancing the local potential distribution, induced electrons will flow between the two electrodes. This mode shows excellent performance and has no masking effect. In general, each work mode has its own structural characteristics and application advantages.

8.3 Nanofiber-based self-powered electronic skins

8.3.1 Individual sensing system

Generally, it is difficult for the electronic skin integrated with the single sensor to mimic the basic characteristics of biological skin in sensing external stimuli while maintaining self-powered functionality. One of the promising solutions is the construction of sensors based on piezoelectric, triboelectric, and thermoelectric effects.

Lou *et al* [39] reported a triboelectric all-fiber structured pressure sensor composed of the polyvinylidene fluoride/Ag nanowire (PVDF/Ag NW) and ethyl cellulose nanofibrous membrane (EC NFM) with hierarchically rough structure fabricated using the electrospinning method (figures 8.4(a) and (b)). During the fabrication of the negative triboelectric layer, the stretching with high electric field and the incorporation of Ag NWs were supposed to promote the formation of highly

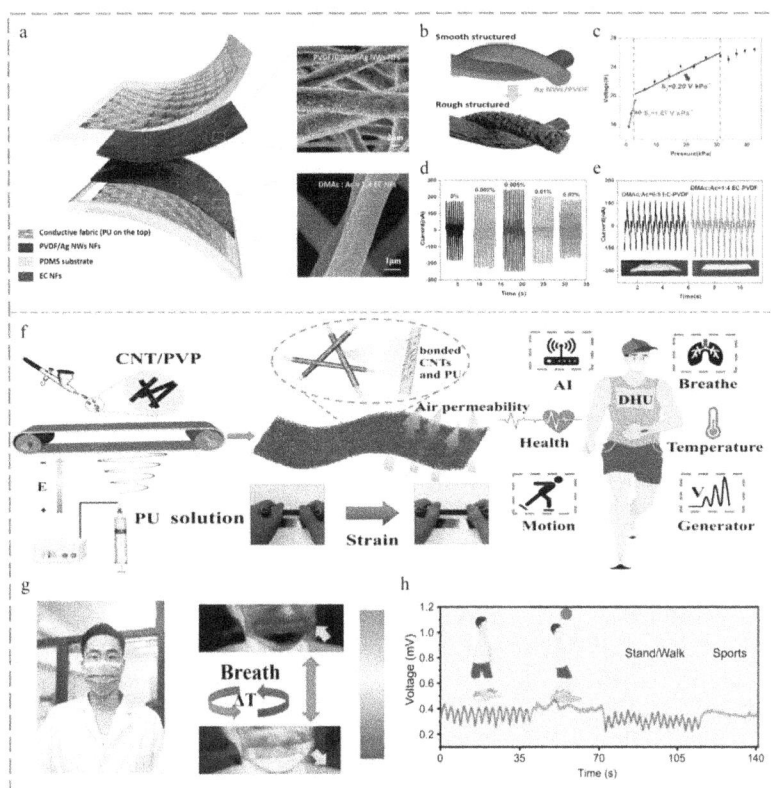

Figure 8.4. Representative electronic skins integrated with individual sensing system. (a) Schematic design of the pressure sensor textile. (b) Comparison of morphological differences between smooth and rough surfaces. (c) Electrical signals in terms of output voltage as a function of the applied pressure. (d) Comparison of the output current of the pressure sensor textile constructed of the PVDF NFM and EC NFM with different solvent ratios. (e) Comparison of output current of the pressure sensor textile constructed of the EC NFM and PVDF/Ag NW NFMs with different contents of Ag NWs. Reproduced from [39]. Copyright 2020, American Chemical Society. (f) The stretchable CNT/PVP/PU composite fabric was prepared using a typical electro-spinning process combined with spraying technology and can be applied in different fields. (g) Physical image of the functional mask for respiratory monitoring. (h) Output thermal voltage for different states of motion. Reproduced from [40] CC BY 4.0.

oriented β-phase crystals inside PVDF nanofibers. Moreover, the contact area of the friction layer surface was further increased due to the formation of rough surfaces induced by the phase separation process (figures 8.4(d) and (e)). The above strategies significantly improve the sensing performance of the self-powered electronic skin based on the contact/separation mode of the TENG. The outstanding sensitivity of the product was 1.67 and 0.20 V · kPa^{-1}, respectively, in the pressure range of 0–3 and 3–32 kPa (figure 8.4(c)). It was capable of detecting and quantifying various body movements associated with joints such as elbows, knees, ankles and necks without the need for additional external power, showing great promise for application.

On the other hand, He *et al* [40] proposed an advanced preparation strategy combining electrostatic spinning and spraying techniques (figure 8.4(f)). The novel composite thermoelectric skin was prepared by first preparing polyurethane (PU) nanofibers as a stretchable backbone and then uniformly spraying carbon nanotubes (CNTs)/polyvinyl pyrrolidone (PVP)/ethanol on it. PVP was considered as a promising candidate for the preparation of composite materials by virtue of its fine cohesiveness and solubilization. Its presence not only promoted the dispersion of the thermoelectric material CNTs, but also improved the stability of the whole skin. The temperature difference between the internal and external environment was utilized as a source of energy for this electronic skin, enabling the multi-scenario applications (figures 8.4(g) and (h)). In addition, an amazing preparation strategy was also explored by the team to enable efficient and continuous fabrication of CNTs/poly(3,4-ethylenedioxythiophene):poly(styrenesulfonate)   (CNT/PEDOT: PSS) thermoelectric nanofiber yarns [41]. Instead of a simple coating process, the thermoelectric material was loaded inside the yarn and exhibited extremely high mechanical stability. The results in this study showed that the thermoelectric nanofiber yarns have high Seebeck coefficient (44 μV K^{-1}) and ~350% tensile properties. It had been integrated into some textiles such as gloves and masks. Based on the thermoelectric effect, it could produce 1.1 mV output voltage at room temperature, which can realize the functions of hot/cold object recognition and human respiration monitoring.

8.3.2 Multifunctional sensing systems

The electronic skins constructed of the single sensor have the limitation of the function, which hinders their application in multiple scenarios. In contrast, by integrating different sensors together, the perception of multiple parameters (such as temperature, humidity, etc) can be achieved through reasonable material selection and structural design. This can greatly enrich the sensing system [42–45]. And more recently, some researchers have gone into fabricating self-powered electrical skins based on multiple sensors.

For example, Zhu *et al* [46] proposed a mature method to prepare multi-modal self-powered electronic skin (figure 8.5(a)). In the whole sensing system, the multi-walled CNTs/polyvinylidene fluoride nanofibrous membrane (MWCNTs/PVDF NFM) prepared by electrospinning method acted as the piezoelectric layer. And the authors leveraged lotus leaf as the template to construct triboelectric layer, in this case depositing the homogeneous solution that fabricated with PDMS prepolymer and cross-linker onto the surface of the lotus leaf, and then washing and drying it for future use. The two sensing layers mentioned above were ultimately integrated vertically with the electrodes and substrate into the electronic skin. The sensing performance of the as-prepared sample had been improved to a certain extent by virtue of the introduction of the MWCNTs and the construction of porous structures. The sensitivity of pressure in the detection range of 0–80 and 80–240 kPa was 9.80 and 54.37 mV kPa^{-1}, respectively (figure 8.5(b)). At the same time, the electronic skin could withstand 14,000 cycles or 14 days of continuous work, showing remarkable

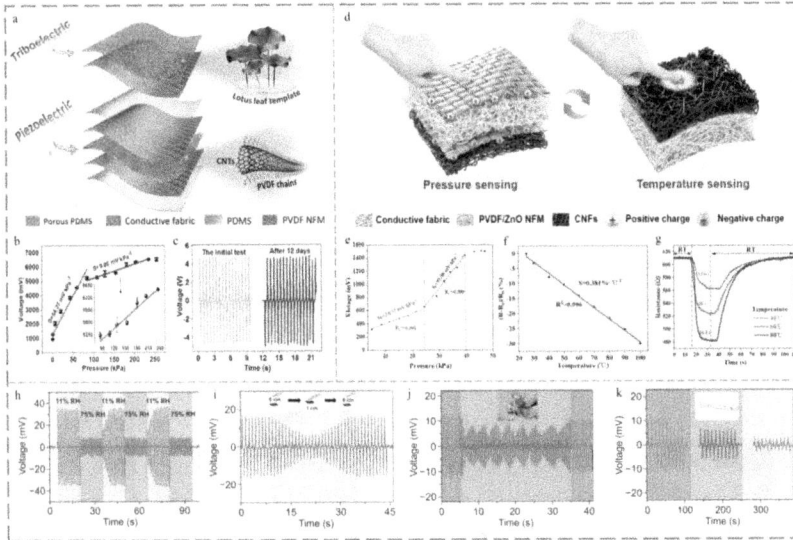

Figure 8.5. Representative electronic skins integrated with multifunctional sensing systems. (a) Schematic diagram of the structure of the electronic skin. (b) Sensitivity characterization of electronic skin. (c) The durability and stability of the sample. Reproduced from [46]. Copyright 2020, with permission from Elsevier. (d) Conceptual diagram of the dual-model textile. (e) Output voltage of pressure sensing layer. (f) Relative resistance changes of the as-prepared sample. (g) Time-resolved temperature responses of the fabric at different temperatures. Reproduced from [47]. Copyright 2021, with permission from Elsevier. (h) Repeatability of flexible humidity sensing device. (i) An image of the output voltage as finger slowly approaches the sensor. (j) Samples for detecting human respiratory rate. (k) Results of arm skin surface humidity detection after different number of exercises. Reproduced from [48] CC BY 4.0.

durability (figure 8.5(c)). Interestingly, the triboelectric sensor and the piezoelectric sensor had complementary effects that respond to the entire contact, separation, deformation, and recovery process. The skins fabricated in this work could be used for both contact and non-contact applications, such as monitoring human pulse, breathing, vocal cord vibration, etc.

Wang *et al* [47] presented a procedure to obtain the all-in-one self-powered electronic skin based on the piezoelectric polyvinylidene fluoride nanofibrous membrane doped with zinc oxide nanoparticles (PVDF/ZnO NFM) as pressure sensing layer and thermal-resistance carbon nanofibers (CNFs) with localized graphitic structures as the temperature sensing layer (figure 8.5(d)). In terms of pressure sensing, the prominent sensitivity of 15.75 mV kPa^{-1} could be gained in the range of 4.9–25 kPa, and the corresponding value within the range of 25–45 kPa was 52.09 mV kPa^{-1} (figure 8.5(e)). Additionally, for temperature sensing, it exhibited the temperature resolution of 0.381% per centigrade in the wide detection range of 25 °C to 100 °C (figure 8.6(f)). The results demonstrated that the electronic skin could accurately sense the changes of pressure and temperature at the same time, and then made rapid response, which could realize the environmental temperature detection, tactile space mapping and other functions (figure 8.6(g)).

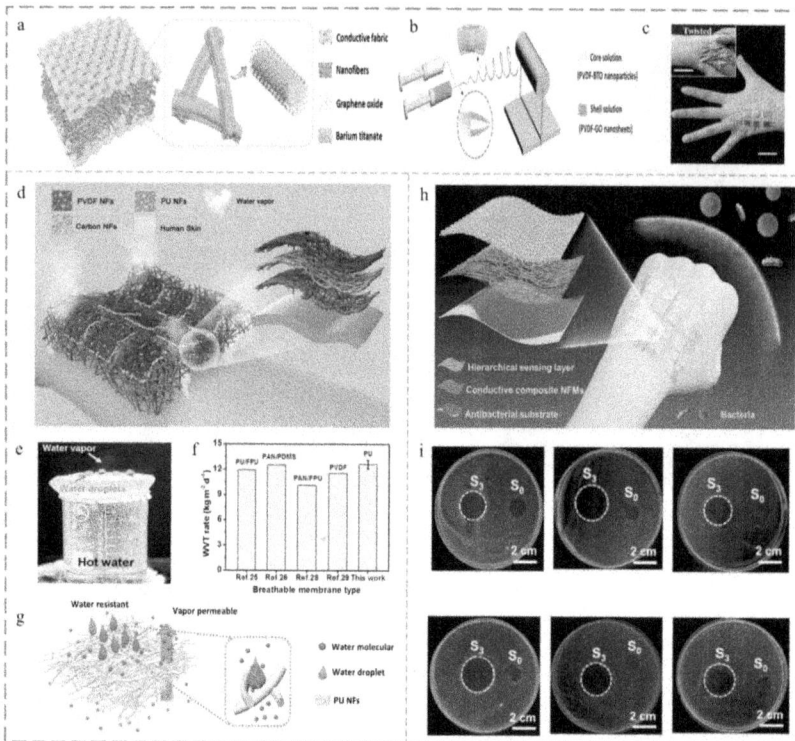

Figure 8.6. Representative properties of self-powered electronic skins. (a) Schematic diagram of the as-prepared electronic skin. (b) Electrostatic spinning experimental equipment. (c) Optical photograph of electronic skin attached to the back of the hand. Reproduced from [56]. Copyright 2020, with permission from Elsevier. (d) Schematic diagram of electronic skin (3 × 3 pixels). (e) Characterization of waterproofing and breathability. (f) Comparison of water vapour transmission rates of various membranes. (g) Mechanistic diagram of the breathability and waterproofness of e-skin. Reproduced from [64] John Wiley & Sons, copyright 2019 WILEY-VCH Verlag GmbH & Co. KGaA, Weinheim. (h) Diagrammatic sketch of the electronic skin. Zone of inhibition of *Escherichia coli* (i) and *Bacillus subtilis* (j) after the antibacterial membrane was placed for 10 days. Reproduced from [65]. Copyright 2021, with permission from Elsevier.

Wang *et al* [48] constructed a self-powered flexible humidity sensing device. The single layer of molybdenum diselenide ($MoSe_2$) was deposited on the polyethylene terephthalate (PET) substrate by atmospheric pressure chemical vapor deposition (APCVD) technique. Benefiting from the piezoelectric effect, the component could harvest energy generated in different parts of the body and convert it into electrical energy, with a peak output of 35 mV and a power density of 42 $mW\,m^{-2}$. Furthermore, the authors prepared the humidity sensing layer by integrating the PVA/MXene nanofibers on interdigital electrodes (IDEs). It had the compelling characteristic advantages, such as high response ($\sim$40), fast response/recovery time (0.9/6.3 s), low hysteresis (1.8%), and favorable repeatability (figures 8.5(h)–(k)). The electronic skin could detect the mechanical movement of the human body, skin humidity and environmental humidity. In summary, these emerging hybrid electronic skins will play an irreplaceable role in various fields.

8.4 Corresponding properties of the self-powered electronic skins

Based on the complex and changeable application environments, the performance of self-powered electronic skin constructed of electrospun nanofibers has put forward new requirements. Considering the characteristics and functions of human skin, strategies such as material functionalization and structural innovation are often used to endow electronic skin flexibility, air permeability, self-healing, antibacterial property, biodegradability, etc [49–52]. This is undoubtedly of strategic importance for the sustainable development of electronic skin and the expansion of its applications. Therefore, this section provides a comprehensive overview of some of the properties that the electronic skin should possess.

8.4.1 Flexibility/shape adaptability

The traditional preparation method of integrating rigid and brittle electronic components on a flexible plastic or elastic rubber substrate, is obviously no longer suitable for the use of electronic skin in the modern life of stretching, bending, twisting and other situations. Therefore, whether for the consideration of wearing comfort or practical application requirements, the electronic skin with the basic characteristics of outstanding flexibility, stretchability and shape adaptability, that can carry out seamless and good contact with the carrier is even more critical [53–55]. As material with great potential, electrospun nanofibers offer a novel approach to the construction of the highly flexible as well as shape-adaptive electronic skins and are gaining widespread interest.

The development of flexible and stretchable conductive materials or sensing elements can increase the flexibility and shape adaptability of electronic skins. For instance, Zhu *et al* [56] developed an electronic skin consisting of three layers of materials with excellent flexibility and stretchability (figure 8.6(a)). For the key sensing layer, the authors utilized polyvinylidene difluoride/graphene oxide (PVDF/GO) as the shell layer and polyvinylidene difluoride/barium titanate (PVDF/BTO) as the core layer to fabricate a novel coaxial piezoelectric fiber via electrostatic spinning method (figure 8.6(b)). The electronic skin prepared by combining the fiber with conductive fabric and elastic polyurethane film exhibited excellent flexibility and shape adaptability while having ultra-high sensitivity. It could fit seamlessly into the skin, even in twisted situations (figure 8.6(c)). Furthermore, Narendar *et al* [57] prepared an electronic tattoo sticker composed of CNTs and nanofilament fibers (SNFS). Based on van der Waals force, the sticker can be easily, seamlessly and stably attached to human skin. The above designs are conducive to the improvement of the wearability of electronic skin.

8.4.2 Air permeability/antibacterial properties

Air permeability is a major reference factor for regulating and maintaining the balance of temperature and humidity of humans, and realizing the exchange of air and water with the external environment [58, 59]. Electronic skins with excellent air permeability show a lower tendency to cause itching or inflammation in the wearers [60–63].

The interconnected porous structure of the nanofibrous membranes can effectively improve the water vapour permeability, thus providing a solution to the issue of long-lasting wearing discomfort. Li et al [64] developed an all-fiber structured electronic skin through the novel electrospinning fabrication technique (figure 8.6(d)). In this study, PVDF (sensing layer), carbon (electrode layer), and polyurethane (PU, substrate layer) nanofibers were properly integrated, and the as-prepared electronic skin exhibited remarkable air permeability, which was up to 10.2 $kg\,m^{-2}d^{-1}$ (figures 8.6(e) and (f)). This could be attributed to the Fickian diffusion of water vapor within the porous structure of the nanofiber membranes, which was closely related to the aperture effect of the polyurethane substrate. Furthermore, the convincing hydrophobicity, flexibility and stretchability all contributed to improving the authenticity of the skin simulation (figure 8.6(g)).

Besides, the prolonged contact between the human body and the external environment makes it inevitable that, in some specific circumstances, the electronic skin will become a suitable carrier for the growth of microorganisms, thus causing irreversible damage to the human body. Therefore, it is urgent to explore electronic skins with antibacterial properties for the purpose of inhibiting bacterial growth and preventing bacterial infections. Currently, some proven strategies have already been reported. Zhu et al [65] added the commercial antimicrobial agent (glycerin mono-laurate composite) to the polydimethylsiloxane (PDMS) substrate to confer anti-bacterial properties on the entire electronic skin (figure 8.6(h)). And the authors had confirmed that the substrate still showed excellent antibacterial properties after ten days of placement (figures 8.6(i) and (j)). This would effectively prevent secondary damage to the skin from bacteria and greatly improve the safety of wearing the electronic skin. It is worth mentioning that silver is a promising and emerging antimicrobial agent, that has attracted widespread attention for its broad-spectrum killing properties against specific bacteria, fungi and viruses [66]. Peng et al [67] prepared a silver nanowire fiber membrane with intrinsic antibacterial properties, and sandwiched the membrane between polylactic acid glycolic acid (PLGA) and polyvinyl alcohol (PVA) to fabricate the electronic skin with 3D micro–nano porous laminar structure, showing outstanding antibacterial properties. Meanwhile, the product also had satisfying biodegradability, air permeability, etc, that showed great potential in healthcare, human–computer interaction and so forth.

8.4.3 Self-healing

Inspired by the self-repair process of organisms, endowing electronic skins with self-healing properties to maintain the stability as well as safety of their mechanical properties and functions is of great significance for sustainable development [68–72]. Combining exogenous healing agents with materials in special forms, such as microcapsules and hollow tubes, is considered to be an effective way to extend the service life of electronic skin. In cases where the material is damaged by external forces, the self-healing agent will be released to activate the healing process and seal the damaged area. However, the limited healing period and the susceptibility to loss of the original fiber pattern significantly limit the application of this method.

In contrast, the strategy of constructing materials with intrinsic self-healing properties that enable self-healing through polymer chain diffusion and reversible dynamic bond reconstruction is more favored by researchers.

Recently, Zhu *et al* [73] developed an unprecedented strategy for the fabrication of inherently self-healing electronic skins with bionic confined protective structure (figure 8.7(a)). The pristine self-healing thermochromic fibrous membranes (STFMs) were prepared via electrospinning method by means of the transparent homogeneous spinning precursor consisting of isophorone diisocyanate (IPDI) and amino-propyl terminated poly(dimethylsiloxane) (NH_2-PDMS-NH_2, ATPDMS). Based on the rich dynamic hydrogen bonding interactions of the urea molecules within the fibers, the pristine STFMs exhibited excellent self-healing properties. However, it was worth noting that supramolecular interaction between single fibers could cause fiber fusion, thus affecting the permeability of electronic skins (figure 8.7(b)). Therefore, by self-assembling polyacrylic acid (PAA) and branched polyethylenei-mine (bPEI) layers on the surface of pristine STFMs to form a closed protective layer, the above problems could be solved without compromising the self-healing properties (figure 8.7(c)). Moreover, from the point of view of simulating biological skin, the self-healing mechanism of the electronic skin was further explored

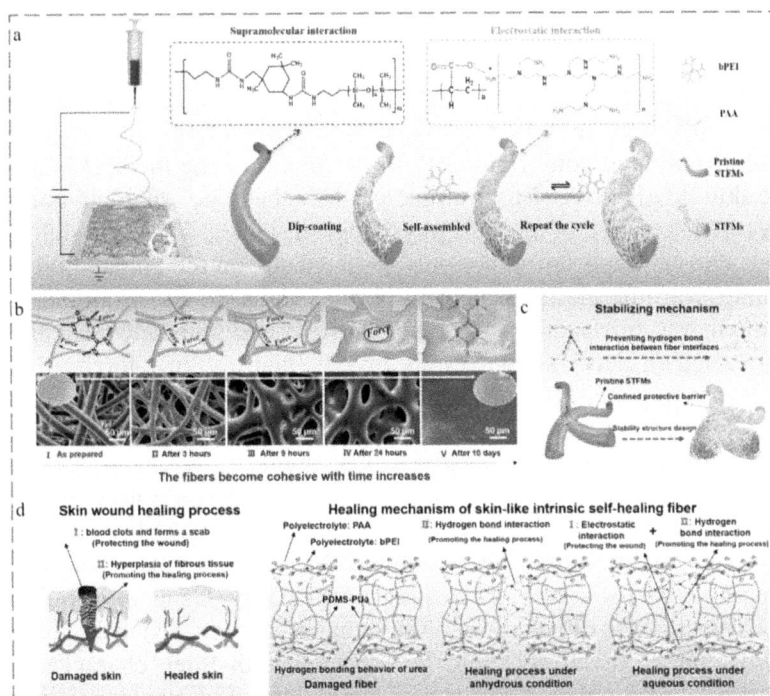

Figure 8.7. The typical example of electronic skin with self-healing properties. (a) Flow chart for the preparation of STFMs. (b) Mechanism of the instability of pristine STFMs. (c) Stability mechanism of the as-prepared STFMs. (d) Schematic illustration of the skin healing process and healing mechanism of STFMs. Reproduced from [73] John Wiley & Sons. Copyright 2022 Wiley-VCH GmbH.

(figure 8.7(d)). The authors had demonstrated that damaged electronic skin, which was not infected by external factors such as sweat, could heal in a relatively short time while maintaining good performance characteristics. This work provides new directions for the field of bionic and advanced flexible electronic skins.

8.5 Applications of self-powered electronic skins

With the continuous iteration and update of wireless communication, multimedia, artificial intelligence and other technologies, electronic skins constructed of electrospun nanofibers act as a carrier of multi-technology integration, and their functional advancements have been greatly enhanced. In this section, the main application scenarios for electronic skin will be described in detail in terms of health monitoring, human–computer interaction and brain–computer interfaces [74–77].

8.5.1 Health monitoring

In recent years, the field of healthcare has gained increasing attention. Compared with traditional bulky or invasive monitoring instruments, the novel health monitoring products based on flexible electronics, with their portability and intelligence, can be flexibly used for information detection of various health indicators, status identification and risk warning [78–81].

Peng *et al* [67] proposed an all-nanofiber-based self-powered electronic skin. Benefiting from the multi-layer interlacing fiber network and layered porous structure, the fiber specific surface area and the capillary channels of thermal-moisture transfer were increased, and the skin showed brilliant air permeability, pressure sensitivity and conformability. With the aid of the medical bandages, the electronic skin fitted perfectly to multiple parts of the body, enabling non-invasive, real-time, rapid detection of multiple physiological signals. These physiological data are of great clinical importance, not only for feedback on the human body, but also for screening, warning and treatment of chronic diseases. For example, paralysed patients often rely on facial micro-expressions to communicate with the outside world, so placing an electronic skin on the eyelids (figure 8.8(a)), forehead (figure 8.8 (b)) or the vent of a mask (figure 8.8(c)) could identify and monitor a relaxed or furrowed brow, a normal or rapid blink, and an inhale or exhale, respectively. This would undoubtedly help patients express their emotions better. Further, the four different states of normal, slow, fast and deep breathing could also be recorded by the movement of the chest and abdomen (figure 8.8(d)), which would be a more comprehensive way of monitoring breathing [82]. Heart rate and pulse are also key physical parameters for assessing a patient's physiology. Using electronic skin as a wrist pulse detector, the radial artery pulse signal was monitored in real time and the normal wrist pulse waveform could be collected with three characteristic peaks including shock, tidal and diastolic wave (figure 8.8(f)). The jugular vein pulse (JVP) tended to supply more precious diagnostic information for patients with heart disease than the former. The test results showed that the JVP waveform consisted of three peaks, namely 'a' (atrial systole), 'C' (tricuspid valve dilatation) and 'V' (atrial systolic filling), and two descending peaks containing 'X' (atrial diastole) and 'Y'

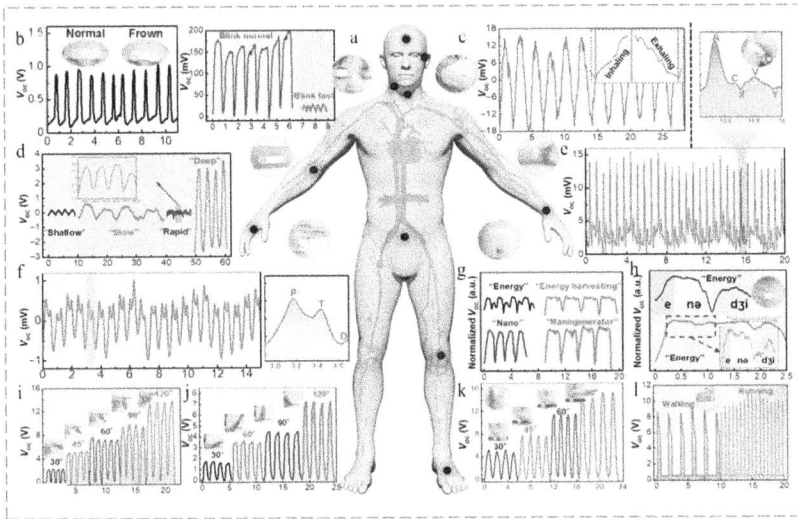

Figure 8.8. Physical health monitoring by attaching electronic skin to the forehead (a), eyelid (b), dust mask (c), belly (d), neck (e), wrist (f), throat (g), knuckle (i), elbow (j), knee (k), heel (l). (h) Voltage signal amplification diagram for (g). From [67], reprinted with permission from AAAS.

(early ventricular filling) (figure 8.8(e)). The electronic skin attached to the throat could be used to distinguish between different words and phrases, exercise laryngeal muscles in patients with damaged vocal cords, and restore acoustics (figures 8.8(g) and (h)). On the other hand, the electronic skin could also track tiny bending movements of joints such as elbows, knees and ankles (figures 8.8(i)–(l)). As a result, people could better understand their condition and develop better health management plans.

Moreover, it is well known that adequate sleep is very important, but due to the fast pace and pressure of life in today's society, the proportion of people suffering from insomnia is increasing. Yue *et al* [83] designed a dermal papillae-bioinspired self-powered multifunctional electronic skin to scientifically evaluate people's sleep quality and comfort. The electronic skin could distinguish between different sleep stages by monitoring respiratory rate (RR) and heart rate (HR) and prevent the onset of sleep disorders, including obstructive sleep apnea-hypopnea syndrome diagnosing. In the meantime, the skin could take into account the assessment of the humidity of the sleeping environment. Additionally, a number of electronic skins for monitoring blood glucose, lactate, alcohol, etc are very widely used. In the future, the health monitoring system based on electronic skin will play an irreplaceable role in the safety management of the human body.

8.5.2 Human–computer interaction

Human–computer interaction system is the result of multidisciplinary integration. Firstly, the high-dimensional, irregular multi-modal data captured by the sensors, containing some physiological and biochemical signals such as pressure, temperature, metabolism, etc, are collected, cleaned and normalized together with other discrete data, time series data and image data in a raw data set. Subsequently, thanks to the current advances in big data and algorithmic optimization, the machine learning (ML), especially deep learning techniques, can be utilized to analyze and interpret data, then generate intelligent decisions, and specific actions are performed by machines [84, 85]. Considering the sensors as one of the fundamental elements in the construction of the intelligent system, their multifunctional characteristics have been given at the level of flexible material selection and special structural design, resulting in an increase in the quantity and quality of data collection, which has a great influence on the final prediction results. In addition, replacing the previous rigid and heavy robots, it is now more likely for them to be combined with soft robots. These make the human–computer interaction system even more functional and have a wider range of applications [86–89].

Typically, humans can work with robots through tactile perception to recognize gestures and movements, manipulate objects as well as translate symbols to speech. This is certainly exciting as these will be utilized to solve key problems in specific areas such as digital healthcare, prosthetic control, virtual communication and social networking. For instance, Guo et al [90] modified polyvinylidene-trifluoro-ethylene nanofibers using MoS_2 and PEDOT:PSS through hydrothermal and vacuum filtration methods, which were used as wet-sensitive and pressure-sensitive membranes, respectively. The two layers integrated with universal electrodes into dual-function flexible sensors. The sensor was capable of monitoring a variety of human physiological signals. Subsequently, the intelligent system was constructed by combining flexible sensor, multi-channel signal acquisition module, data analysis module and display module. The authors had demonstrated that the volunteer's gestures can be accurately displayed on the interface in real time, and that the virtual hand changes in response to the human gestures. This is expected to replace the currently available vertical live-action grips and give users a higher quality of experience in areas such as VR gaming. Moreover, the sensor could be attached to the robot arm, so that when the arm came into contact with other objects or the environment was too humid, the arm would stop working and retracted to its initial position, achieving both contact (pressure) and non-contact (humidity) risk avoidance, extending the life cycle of the robot arm and avoiding accidental injuries to the human body from mechanical operations.

Lu et al [91] obtained polyacrylonitrile-based branched rod-like carbon nanofibers by a triple process of electrostatic spinning, carbonization and ultrasonication. It was then blended with carbon black and deposited on thermoplastic polyurethane (TPU) films to produce strain sensors with high sensitivity (GF of 18.7), fast response time (67 ms) and wide operating range (0.1%–1000%). With the integration of the intelligent system, the sensor was attached to the human body, and when the volunteer made a specific gesture, the robotic hand would made the corresponding gesture. There are many examples of manipulation of manipulators with this strategy, and these have undoubtedly brought great convenience to our lives.

Take a realistic example, during the epidemic, direct human contact greatly increases the risk of transmission, while the human–computer interaction model allows for contact-free sampling, improving safety and ensuring work efficiency at the same time. All in all, this application field is extremely promising.

8.5.3 Brain–computer interfaces

The convergence of brain science and artificial intelligence has made brain–machine interface (BMI) technology one of the most hotly debated high-tech topics in the public domain. The brainwave data captured by the electronic skin is controlled through a complete closed loop of neural interface, neural decoding, neural control and neural feedback, allowing for efficient human–machine collaboration, bypassing the senses and muscles [92–96].

This makes the sci-fi scenario of controlling objects with the mind, commonly seen in movies, become reality. These technologies will have a disruptive impact on existing human life. The combination of brain–computer interaction and electronic skin can help patients to repair, improve and even compensate for damaged parts of the nerves. The use of brain–computer interfaces to control prosthetic limbs can act as a real-life extension of the missing body parts of people with disabilities, helping them to perform sports, learning and games in ways that they would not otherwise be able to experience on their own. However, as a precise human component, the brain's expansion and contraction characteristics give it very high requirements for the material selection, quantity, compatibility and other parameters of invasive or non-invasive sensor parts. In the future, it is expected that nanomaterial-based technologies will be used to prepare electronic skins that cater for neuronal interfaces. It is convincing that a great deal of effort will be devoted to the development of electrospun carbon nanofiber-based electronic skins.

8.6 Conclusions and perspectives

The invention of self-powered electronic skin constructed of electrospun nanofibers has been an important breakthrough in the field of wearable electronics. It provides a green and sustainable way to harvest the energy that is easily overlooked from the human body and even the surrounding environment, and convert them into electricity for driving electronic devices. While ensuring superior sensing characteristics, the tailored structure and properties of nanofibers can further endow the electronic skin with particular properties such as breathability, antimicrobial and self-healing performances to meet the needs of daily use in complex environments. Furthermore, the integration of multiple sensors and the effective combination with the Internet of Things (IoT), artificial intelligence and other emerging technologies have expanded the application scope of electronic skin. Currently, electronic skin is showing excellent development prospects in medical monitoring, human–computer interaction, brain–computer interface and other fields. We expect that self-powered electronic skins based on electrospun nanofibers will perform an increasing role in the future of smart textiles and a new generation of wearable electronics.

Acknowledgments

This work was financially supported by the National Natural Science Foundation of China (52373054, 52073051), Natural Science Foundation of Shanghai (23ZR1400900), the Fundamental Research Funds for the Central Universities (2232023Y-01, 2232022A-04), the 'DHU Distinguished Young Professor Program' (23D210102).

References

[1] Kim S-R, Lee S and Park J-W 2022 A skin-inspired, self-powered tactile sensor *Nano Energy* **101** 107608

[2] Cai S, Xu C, Jiang D, Yuan M, Zhang Q, Li Z and Wang Y 2022 Air-permeable electrode for highly sensitive and noninvasive glucose monitoring enabled bygraphene fiber fabrics *Nano Energy* **93** 106904

[3] Wei X, Zhu M, Li J, Liu L, Yu J, Li Z and Ding B 2021 Wearable biosensor for sensitive detection of uric acid in artificial sweat enabled by a fiber structured sensing interface *Nano Energy* **85** 106031

[4] Su L, Xiong Q, Wang H and Zi Y 2022 Porous-structure-promoted tribo-induced high-performance self-powered tactile sensor toward remote human-machine interaction *Adv. Sci.* **9** 2203510

[5] Lv X, Liu Y, Yu J, Li Z and Ding B 2023 Smart fibers for self-powered electronic skins *Adv. Fiber Mater.* **5** 401–28

[6] Wang W *et al* 2023 Neuromorphic sensorimotor loop embodied by monolithically integrated, low-voltage, soft e-skin *Science* **380** 735–42

[7] Song Y, Tay R Y, Li J, Xu C, Min J, Shirzaei Sani E, Kim G, Heng W, Kim I and Gao W 2023 3D-printed epifluidic electronic skin for machine learning–powered multimodal health surveillance *Sci. Adv.* **9** eadi6492

[8] Chou H-H, Nguyen A, Chortos A, To J W F, Lu C, Mei J, Kurosawa T, Bae W-G, Tok J B-H and Bao Z 2015 A chameleon-inspired stretchable electronic skin with interactive colour changing controlled by tactile sensing *Nat. Commun.* **6** 8011

[9] Guan S, Yang Y, Wang Y, Zhu X, Ye D, Chen R and Liao Q 2024 A dual-functional MXene-based bioanode for wearable self-charging biosupercapacitors *Adv. Mater.* **36** 2305854

[10] Zhou Y, Jia X, Pang D, Jiang S, Zhu M, Lu G, Tian Y, Wang C, Chao D and Wallace G 2023 An integrated Mg battery-powered iontophoresis patch for efficient and controllable transdermal drug delivery *Nat. Commun.* **14** 297

[11] Su Y *et al* 2023 Soft–rigid heterostructures with functional cation vacancies for fast-charging and high-capacity sodium storage *Adv. Mater.* **35** 2305149

[12] Gupta S, Yang X and Ceder G 2023 What dictates soft clay-like lithium superionic conductor formation from rigid salts mixture *Nat. Commun.* **14** 6884

[13] Xiong R, Hu K, Grant A M, Ma R, Xu W, Lu C, Zhang X and Tsukruk V V 2016 Ultrarobust transparent cellulose nanocrystal-graphene membranes with high electrical conductivity *Adv. Mater.* **28** 1501–9

[14] Wu H, Su Z, Shi M, Miao L, Song Y, Chen H, Han M and Zhang H 2018 Self-powered noncontact electronic skin for motion sensing *Adv. Funct. Mater.* **28** 1704641

[15] Ma M, Zhang Z, Liao Q, Yi F, Han L, Zhang G, Liu S, Liao X and Zhang Y 2017 Self-powered artificial electronic skin for high-resolution pressure sensing *Nano Energy* **32** 389–96

[16] Liu R, Lai Y, Li S, Wu F, Shao J, Liu D, Dong X, Wang J and Wang Z L 2022 Ultrathin, transparent, and robust self-healing electronic skins for tactile and non-contact sensing *Nano Energy* **95** 107056

[17] Zhang X, Li Z, Du W, Zhao Y, Wang W, Pang L, Chen L, Yu A and Zhai J 2022 Self-powered triboelectric-mechanoluminescent electronic skin for detecting and differentiating multiple mechanical stimuli *Nano Energy* **96** 107115

[18] Zhou M, Xu F, Ma L, Luo Q, Ma W, Wang R, Lan C, Pu X and Qin X 2022 Continuously fabricated nano/micro aligned fiber based waterproof and breathable fabric triboelectric nanogenerators for self-powered sensing systems *Nano Energy* **104** 107885

[19] Kanokpaka P, Chang L-Y, Wang B-C, Huang T-H, Shih M-J, Hung W-S, Lai J-Y, Ho K-C and Yeh M-H 2022 Self-powered molecular imprinted polymers-based triboelectric sensor for noninvasive monitoring lactate levels in human sweat *Nano Energy* **100** 107464

[20] Chen C, Feng J, Li J, Guo Y, Shi X and Peng H 2023 Functional fiber materials to smart fiber devices *Chem. Rev.* **123** 613–62

[21] Wang Y, Xu Y, Zhai W, Zhang Z, Liu Y, Cheng S and Zhang H 2022 In-situ growth of robust superlubricated nano-skin on electrospun nanofibers for post-operative adhesion prevention *Nat. Commun.* **13** 5056

[22] Zhang J-H *et al* 2022 Versatile self-assembled electrospun micropyramid arrays for high-performance on-skin devices with minimal sensory interference *Nat. Commun.* **13** 5839

[23] Shi X-L, Chen W-Y, Zhang T, Zou J and Chen Z-G 2021 Fiber-based thermoelectrics for solid, portable, and wearable electronics *Energy Environ. Sci.* **14** 729–64

[24] Zhang X and Zhao L-D 2015 Thermoelectric materials: energy conversion between heat and electricity *J. Materiomics* **1** 92–105

[25] Chen X, Zhou Z, Lin Y-H and Nan C 2020 Thermoelectric thin films: promising strategies and related mechanism on boosting energy conversion performance *J. Materiomics* **6** 494–512

[26] Gokana M R, Wu C-M, Motora K G, Qi J Y and Yen W-T 2022 Effects of patterned electrode on near infrared light-triggered cesium tungsten bronze/poly(vinylidene)fluoride nanocomposite-based pyroelectric nanogenerator for energy harvesting *J. Power Sources* **536** 231524

[27] Alagumalai A, Mahian O, Aghbashlo M, Tabatabaei M, Wongwises S and Wang Z L 2021 Towards smart cities powered by nanogenerators: bibliometric and machine learning–based analysis *Nano Energy* **83** 105844

[28] Korkmaz S and Kariper İ A 2021 Pyroelectric nanogenerators (PyNGs) in converting thermal energy into electrical energy: Fundamentals and current status *Nano Energy* **84** 105888

[29] Bi D, Xu B, Gao P, Sun L, Grätzel M and Hagfeldt A 2016 Facile synthesized organic hole transporting material for perovskite solar cell with efficiency of 19.8% *Nano Energy* **23** 138–44

[30] Li H *et al* 2023 2D/3D heterojunction engineering at the buried interface towards high-performance inverted methylammonium-free perovskite solar cells *Nat. Energy* **8** 946–55

[31] Sharma D, Nicoara N, Jackson P, Witte W, Hariskos D and Sadewasser S 2024 Charge-carrier-concentration inhomogeneities in alkali-treated Cu(In,Ga)Se2 revealed by conductive atomic force microscopy tomography *Nat. Energy* **9** 163–71

[32] Sun Q *et al* 2024 Charge dispersion strategy for high-performance and rain-proof tribo-electric nanogenerator *Adv. Mater.* **36** 2307918

[33] Chen C, Zhao S, Pan C, Zi Y, Wang F, Yang C and Wang Z L 2022 A method for quantitatively separating the piezoelectric component from the as-received 'Piezoelectric' signal *Nat. Commun.* **13** 1391

[34] Cao X, Xiong Y, Sun J, Zhu X, Sun Q and Wang Z L 2021 Piezoelectric nanogenerators derived self-powered sensors for multifunctional applications and artificial intelligence *Adv. Funct. Mater.* **31** 2102983

[35] Wu C, Wang A C, Ding W, Guo H and Wang Z L 2019 Triboelectric nanogenerator: a foundation of the energy for the new era *Adv. Energy Mater.* **9** 1802906

[36] Zhou K, Zhao Y, Sun X, Yuan Z, Zheng G, Dai K, Mi L, Pan C, Liu C and Shen C 2020 Ultra-stretchable triboelectric nanogenerator as high-sensitive and self-powered electronic skins for energy harvesting and tactile sensing *Nano Energy* **70** 104546

[37] Chen H, Song Y, Cheng X and Zhang H 2019 Self-powered electronic skin based on the triboelectric generator *Nano Energy* **56** 252–68

[38] Dong K, Peng X and Wang Z L 2020 Fiber/fabric-based piezoelectric and triboelectric nanogenerators for flexible/stretchable and wearable electronics and artificial intelligence *Adv. Mater.* **32** 1902549

[39] Lou M, Abdalla I, Zhu M, Yu J, Li Z and Ding B 2020 Hierarchically rough structured and self-powered pressure sensor textile for motion sensing and pulse monitoring *ACS Appl. Mater. Interfaces* **12** 1597–605

[40] He X, Shi J, Hao Y, He M, Cai J, Qin X, Wang L and Yu J 2022 Highly stretchable, durable, and breathable thermoelectric fabrics for human body energy harvesting and sensing *Carbon Energy* **4** 621–32

[41] He X, Gu J, Hao Y, Zheng M, Wang L, Yu J and Qin X 2022 Continuous manufacture of stretchable and integratable thermoelectric nanofiber yarn for human body energy harvesting and self-powered motion detection *Chem. Eng. J.* **450** 137937

[42] Pang Y, Xu X, Chen S, Fang Y, Shi X, Deng Y, Wang Z-L and Cao C 2022 Skin-inspired textile-based tactile sensors enable multifunctional sensing of wearables and soft robots *Nano Energy* **96** 107137

[43] Wu W *et al* 2017 Free-standing and eco-friendly polyaniline thin films for multifunctional sensing of physical and chemical stimuli *Adv. Funct. Mater.* **27** 1703147

[44] Deng C *et al* 2019 Ultrasensitive and highly stretchable multifunctional strain sensors with timbre-recognition ability based on vertical graphene *Adv. Funct. Mater.* **29** 1907151

[45] Guo Y, Wei X, Gao S, Yue W, Li Y and Shen G 2021 Recent advances in carbon material-based multifunctional sensors and their applications in electronic skin systems *Adv. Funct. Mater.* **31** 2104288

[46] Zhu M, Lou M, Yu J, Li Z and Ding B 2020 Energy autonomous hybrid electronic skin with multi-modal sensing capabilities *Nano Energy* **78** 105208

[47] Wang Y, Zhu M, Wei X, Yu J, Li Z and Ding B 2021 A dual-mode electronic skin textile for pressure and temperature sensing *Chem. Eng. J.* **425** 130599

[48] Wang D, Zhang D, Li P, Yang Z, Mi Q and Yu L 2021 Electrospinning of flexible poly(vinyl alcohol)/MXene nanofiber-based humidity sensor self-powered by monolayer molybdenum diselenide piezoelectric nanogenerator *Nano-Micro Lett.* **13** 57

[49] Lin X, Bing Y, Li F, Mei H, Liu S, Fei T, Zhao H and Zhang T 2022 An all-nanofiber-based, breathable, ultralight electronic skin for monitoring physiological signals *Adv. Mater. Technol.* **7** 2101312

[50] Niu Q, Wei H, Hsiao B S and Zhang Y 2022 Biodegradable silk fibroin-based bio-piezoelectric/triboelectric nanogenerators as self-powered electronic devices *Nano Energy* **96** 107101

[51] Huang J, Xu Z, Qiu W, Chen F, Meng Z, Hou C, Guo W and Liu X Y 2020 Stretchable and heat-resistant protein-based electronic skin for human thermoregulation *Adv. Funct. Mater.* **30** 1910547

[52] Zarei M, Lee G, Lee S G and Cho K 2023 Advances in biodegradable electronic skin: material progress and recent applications in sensing, robotics, and human–machine interfaces *Adv. Mater.* **35** 2203193

[53] Choi S, Lee H, Ghaffari R, Hyeon T and Kim D-H 2016 Recent advances in flexible and stretchable bio-electronic devices integrated with nanomaterials *Adv. Mater.* **28** 4203–18

[54] Wang S *et al* 2018 Skin electronics from scalable fabrication of an intrinsically stretchable transistor array *Nature* **555** 83–8

[55] Larson C, Peele B, Li S, Robinson S, Totaro M, Beccai L, Mazzolai B and Shepherd R 2016 Highly stretchable electroluminescent skin for optical signaling and tactile sensing *Science* **351** 1071–4

[56] Zhu M, Lou M, Abdalla I, Yu J, Li Z and Ding B 2020 Highly shape adaptive fiber based electronic skin for sensitive joint motion monitoring and tactile sensing *Nano Energy* **69** 104429

[57] Gogurla N and Kim S 2021 Self-powered and imperceptible electronic tattoos based on silk protein nanofiber and carbon nanotubes for human–machine interfaces *Adv. Energy Mater.* **11** 2100801

[58] Qiu Q, Zhu M, Li Z, Qiu K, Liu X, Yu J and Ding B 2019 Highly flexible, breathable, tailorable and washable power generation fabrics for wearable electronics *Nano Energy* **58** 750–8

[59] Lou M, Abdalla I, Zhu M, Wei X, Yu J, Li Z and Ding B 2020 Highly wearable, breathable, and washable sensing textile for human motion and pulse monitoring *ACS Appl. Mater. Interfaces* **12** 19965–73

[60] Wu F, Lan B, Cheng Y, Zhou Y, Hossain G, Grabher G, Shi L, Wang R and Sun J 2022 A stretchable and helically structured fiber nanogenerator for multifunctional electronic textiles *Nano Energy* **101** 107588

[61] Sun G, Wang P, Jiang Y, Sun H, Liu T, Li G, Yu W, Meng C and Guo S 2023 Bioinspired flexible, breathable, waterproof and self-cleaning iontronic tactile sensors for special underwater sensing applications *Nano Energy* **110** 108367

[62] Zhao C *et al* 2023 Ultrathin Mo2S3 nanowire network for high-sensitivity breathable piezoresistive electronic skins *ACS Nano* **17** 4862–70

[63] Zheng X, Zhang S, Zhou M, Lu H, Guo S, Zhang Y, Li C and Tan S C 2023 MXene functionalized, highly breathable and sensitive pressure sensors with multi-layered porous structure *Adv. Funct. Mater.* **33** 2214880

[64] Li Z, Zhu M, Shen J, Qiu Q, Yu J and Ding B 2020 All-fiber structured electronic skin with high elasticity and breathability *Adv. Funct. Mater.* **30** 1908411

[65] Zhu M, Wang Y, Lou M, Yu J, Li Z and Ding B 2021 Bioinspired transparent and antibacterial electronic skin for sensitive tactile sensing *Nano Energy* **81** 105669

[66] Shi Y, Wei X, Wang K, He D, Yuan Z, Xu J, Wu Z and Wang Z L 2021 Integrated all-fiber electronic skin toward self-powered sensing sports systems *ACS Appl. Mater. Interfaces* **13** 50329–37

[67] Peng X, Dong K, Ye C, Jiang Y, Zhai S, Cheng R, Liu D, Gao X, Wang J and Wang Z L 2020 A breathable, biodegradable, antibacterial, and self-powered electronic skin based on all-nanofiber triboelectric nanogenerators *Sci. Adv.* **6** eaba9624

[68] Li C, Guo H, Wu Z, Wang P, Zhang D and Sun Y 2023 Self-healable triboelectric nanogenerators: marriage between self-healing polymer chemistry and triboelectric devices *Adv. Funct. Mater.* **33** 2208372

[69] Huynh T-P, Sonar P and Haick H 2017 Advanced materials for use in soft self-healing devices *Adv. Mater.* **29** 1604973

[70] Choi S, Eom Y, Kim S-M, Jeong D-W, Han J, Koo J M, Hwang S Y, Park J and Oh D X 2020 A self-healing nanofiber-based self-responsive time-temperature indicator for securing a cold-supply chain *Adv. Mater.* **32** 1907064

[71] Wang C *et al* 2022 Ultra-stretchable and fast self-healing ionic hydrogel in cryogenic environments for artificial nerve fiber *Adv. Mater.* **34** 2105416

[72] Liu J, Zhang L, Wang N, Zhao H and Li C 2022 Nanofiber-reinforced transparent, tough, and self-healing substrate for an electronic skin with damage detection and program-controlled autonomic repair *Nano Energy* **96** 107108

[73] Zhu M, Li J, Yu J, Li Z and Ding B 2022 Superstable and intrinsically self-healing fibrous membrane with bionic confined protective structure for breathable electronic skin *Angew. Chem. Int. Ed.* **61** e202200226

[74] Du W, Li Z, Zhao Y, Zhang X, Pang L, Wang W, Jiang T, Yu A and Zhai J 2022 Biocompatible and breathable all-fiber-based piezoresistive sensor with high sensitivity for human physiological movements monitoring *Chem. Eng. J.* **446** 137268

[75] Chao M, He L, Gong M, Li N, Li X, Peng L, Shi F, Zhang L and Wan P 2021 Breathable Ti$_3$C2T$_x$ MXene/protein nanocomposites for ultrasensitive medical pressure sensor with degradability in solvents *ACS Nano* **15** 9746–58

[76] Yao H, Sun T, Chiam J S, Tan M, Ho K Y, Liu Z and Tee B C K 2021 Augmented reality interfaces using virtual customization of microstructured electronic skin sensor sensitivity performances *Adv. Funct. Mater.* **31** 2008650

[77] Liu H, Li H, Wang Z, Wei X, Zhu H, Sun M, Lin Y and Xu L 2022 Robust and multifunctional Kirigami electronics with a tough and permeable aramid nanofiber framework *Adv. Mater.* **34** 2207350

[78] Lee G, Wei Q and Zhu Y 2021 Emerging wearable sensors for plant health monitoring *Adv. Funct. Mater.* **31** 2106475

[79] Gao Y, Yu L, Yeo J C and Lim C T 2020 Flexible hybrid sensors for health monitoring: materials and mechanisms to render wearability *Adv. Mater.* **32** 1902133

[80] Han F, Wang T, Liu G, Liu H, Xie X, Wei Z, Li J, Jiang C, He Y and Xu F 2022 Materials with tunable optical properties for wearable epidermal sensing in health Monitoring *Adv. Mater.* **34** 2109055

[81] Zhang S, Zhou Z, Zhong J, Shi Z, Mao Y and Tao T H 2020 Body-integrated, enzyme-triggered degradable, silk-based mechanical sensors for customized health/fitness monitoring and *in situ* treatment *Adv. Sci.* **7** 1903802

[82] Peng X, Dong K, Ning C, Cheng R, Yi J, Zhang Y, Sheng F, Wu Z and Wang Z L 2021 All-nanofiber self-powered skin-interfaced real-time respiratory monitoring system for obstructive sleep apnea-hypopnea syndrome diagnosing *Adv. Funct. Mater.* **31** 2103559

[83] Yue O, Wang X, Hou M, Zheng M, Bai Z, Cui B, Cha S and Liu X 2022 Skin-inspired wearable self-powered electronic skin with tunable sensitivity for real-time monitoring of sleep quality *Nano Energy* **91** 106682

[84] Wang M, Wang T, Luo Y, He K, Pan L, Li Z, Cui Z, Liu Z, Tu J and Chen X 2021 Fusing stretchable sensing technology with machine learning for human–machine interfaces *Adv. Funct. Mater.* **31** 2008807

[85] Yang W, Gong W, Gu W, Liu Z, Hou C, Li Y, Zhang Q and Wang H 2021 Self-powered interactive fiber electronics with visual–digital synergies *Adv. Mater.* **33** 2104681

[86] Zhang R *et al* 2019 Sensing body motions based on charges generated on the body *Nano Energy* **63** 103842

[87] Zhang R, Hummelgård M, Örtegren J, Olsen M, Andersson H, Yang Y, Olin H and Wang Z L 2022 Utilising the triboelectricity of the human body for human–computer interactions *Nano Energy* **100** 107503

[88] Tan P *et al* 2022 Self-powered gesture recognition wristband enabled by machine learning for full keyboard and multicommand input *Adv. Mater.* **34** 2200793

[89] Zhai K, Wang H, Ding Q, Wu Z, Ding M, Tao K, Yang B-R, Xie X, Li C and Wu J 2023 High-performance strain sensors based on organohydrogel microsphere film for wearable human–computer interfacing *Adv. Sci.* **10** 2205632

[90] Guo K *et al* 2023 A P(VDF-TrFE) nanofiber composites based multilayer structured dual-functional flexible sensor for advanced pressure-humidity sensing *Chem. Eng. J.* **461** 141970

[91] Lu X, Qin Y, Chen X, Peng C, Yang Y and Zeng Y 2022 An ultra-wide sensing range film strain sensor based on a branch-shaped PAN-based carbon nanofiber and carbon black synergistic conductive network for human motion detection and human–machine interfaces *J. Mater. Chem.* C **10** 6296–305

[92] Lei K *et al* 2023 Environmentally adaptive polymer hydrogels: maintaining wet-soft features in extreme conditions *Adv. Funct. Mater.* **33** 2303511

[93] Robinson D A *et al* 2023 Tunable intervalence charge transfer in ruthenium prussian blue analog enables stable and efficient biocompatible artificial synapses *Adv. Mater.* **35** 2207595

[94] Yan Z *et al* 2017 Thermal release transfer printing for stretchable conformal bioelectronics *Adv. Sci.* **4** 1700251

[95] Sussillo D, Stavisky S D, Kao J C, Ryu S I and Shenoy K V 2016 Making brain–machine interfaces robust to future neural variability *Nat. Commun.* **7** 13749

[96] Proix T *et al* 2022 Imagined speech can be decoded from low- and cross-frequency intracranial EEG features *Nat. Commun.* **13** 48

IOP Publishing

Energy Harvesting Properties of Electrospun Nanofibers
(Second Edition)

Jian Fang and Tong Lin

Chapter 9

Electrospinning of functional nanofibers: a pathway to flexible piezoelectric and triboelectric wearable devices

Shayan Abrishami, Armineh Shirali, Mahshid Sadeghi, Amin Sarmadi and Roohollah Bagherzadeh

Elaborating on the significance of flexibility and stretchability, this work highlights their pivotal role in the production of wearable electronics. The growth of wearable electronics based on piezoelectric and triboelectric materials is evident, with particular emphasis on the enhancing applications of energy harvesting systems and sensors. Energy harvesting systems, capable of converting environmental mechanical energy into electrical energy, have emerged as vital components. Notably, flexible electrodes are essential for the development of wearable energy harvesters. Among the best ways to produce energy for wearable harvesters, the electrospinning method is a method that has received a lot of attention due to its unique features. The electrospinning method has significantly impacted the production of wearable electronics by enabling the creation of nano-sized fibers from piezoelectric and triboelectric materials. By controlling and optimizing various factors influencing the electrospinning process, desirable properties can be achieved in piezoelectric or triboelectric layers. The expanding application of electrospun layers with piezoelectric properties in body movement sensors, vehicle-related technologies, and storage is noteworthy. However, further optimizations are required to enhance the output of devices produced using this method, focusing on parameters like piezoelectric materials, structural design, and substrate selection. This work emphasizes the exceptional characteristics and potential for integrating fibrous wearable energy harvesting devices with other materials and technologies, positioning them as promising components in the future of wearable technology.

doi:10.1088/978-0-7503-5487-5ch9

9.1 Introduction

In recent years, energy harvester devices have gained significant popularity for capturing and converting ambient energy sources like solar, thermal, and mechanical energy into usable electric power. These innovative devices offer tremendous potential for powering a variety of electronic devices, providing an efficient, eco-friendly solution that benefits both consumers and the environment. Mechanical energy, a prevalent and widely used source due to its abundance in everyday activities and motion-related applications, can provide sufficient power to run various machines and devices. This highly efficient, cost-effective, renewable, and clean source of energy can be harnessed with minimal effort. Rain drops, gusts of wind, ocean waves, and human body movements are all examples of how vibrational energy is readily available around us. The human body itself is an incredibly powerful energy source that can be tapped into for various tasks. Our natural body heat and motion from walking, running, or even simple everyday activities can be harvested to generate electricity. This previously wasted energy can be utilized in innovative ways to power wearable devices or even contribute to charging portable electronics [1–4]. Furthermore, vibration is a ubiquitous type of mechanical motion that exists in our daily lives, ranging from small to large scales, and taking on a variety of forms. The frequency of the vibration can be found at around a few hundred Hertz, which is incredibly low and also changes dynamically over time, making it difficult for current technologies to capture and harvest this energy effectively [5]. Also, kinetic energy from any kind of motion like vibration, compression or torsion that triggers dynamic strain on piezo-electric materials can be collected and transformed into electrical energy with remarkable efficiency [6, 7]. In order to convert mechanical energy into electrical energy, various structures with piezoelectric and triboelectric properties have been used [8, 9]. Compared with traditional mechanical energy generators such as hydraulic generators and wind power generators, nanogenerators and energy harvesters are significantly superior in many ways. They have a much smaller form factor, are cost-effective to manufacture at scale, require less intricate fabrication processes, and they can be easily transported due to their portability [10, 11].

Piezoelectric nanogenerators (PENGs) and triboelectric nanogenerators (TENGs) are emerging technologies that harvest ambient energy, particularly vibrations, to generate electricity. These innovative devices, based on the piezo-electric and triboelectric effects, hold immense potential to revolutionize how we power electronics in daily life. Beyond energy harvesting, PENGs and TENGs have found applications in diverse fields. From innovative engineering solutions like sensors for monitoring and soft robotics to life science advancements like electronic skin, these technologies offer enhanced capabilities that can transform various industries [12]. Piezoelectric materials are substances that generate an electrical charge when subjected to physical pressure. This charge can accumulate on the surface of the material and be used for various purposes [13]. Of the numerous different types of piezoelectric materials that have been discovered and studied up to this point, only a select few have had the opportunity to be used successfully in technical applications. Piezoceramics, piezopolymers, and piezocomposites are three of the most commonly used materials when it comes to the manufacture of generators, actuator systems, and sensors [14]. Piezoelectric ceramics are immensely functional ceramic materials that are able to perform a variety of operations by exploiting the piezoelectric effect. These ceramics can be used in a wide range of applications and can be found in numerous products such as medical devices, automotive parts and smart home devices. The first generation of piezoceramics were discovered and utilized in the 1950s; among the most prominent and widely used piezoceramics are barium titanate, lead zirconate, lead titanate and lead zirconate titanate [14, 15]. Despite their large piezoelectric coefficient, traditional piezoceramics are known to be brittle and have significantly less mechanical stability when compared to their more flexible counterparts, piezopolymers. Since many of the final applications of piezoelectric-based systems require appropriate mechanical properties, piezopolymers have become a proper option to choose with acceptable mechanical properties like flexibility [14, 16]. Piezoelectric polymers such as polyvinylidene fluoride (PVDF) and its copolymers have been demonstrated to possess remarkable piezoelectric properties, particularly at the nanoscale, and are thus especially suitable for use in the development of piezoelectric nanogenerators [17]. They are incredibly flexible, robust, and lightweight yet surprisingly affordable, making them ideal for a wide range of applications. Polymers have become increasingly important components of triboelectric energy harvesters due to exceptional mechanical robustness, dielectric properties and ease of fabrication, so all of these properties make them ideal for use in some applications [18, 19]. Specifically, Polymer-based energy harvesting systems that take advantage of multiple physical phenomena such as piezoelectric, pyroelectric, and triboelectric effects have recently become increasingly popular and relevant for self-powered nanogenerators [20].

There are different fabrication methods for polymeric piezoelectric wearable electronics. The fabrication method of a piezoelectric device must be carefully chosen since it is dependent on the selected raw materials, the exact dimensions required and the ultimate application of the device. A number of piezopolymers fabrication methods are being utilized today such as photolithography [21, 22], printing techniques [23, 24], casting [25] laser printing [26], coating [27] and electrospinning [17, 28–30]. Electrospinning is one of the most widely utilized fabrication methods due to its uncomplicated operation, ability to produce continuous nanofibers, adjustable porosity of the electrospun structure, and flexibility in creating various shapes and sizes [17, 31]. Meanwhile, in the conditions of high electrical fields that are present during the electrospinning process, electric poling of piezoelectric materials can be spontaneously implemented, which has a profound effect on the material's properties [17]. Electrospun piezoelectric nanofiber webs have proven to be extremely advantageous compared to other scaffolding techniques due to their high surface-to-volume ratios and high porosity, yet reduced pore size. These features make the nanofiber webs an attractive choice for many applications that require extreme precision, durability, and flexibility. In addition, due to biocompatibility of electrospun piezoelectric fiber, it is an ideal candidate for broad range of biological sensor and biomedical applications. Their unique properties provide a high degree of accuracy, sensitivity, and durability that make them an attractive option for numerous industries [31].

Different materials have been used for textiles that possess distinct characteristics. Combining these materials to create a unique outcome has been a popular method of maximizing textile properties. With the right selection of materials, it is now possible to create textiles with intelligent functionalities. Smart textiles react to various stimuli in the environment such as chemical, electric, magnetic, and biological influences. These textiles can alter properties in response to some external conditions. Smart textile is defined as a textile that can detect and react to external triggers. It could detect changes in its environment and respond accordingly. In fact, electronic textiles (E-Textiles) are a subclass of smart textiles. Although all e-textiles are smart textiles, not all smart textiles can be classified as e-textiles [14]. Wearable electronics and smart textiles are quickly becoming important advances in modern technology, offering unmatched capabilities for providing fabrics and apparel with sensing, actuating, and adaptive functions that can respond to a wide range of different stimuli. Textile-based sensors can have the ability to detect pressure, strain, temperature, chemicals, humidity, obstacles, and location with remarkable accuracy and convert the corresponding stimuli into electrical signals that can be easily recorded and processed by a variety of digital devices. Textile-based piezoelectric devices are an increasingly popular trend in the energy industry as they rely on technologically advanced polymers, composites, and piezoelectric nanogenerators to convert mechanical energy into electrical energy [32, 33]. Smart textiles have the potential to make an alteration in the world of clothing by combining electronic components with fabric, creating a new generation of multifunctional, adaptive, flexible, and comfortable structures to produce e-textiles. Wearable sensors must be designed with flexibility in mind to allow them to integrate seamlessly into textiles. Textile-based sensor technology is rapidly advancing and becoming more widely used nowadays. Ideally, wearable sensors should create a sensory system that can respond to a wide range of mechanical motions from various body parts, such as compression, bending, stretching, and sliding [27, 34–36]. Therefore, textile-based piezoelectric sensors are the way of the future for wearables, paving the way for more advanced electronic capabilities [37].

9.2 Smart materials

9.2.1 Piezoelectric energy harvesting systems

Piezoelectric materials contain dielectric substances with asymmetric crystals. By using an external electric field, cations and anions tend to become displaced and orient asymmetrically, resulting in deformation of the crystal lattice structure. When piezoelectric materials are deformed or strained by external forces like pressure, their underlying crystal structures get distorted. As a result, electrical dipoles within the crystal structure become oriented in such a way that unidentified charges appear on either side of the material. This creates an electric field that propagates throughout the entire material and is called the direct piezoelectric effect. On the other hand, the piezoelectric indirect effect refers to a vice-versa process [14, 38, 39]. Piezoelectric charge constant (d_{33}) is a measure of the piezoelectric effect in a material, which is the ability of a material to generate an electric charge in response to mechanical stress. The d_{33} value is commonly used to characterize piezoelectric materials such as quartz, ceramics, and polymers, and is an important parameter in the design of piezoelectric devices such as sensors, actuators, and transducers [40–44].

There are various types of piezo materials such as piezoceramics, piezopolymers and piezocomposites. Piezoceramics are created by a specific thermal treatment of a formulated combination of carbonates, oxides, and salts of various metals. This process enables engineers to design the material to accommodate the desired purpose needs for their specific application and they can be used in sensors, transducers, actuators, and electronic materials. The most efficient piezoceramics are based on lead-containing substances like lead zirconate and lead titanate. The utilization, recycling, and disposal of lead-based materials has continuously posed an environmental risk for many years. These materials are toxic and can linger in the environment for extended periods of time, leading to a variety of dangerous consequences if not managed properly [45]. As mentioned earlier, piezoceramics are efficient for use in devices, but are not flexible. Piezopolymers are becoming increasingly popular as a preferred choice for applications requiring a high level of performance. Piezopolymers such as, PVDF and its copolymers, cellulose and their derivatives, polylactic acids (PLA), polyurethanes (PU), polyimides (PI) have revolutionized the manufacturing industry due to their superior mechanical properties, flexibility, and faster processing times compared to conventional materials [40, 46]. In comparison with piezoceramics, piezopolymers have this ability to be created and processed at low temperatures. Additionally, they can be molded into various shapes [47, 48]. Piezocomposites are very popular materials that can have better piezoelectric and mechanical properties than individual components by mixing matrix and filler. The selection of materials and structural design options enable a wide range of mechanical and piezoelectric properties to be developed, allowing engineers to focus on meeting their desired levels of anisotropic performance [49, 50]. The mechanical properties of piezocomposites are highly reliant on the volume fraction of the components, the type and design of their connection, and their production method [14, 51].

Piezoelectric materials can be used in some devices such as nanogenerators, energy harvesters and sensors for various applications. PENGs work based on the piezoelectric effect and receive a lot of attention these days. Efficient PENGs are composed of piezoelectric layer (such as PVDF, PDMS, etc) and electrodes for harvesting the generated electrical energy [52]. They have efficient reversible conversion between mechanical and electrical energy, with an effective electro-mechanical coupling, while also requiring simple engineered structures. PENGs act not only as an energy harvester but also have sensing abilities in the shape of acoustic sensors, stress–strain sensors, or tactile sensors. Additionally, PENGs can attach to the human body and monitor the weak physiological signals that come from human activities such as respiration, blood flow, and heartbeat. In recent years,

PENGs have been utilized broadly in drug delivery systems and biomedical devices [53–56]. These devices can be placed on human fingers or elbow joints and the movement of them can create a bending force, which in turn generates electric signals that can detect the various movements of the human's hands and arms. Joints can be set into various positions which will produce different bending forces, thus generating a range of electric signal strengths [57]. Piezoelectric energy harvesters have become so popular for the variety of their applications. He *et al* [58] developed a piezoelectric energy harvester structure consisting of double-layer squeezing structures and a piezoelectric beam array. Throughout the experiment a person, whose weight was 60 kg, stepped on and off the structure with various frequencies. As a result, a single piezoelectric beam was able to achieve an output power of 134.2 μW when subjected to a step frequency of 1.81 Hz. However, they estimated that the potential output power could reach up to 5.368 mW with the employment of 40 piezoelectric beams installed inside the floor structure. Another study by Zhang *et al* [59] designed a piezoelectric energy harvester device that can detect the mechanical energy of the foot by integrating a spring with an energy generator. It was placed near the heel of a shoe. When the human's foot touches the ground, a pedal will be compressed by the heel and a piezoelectric beam is bent, so electrical energy will be produced. When the foot leaves the ground, the potential energy stored by the compressed spring turns into kinetic energy and the piezoelectric layer bends again. So, a person can get energy of 235.2 mJ from each step. Cho *et al* [60] designed an ingenious road-compatible piezoelectric energy harvester using two piezoelectric transducers that are fixed onto both ends of the device. The stress applied from the movement of the vehicle is then converged towards the center of the device via a rigid bar which enables it to harvest energy efficiently. The harvester was tested in real-life road conditions for five months. It was stressed when vehicles traveled at speeds between 10 and 50 km h^{-1}. At 50 and 10 km h^{-1} the output power of piezoelectric energy harvester was 2.381 W and 576 mW, respectively. The produced energy was effectively utilized to power the LED indicators.

In general, piezoelectric sensors are becoming increasingly popular due to their remarkable characteristics, including high sensitivity, broad sensing range, fast response time, flexibility, as well as low-to-zero power consumption [61]. Piezoelectric sensors can play a key role in biomedical applications and health detecting. A flexible self-powered piezoelectric sensor was developed based on cowpea-structured PVDF/ZnO nanofibers and this sensor showed an impressive level of sensitivity up to 0.33 V kPa^{-1} during pressing and bending operations, making it a wonderful device for physical signal monitoring and gesture sensing applications of humans [62, 63]. Furthermore, electronic skin (e-skin) has been expanded in health detecting systems for diagnosis of diseases [64, 65]. Pressure sensors are capable of interacting with e-skin [66]. Piezoelectric artificial skin is fabricated by PVDF and reduced graphene oxide (rGO). Moreover, pressure sensing applications often use PVDF/Au biosensors. Gold films act as electrodes, while a silicon substrate increases flexibility. These piezoelectric sensors can be placed in various places within the human body for health monitoring and sensing purposes such as muscle movement and level of respiration of patients [67–70].

9.2.2 Triboelectric energy harvesting systems

The triboelectric effect is the most fundamental topic of study in the field of tribology. This phenomenon occurs when two different materials are rubbed together, causing an electrical charge to transfer between them and generating an energy exchange [71]. The produced triboelectric signals are not only useful for directly operating electric devices, but they also provide effective monitoring of the mechanical or chemical stimuli that occur on the devices. Triboelectric devices are increasingly becoming popular due to their ease of fabrication, cost-effectiveness, impressive output performance and flexibility in comparison with other technologies. These qualities make triboelectric devices suitable for self-powered wearable applications [72–74]. Nowadays, most of the triboelectric materials that are used are available polymers with known triboelectric charge densities, dielectric properties, and various other relevant parameters like coefficient of friction [75]. Some polymers such as polyvinylidene fluoride (PVDF) [76], polytetrafluoroethylene (PTFE) [77], polydimethylsiloxane (PDMS) [78], due to having fluorine in their structure have strong electron affinity. Despite their strong triboelectric properties, these polymers can be environmentally hazardous. Additionally, there are other materials that can have more efficient performance, such as regenerated cellulose [79]. Regenerated cellulose is produced with only green materials that were not harmful to the environment, and all of the mechanism of production ias environmentally friendly but shows impressive triboelectricity to the same order as fluoropolymer-based TENGs which can be used for powering LED lamps and emergency indicators [80]. Lignin [81], chitosan [82], fabricated nanofibers [83] and fish gelatins [84] are the other biopolymers that can be used in TENGs.

Triboelectric materials can be utilized in nanogenerators, energy harvesters and sensors with a wide range of applications. TENGs have become popular due to their simple operational principles, wide range of structurally diverse designs, miniaturization possibilities and incredibly high energy conversion efficiency achieved by coupling friction electrification and electrostatic induction between materials with different triboelectric polarities. A TENG is a combination of three essential components; a positive triboelectric layer, a negative triboelectric layer and conductive electrodes [85]. The fundamental principle of triboelectricity is based on the different polarities of the two layers, when they encounter each other, negative and positive electrostatic charge will be generated on the surface of each layer. By applying an external force to separate the layers, the electrostatic charges will be separated, consequently, the potential difference will be generated between electrodes. The generated potential difference causes the flow of electrons in the external circuit [86].

TENGs can play a key role in healthcare and biomedical applications. Cardiac pacemakers help regulate the rhythm of the heartbeats by sending an electrical pulse to the heart muscles. The heart rate can diminish due to several reasons, such as dizziness and angina. Pacemakers help the heart to perform better and increase the heartbeat [87]. Zeng *et al* [88] designed the first self-powered pacemaker based on an implantable TENG. The innovative implantable triboelectric nanogenerator (iTENG) was fabricated with a pyramid patterned polydimethylsiloxane (PDMS) array on Kapton, combined with a micro–nano structured aluminum (Al) foil as the opposite layer. Kapton was used as a substrate, and gold (Au) was utilized as an additional electrode to be deposited on it and a thick polyethylene terephthalate (PET) was used as a spacer, and a thick PDMS as encapsulation layer was used and this device was packaged. The pacemaker was put in the chest skin and generated output of 3.73 V and 0.14 μA in *in vivo* conditions (breathing rate of 50 min^{-1}). In addition, TENG can be helpful in healthcare monitoring. In cases of various diseases, it is necessary to keep track of physiological parameters like respiratory rate, heartbeat, and blood pressure on a regular basis. Sudden shifts in these parameters can be a life-threatening condition monitoring device [89]. Zhao *et al* [90] designed a wash durable textile-based TENG. Fabrication of TENG is done with a weaving process for coated copper polyester stands as the weft yarn and the warp yarn. The textile-based TENG was placed on the chest to show its sensitivity to human respiration monitoring the output voltage of textile-based TENG was 4.98 V when tapped at 10 cm s^{-1}.

Furthermore, triboelectric energy harvesting is a mechanical process that relies on rubbing contact and electron transfer. It has several advantageous, such as flexibility, high power density, high conversion efficiency, which make it a suitable choice for wearable applications [91, 92]. Zhong *et al* [93] developed a cost-effective fiber-based triboelectric nanogenerator that can harvest biomechanical energy from human motion and vibration of the body. This wearable TENG converted it to electric power by utilizing polytetrafluoroethylene aqueous suspension, cotton threads and carbon nanotubes (CNTs). The power density can reach levels of approximately 0.1 μW cm^{-2}. In addition, Yi *et al* [94] designed a TENG with sandwich shaped structure and checker like electrodes for harvesting mechanical energy in various directions. The TENG device demonstrated remarkable performance in both sliding directions while generating voltage output of up to 210 V. Moreover, they were able to light up LEDs using the mouse operation energy when the sliding mode TENG was integrated into a mouse pad or other sliding panel. Wang *et al* [95] produced a fiber-based TENG that is designed to continuously power wearable electronics, just through the human motion. It was made by using elastomeric materials and a helix inner electrode that is attached to the tube with a dielectric layer and outer electrode. It showed properties, such as great flexibility, an excellent level of stretchability, isotropy, weavability as well as water-resistance and an incredibly high surface charge density of up to 250 mC m^{-2}. By utilizing only, the energy extracted from walking or jogging, the TENG that is embedded in the soles of shoes can be utilized to power a variety of wearable electronics such as an electronic watch and fitness tracker.

Triboelectric sensors are the other types of triboelectric materials that can be used in various devices with different functions. TENG-based sensor data can be used for early detection of health conditions. Lin *et al* [96] designed a TENG-based pressure sensor with electrospinning technique that can monitor the human heart rate through the wrist. The dielectric layers of the sensor are composed of core–shell PDMS ion gel/poly (vinylidene fluoride)-co-hexafluoropropylene elastic nanofiber mats. Cu was utilized as an electrode layer, providing a capacitive layer that wraps around the core–shell fiber mats. Kapton was used as the triboelectric negative layer for the structure. This device showed sensitivity of 0.068 V kPa^{-1} from 100 to 700 kPa and a maximum power density of 0.9 W m^{-2} in experimental tests on living subjects. Cui *et al* [97] colleagues produced a wearable pulse sensor with trench structure in which the PDMS film is in contact with the human skin as a triboelectric layer and the indium tin oxide (ITO) film acts as an electrode. This sensor is cost-effective and flexile as a wearable sensor. The negatively charged PDMS layer would bend and contact the positively charged ITO layer, When the PDMS layer is compressed and released alternately, the distance between the two layers' changes and then the number of charges also changes. The designed trench measuring 25 mm in length, 7 mm in width and 0.5 mm in depth, was able to provide a current output of 0.97 nA that was optimal for this wearable pulse sensor. In 2018, Cao *et al* [98] designed a nanofiber-based triboelectric sensor (SNTS). This sensor with arch structure composed of an electrospun polyvinylidene fluoride (PVDF) nanofiber membrane and a screen-printed Ag nanoparticles (AgNPs) electrode. according to the various tribo-polarities of PVDF nanomembrane and AgNPs, when these two tribolayers contact with each other, an equal amount of both positive and negative triboelectric charges will be formed on their surfaces. The AgNPs electrode is connected to the ground through an external load, from which the sensing signal is obtained. The SNTS, encased in a protective mask, can accurately monitor, and analyze the breathing condition of a person.

9.2.3 Pyroelectric

Today, with the increase of energy crises and environmental problems of many energy sources, there has been a widespread interest in new sources. One of these sources, which are mainly known as wasted energy, is heat energy produced by the human body, car exhausts, and solar radiation. Pyroelectrics are materials that can convert thermal energy into electrical energy by spontaneously changing internal polarity when there is no applied electric field. This electrical response to thermal energy is due to the internal oscillations of the pyroelectric material with respect to temperature changes [99–101]. The performance of pyroelectric materials is based on the pyroelectric effect which refers to the alteration of polarization caused by temperature fluctuation. Heating of pyroelectric materials leads to the creation of thermal vibrations, as a result of which a disturbance in the bipolar balance occurs. As a result, it causes a change in electrical polarization. If this change is significant, it can lead to the separation of bound charge carriers and the placing of charges on the surface of the electrodes [102].

Many pyroelectric materials have been reported so far for various applications including energy harvesting and sensors [103–107]. These materials have been introduced and consumed in different forms such as triglycine sulfate, polymer, ceramics and single crystals [108]. Each of these forms has its own characteristics that have made them very popular in a specific function. For example, although triglycine sulfate has a very high pyroelectric coefficient, it is not a suitable option for energy harvesting due to its brittleness, instability in high humidity and solubility in water. These materials are one of the most widely used pyroelectric materials in sensors. Pyroelectric polymers such as PVDF and composites made from it are widely used in sensors and nanogenerators due to their ease of production and flexibility. Compared to triglycine sulfate, polymers have more stability, more suitable mechanical properties and lower pyroelectric coefficient [102]. Pyroelectric ceramic structures are more resistant and thermally stable than the polymers and triglycine sulfate, which has made them popular and widely used. Additionally, These materials have high pyroelectric coefficient and low dielectric loss, which makes them one of the most suitable choices for energy harvesters [109].

Among the wide applications of pyroelectric materials, energy harvesters and nanogenerators have grown greatly in recent years. Polymers are used in this field due to their low cost and ease of production process, for instance, PVDF is one of the light polymers, which besides pyroelectric properties, also has piezoelectric properties, which is why it is used in hybrid applications. By presenting a nonwoven layer composed of PVDF nanofibers, You $et\ al$ [110] developed a flexible hybrid nanogenerator with piezoelectric–pyroelectric properties. In another study, Zhang $et\ al$ [111] produced a flexible pyroelectric nanogenerator using PVDF thin film, the measured output voltage of which was 8.2 V. Also, its maximum output power was equal to 2.2 μW. Ceramics, lead-based and lead-free, are among the most widely used materials in the field of pyroelectric harvesting energy applications. Ko $et\ al$ [112] placed a thin layer of lead zirconate titanate film on a nickel chrome foil layer and using lanthanum nickel oxide electrodes, which enables the high temperature growth of the film. The PZT film had high piezoelectric and pyroelectric coefficients ($50\ \mathrm{nC\ cm^{-2} \cdot K^{-1}}$) at room temperature.

9.2.4 Photovoltaic energy harvesting system

Solar energy holds great potential as a renewable source of power, as it is the most bounteous exploitable renewable energy source. A photovoltaic process using semiconductor materials with the photovoltaic effect can be employed to convert solar radiation into electricity. However, despite the enormous capacity of solar energy, photovoltaic technology is responsible for a trivial amount of electricity generation so far. However, the growth of photovoltaic technology is at a high rate, and innovative technologies are being developed to increase the efficiency of photovoltaic cells at lower costs [113–116]. Nanotechnology has enabled researchers to develop advanced materials like electrospun polymer and ceramic nanofibers that can overcome the limitations associated with traditional materials [116, 117]. Moreover, these advanced materials can be used to create flexible and stretchable solar energy harvesters. For instance, flexible electrospun TiO_2–SiO_2 mats can achieve a 1.3–3.4 mm radius of curvature even after heat treatment [118]. However, the fabrication of composite nanofibers using ceramic nanoparticles like hydroxyapatite randomly decorated on polyamide 6 (PA6) fibers without agglomeration by electrospinning methods has its limitations, since the resulting energy harvesters are not that flexible due to the fragility and rigidity of the composite material [119].

Electrospun nanofibers have been gaining increasing attention as effective materials for application in photovoltaic systems. These nanofibers offer unique features such as high surface area-to-volume ratio, inherent porosity, tunable morphology, and high surface energy, which make them attractive for use in photovoltaic devices. The use of electrospun nanofibers in photovoltaic systems provides various benefits such as enhanced light absorption, increased charge carrier separation efficiency, and improved device performance. For example, electrospun titanium dioxide nanofibers have been shown to improve the power conversion efficiency of dye-sensitized solar cells due to their high surface area and high charge carrier mobility [120].

9.3 Electrospun nanofibers

Diminishing the size of materials and reaching their dimensions to the nanoscale allows unique features to be seen in various materials, which is the reason why nanomaterials have developed tremendously in recent years. In fact, nanofibers are fibers whose dimensions are on the nano scale, and according to the nanotechnologies standards, fibers with a diameter below 1000 nm are considered as nanofibers [121]. When it comes to fibers, decreasing their diameter primarily results in a greater amount of specific surface area. Nanofibers with high specific surface area can lead to changes in properties like bioactivity, electroactivity, and strength. Nanofibers have numerous applications including filtration, composites, electronic technology, biomedical technology [122–128]. There are various methods for producing fibers with a diameter in the nanometer scale, electrospinning is one of the most widely used and popular methods, because of ease of operation, continuous nanofiber formation, adjustable porosity of electrospun structure [129]. Electrospinning is a method for producing ultrafine fibers with diameters in the range of a few nanometers to microns in diameter. This technique utilizes an electric field to draw a polymer solution or melt into a fiber-forming jet that is then collected on a surface. Recent advancements in electrospinning technology have led to the development of new techniques for fabricating complex nanofiber structures, including core–shell nanofibers and composite nanofibers. The process has gained significant attention in recent years due to its versatility, low cost, ease of operation, and ability to produce nanofibers with varied properties for various applications in engineering, biomedical devices, electronic devices, and tissue engineering [130–134].

The basis of the electrospinning method is not complex, and its operation is also uncomplicated, and this has become one of the factors of the increasing development of this method. To produce electrospun nanofibers, a conductive surface is used as a collector, and an electric field is applied between it and the physical zone of jet emission (such as the tip of a nozzle) in a polymer or solvent. The fluid takes on a specific shape known as the Taylor cone when a strong electric field is produced by a high voltage. This is because the electric field stretches the fluid. If the voltage is higher than critical amount, a polymer jet is created and pulled towards a collector because the surface tension of the fluid can no longer keep the cone shape intact. The process of melt electrospinning involves heating the polymer above its melting point so that it can flow through the nozzle and then undergo elongational stretching to form a fiber during flying time. This technique is typically used when finding an appropriate solvent for the polymer at room temperature poses a challenge. Although melt electrospinning is highly efficient, the mean diameter of the fibers produced is typically larger than those obtained through solution electrospinning. Additionally, this process requires specialized equipment and high voltages due to the low electrical conductivity and high viscosity of molten polymers.

The schematic principle of a solution electrospinning can be seen in figure 9.1. To create nanofibers, a polymer solution is loaded into a syringe and a syringe pump is used to control the flow rate. The needle and collector are given a specific voltage, and the process parameters are adjusted to ensure that a Taylor cone forms at the tip of the needle. When electrostatic forces exceed surface tension and viscous forces, a fine jet is extruded from the cone. During flight, solvent evaporation occurs, resulting in solid nanofibers that collect on the surface of the collector. Each of the mentioned parameters has an optimal value, which decreases or increases these values, causing a change in the properties of nanofibers. Additionally, there are

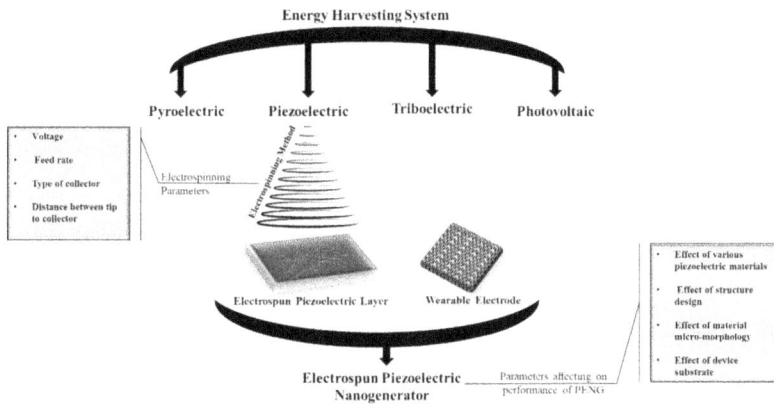

A summary of basic structure of electrospun piezoelectric nanogenerators and the parameters affecting on electrospinning process and performance of PENGs.

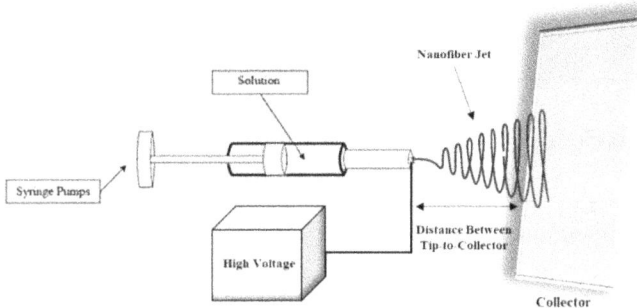

Figure 9.1. Schematic diagram of an electrospinning process.

various types of collector which have unique properties and the final product can be under their influence [17]. In general, three categories of factors can have a direct effect on the final properties of electrospun nanofibers, these three categories of factors are electrospinning parameters, polymer solution parameters, and ambient parameters. Here, these factors and their effects will be discussed in detail.

As mentioned earlier, one of the applications of electrospun nanofluids is in electronic, biomedical and energy systems. Electrospinning makes it possible to produce many polymers with piezoelectric properties in the form of nanofibers and use them in piezoelectric devices for the mentioned applications. PVDF is one of the most widely used polymers prepared by this method in the form of nanofibers. This polymer has good piezoelectric and triboelectric properties due to its crystal network and the arrangement of its molecules. In fact, this polymer is a linear semicrystalline that has five different crystal phases. One of these crystalline phases is the β-phase, which has the highest spontaneous polarity due to the arrangement of atoms in the polymer chain, which leads to the emergence of piezoelectric properties. In other words, the higher the β-phase in the structure of this polymer, the higher its piezoelectric property [28, 129, 135–137]. Each of the categories of parameters affecting the properties of nanofibers have a direct effect on the amount of β-phase in the PVDF crystal lattice. This crystal network is not unique to this polymer and other polymers also have this network.

9.3.1 Electrospinning parameters

The characteristics of nanofibers are significantly influenced by various electro-spinning factors such as the voltage applied, the distance between the tip and collector, the flow rate, and the type of collector (static and dynamic collectors). In this section, we are going to examine each of the electrospinning parameters and point out their effects on the produced nanofibers, and in figure 9.2 all the electrospinning parameters are summarized along with their effects on the electro-spun nanofibers.

9.3.1.1 Voltage
In the process of electrospinning, the structure and morphology of nanofibers are greatly affected by the electric field. Increasing the applied voltage results in a decrease in the average diameter of fibers until it reaches a certain value, usually known as critical value or optimal voltage. This can be attributed to stronger electrostatic forces acting on the polymer solution and increased drawing stress. However, if the voltage is higher than optimal voltage, the fibers will become beaded and eventually break due to excessive drawing stress. Therefore, it is challenging to determine optimal values for the Taylor cone generation and control fiber formation. By applying optimal voltage, the stretching of polymer jets encourages orientation of the molecular chain dipoles. For example, in electrospinning of PVDF, optimal voltage can lead to the α crystalline phase to β crystalline phase. Therefore, increasing β crystalline phase is equal to increasing piezoelectric properties of the electrospun layer [138–140].

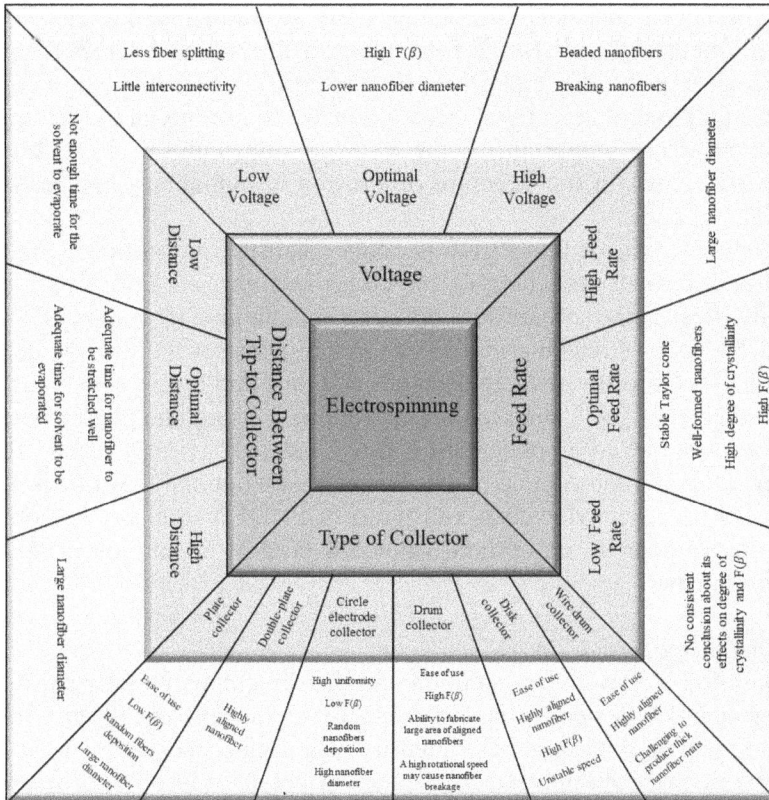

Figure 9.2. The electrospinning parameters and their effects on electrospun nanofibers.

There are several studies on how voltage affects the properties of electrospun nanofibers. For example, Jiyong *et al* [139] concluded in study that the most optimal applied voltage value for PVDF polymer electrospinning is about 20 kV. In the study conducted by Damarajo *et al* [140], the applied voltage was increased from 12 to 25 kV and it was observed that the β crystalline phase increased from 67.8% to 72.4%, with a decrease in average diameter of PVDF electrospun nanofiber from 295 to 177 nm. If the applied voltage is increased, more charge will accumulate on the surface of the jet, which will cause the jet to move faster and elongate more. This will result in a reduction in the fiber diameter, as the voltage is increased. Using high voltage can ensure that the electric dipoles in the PVDF solution become aligned during the electrospinning process. It should be noted that there is an optimal value for applied voltage and if the voltage exceeds the optimal value, the flying time is reduced and polymer chains do not have enough time to be well-oriented [17].

9.3.1.2 Feed rate
The feed rate has a direct effect on the structure of nanofibers and Taylor cone. If the feeding rate is low, a vacuum is created inside the nozzle, while if the feeding rate is

too high, the Taylor cone is not formed properly. Additionally, if the feeding rate is more than optimal, it leads to the production of fibers with a larger diameter. In order to achieve a durable and stable Taylor cone, the feeding rate needs to be optimal. In addition to the optimal feeding rate, the uniformity of the feeding rate at every moment is also important. If the feeding is not uniform, it can lead to the formation of willows in the structure of electrospun nanofibers and results in an unstable Taylor cone.

Several studies indicate that PVDF nanofiber diameter increases with higher flow rates due to decreased stretching of the electrospinning jet. This results in lower crystallinity because the solvent does not have enough time to evaporate. Low flow rates lead to higher stretching of the jet, which increases the β crystalline phase content [28, 141]. However, there are contradictory findings in some studies. For instance, Singh *et al* [141] observed an increase in β crystalline phase content with higher flow rates due to amplified shear force, but only up to a certain threshold level, after which the content decreased. Therefore, an optimal flow rate is necessary to produce a proper Taylor cone and uniform PVDF nanofibers without beads. When the feed rate is set on optimal value, the level of orientation of electrospun nanofibers is higher.

9.3.1.3 Distance between tip and collector
The evaporation of solvent in a solution and the stretching of a polymer molecular chain are primarily affected by the distance between the tip and collector, but it also leads to a reduction in electric field intensity. The connection between the distance between the tip and collector and the β-phase of polymers like PVDF nanofibers is not a simple linear relationship. Researchers have found that changes in the distance between the tip and collector do not significantly impact the fiber diameter and β-phase of PVDF nanofibers. This is due to the competing effects of distance on the formation of the β-phase, where an increase in distance allows the nanofibers more time to stretch and solvents to evaporate, leading to the formation of the β-phase. However, as the distance between the tip and collector increases, the electric field intensity decreases, and the instability of the jet increases, which is detrimental to the formation of β-phase [141–146].

9.3.1.4 Type of collector
The last part of the electrospinning process on which the nanofibers are collected is the collector, and the type of this collector can affect the properties of the nanofibers. In general, collectors can be divided into two categories: static and dynamic, and each of these two categories has sub-categories. When a static collector is used to bind nanofibers, the only dominant force is the Coulomb force caused by the application of voltage, while when using an electrodynamic collector, a mechanical force is also applied in addition to the Coulomb force. This is the reason why nanofibers are more aligned and orderly when using the dynamic collector, therefore, the dynamic collector leads to more arrangement and increase of β-phase in the polymer chain [17]. In figure 9.3, the types of collectors used in the electrospinning process can be seen.

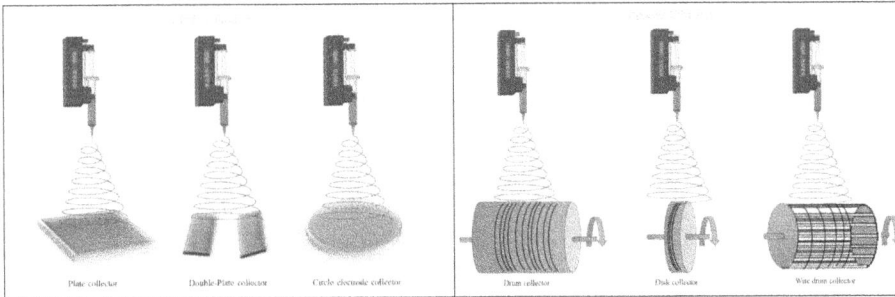

Figure 9.3. Schematic illustration of various types of collectors.

Using a fixed metal plate as a collector is one of the most common collectors used in the electrospinning process. The fixed metal plate is placed at a certain distance from the tip of the nozzle, when the electrospun polymer comes out of the nozzle, it is placed on this static plate without any other mechanical force. In this case, the β-phase in the electrospun polymer is not high and the decoration is relatively low. Instead of using a single fixed plate, you can use a double fixed plate. This change causes a 10%–15% increase in the β-phase to be observed, the homogeneity also increases to some extent. The reason for this occurrence is associated with the single plate collector lacking a specific direction for the electric field. Conversely, when utilizing the double-metal plate collector, the electric field vectors generate a stretching force in the interval between the two metal bars. In addition to the previous two static collectors, a circular static collector is also used, which has a higher productivity due to the similarity of its cross-section to the end of a Taylor cone. However, the percentage of β-phase is not high enough and the distribution of nanofibers is random [147–152].

Dynamic collectors are another category of collectors that apply a mechanical force by rotating and as a result lead to more formation and increase in β-phase percentage. The speed of these collectors is related to the applied mechanical force. Among the types of dynamic collectors, rotating disk compared to drums leads to the production of higher quality nanofibers, while the biggest problem is its limited surface, which leads to not being able to produce on a large surface. Many studies compared the quality of nanofibers with static and dynamic collectors [152]. For instance, Abolhasani *et al* [153], compared the crystalline structures of PVDF nanofibers produced using a static collector and a rotating collector at 2500 rpm. They found that the PVDF nanofibers obtained using the static collector had a β-phase content ranging from 50% to 60% and were distributed randomly. In contrast, those produced using the rotating collector had an $F(\beta)$ ranging from 70% to 80% and were well aligned. Additionally, the nanofibers collected on the rotating drum were thinner than those collected on the static collector, but the total crystallinity values of the PVDF nanofibers produced were not significantly affected. The rotation speed of dynamic collectors has an effect on the final quality of electrospun nanofibers. According to the results reported in the studies, increasing the speed increases the percentage of β-phase in the structure of nanofibers, while increasing the average diameter slightly increases with increasing speed [154].

9.3.2 Polymer solution parameters

Optimization of electrospinning process parameters is critical for controlling the fiber morphology, diameter, and uniformity. The polymer viscosity is an important parameter in electrospinning, which directly affects the fiber diameter. High viscosity solutions produce thicker fibers, while low viscosity solutions produce thinner fibers. The high viscosity of the polymer is the result of its high concentration in the solvent, which can lead to a decrease in the percentage of the β crystalline phase in the electrospun structure. In addition, at low viscosity, the beads increase and reduce the uniformity of the production layer. Each polymer has its own optimum value according to its properties and structure. For example, for PVDF polymer, which is one of the most widely used polymers for the production of electrospun piezoelectric layers, its optimal concentration is between 10% and 25% by weight. This range is determined according to the molecular weight of this polymer, the lower the molecular weight of this polymer, the higher the concentration needed [28].

In electrospinning, the polymer solution viscosity determines the jet formation as mentioned above, which significantly affects the fiber diameter. As molecular weight increases, the viscosity of polymer solutions increases, resulting in higher surface tension, slower evaporation rate, and larger fiber diameters [155]. However, this behavior is also dependent on the molecular weight distribution, with polymers having a narrow molecular weight distribution showing less sensitivity to the changes in molecular weight, and therefore less sensitivity to changes in fiber diameter. Polymer molecular weight also influences the spinnability of the polymer solution. High molecular weight polymers tend to have high chain entanglements, resulting in high solution viscosities, which can make it difficult to spin uniform fibers. In contrast, low molecular weight polymers exhibit lower solution viscosities and may require the addition of solvent or surfactants to improve electrospinnability [156]. Moreover, polymer molecular weight can impact the mechanical properties of electrospun nanofibers. In general, high molecular weight polymers can result in stronger fibers with higher tensile strength and elongation at break, while low molecular weight polymers may produce fibers with lower strength and flexibility [157].

The solvent used in electrospinning plays a vital role in the process as it not only affects the fiber diameter, but also influences the physical and chemical properties of the resultant electrospun nanofibers. The fundamental principle of electrospinning involves the formation of a jet of polymer solution/melt under the influence of an electric field that leads to the evaporation of solvent, causing solidification of the fiber. The choice of solvent is crucial in determining the physical properties of the resultant nanofibers, including morphology, viscosity, and surface tension. The polarity and dielectric constant of the solvent can affect the ionization of the polymer solution that further prompts the formation of Taylor cones, the initiation of the jet, and the formation of fibers. Furthermore, the solvent properties can also influence the surface structure and molecular orientation of the polymer nanofibers. solvents with high dielectric constants are known to exert a higher electrostatic force on the polymer solution, which enhances the chain alignment and promotes the formation of highly aligned nanofibers. The solubility of the polymer in the solvent also plays a crucial role as excess solvent can cause residual solvent toxicity and affect the mechanical and thermal stability of the resultant nanofibers. The capability of the solvent to establish intermolecular forces can also influence the chemical properties of the nanofibers. For instance, solvents that are highly polar and have high surface tension can induce the aggregation of the polymer chain, leading to the formation of nanofibers with a higher degree of crosslinking [17, 28, 158, 159].

9.3.3 Ambient parameters

The properties of the nanofiber produced depend on various parameters, including ambient properties such as temperature and relative humidity. Temperature is one of the critical factors influencing the electrospinning process, as it regulates the solvent evaporation rate and polymer chain relaxation. The evaporation rate of the solvent increases with an increase in temperature, thus reducing the time for the solvent to reach the surface and increasing the solvent concentration in the air around the jet. This concentration can cause a non-uniform distribution of solute, leading to a fiber's beading or irregular diameter distribution. Moreover, a higher temperature decreases the viscosity of the polymer solution, decreasing the electrospinning process's stability. The solution's surface tension decreases while solvent volatilization rate increases as the working temperature rises. Research has suggested that higher temperatures can result in the creation of nanofibers with narrower diameters [17, 28, 152, 160].

Relative humidity is another critical ambient factor that affects the electrospun fiber morphology. Relative humidity refers to the amount of water vapor present in the air compared to its maximum saturation level. High relative humidity can cause moisture absorption by the polymer solution, leading to an increase in polymer chain mobility and viscosity. Increased humidity can also lead to the formation of beads or globules on the electrospun fibers, leading to reduced fiber alignment. The level of moisture in the air has a significant impact on how quickly solvents evaporate, and this in turn affects the formation of the β-phase and the appearance of the fiber surfaces. When humidity is low, solvents evaporate rapidly, which can cause problems with the fiber's structure, while higher humidity, around 60%–70%, can result in thinner and smoother nanofibers that contain more β-phase material [17, 28].

9.3.4 Parameters affecting performance of PENG

The primary purpose of PENGs is to amplify both the output current and voltage while simultaneously increasing the device's flexibility. There are numerous techniques to enhance performance of these PENGs. Innovative methods to improve the performance include the use of different piezoelectric materials, structural design of commonly used piezoelectric materials, changing the microscopic morphology of existing piezoelectric, selecting appropriate substrate for enhanced electrical properties. These are described in detail in this section and they aim to reach flexible PENGs with increased output current and voltage [161].

9.3.4.1 Effect of various piezoelectric materials

One of the vital factors that have an impressive impact on performance of PENGs is the type of piezoelectric materials. Various piezoelectric materials, such as $BaTio_3$, PVDF, PZT and ZnO show different output performance. ZnO as an earliest piezoelectric material has some unique properties, like acceptable semiconducting and piezoelectric properties, that make it suitable for PENGs. Also, it has simple structure and all its variable factors such as purity and size can be controlled. The other properties of ZnO nanowires are its bendability and flexibility. Recently, researchers have put significant effort into finding the optimal structure and improving the piezoelectric properties and preparation of nanowires for PENGs [161]. Some ZnO designs were used to convert heartbeat and finger movement into electrical energy and this can have positive impact on biomedical devices. Such a design was developed by fixing a single ZnO nanowire at both ends to electrodes and laterally packaging it on a substrate that is flexible. When the substrate is bent, the wire becomes stretched, causing a generation of electric charge along its length. This, in turn, results in the production of a current in an external circuit [162, 163]. Various distribution densities of ZnO nanowires generate different output currents and will be prepared for different applications. For example, low density ZnO nanowires produced direct current, and high density ZnO nanowires generated current that was alternating [164]. Although piezoelectric materials have the ability of generating current and voltage, this ability is not enough to be used in devices. Therefore, alternative materials need to be utilized. PZT-based ceramic materials can be one of them that can be used in PENGs. They have excellent features, such as considerable piezoelectricity, high dielectric constant, and perfect crystallinity. PZT single crystal nanowire was prepared with hydrothermal synthesis. The output of PZT-based PENG was in open circuit voltages up to 0.12 V and short-circuit currents up to 1.1 nA [165]. In another example, flexible PENG was developed with highly crystalline PZT piezoelectric nanoribbons on a MgO substrate with radio frequency sputtering method (RF-sputtering method). The PZT piezoelectric nanoribbons were highly crystalline and were transferred onto PDMS silica gel after being separated from the substrate. Generally, the open circuit voltages of the small PENG with printed PZT piezoelectric nanoribbons was 0.25 V and short-circuit current was reported as 40 nA [166]. The laser lift-off process can create a PZT thin film nanogenerator with properties such as light weight, flexibility and high efficiency. This nanogenerator was designed by transferring the PZT thin film onto a flexible plastic substrate without any mechanical destruction. The PENG was periodically bent and released, causing an output voltage of 200 V and a current of 105 μA [167]. In general, PZT-based PENGs showed better performance than ZnO-based PENGs. However, due to toxic effect of lead element in PZT, it cannot be widely used in microdevices [168]. As a result of this, materials such as $BaTiO_3$, which were lead-free, were used as replacements. Additionally, other alternatives like sodium potassium bismuth titanate (BNKT) and bismuth titanate ($Bi_4Ti_3O_{12}$) and potassium sodium niobate (KNN) became suitable options that will be used as a piezoelectric layer of PENGs [161]. Huan *et al* [169] designed a PENG based on KNN nanoparticles and multi-walled carbon nanotubes (MWCNTs). The output voltage of this device was 3.5 V for open circuit voltage and 0.3 μA for short-circuit current.

The process of electrospinning was utilized to produce layers of $BaTiO_3$, which can be used in flexible PENGs. The $BaTiO_3$ nanowires were moved onto a polyethylene terephthalate (PET) substrate. The two ends of the nanowires were adhered to the flexible substrate using silver paste, and cooper wire electrodes were connected to the current and voltmeter. Eventually, the device was filled with PDMS material [168]. $BaTiO_3$ with high piezoelectric properties and cost-effectiveness, does not show acceptable flexibility. However, polymer-based piezoelectric materials like PVDF with high flexibility, high chemical resistance and ease of production could be used in sensors and energy harvesters. PVDF nanofibers should be put on a flexible substrate with electrospinning technique for preparation of polymer-based PENGs [170]. Additionally, PENGs made from the organic polymer PVDF can have better flexibility but experienced lower value of d_{33} in comparison with inorganic ones, reducing the performance of piezoelectric devices [161, 171].

9.3.4.2 Effect of structure design
Structure design is another factor that can effect performance of PENGs. This factor has direct influence on mechanical and piezoelectric properties of devices [172]. There are various types of structural designs such as coaxial nanofibers, hierarchical structure and coated nanofibers [17].

9.3.5 Coaxial nanofibers

Core–sheath nanofibers can be produced effectively by utilizing the coaxial electro-spinning method [172–174]. In this method two needles with different polymer solutions were used, with uniaxial electrospinning that utilized only one nozzle. With the combination of specially crafted and complementary polymers in the core and shell layers, it is possible to create nanofibers that possess superior mechanical and piezoelectric properties [174]. Crafting core–shell high quality nanofiber requires controlling the interfacial tension and viscosity of both solution components [175]. Polymers in coaxial core–shell structures have high density in comparison with the neat nanofibers that can lead to better interaction between core and shell layer. The exceptional tensile properties of coaxial core–shell structures can be attributed to factors such as their high level of alignment and tightly packed structures [96, 172]. Han *et al* [172] produced a core–shell nanofiber which contained poly (vinylidene fluoride) trifluoroethylene (PVDF-TrFE) as a shell and PEDOT: PSS as a core for comparing with the neat PVDF-TrFE. They examined various thicknesses for shell layer and the result showed that the thinnest shell layer showed output voltage of 8.76 V which was 10 times higher than neat PVDF-TrFE. In another study, PVDF-TrFE and poly (vinylidene fluoride-trifluoroethylene-chlorotrifluoroethylene)

(PVDF-TrFE-CTFE) can act as core and shell, respectively. PVDF-TrFE-CTFE can show high piezoelectricity in comparison with PVDF-TrFE [176]. Furthermore, coaxial electrospinning can be utilized for fabrication of core–shell nanocomposite nanofibers. Zhu et al [177] fabricated a coaxial core–shell nanofiber with PVDF/BaTiO$_3$as core and PVDF/graphene oxide (GO) as shell. The results showed a high level of piezoelectricity in comparison with neat PVDF.

9.3.6 Other hierarchical structures

Several designs which feature electrospun nanofibers as the primary sensing elements have also been made, exhibiting a hierarchical structure. One proper design for smart fabrics in wearable electronics is to transform the electrospun nanofiber mat into a yarn structure through twinning [178–181]. Ji et al [178] produced a wearable energy harvester based on piezoelectric core–shell yarn. By electrospinning method PVDF-TrFE and BNT-ST ($0.78Bi_{0.5}Na_{0.5}TiO_3$–$0.22SrTiO_3$) were produced. The nanofibrous layer twined around a conductive thread as an inner electrode. After producing the core–shell yarn structure, a conductive yarn was added to serve as the external electrode. The braided yarn was then sewn onto fabric pieces in the last step. The outcome demonstrated that a 15 cm stitched yarn produced 19.1 V when bent. Gao et al [179] fabricated a special type of yarn that contains a conductive core made of silver-coated PA6. The core serves as an inner electrode and is surrounded by nanofibers of PVDF. Finally, a layer of silver is added to the outside of the yarn to act as an external electrode. This innovative yarn generates an average voltage of 0.52 V and a current of 18.76 nA. Additionally, to make piezoelectric fabrics, the piezoelectric nanofiber mat can be twisted into yarns. Ji et al [182] fabricated a three-layered structure consisting of two layers of piezoelectric energy harvesters at the top and bottom, and a middle layer of triboelectric energy harvester. This package demonstrated outstanding voltage output of 253 V, as well as exceptional durability and efficient charging speed.

9.3.7 Coated nanofibers

Many nanofillers are successfully applied into the polymer matrix of nanofibers through a direct approach, by being uniformly dispersed in the polymer solution before electrospinning. Some research has been done on fillers in order to increasing the piezoelectric properties [183]. Coating a polymer nanofibers mat with nano-particles is an effective technique. Kim and Fan [184] made a comparison between the output voltage of PVDF nanofibers mat that was electrosprayed with ZnO nanorods and dispersed ZnO nanorods into the PVDF solution before electro-spinning and the results were 245.63 and 5.69 nW cm^{-2}, respectively. In a research by Li et al [185], poly (vinylidene fluoride-co-hexafluoropropylene) (PVDF-HFP) nanofibers were coated with MWCNTs as collector and a copper frame was used. At the final stage the output voltage was reported 0.62 V with excellent properties such as washability and breathability. Furthermore, by increasing the contact area between nanofibers and electrodes better charge transfer will be achieved [186].

9.3.8 Effect of material micro-morphology

During the process of materials preparation some parameters such as reaction temperature can be controlled, and the same materials with various micro-morphologies will be produced like nanotubes, nanofibers, nanorods and nanowires. Therefore, various types of micro-morphologies can directly affect piezoelectric properties in PENGs. For example, PENGs based on ZnO nanowires showed better piezoelectricity compared with ZnO nanorods. Two distinct micro-morphologies of ZnO were utilized to produce PENG. The initial PENG was produced by growing ZnO nanorods arrays on silver-coated cotton fabric using an aqueous chemical growth technique at subzero temperatures. The second PENG was fabricated using ZnO nanowires and a 3D printing method, resulting in two different output voltages of 9.5 mV and 2.03 V, respectively [161, 187–189]. The PENGs based on 2D ZnO nanowires showed better output voltage compared with 1D ZnO nanowire under similar conditions [190]. Besides that, piezoelectric materials based on $BaTiO_3$ can be produced in the shape of different micro-morphologies such as nanowires, nanotubes and nanopillars, and this variety in the morphologies of $BaTiO_3$ based materials can lead to different piezoelectricity, efficiency of devices and materials [191]. The output voltage of PENGs based on $BaTiO_3$ nanowires is higher than ones with $BaTiO_3$ nanotubes. One of the most popular nanotubes that was utilized in PENGs is lead-free $BaTiO_3$ most of which was produced with hydrothermal method. Using a mixture of $BaTiO_3$ nanotubes and PDMS, researchers have been able to produce a highly flexible and transparent nanogenerator that has excellent properties with the current and output voltage of 350 nA and 5.5 V, respectively [192]. Moreover, the other micro-morphology of $BaTiO_3$ is nanopillar arrays which was produced through a sol–gel method and this kind of PENGs showed output voltage of 6 V and current of 0.8 µA [193]. PZT-based materials with different morphologies can be used in PENGs. But the most piezoelectric performance devoted to PZT is with nanorods arrays. The output voltage of thin films PENGs based on PZT nanorods arrays was 1.63 V [194, 195]. In general, the same materials with different micro-morphologies result in different piezoelectricity. The nanowire form of ZnO and PZT illustrates a high level of piezoelectricity compared with other micro-morphologies. Moreover, micro-morphology of the matrix of piezoelectric materials such as 1D, 2D, 3D bulks and spherical particles is the other influential factor for piezoelectric properties. These matrixes will be used in PENGs with high output voltages. The 1D and 2D materials are highly beneficial in increasing the strength of PENGs. Moreover, the 3D bulk particles and spherical particles can also offer outstanding output performances, which further enhance their overall efficiency [161].

9.3.9 Effect of device substrate

The types of substrates play a key role in performance of PENGs and by altering them the piezoelectric properties will be changed [196]. Nour *et al* [196] fabricated a flexible PENG based on nanowires of ZnO and PVDF polymer hybrid structure which was grown on a substrate of paper. For measuring the effect of different substrates on performance of PENG, a comparison was made between plastic and paper. The results showed that the output voltage of PENG with paper substrate was about 4.8 V which is much higher than the one with plastic substrate that reported 0.04 V. This research demonstrated the importance of substrate and porosity in improvement of piezoelectric properties of PENGs. On the other hand, utilizing PET as substrate in PENGs illustrated outstanding output voltage due to its perfect flexibility when used in equipment that is put under the weak mechanical force to convert mechanical to electrical energy. Besides, PDMS provided high piezoelectric performance because of its ability in the transmission process which can lose less electrical energy. Using PDMS substrate, not only can one protect the piezoelectric materials from air and water but one can also help prevent mechanical damage. This makes it an ideal choice for protecting and preserving these materials in varying ambient conditions. In general, PDMS and PET are the two most popular substrates that can be used effectively in flexible devices and enhance the efficiency of PENGs [161, 190, 197–199].

9.4 Wearable electrodes

Energy harvesters and nanogenerators are composed of at least two components. The initial piece is the nanoconverter, made from either piezoelectric, triboelectric, or pyroelectric materials. It transforms other types of energy into electrical energy and can include multiple components. The other one is the electrode. Electrodes in these devices are responsible for transferring the electricity from the converter layer to other parts, including a battery or an electrical consumer. This makes them a critical part of maintaining stable power delivery [14]. Electrodes used in energy harvesters, sensors, and nanogenerators are made of various materials, among the most widely used of which are carbon-based, metal-based, polymer-based, and ink-based electrodes. It should be noted that the electrodes can be multi-component like other components of these devices [200–206].

Wearable and flexible electronics have advanced a lot in recent years, and energy harvesters are no exception. To achieve flexible and wearable electronics with suitable mechanical properties, it is necessary that all the components of that device have desirable mechanical properties. As a result, to have a wearable energy harvester, flexible electrodes are needed. Various techniques are used to obtain flexible electrodes. One of these methods is the use of conductive inks that can be converted into flexible electrodes by being printed on the desired surface. Inks can mix other additives such as polymers or metals and print on surfaces under certain conditions with a special solvent. One of the advantages of this method is the possibility of mixing other materials such as metals and carbon with the printed ink, which can lead to the improvement of the electrical properties and conductivity of the electrode [14]. Rajala *et al* [207] used three specific printing inks to create flexible electrodes that had great performance. Two of these inks were based on graphene, while the other was formulated from CNTs. They compared the output of new electrodes with a copper electrode on a piezoelectric of PVDF as reference. Sensors made of three printed electrodes proved to be much more sensitive than copper electrodes. Compared to the two graphene samples, the electrical conductivity of the sample made with carbon nanotube-based ink was much superior. Bandodkar *et al* [208] created a stretchable, flexible and conductive electrode for their printed electrochemical sensor using an ink containing polyurethane and CNTs. This enabled them to produce a sensor with enhanced stretchability and performance. The printed layer had amazing endurance, being able to withstand strain of up to 500% with little impact on its stability.

Another method is to use advanced and finished textiles as electrodes. Since textiles inherently have flexibility and breathability properties, they can be a suitable option for wearable electrodes after performing conductive processes. To make textiles conductive and use them as electrodes in nanogenerators and energy harvesters and sensors, methods such as immersing [209], deep coating [210], *in situ* synthesis [211], sputtering [212] and printing [213] have been reported. Using these methods, it is possible to place materials such as graphene, polyaniline, silver, gold, and conductive ink on fabrics. For instance, Mao *et al* [212] developed an e-textile by coating it with a graphene oxide to sense physiological changes in the human body. It provides a convenient way to detect and monitor health conditions through its wearable design feature. By employing a sputtering method, gold is applied to the surface of this coated fabric and *in situ* ZnO nanorods are synthesized which will serve as an electrode. By using this fabric, human arterial pulse and respiration can be monitored. This fabric was evaluated for its output and was found to have an average output voltage of 0.2 V.

Polymers generally have unique mechanical properties that are hard to find in other materials. Polymers in the rubbery state can have high mechanical elasticity due to the slippage of the polymer chains. There are a group of polymers that, in addition to proper mechanical properties, have conductive properties, so they have become a suitable option for wearable electronics. These conductive polymers can be used as flexible and sometimes transparent electrodes in various electronic systems such as energy harvesters, sensors, and optoelectronics. The number of the conductive polymers is large, and each of them is unique in its properties; among the most widely used are polyacetylene, polyaniline, polypyrrole, polythiophene, and poly (3,4 ethylenedioxythiophene) (PEDOT) [214]. There are various methods to apply these conductive polymers in the desired interference in wearable electronic systems. One of these methods is *in situ* synthesis or polymerization, by which smart clothing can be obtained. For example, in order to obtain a wearable nanogenerator, Mule *et al* [215] performed the *in situ* synthesis process of polypyrrole polymer on cotton fabric so that they could use it as a nanogenerator electrode. One of the features that make conductive polymers very popular is their ability to be used with other conductive materials to improve their conductivity properties. Many studies have been conducted in the field of simultaneous use of non-polymer conductive materials such as silver, carbon, etc along with conductive polymers. In one of the studies, it has been reported that by adding polymer PEDOT; PSS to an electrode made of silver nanowires, up to 20% increase in the output voltage of the nanogenerator was recorded [216]. Production of multi-component composites with conductive polymers and conductive fillers such as silver nanowires, CNTs and metal nanoparticles are very common. Kim *et al* [217] used PEDOT polymer and silver nanowires to produce a conductive and multi-component electrode. Table 9.1 provides a brief summary of

Table 9.1. A summary of various types of electrodes, fabrication methods and applications.

Type of electrode	Materials	Fabrication method	Application	Output	References
Printed electrode	Solvent: poly (gelycole sebacate) additive: carbon nanotubes	Printing	TENG	5.93 $W\,m^{-2}$	[249]
Printed electrode	Solvent: dimethylformamide (DMF)/ tetrahydrofuran (THF)/thermoplastic polyurethan (TPU) additive: silver flakes	Printing	Sensor	—	[250]
Printed electrode	Solvent: DMF/methyl ethyl ketone (MEK) additive: BaTiO$_3$/PVDF	Printing	PENG	0.014 $W\,m^{-2}$	[251]
Printed electrode	Solvent: polyurethane (PU) additive: CNTs	Printing	Sensor	—	[208]
Printed electrode	Solvent: DMF/TPU additive: carbon black/ nano clay	Printing	Sensor	—	[252]
Textile electrode	Cotton-PA6–spandex fabric /copper and silver	In situ synthesized	Energy harvester	40 V	[253]
Textile electrode	polyester fabric/nickel chrome-plated	Commercial fabric	TENG	0.25 Ω	[254]
Textile electrode	Cotton fabric/polyaniline	Dip coating	TENG	350 V	[210]
Textile electrode	PA6 fabric/graphene	Immersing	TENG	213.75 V	[209]
Textile electrode	Chitosan fabric/silver	In situ synthesized	Smart garment	0.0332 Ω/sq	[211]
Textile electrode	Cotton fabric/polypyrrole	In situ polymerization	TENG	~180–200 V	[215]
Polymer-based electrode	PEDOT; PSS	Spraying	Generator	1.2 V	[255]
Polymer-based electrode	Silver nanowires/PEDOT: PSS	Compositing	Nanogenerator	750 mV	[256]
Polymer-based electrode	Silver nanowire/PEDOT: PSS	Compositing	PENG	2.68 V	[216]

some examples of various electrodes with their fabrication techniques, applications, and performance.

Piezoelectric devices rely heavily on electrodes to function properly. Without them, the device cannot operate effectively. Stretchable piezoelectric devices require electrodes that have superior electrical conductivity, flexibility, and the ability to maintain constant resistance under dynamic and static tensile strain. All these factors are essential for their successful application in practical scenarios. Stretchable electrodes are designed using two main approaches: incorporating stretchable materials and structural designs. Unlike conductive materials such as metals, traditional ones are rigid and fragile. Stretchable structures are therefore frequently utilized to increase their stretchability and make them more versatile [17]. A few popular structures that often come into play are serpentine [218], wavy [219, 220], coiled [221], kirigami [222–224]. Modifying the electrode design helps maintain the electrical conductivity when being stretched in this approach. Yet, the complexity of techniques like photolithography hinders their use in electronic devices as it is needed for patterning [225]. An approach that has seen some success is to blend conductive nanomaterials (carbon or metal nanomaterials), conducting polymers and liquid metals with flexible, stretchable materials like rubber or thermoplastic polyurethane [17].

Despite the remarkable progress in the field of production and design of flexible and stretchable electrodes, which are known for their use in wearable electronics, several problems and challenges remain to be addressed before they can be effectively used in practical applications. There is still a need for further research to explore methods of improving the reliability, durability, and productivity of these electrodes so that they can be implemented more widely. One of the major issues that needs to be addressed when designing a wearable electrode is its lower electrical conductivity when compared to rigid and conventional electrodes such as metal plates. This can severely limit the performance of the device, and thus it becomes incredibly important to optimize the electrical conductivity of these electrodes in order to make them more efficient and reliable. In addition to the mechanical instability caused by the structural components, another major issue leading to functional instability of these structures is significant changes in electrical conductivity under deformation. These fluctuations occur due to the strain experienced by the materials during their deformation, causing them to lose their initial electrical properties and thus leading to unwanted electric activity. The other major challenge that arises from the development of stretchable electronics is the difficulty in effectively combining the stretchable electrodes with the active polymer nanofiber sensing layer. This requires a precise and meticulous integration process, as any misalignment between the two components could lead to significant performance losses or even complete functionality failure. Furthermore, it can be quite tricky to achieve a seamless combination due to their dissimilar physical properties [226–229].

9.5 Recent progress and development in wearable energy harvesters

Piezoelectric nanofibers have the ability to gather the biological energy resulting from human body movements, such as walking [228, 230, 231], breathing [232, 233], and heartbeat [234–236]. Moreover, other mechanical energy types such as airflow and vibration can be extracted through these nanofibers [162]. PENGs have been utilized for detecting water velocity and pH [237], monitoring deformation [238], and tracking human motion in self-powered systems [239–241]. Shi *et al* [242] utilized a flexible (PENG) for energy harvesting. The PENG was made of electrospun fiber mats that contained 15 wt.% $BaTiO_3$ nanoparticles, 0.15 wt.% graphene nanosheets, and PVDF. The PENG generated a peak voltage of 112 V when the finger was pressed and released at fast strain rates. It was also able to generate energy from human movements, such as finger tapping, wrist flexing, and foot stepping. When tested under wrist flexing and finger tapping conditions, the output voltage reached 7.7 and 7.5 V, respectively. Additionally, when the PENG was located under the foot heel and toe, it produced a maximum voltage of 7.8 and 2.8 V, respectively. The heel produced a higher output voltage because it applied more pressure compared to the toe. Choudhry *et al* [243] conducted research on producing electrical energy through bodily movements, specifically running. They utilized two distinct composites: one with impressive piezoelectric characteristics and the other with silicon and MWCNTs, serving as an electrode. In their investigation, they also experimented with a PZT–silicon–graphene composite structure for an energy picker located in a shoe's shaft and managed to achieve an output voltage of 27 V. A very adaptable deformation PENG was developed using a very thin sheet of aluminum foil. This nanogenerator was designed to detect wrinkles on the human face and track eye movement. The PENG used ZnO NWs and was positioned close to an eye on human skin. When the eye blinked, the PENG adapted to the dynamic wrinkles, deforming as it was highly flexible and conformable. The output voltage and current measured approximately 0.2 V and 2 nA during eye blinking motion.

One of the application areas of nanogenerators is in applications related to vehicles. Various energies are created and wasted in the process of moving a vehicle, so this can be a suitable source to be used as input energy for a nanogenerator. In a study by Ma *et al* [244], a flexible nanogenerator was developed that has adaptable shape and can fit the curvature of a bicycle tire. The nanogenerator was created by combining ferroelectricity and piezoelectricity, adding PZT and salt to PDMS, and polarization and electrode preparation. The nanogenerator had small dimensions and generated an open circuit voltage of 29 V and short-circuit current of 116 nA. This energy harvesting method is similar to that of automobile tires. Bicycles equipped with nanogenerators have great potential for monitoring their own status and transmitting signals, which could lead to intelligent development. In a study by Dudem *et al* [245], they developed a PENG by combining $BaTiO_3$ micro stones and Ag nanowires in a PVDF matrix. This PENG was able to effectively harvest energy from various types of vehicular motion, such as bicycles, motorcycles, and cars. The impact force generated by the motion of these vehicles was efficiently converted into electricity by this PENG.

In addition, nanogenerators have a great potential to utilize wasted low energy as an input to generate electrical energy. For example, the energy of sound vibration is usually assumed as wasted energy and mostly does not get attention, while some nanogenerators can use it. Park *et al* [246] created an energy collection tool that utilized graphene as a clear and flexible electrode. They tested the tool and found that sound vibrations were primarily responsible for generating energy. When exposed to acoustic vibrations of 105 dB, the device produced an output voltage of approximately 7.6 V. Harvested energy is a promising renewable energy source that can replace traditional batteries. For instance, He *et al* [247] introduced a wearable self-powered electrochromic supercapacitor that is driven by piezoelectricity. However, since piezoelectric materials have low power output, rechargeable batteries or supercapacitors are necessary to store the harvested energy. To increase energy accumulation, storage cells can be connected in series to expand the voltage range. The voltage of the storage device is crucial for efficient energy harvesting, and a rectification circuit is required to connect the piezoelectric device to the storage cell. A diode bridge rectifier is commonly used as an AC–DC rectifier because the current source is alternating [248]. Ji *et al* [182] developed a piezo-triboelectric device using BNT-ST/PVDF-TrFE nanofibers module that includes piezoelectric and triboelectric layers. This device can charge a 10 µF capacitor up to 6.5 V in 100 s.

9.6 Conclusion

Wearable electronics based on piezoelectric and triboelectric materials has gained significant traction in recent years, particularly for energy harvesting systems and sensors due to their flexibility and stretchability. Among the various types of wearable electronics, energy harvesting systems, and sensors have improved day by day due to their wide applications. Energy harvesting systems can convert the mechanical energy of the surrounding environment, which is usually considered wasted energy, into electrical energy. One of the most important components of any wearable electronics is the electrode, which needs to be compatible with the mechanical properties of the wearable product in terms of elasticity. In order to achieve flexible and wearable electronics with suitable mechanical properties, it is necessary that all the components of that device have desirable mechanical properties. As a result, to have a wearable energy harvester, flexible electrodes are needed.

The electrospinning method offers a revolutionary approach for producing wearable electronics by enabling the creation of nano-sized fibers from piezoelectric and triboelectric materials. Various factors influence the final product. By carefully controlling and optimizing these parameters, researchers can achieve piezoelectric or triboelectric layers with desired properties. The controllability of these effective parameters has led to the increasing popularity of this method. Electrospun layers with piezoelectric properties have found many applications in the fields related to body movement sensors, vehicle applications, storage, etc. It should be noted that in order to improve the output of production devices with this method, more optimizations are needed. There are various parameters which can be controlled or changed to enhance the output, some of which include piezoelectric materials, structural design of commonly used piezoelectric materials, changing the microscopic morphology of existing piezoelectric, selecting an appropriate substrate. The exceptional characteristics and potential for integration with other materials and technologies make fibrous wearable energy harvesting devices a promising component in the future of wearable technology.

Declaration of conflicting interests

The authors declare that there is no conflict of interest regarding the publication of this article.

Data availability statement

The data that support the findings of this study are available on request from the corresponding authors.

References

[1] Huang L, Lin S, Xu Z, Zhou H, Duan J, Hu B and Zhou J 2020 Fiber-based energy conversion devices for human-body energy harvesting *Adv. Mater.* **32** 1902034

[2] Mahmud M P, Bazaz S R, Dabiri S, Mehrizi A A, Asadnia M, Warkiani M E and Wang Z L 2022 Advances in mems and microfluidics-based energy harvesting technologies *Adv. Mater. Technol.* **7** 2101347

[3] Mohammadnia A, Rezania A, Ziapour B M, Sedaghati F and Rosendahl L 2020 Hybrid energy harvesting system to maximize power generation from solar energy *Energy Convers. Manage.* **205** 112352

[4] Ryu H and Kim S W 2021 Emerging pyroelectric nanogenerators to convert thermal energy into electrical energy *Small* **17** 1903469

[5] Mahapatra B, Patel K K and Patel P K 2021 A review on recent advancement in materials for piezoelectric/triboelectric nanogenerators *Mater. Today Proc.* **46** 5523–9

[6] Wu N, Bao B and Wang Q 2021 Review on engineering structural designs for efficient piezoelectric energy harvesting to obtain high power output *Eng. Struct.* **235** 112068

[7] Wu M, Yao K, Li D, Huang X, Liu Y, Wang L, Song E, Yu J and Yu X 2021 Self-powered skin electronics for energy harvesting and healthcare monitoring *Mater. Today Energy* **21** 100786

[8] Wang Z L and Song J 2006 Piezoelectric nanogenerators based on zinc oxide nanowire arrays *Science* **312** 242–6

[9] Fan F-R, Tian Z-Q and Wang Z L 2012 Flexible triboelectric generator *Nano Energy* **1** 328–34

[10] Zhang J, He Y, Boyer C, Kalantar-Zadeh K, Peng S, Chu D and Wang C H 2021 Recent developments of hybrid piezo–triboelectric nanogenerators for flexible sensors and energy harvesters *Nanoscale Adv.* **3** 5465–86

[11] Peng W and Du S 2023 The advances in conversion techniques in triboelectric energy harvesting: a review *IEEE Trans. Circuits Syst.* I **70** 3049–62

[12] Jiao P 2021 Emerging artificial intelligence in piezoelectric and triboelectric nanogenerators *Nano Energy* **88** 106227

[13] Ji G and Huber J 2022 Recent progress in acoustic metamaterials and active piezoelectric acoustic metamaterials-a review *Appl. Mater. Today* **26** 101260

[14] Bagherzadeh R, Abrishami S, Shirali A and Rajabzadeh A R 2022 Wearable and flexible electrodes in nanogenerators for energy harvesting, tactile sensors, and electronic textiles: Novel materials, recent advances, and future perspectives *Mater. Today Sustain.* **20** 100233

[15] Panda P and Sahoo B 2015 PZT to lead free piezo ceramics: a review *Ferroelectrics* **474** 128–43

[16] Majumder S, Sagor M M H and Arafat M T 2022 Functional electrospun polymeric materials for bioelectronic devices: a review *Mater. Adv.* **3** 6753–72

[17] Mirjalali S, Mahdavi Varposhti A, Abrishami S, Bagherzadeh R, Asadnia M, Huang S, Peng S, Wang C H and Wu S 2023 A review on wearable electrospun polymeric piezoelectric sensors and energy harvesters *Macromol. Mater. Eng.* **308** 2200442

[18] Jing Q and Kar-Narayan S 2018 Nanostructured polymer-based piezoelectric and triboelectric materials and devices for energy harvesting applications *J. Phys. D: Appl. Phys.* **51** 303001

[19] Costa P, Nunes-Pereira J, Pereira N, Castro N, Gonçalves S and Lanceros-Mendez S 2019 Recent progress on piezoelectric, pyroelectric, and magnetoelectric polymer-based energy-harvesting devices *Energy Technol.* **7** 1800852

[20] Rodrigues-Marinho T, Perinka N, Costa P and Lanceros-Mendez S 2023 Printable lightweight polymer-based energy harvesting systems: materials, processes, and applications *Mater. Today Sustain.* **21** 100292

[21] Mohammadpourfazeli S, Arash S, Ansari A, Yang S, Mallick K and Bagherzadeh R 2023 Future prospects and recent developments of polyvinylidene fluoride (PVDF) piezoelectric polymer; fabrication methods, structure, and electro-mechanical properties *RSC Adv.* **13** 370–87

[22] Shin J, Jeong B, Kim J, Nam V B, Yoon Y, Jung J, Hong S, Lee H, Eom H and Yeo J 2020 Wearable Temperature sensors: sensitive wearable temperature sensor with seamless monolithic integration *Adv. Mater.* **32** 2070014

[23] Aliqué M, Simão C D, Murillo G and Moya A 2021 Fully-printed piezoelectric devices for flexible electronics applications *Adv. Mater. Technol.* **6** 2001020

[24] Megdich A, Habibi M and Laperrière L 2023 A Review on 3D printed piezoelectric energy harvesters: materials, 3D printing techniques, and applications *Mater. Today Commun.* **35** 105541

[25] Yan M, Xiao Z, Ye J, Yuan X, Li Z, Bowen C, Zhang Y and Zhang D 2021 Porous ferroelectric materials for energy technologies: current status and future perspectives *Energy Environ. Sci.* **14** 6158–90

[26] Zeng Y, Jiang L, He Q, Wodnicki R, Yang Y, Chen Y and Zhou Q 2021 Recent progress in 3D printing piezoelectric materials for biomedical applications *J. Phys. D: Appl. Phys.* **55** 013002

[27] Rashid A, Zubair U, Ashraf M, Javid A, Abid H A and Akram S 2023 Flexible piezoelectric coatings on textiles for energy harvesting and autonomous sensing applications: a review *J. Coat. Technol. Res.* **20** 141–72

[28] Kalimuldina G, Turdakyn N, Abay I, Medeubayev A, Nurpeissova A, Adair D and Bakenov Z 2020 A review of piezoelectric PVDF film by electrospinning and its applications *Sensors* **20** 5214

[29] Zhao B, Chen Z, Cheng Z, Wang S, Yu T, Yang W and Li Y 2022 Piezoelectric nanogenerators based on electrospun PVDF-coated mats composed of multilayer polymer-coated $BaTiO_3$ nanowires *ACS Appl. Nano Mater.* **5** 8417–28

[30] Rasoolzadeh M, Sherafat Z, Vahedi M and Bagherzadeh E 2022 Structure dependent piezoelectricity in electrospun PVDF-SiC nanoenergy harvesters *J. Alloys Compd* **917** 165505

[31] Azimi B, Milazzo M, Lazzeri A, Berrettini S, Uddin M J, Qin Z, Buehler M J and Danti S 2020 Electrospinning piezoelectric fibers for biocompatible devices *Adv. Healthcare Mater.* **9** 1901287

[32] Dolez P I 2021 Energy harvesting materials and structures for smart textile applications: Recent progress and path forward *Sensors* **21** 6297

[33] Maestri G, Ferreira L B, Bachmann P, Paim A A, Merlini C and Steffens F 2023 Recent advances in piezoelectric textile materials: a brief literature review *J. Eng. Fibers Fabrics* **18** 15589250231151242

[34] He T, Shi Q, Wang H, Wen F, Chen T, Ouyang J and Lee C 2019 Beyond energy harvesting-multi-functional triboelectric nanosensors on a textile *Nano Energy* **57** 338–52

[35] Abrishami S, Shirali A, Razbin M, Ziaee S, Xiaolan Q and Bagherzadeh R 2023 Nickel/ silver *in situ* modification of wearable and stretchable electrodes for textile-based piezo-electric applications *J. Text. Inst.* **115** 2363–76

[36] Vahdani M, Mirjalali S, Razbin M, Moshizi S A, Payne D, Kim J, Huang S, Asadnia M, Peng S and Wu S 2024 Bio-disintegrable elastic polymers for stretchable piezoresistive strain sensors *Adv. Sustain. Syst.* **8** 2300482

[37] Cheng Y, Wang R, Sun J and Gao L 2015 A stretchable and highly sensitive graphene-based fiber for sensing tensile strain, bending, and torsion *Adv. Mater.* **27** 7365–71

[38] Dineva P, Gross D, Müller R, Rangelov T, Dineva P, Gross D, Müller R and Rangelov T 2014 *Piezoelectric Materials* (Berlin: Springer)

[39] Parali L and Sari A 2017 Vibration modelling of piezoelectric actuator (PEA) using Simulink software *2017 4th Int. Conf. on Electrical and Electronic Engineering (ICEEE)* (Piscataway, NJ: IEEE) pp 153–7

[40] Mishra S, Unnikrishnan L, Nayak S K and Mohanty S 2019 Advances in piezoelectric polymer composites for energy harvesting applications: a systematic review *Macromol. Mater. Eng.* **304** 1800463

[41] Van den Ende D, Bory B, Groen W and Van der Zwaag S 2010 Improving the d_{33} and g_{33} properties of 0–3 piezoelectric composites by dielectrophoresis *J. Appl. Phys.* **107** 024107

[42] Yan Y, Geng L D, Zhu L F, Leng H, Li X, Liu H, Lin D, Wang K, Wang Y U and Priya S 2022 Ultrahigh piezoelectric performance through synergistic compositional and micro-structural engineering *Adv. Sci.* **9** 2105715

[43] Guo Q, Cao G and Shen I 2013 Measurements of piezoelectric coefficient d_{33} of lead zirconate titanate thin films using a mini force hammer *J. Vib. Acoust.* **135** 011003

[44] Sun H, Zarkadoula E, Crespillo M L, Weber W J, Rathod V, Zinkle S J and Ramuhalli P 2022 Laser Doppler vibrometry for piezoelectric coefficient (d_{33}) measurements in irradi-ated aluminum nitride *Sens. Actuators,* A **347** 113886

[45] Pinheiro E and Deivarajan T 2019 A concise review encircling lead free porous piezoelectric ceramics *Acta Phys. Pol.* A **136** 555–65

[46] Mahapatra S D, Mohapatra P C, Aria A I, Christie G, Mishra Y K, Hofmann S and Thakur V K 2021 Piezoelectric materials for energy harvesting and sensing applications: Roadmap for future smart materials *Adv. Sci.* **8** 2100864

[47] Surmenev R A, Chernozem R V, Pariy I O and Surmeneva M A 2021 A review on piezo- and pyroelectric responses of flexible nano-and micropatterned polymer surfaces for biomedical sensing and energy harvesting applications *Nano Energy* **79** 105442

[48] Klimiec E, Zaraska K, Zaraska W and Kuczyński S 2012 Micropower generators and sensors based on piezoelectric polypropylene PP and polyvinylidene fluoride PVDF films-energy harvesting from walking *Appl. Mech. Mater.* **110** 1245–51

[49] Krishnaswamy J A, Buroni F C, Melnik R, Rodriguez-Tembleque L and Saez A 2020 Design of polymeric auxetic matrices for improved mechanical coupling in lead-free piezocomposites *Smart Mater. Struct.* **29** 054002

[50] Khazaee M, Rezaniakolaie A and Rosendahl L 2020 A broadband macro-fiber-composite piezoelectric energy harvester for higher energy conversion from practical wideband vibrations *Nano Energy* **76** 104978

[51] Liu Y, Khanbareh H, Halim M A, Feeney A, Zhang X, Heidari H and Ghannam R 2021 Piezoelectric energy harvesting for self-powered wearable upper limb applications *Nano Select* **2** 1459–79

[52] Lu L, Ding W, Liu J and Yang B 2020 Flexible PVDF based piezoelectric nanogenerators *Nano Energy* **78** 105251

[53] Deng W, Zhou Y, Libanori A, Chen G, Yang W and Chen J 2022 Piezoelectric nanogenerators for personalized healthcare *Chem. Soc. Rev.* **51** 3380–435

[54] Abolhasani M M, Naebe M, Hassanpour Amiri M, Shirvanimoghaddam K, Anwar S, Michels J J and Asadi K 2020 Hierarchically structured porous piezoelectric polymer nanofibers for energy harvesting *Adv. Sci.* **7** 2000517

[55] Sun Y, Chen J, Li X, Lu Y, Zhang S and Cheng Z 2019 Flexible piezoelectric energy harvester/sensor with high voltage output over wide temperature range *Nano Energy* **61** 337–45

[56] Shepelin N A, Glushenkov A M, Lussini V C, Fox P J, Dicinoski G W, Shapter J G and Ellis A V 2019 New developments in composites, copolymer technologies and processing techniques for flexible fluoropolymer piezoelectric generators for efficient energy harvesting *Energy Environ. Sci.* **12** 1143–76

[57] Khan H, Mahmood N, Zavabeti A, Elbourne A, Rahman M A, Zhang B Y, Krishnamurthi V, Atkin P, Ghasemian M B and Yang J 2020 Liquid metal-based synthesis of high performance monolayer SnS piezoelectric nanogenerators *Nat. Commun.* **11** 3449

[58] He M, Wang S, Zhong X and Guan M 2019 Study of a piezoelectric energy harvesting floor structure with force amplification mechanism *Energies* **12** 3516

[59] Zhang Y-H, Lee C-H and Zhang X-R 2019 A novel piezoelectric power generator integrated with a compliant energy storage mechanism *J. Phys. D: Appl. Phys.* **52** 455501

[60] Cho J Y, Kim K-B, Hwang W S, Yang C H, Ahn J H, Do Hong S, Jeon D H, Song G J, Ryu C H and Woo S B 2019 A multifunctional road-compatible piezoelectric energy harvester for autonomous driver-assist LED indicators with a self-monitoring system *Appl. Energy* **242** 294–301

[61] Duan S, Wu J, Xia J and Lei W 2020 Innovation strategy selection facilitates high-performance flexible piezoelectric sensors *Sensors* **20** 2820

[62] Yang T, Pan H, Tian G, Zhang B, Xiong D, Gao Y, Yan C, Chu X, Chen N and Zhong S 2020 Hierarchically structured PVDF/ZnO core–shell nanofibers for self-powered physiological monitoring electronics *Nano Energy* **72** 104706

[63] Yang Y, Pan H, Xie G, Jiang Y, Chen C, Su Y, Wang Y and Tai H 2020 Flexible piezoelectric pressure sensor based on polydopamine-modified BaTiO$_3$/PVDF composite film for human motion monitoring *Sens. Actuators* A **301** 111789

[64] Kim J, Campbell A S, de Ávila B E-F and Wang J 2019 Wearable biosensors for healthcare monitoring *Nat. Biotechnol.* **37** 389–406

[65] Rodrigues D, Barbosa A I, Rebelo R, Kwon I K, Reis R L and Correlo V M 2020 Skin-integrated wearable systems and implantable biosensors: a comprehensive review *Biosensors* **10** 79

[66] Zang Y, Zhang F, Di C-A and Zhu D 2015 Advances of flexible pressure sensors toward artificial intelligence and health care applications *Mater. Horizons* **2** 140–56

[67] Kishore R A 2021 Harvesting thermal energy with ferroelectric materials *Ferroelectric Materials for Energy Harvesting and Storage* (Amsterdam: Elsevier) pp 85–106

[68] Zou Y, Raveendran V and Chen J 2020 Wearable triboelectric nanogenerators for biomechanical energy harvesting *Nano Energy* **77** 105303

[69] Liu Z, Zheng Q, Shi Y, Xu L, Zou Y, Jiang D, Shi B, Qu X, Li H and Ouyang H 2020 Flexible and stretchable dual mode nanogenerator for rehabilitation monitoring and information interaction *J. Mater. Chem.* B **8** 3647–54

[70] Roy K, Ghosh S K, Sultana A, Garain S, Xie M, Bowen C R, Henkel K, Schmeiβer D and Mandal D 2019 A self-powered wearable pressure sensor and pyroelectric breathing sensor based on GO interfaced PVDF nanofibers *ACS Appl. Nano Mater.* **2** 2013–25

[71] Pan S and Zhang Z 2019 Fundamental theories and basic principles of triboelectric effect: a review *Friction* **7** 2–17

[72] Luo J, Xu L, Tang W, Jiang T, Fan F R, Pang Y, Chen L, Zhang Y and Wang Z L 2018 Direct-current triboelectric nanogenerator realized by air breakdown induced ionized air channel *Adv. Energy Mater.* **8** 1800889

[73] Liu J, Goswami A, Jiang K, Khan F, Kim S, McGee R, Li Z, Hu Z, Lee J and Thundat T 2018 Direct-current triboelectricity generation by a sliding Schottky nanocontact on MoS2 multilayers *Nat. Nanotechnol.* **13** 112–6

[74] Kim M P, Ahn C W, Lee Y, Kim K, Park J and Ko H 2021 Interfacial polarization-induced high-k polymer dielectric film for high-performance triboelectric devices *Nano Energy* **82** 105697

[75] Zou H, Zhang Y, Guo L, Wang P, He X, Dai G, Zheng H, Chen C, Wang A C and Xu C 2019 Quantifying the triboelectric series *Nat. Commun.* **10** 1427

[76] Im J-S and Park I-K 2018 Mechanically robust magnetic Fe3O4 nanoparticle/polyvinylidene fluoride composite nanofiber and its application in a triboelectric nanogenerator *ACS Appl. Mater. Interfaces* **10** 25660–5

[77] Kang D, Lee H Y, Hwang J-H, Jeon S, Kim D, Kim S and Kim S-W 2022 Deformation-contributed negative triboelectric property of polytetrafluoroethylene: a density functional theory calculation *Nano Energy* **100** 107531

[78] Liu Y, Sun N, Liu J, Wen Z, Sun X, Lee S-T and Sun B 2018 Integrating a silicon solar cell with a triboelectric nanogenerator via a mutual electrode for harvesting energy from sunlight and raindrops *ACS Nano* **12** 2893–9

[79] Zhang R and Olin H 2020 Material choices for triboelectric nanogenerators: a critical review *EcoMat* **2** e12062

[80] Zhang R, Dahlström C, Zou H, Jonzon J, Hummelgård M, Örtegren J, Blomquist N, Yang Y, Andersson H and Olsen M 2020 Cellulose-based fully green triboelectric nanogenerators with output power density of 300 W m^{-2} *Adv. Mater.* **32** 2002824

[81] Bao Y, Wang R, Lu Y and Wu W 2017 Lignin biopolymer based triboelectric nano-generators *APL Mater.* **5** 074109

[82] Kim J-N, Lee J, Go T W, Rajabi-Abhari A, Mahato M, Park J Y, Lee H and Oh I-K 2020 Skin-attachable and biofriendly chitosan-diatom triboelectric nanogenerator *Nano Energy* **75** 104904

[83] An S, Sankaran A and Yarin A L 2018 Natural biopolymer-based triboelectric nano-generators via fast, facile, scalable solution blowing *ACS Appl. Mater. Interfaces* **10** 37749–59

[84] Han Y, Han Y, Zhang X, Li L, Zhang C, Liu J, Lu G, Yu H-D and Huang W 2020 Fish gelatin based triboelectric nanogenerator for harvesting biomechanical energy and self-powered sensing of human physiological signals *ACS Appl. Mater. Interfaces* **12** 16442–50

[85] Wu C, Wang A C, Ding W, Guo H and Wang Z L 2019 Triboelectric nanogenerator: a foundation of the energy for the new era *Adv. Energy Mater.* **9** 1802906

[86] Li X, Jiang C, Ying Y and Ping J 2020 Biotriboelectric nanogenerators: materials, structures, and applications *Adv. Energy Mater.* **10** 2002001

[87] Khandelwal G, Raj N P M J and Kim S-J 2020 Triboelectric nanogenerator for healthcare and biomedical applications *Nano Today* **33** 100882

[88] Zheng Q, Shi B, Fan F, Wang X, Yan L, Yuan W, Wang S, Liu H, Li Z and Wang Z L 2014 *In vivo* powering of pacemaker by breathing-driven implanted triboelectric nanogenerator *Adv. Mater.* **26** 5851–6

[89] Chen X, Xie X, Liu Y, Zhao C, Wen M and Wen Z 2020 Advances in healthcare electronics enabled by triboelectric nanogenerators *Adv. Funct. Mater.* **30** 2004673

[90] Zhao Z, Yan C, Liu Z, Fu X, Peng L M, Hu Y and Zheng Z 2016 Machine-washable textile triboelectric nanogenerators for effective human respiratory monitoring through loom weaving of metallic yarns *Adv. Mater.* **28** 10267–74

[91] Zhang R, Hummelgård M, Örtegren J, Olsen M, Andersson H, Yang Y, Zheng H and Olin H 2021 The triboelectricity of the human body *Nano Energy* **86** 106041

[92] Covaci C and Gontean A 2020 Piezoelectric energy harvesting solutions: a review *Sensors* **20** 3512

[93] Zhong J, Zhang Y, Zhong Q, Hu Q, Hu B, Wang Z L and Zhou J 2014 Fiber-based generator for wearable electronics and mobile medication *ACS Nano* **8** 6273–80

[94] Yi F, Wang X, Niu S, Li S, Yin Y, Dai K, Zhang G, Lin L, Wen Z and Guo H 2016 A highly shape-adaptive, stretchable design based on conductive liquid for energy harvesting and self-powered biomechanical monitoring *Sci. Adv.* **2** e1501624

[95] Wang J, Li S, Yi F, Zi Y, Lin J, Wang X, Xu Y and Wang Z L 2016 Sustainably powering wearable electronics solely by biomechanical energy *Nat. Commun.* **7** 12744

[96] Lin M-F, Xiong J, Wang J, Parida K and Lee P S 2018 core–shell nanofiber mats for tactile pressure sensor and nanogenerator applications *Nano Energy* **44** 248–55

[97] Cui X, Zhang C, Liu W, Zhang Y, Zhang J, Li X, Geng L and Wang X 2018 Pulse sensor based on single-electrode triboelectric nanogenerator *Sens. Actuators* A **280** 326–31

[98] Cao R, Wang J, Zhao S, Yang W, Yuan Z, Yin Y, Du X, Li N-W, Zhang X and Li X 2018 Self-powered nanofiber-based screen-print triboelectric sensors for respiratory monitoring *Nano Res.* **11** 3771–9

[99] Korkmaz S and Kariper İ A 2021 Pyroelectric nanogenerators (PyNGs) in converting thermal energy into electrical energy: Fundamentals and current status *Nano Energy* **84** 105888

[100] Zhang K, Wang Y, Wang Z L and Yang Y 2019 Standard and figure-of-merit for quantifying the performance of pyroelectric nanogenerators *Nano Energy* **55** 534–40

[101] Mondal R, Hasan M A M, Baik J M and Yang Y 2023 Advanced pyroelectric materials for energy harvesting and sensing applications *Mater. Today* **66** 273–301

[102] Thakre A, Kumar A, Song H-C, Jeong D-Y and Ryu J 2019 Pyroelectric energy conversion and its applications—flexible energy harvesters and sensors *Sensors* **19** 2170

[103] Zhang Y, Hopkins M A, Liptrot D J, Khanbareh H, Groen P, Zhou X, Zhang D, Bao Y, Zhou K and Bowen C R 2020 Harnessing plasticity in an amine-borane as a piezoelectric and pyroelectric flexible film *Angew. Chem. Int. Ed.* **59** 7808–12

[104] Cook F, Lord R, Sitbon G, Stephens A, Rust A and Schwarzacher W 2020 A pyroelectric thermal sensor for automated ice nucleation detection *Atmos. Meas. Tech.* **13** 2785–95

[105] Navid A, Vanderpool D, Bah A and Pilon L 2010 Towards optimization of a pyroelectric energy converter for harvesting waste heat *Int. J. Heat Mass Transfer* **53** 4060–70

[106] Luo L, Jiang X, Zhang Y and Li K 2017 Electrocaloric effect and pyroelectric energy harvesting of $(0.94 - x)$ $Na_{0.5}Bi_{0.5}TiO_3$-$0.06BaTiO_3$-$xSrTiO_3$ ceramics *J. Eur. Ceram. Soc.* **37** 2803–12

[107] Yu C, Park J, Youn J R and Song Y S 2022 Integration of form-stable phase change material into pyroelectric energy harvesting system *Appl. Energy* **307** 118212

[108] Sebald G, Lefeuvre E and Guyomar D 2008 Pyroelectric energy conversion: optimization principles *IEEE Trans. Ultrason. Ferroelectr. Freq. Control* **55** 538–51

[109] Bowen C R, Taylor J, LeBoulbar E, Zabek D, Chauhan A and Vaish R 2014 Pyroelectric materials and devices for energy harvesting applications *Energy Environ. Sci.* **7** 3836–56

[110] You M-H, Wang X-X, Yan X, Zhang J, Song W-Z, Yu M, Fan Z-Y, Ramakrishna S and Long Y-Z 2018 A self-powered flexible hybrid piezoelectric–pyroelectric nanogenerator based on non-woven nanofiber membranes *J. Mater. Chem.* A **6** 3500–9

[111] Zhang H, Xie Y, Li X, Huang Z, Zhang S, Su Y, Wu B, He L, Yang W and Lin Y 2016 Flexible pyroelectric generators for scavenging ambient thermal energy and as self-powered thermosensors *Energy* **101** 202–10

[112] Ko Y J, Kim D Y, Won S S, Ahn C W, Kim I W, Kingon A I, Kim S-H, Ko J-H and Jung J H 2016 Flexible Pb $(Zr_{0.52}Ti_{0.48})O_3$ films for a hybrid piezoelectric-pyroelectric nanogenerator under harsh environments *ACS Appl. Mater. Interfaces* **8** 6504–11

[113] Bandaru S H, Becerra V, Khanna S, Radulovic J, Hutchinson D and Khusainov R 2021 A review of photovoltaic thermal (PVT) technology for residential applications: performance indicators, progress, and opportunities *Energies* **14** 3853

[114] Green M A 2019 Photovoltaic technology and visions for the future *Prog. Energy* **1** 013001

[115] Yao H, Wang J, Xu Y, Zhang S and Hou J 2020 Recent progress in chlorinated organic photovoltaic materials *Acc. Chem. Res.* **53** 822–32

[116] Hettiarachchi M, Dissanayake M, Senadeera G and Umair K 2021 Optimization of photovoltaic performance of electrospun PVdF-HFP nanofiber membrane based dye sensitized solar cells with membrane thickness *5th Int. Research Conf. of Uva Wellassa University, IRCUWU2021* Paper ID: IRCUWU2021-311

[117] Zhang N and Xiang D 2021 Self-assembling of versatile $Si_3N_4@SiO_2$ nanofibre sponges by direct nitridation of photovoltaic silicon waste *J. Hazard. Mater.* **419** 126385

[118] Biswas A, Park H and Sigmund W M 2012 Flexible ceramic nanofibermat electrospun from TiO_2-SiO_2 aqueous sol *Ceram. Int.* **38** 883–6

[119] Esfahani H, Prabhakaran M P, Salahi E, Tayebifard A, Rahimipour M R, Keyanpour-Rad M and Ramakrishna S 2016 Electrospun nylon 6/zinc doped hydroxyapatite membrane for protein separation: mechanism of fouling and blocking model *Mater. Sci. Eng.* C **59** 420–8

[120] Du P, Song L, Xiong J, Li N, Wang L, Xi Z, Wang N, Gao L and Zhu H 2013 Dye-sensitized solar cells based on anatase TiO_2/multi-walled carbon nanotubes composite nanofibers photoanode *Electrochim. Acta* **87** 651–6

[121] Murashov V and Howard J 2011 *Nanotechnology Standards* (Springer Science & Business Media)

[122] Norrrahim M N F, Kasim N A M, Knight V F, Halim N A, Shah N A A, Noor S A M, Jamal S H, Ong K K, Yunus W M Z W and Farid M A A 2021 Performance evaluation of cellulose nanofiber reinforced polymer composites *Funct. Compos. Struct.* **3** 024001

[123] Zaarour B and Alhinnawi M F 2022 A comprehensive review on branched nanofibers: preparations, strategies, and applications *J. Ind. Text.* **51** 1S–35S

[124] Zaarour B, Tina H, Zhu L and Jin X 2022 Branched nanofibers with tiny diameters for air filtration via one-step electrospinning *J. Ind. Text.* **51** 1105S–17S

[125] Zhao Y, Yan J, Yu J and Ding B 2023 Electrospun nanofiber electrodes for lithium-ion batteries *Macromol. Rapid Commun.* **44** 2200740

[126] Yan B, Zhang Y, Li Z, Zhou P and Mao Y 2022 Electrospun nanofibrous membrane for biomedical application *SN Appl. Sci.* **4** 172

[127] Chen Y, Dong X, Shafiq M, Myles G, Radacsi N and Mo X 2022 Recent advancements on three-dimensional electrospun nanofiber scaffolds for tissue engineering *Adv. Fiber Mater.* **4** 959–86

[128] Mirjalali S, Bagherzadeh R, Abrishami S, Asadnia M, Huang S, Michael A, Peng S, Wang C H and Wu S 2023 Multilayered electrospun/electrosprayed polyvinylidene fluoride + zinc oxide nanofiber mats with enhanced piezoelectricity *Macromol. Mater. Eng.* **308** 2300009

[129] Zaarour B, Zhu L, Huang C, Jin X, Alghafari H, Fang J and Lin T 2021 A review on piezoelectric fibers and nanowires for energy harvesting *J. Ind. Text.* **51** 297–340

[130] Zhou Y, Liu Y, Zhang M, Feng Z, Yu D-G and Wang K 2022 Electrospun nanofiber membranes for air filtration: a review *Nanomaterials* **12** 1077

[131] Akdere M and Schneiders T 2021 Modeling of the electrospinning process *Advances in Modeling and Simulation in Textile Engineering* (Amsterdam: Elsevier) pp 237–53

[132] Dehaghi N G and Kokabi M 2023 Polyvinylidene fluoride/barium titanate nanocomposite aligned hollow electrospun fibers as an actuator *Mater. Res. Bull.* **158** 112052

[133] Hao Y, Hu F, Chen Y, Wang Y, Xue J, Yang S and Peng S 2022 Recent progress of electrospun nanofibers for zinc–air batteries *Adv. Fiber Mater.* **4** 185–202

[134] Pattnaik S, Swain K and Ramakrishna S 2023 Optimal delivery of poorly soluble drugs using electrospun nanofiber technology: challenges, state of the art, and future directions *Wiley Interdiscip. Rev. Nanomed. Nanobiotechnol.* **15** e1859

[135] Wang X, Sun F, Yin G, Wang Y, Liu B and Dong M 2018 Tactile-sensing based on flexible PVDF nanofibers via electrospinning: a review *Sensors* **18** 330

[136] Xu Q, Wen J and Qin Y 2021 Development and outlook of high output piezoelectric nanogenerators *Nano Energy* **86** 106080

[137] Mokhtari F, Cheng Z, Raad R, Xi J and Foroughi J 2020 Piezofibers to smart textiles: a review on recent advances and future outlook for wearable technology *J. Mater. Chem.* A **8** 9496–522

[138] Ibrahim H M and Klingner A 2020 A review on electrospun polymeric nanofibers: Production parameters and potential applications *Polym. Test.* **90** 106647

[139] Jiyong H, Yinda Z, Hele Z, Yuanyuan G and Xudong Y 2017 Mixed effect of main electrospinning parameters on the β-phase crystallinity of electrospun PVDF nanofibers *Smart Mater. Struct.* **26** 085019

[140] Damaraju S M, Wu S, Jaffe M and Arinzeh T L 2013 Structural changes in PVDF fibers due to electrospinning and its effect on biological function *Biomed. Mater.* **8** 045007

[141] Singh R K, Lye S W and Miao J 2019 PVDF nanofiber sensor for vibration measurement in a string *Sensors* **19** 3739

[142] Bokka S, Li Y, Reneker D H and Chase G G 2023 Achievement of high surface charge in poly (vinylidene fluoride) fiber yarns through dipole orientation during fabrication *J. Appl. Polym. Sci.* **140** e53265

[143] Gade H, Nikam S, Chase G G and Reneker D H 2021 Effect of electrospinning conditions on β-phase and surface charge potential of PVDF fibers *Polymer* **228** 123902

[144] Nasir M, Matsumoto H, Danno T, Minagawa M, Irisawa T, Shioya M and Tanioka A 2006 Control of diameter, morphology, and structure of PVDF nanofiber fabricated by electrospray deposition *J. Polym. Sci.* B **44** 779–86

[145] Bae J, Baek I and Choi H 2017 Efficacy of piezoelectric electrospun nanofiber membrane for water treatment *Chem. Eng. J.* **307** 670–8

[146] Motamedi A S, Mirzadeh H, Hajiesmaeilbaigi F, Bagheri-Khoulenjani S and Shokrgozar M 2017 Effect of electrospinning parameters on morphological properties of PVDF nanofibrous scaffolds *Prog. Biomater.* **6** 113–23

[147] Shehata N, Elnabawy E, Abdelkader M, Hassanin A H, Salah M, Nair R and Ahmad Bhat S 2018 Static-aligned piezoelectric poly (vinylidene fluoride) electrospun nanofibers/MWCNT composite membrane: facile method *Polymers* **10** 965

[148] Kang D H and Kang H W 2018 Advanced electrospinning using circle electrodes for freestanding PVDF nanofiber film fabrication *Appl. Surf. Sci.* **455** 251–7

[149] Hansen B J, Liu Y, Yang R and Wang Z L 2010 Hybrid nanogenerator for concurrently harvesting biomechanical and biochemical energy *ACS Nano* **4** 3647–52

[150] Mokhtari F, Shamshirsaz M and Latifi M 2016 Investigation of β phase formation in piezoelectric response of electrospun polyvinylidene fluoride nanofibers: LiCl additive and increasing fibers tension *Polym. Eng. Sci.* **56** 61–70

[151] Andrew J and Clarke D 2008 Effect of electrospinning on the ferroelectric phase content of polyvinylidene difluoride fibers *Langmuir* **24** 670–2

[152] He Z, Rault F, Lewandowski M, Mohsenzadeh E and Salaün F 2021 Electrospun PVDF nanofibers for piezoelectric applications: a review of the influence of electrospinning parameters on the β phase and crystallinity enhancement *Polymers* **13** 174

[153] Abolhasani M M, Azimi S and Fashandi H 2015 Enhanced ferroelectric properties of electrospun poly (vinylidene fluoride) nanofibers by adjusting processing parameters *RSC Adv.* **5** 61277–83

[154] Lins L C, Wianny F, Livi S, Dehay C, Duchet-Rumeau J and Gérard J F 2017 Effect of polyvinylidene fluoride electrospun fiber orientation on neural stem cell differentiation *J. Biomed. Mater. Res.* B **105** 2376–93

[155] Deitzel J M, Kleinmeyer J, Harris D and Tan N B 2001 The effect of processing variables on the morphology of electrospun nanofibers and textiles *Polymer* **42** 261–72

[156] Lee K Y and Mooney D J 2012 Alginate: properties and biomedical applications *Prog. Polym. Sci.* **37** 106–26

[157] Huang Z-M, Zhang Y-Z, Kotaki M and Ramakrishna S 2003 A review on polymer nanofibers by electrospinning and their applications in nanocomposites *Compos. Sci. Technol.* **63** 2223–53

[158] Ma P X and Zhang R 1999 Synthetic nano-scale fibrous extracellular matrix *J. Biomed. Mater. Res.* **46** 60–72

[159] Li D and Xia Y 2004 Electrospinning of nanofibers: reinventing the wheel? *Adv. Mater.* **16** 1151–70

[160] Elnabawy E, Sun D, Shearer N and Shyha I 2023 Electro-blown spinning: new insight into the effect of electric field and airflow hybridized forces on the production yield and characteristics of nanofiber membranes *J. Sci.: Adv. Mater. Devices* **8** 100552

[161] Hu D, Yao M, Fan Y, Ma C, Fan M and Liu M 2019 Strategies to achieve high performance piezoelectric nanogenerators *Nano Energy* **55** 288–304

[162] Fan F R, Tang W and Wang Z L 2016 Flexible nanogenerators for energy harvesting and self-powered electronics *Adv. Mater.* **28** 4283–305

[163] Yang R, Qin Y, Dai L and Wang Z L 2009 Power generation with laterally packaged piezoelectric fine wires *Nat. Nanotechnol.* **4** 34–9

[164] Park H K, Lee K Y, Seo J S, Jeong J A, Kim H K, Choi D and Kim S W 2011 Charge-generating mode control in high-performance transparent flexible piezoelectric nanogenerators *Adv. Funct. Mater.* **21** 1187–93

[165] Bai S, Xu Q, Gu L, Ma F, Qin Y and Wang Z L 2012 Single crystalline lead zirconate titanate (PZT) nano/micro-wire based self-powered UV sensor *Nano Energy* **1** 789–95

[166] Qi Y, Jafferis N T, Lyons K, Lee C M, Ahmad H and McAlpine M C 2010 Piezoelectric ribbons printed onto rubber for flexible energy conversion *Nano Lett.* **10** 524–8

[167] Park K I, Son J H, Hwang G T, Jeong C K, Ryu J, Koo M, Choi I, Lee S H, Byun M and Wang Z L 2014 Highly-efficient, flexible piezoelectric PZT thin film nanogenerator on plastic substrates *Adv. Mater.* **26** 2514–20

[168] Ni X, Wang F, Lin A, Xu Q, Yang Z and Qin Y 2013 Flexible nanogenerator based on single BaTiO$_3$ nanowire *Sci. Adv. Mater.* **5** 1781–7

[169] Huan Y, Zhang X, Song J, Zhao Y, Wei T, Zhang G and Wang X 2018 High-performance piezoelectric composite nanogenerator based on Ag/(K, Na) NbO3 heterostructure *Nano Energy* **50** 62–9

[170] Chang C, Tran V H, Wang J, Fuh Y-K and Lin L 2010 Direct-write piezoelectric polymeric nanogenerator with high energy conversion efficiency *Nano Lett.* **10** 726–31

[171] Persano L, Dagdeviren C, Su Y, Zhang Y, Girardo S, Pisignano D, Huang Y and Rogers J A 1633 High performance piezoelectric devices based on aligned arrays of nanofibers of poly (vinylidenefluoride-co-trifluoroethylene) *Nat. Commun.* **2013** 1

[172] Han J, Kim J H, Choi H J, Kim S W, Sung S M, Kim M S, Choi B K, Paik J H, Lee J S and Cho Y S 2021 Origin of enhanced piezoelectric energy harvesting in all-polymer-based core–shell nanofibers with controlled shell-thickness *Compos. Part B: Eng.* **223** 109141

[173] Zhang X, Lv S, Lu X, Yu H, Huang T, Zhang Q and Zhu M 2020 Synergistic enhancement of coaxial nanofiber-based triboelectric nanogenerator through dielectric and dispersity modulation *Nano Energy* **75** 104894

[174] Huang Y-J, Chen Y-F, Hsiao P-H, Lam T-N, Ko W-C, Luo M-Y, Chuang W-T, Su C-J, Chang J-H and Chung C F 2021 In-situ synchrotron SAXS and WAXS investigation on the deformation of single and coaxial electrospun P (VDF-TrFE)-based nanofibers *Int. J. Mol. Sci.* **22** 12669

[175] Fabiani D, Zucchelli A, Brugo T, Selleri G, Grolli F and Speranza M 2020 Core–shell piezoelectric nanofibers for multifunctional composite materials *2020 IEEE 3rd Int. Conf. on Dielectrics (ICD)* (Piscataway, NJ: IEEE) pp 325–8

[176] Lam T-N, Ma C-Y, Hsiao P-H, Ko W-C, Huang Y-J, Lee S-Y, Jain J and Huang E-W 2021 Tunable mechanical and electrical properties of coaxial electrospun composite nanofibers of P (VDF-TrFE) and P (VDF-TrFE-CTFE) *Int. J. Mol. Sci.* **22** 4639

[177] Zhu M, Lou M, Abdalla I, Yu J, Li Z and Ding B 2020 Highly shape adaptive fiber based electronic skin for sensitive joint motion monitoring and tactile sensing *Nano Energy* **69** 104429

[178] Ji S H, Cho Y-S and Yun J S 2019 Wearable core–shell piezoelectric nanofiber yarns for body movement energy harvesting *Nanomaterials* **9** 555

[179] Gao H, Minh P T, Wang H, Minko S, Locklin J, Nguyen T and Sharma S 2018 High-performance flexible yarn for wearable piezoelectric nanogenerators *Smart Mater. Struct.* **27** 095018

[180] Xue L, Fan W, Yu Y, Dong K, Liu C, Sun Y, Zhang C, Chen W, Lei R and Rong K 2021 A novel strategy to fabricate core-sheath structure piezoelectric yarns for wearable energy harvesters *Adv. Fiber Mater.* **3** 239–50

[181] Latifi M 2021 *Engineered Polymeric Fibrous Materials* (Woodhead Publishing)

[182] Ji S H, Lee W and Yun J S 2020 All-in-one piezo-triboelectric energy harvester module based on piezoceramic nanofibers for wearable devices *ACS Appl. Mater. Interfaces* **12** 18609–16

[183] Jin M-Y, Lin Y, Liao Y, Tan C-H and Wang R 2018 Development of highly-efficient ZIF-8@ PDMS/PVDF nanofibrous composite membrane for phenol removal in aqueous-aqueous membrane extractive process *J. Membr. Sci.* **568** 121–33

[184] Kim M and Fan J 2021 Piezoelectric properties of three types of PVDF and ZnO nanofibrous composites *Adv. Fiber Mater.* **3** 160–71

[185] Li H, Zhang W, Ding Q, Jin X, Ke Q, Li Z, Wang D and Huang C 2019 Facile strategy for fabrication of flexible, breathable, and washable piezoelectric sensors via welding of nanofibers with multiwalled carbon nanotubes (MWCNTs) *ACS Appl. Mater. Interfaces* **11** 38023–30

[186] Tamil Selvan R, Jayathilaka W, Hilaal A and Ramakrishna S 2020 Improved piezoelectric performance of electrospun PVDF nanofibers with conductive paint coated electrode *Int. J. Nanosci.* **19** 1950008

[187] Khan A, Ali Abbasi M, Hussain M, Hussain Ibupoto Z, Wissting J, Nur O and Willander M 2012 Piezoelectric nanogenerator based on zinc oxide nanorods grown on textile cotton fabric *Appl. Phys. Lett.* **101** 193506

[188] Zhu G, Yang R, Wang S and Wang Z L 2010 Flexible high-output nanogenerator based on lateral ZnO nanowire array *Nano Lett.* **10** 3151–5

[189] Serairi L and Leprince-Wang Y 2022 ZnO nanowire-based piezoelectric nanogenerator device performance tests *Crystals* **12** 1023

[190] Wang Q, Yang D, Qiu Y, Zhang X, Song W and Hu L 2018 Two-dimensional ZnO nanosheets grown on flexible ITO-PET substrate for self-powered energy-harvesting nano-devices *Appl. Phys. Lett.* **112**

[191] Aleman C K A, Narvaez J A B, Lopez G D B and Mercado C C 2020 Array pattern effects on the voltage output of vertically aligned BaTiO3 nanotubular flexible piezoelectric nanogenerator *MRS Commun.* **10** 500–5

[192] Lin Z-H, Yang Y, Wu J M, Liu Y, Zhang F and Wang Z L 2012 BaTiO$_3$ nanotubes-based flexible and transparent nanogenerators *J. Phys. Chem. Lett.* **3** 3599–604

[193] Shin S-H, Choi S-Y, Lee M H and Nah J 2017 High-performance piezoelectric nano-generators via imprinted sol–gel BaTiO$_3$ nanopillar array *ACS Appl. Mater. Interfaces* **9** 41099–103

[194] Chen X, Xu S, Yao N and Shi Y 2010 1.6 V nanogenerator for mechanical energy harvesting using PZT nanofibers *Nano Lett.* **10** 2133–7

[195] Yue R, Ramaraj S G, Liu H, Elamaran D, Elamaran V, Gupta V, Arya S, Verma S, Satapathi S and Liu X 2022 A review of flexible lead-free piezoelectric energy harvester *J. Alloys Compd.* **918** 165653

[196] Nour E S, Sandberg M, Willander M and Nur O 2014 Handwriting enabled harvested piezoelectric power using ZnO nanowires/polymer composite on paper substrate *Nano Energy* **9** 221–8

[197] Batra K, Sinha N and Kumar B 2020 Tb-doped ZnO: PDMS based flexible nanogenerator with enhanced piezoelectric output performance by optimizing nanofiller concentration *Ceram. Int.* **46** 24120–8

[198] Arunguvai J and Lakshmi P 2021 Flexible piezoelectric MoS2/P (VDF-TrFE) nano-composite film for vibration energy harvesting *J. Electron. Mater.* **50** 6870–80

[199] Ding Y, Duan Y and Huang Y 2015 Electrohydrodynamically printed, flexible energy harvester using *in situ* poled piezoelectric nanofibers *Energy Technol.* **3** 351–8

[200] Mamlayya V, Maile N and Fulari V 2020 A study on silver nanoleaf-decorated PANI electrodes for improved electrochemical performance *Polym. Bull.* **77** 4587–607

[201] Duan Z, Jiang Y, Huang Q, Wang S, Wang Y, Pan H, Zhao Q, Xie G, Du X and Tai H 2021 Paper and carbon ink enabled low-cost, eco-friendly, flexible, multifunctional pressure and humidity sensors *Smart Mater. Struct.* **30** 055012

[202] Im B, Lee S-K, Kang G, Moon J, Byun D and Cho D-H 2022 Electrohydrodynamic jet printed silver-grid electrode for transparent raindrop energy-based triboelectric nanogen-erator *Nano Energy* **95** 107049

[203] Tiwari S, Purabgola A and Kandasubramanian B 2020 Functionalised graphene as flexible electrodes for polymer photovoltaics *J. Alloys Compd.* **825** 153954

[204] Pace G, Ansaldo A, Serri M, Lauciello S and Bonaccorso F 2020 Electrode selection rules for enhancing the performance of triboelectric nanogenerators and the role of few-layers graphene *Nano Energy* **76** 104989

[205] Sorayani Bafqi M S, Latifi M, Sadeghi A-H and Bagherzadeh R 2022 Expected lifetime of fibrous nanogenerator exposed to cyclic compressive pressure *J. Ind. Text.* **51** 4493S–505S

[206] Latifi M and Bagherzadeh R 2020 Hybrid pizo-triboelectric energy nanogenerator from woollen fabric and PVDF/BaTiO₃ nanofibrous layer *Adv. Mater. New Coat.* **9** 2474–81

[207] Rajala S N K, Mettänen M and Tuukkanen S 2015 Structural and electrical character-ization of solution-processed electrodes for piezoelectric polymer film sensors *IEEE Sens. J.* **16** 1692–9

[208] Bandodkar A J, Jeerapan I, You J-M, Nuñez-Flores R and Wang J 2016 Highly stretchable fully-printed CNT-based electrochemical sensors and biofuel cells: combining intrinsic and design-induced stretchability *Nano Lett.* **16** 721–7

[209] Liu Y, Yiu C, Jia H, Wong T, Yao K, Huang Y, Zhou J, Huang X, Zhao L and Li D 2021 Thin, soft, garment-integrated triboelectric nanogenerators for energy harvesting and human machine interfaces *EcoMat* **3** e12123

[210] Dudem B, Mule A R, Patnam H R and Yu J S 2019 Wearable and durable triboelectric nanogenerators via polyaniline coated cotton textiles as a movement sensor and self-powered system *Nano Energy* **55** 305–15

[211] Qin H, Li J, He B, Sun J, Li L and Qian L 2018 Novel wearable electrodes based on conductive chitosan fabrics and their application in smart garments *Materials* **11** 370

[212] Mao C, Zhang H and Lu Z 2017 Flexible and wearable electronic silk fabrics for human physiological monitoring *Smart Mater. Struct.* **26** 095033

[213] Masihi S, Panahi M, Maddipatla D, Bose A, Zhang X, Hanson A, Palaniappan V, Narakathu B, Bazuin B and Atashbar M 2019 A novel printed fabric based porous capacitive pressure sensor for flexible electronic applications *2019 IEEE Sensors* (Piscataway, NJ: IEEE) pp 1–4

[214] Ouyang J 2021 Application of intrinsically conducting polymers in flexible electronics *SmartMat* **2** 263–85

[215] Mule A R, Dudem B, Patnam H, Graham S A and Yu J S 2019 Wearable single-electrode-mode triboelectric nanogenerator via conductive polymer-coated textiles for self-power electronics *ACS Sustain. Chem. Eng.* **7** 16450–8

[216] Khadtare S, Ko E J, Kim Y H, Lee H S and Moon D K 2019 A flexible piezoelectric nanogenerator using conducting polymer and silver nanowire hybrid electrodes for its application in real-time muscular monitoring system *Sens. Actuators* A **299** 111575

[217] Kim Y, Ryu T I, Ok K H, Kwak M G, Park S, Park N G, Han C J, Kim B S, Ko M J and Son H J 2015 Inverted layer-by-layer fabrication of an ultraflexible and transparent Ag nanowire/conductive polymer composite electrode for use in high-performance organic solar cells *Adv. Funct. Mater.* **25** 4580–9

[218] Liu S, Ha T and Lu N 2019 Experimentally and numerically validated analytical solutions to nonbuckling piezoelectric serpentine ribbons *J. Appl. Mech.* **86** 051010

[219] Kwon H J, Kim G-U, Lim C, Kim J K, Lee S-S, Cho J, Koo H-J, Kim B J, Char K and Son J G 2023 Sequentially coated wavy nanowire composite transparent electrode for stretchable solar cells *ACS Appl. Mater. Interfaces* **15** 13656–67

[220] Tran N-H, Tran P and Lee J-H 2022 Copper nanowire-sealed titanium dioxide/poly (dimethylsiloxane) electrode with an in-plane wavy structure for a stretchable capacitive strain sensor *ACS Appl. Nano Mater.* **5** 7150–60

[221] Choi J H, Noh J H and Choi C 2023 Highly elastically deformable coiled CNT/polymer fibers for wearable strain sensors and stretchable supercapacitors *Sensors* **23** 2359

[222] Kim Y-G, Song J-H, Hong S and Ahn S-H 2022 Piezoelectric strain sensor with high sensitivity and high stretchability based on kirigami design cutting *npj Flexible Electron.* **6** 52

[223] Li X, Zhu P, Zhang S, Wang X, Luo X, Leng Z, Zhou H, Pan Z and Mao Y 2022 A self-supporting, conductor-exposing, stretchable, ultrathin, and recyclable kirigami-structured liquid metal paper for multifunctional E-skin *ACS Nano* **16** 5909–19

[224] Kang C, Kim S W, Kim W, Choi D and Kim H K 2023 Stretchable and flexible snake skin patterned electrodes for wearable electronics inspired by kirigami structure *Adv. Mater. Interfaces* **10** 2202477

[225] Wang Y, Li X, Hou Y, Yin C and Yin Z 2021 A review on structures, materials and applications of stretchable electrodes *Front. Mater. Sci.* **15** 54–78

[226] Park S-H, Lee H B, Yeon S M, Park J and Lee N K 2016 Flexible and stretchable piezoelectric sensor with thickness-tunable configuration of electrospun nanofiber mat and elastomeric substrates *ACS Appl. Mater. Interfaces* **8** 24773–81

[227] Duan Y, Huang Y, Yin Z, Bu N and Dong W 2014 Non-wrinkled, highly stretchable piezoelectric devices by electrohydrodynamic direct-writing *Nanoscale* **6** 3289–95

[228] Siddiqui S, Lee H B, Kim D I, Duy L T, Hanif A and Lee N E 2018 An omnidirectionally stretchable piezoelectric nanogenerator based on hybrid nanofibers and carbon electrodes

for multimodal straining and human kinematics energy harvesting *Adv. Energy Mater.* **8** 1701520

[229] Hou W, Liao Q, Xie S, Song Y and Qin L 2022 Prospects and challenges of flexible stretchable electrodes for electronics *Coatings* **12** 558

[230] Qian F, Xu T-B and Zuo L 2019 Piezoelectric energy harvesting from human walking using a two-stage amplification mechanism *Energy* **189** 116140

[231] Kashfi M, Fakhri P, Amini B, Yavari N, Rashidi B, Kong L and Bagherzadeh R 2022 A novel approach to determining piezoelectric properties of nanogenerators based on PVDF nanofibers using iterative finite element simulation for walking energy harvesting *J. Ind. Text.* **51** 531S–53S

[232] Delnavaz A and Voix J 2013 Ear canal dynamic motion as a source of power for in-ear devices *J. Appl. Phys.* **113**

[233] Delnavaz A and Voix J 2012 Electromagnetic micro-power generator for energy harvesting from breathing *IECON 2012–38th Annual Conf. on IEEE Industrial Electronics Society* (Piscataway, NJ: IEEE) pp 984–8

[234] Dagdeviren C, Yang B D, Su Y, Tran P L, Joe P, Anderson E, Xia J, Doraiswamy V, Dehdashti B and Feng X 2014 Conformal piezoelectric energy harvesting and storage from motions of the heart, lung, and diaphragm *Proc. Natl Acad. Sci.* **111** 1927–32

[235] Panda S, Hajra S, Mistewicz K, In-na P, Sahu M, Rajaitha P M and Kim H J 2022 Piezoelectric energy harvesting systems for biomedical applications *Nano Energy* **100** 107514

[236] Wang Z, Pan X, He Y, Hu Y, Gu H and Wang Y 2015 Piezoelectric nanowires in energy harvesting applications *Adv. Mater. Sci. Eng.* **2015** 165631

[237] Alluri N R, Saravanakumar B and Kim S-J 2015 Flexible, hybrid piezoelectric film (BaTi$_{(1-x)}$Zr$_x$O$_3$)/PVDF nanogenerator as a self-powered fluid velocity sensor *ACS Appl. Mater. Interfaces* **7** 9831–40

[238] Lee S, Hinchet R, Lee Y, Yang Y, Lin Z H, Ardila G, Montès L, Mouis M and Wang Z L 2014 Ultrathin nanogenerators as self-powered/active skin sensors for tracking eye ball motion *Adv. Funct. Mater.* **24** 1163–8

[239] Wang Y, Yu Y, Wei X and Narita F 2022 Self-powered wearable piezoelectric monitoring of human motion and physiological signals for the postpandemic era: a review *Adv. Mater. Technol.* **7** 2200318

[240] Guo W, Tan C, Shi K, Li J, Wang X-X, Sun B, Huang X, Long Y-Z and Jiang P 2018 Wireless piezoelectric devices based on electrospun PVDF/BaTiO$_3$ NW nanocomposite fibers for human motion monitoring *Nanoscale* **10** 17751–60

[241] Liu Y, Wang L, Zhao L, Yu X and Zi Y 2020 Recent progress on flexible nanogenerators toward self-powered systems *InfoMat* **2** 318–40

[242] Shi K, Sun B, Huang X and Jiang P 2018 Synergistic effect of graphene nanosheet and BaTiO$_3$ nanoparticles on performance enhancement of electrospun PVDF nanofiber mat for flexible piezoelectric nanogenerators *Nano Energy* **52** 153–62

[243] Choudhry I, Khalid H R and Lee H-K 2020 Flexible piezoelectric transducers for energy harvesting and sensing from human kinematics *ACS Appl. Electron. Mater.* **2** 3346–57

[244] Ma S W, Fan Y J, Li H Y, Su L, Wang Z L and Zhu G 2018 Flexible porous polydimethylsiloxane/lead zirconate titanate-based nanogenerator enabled by the dual effect of ferroelectricity and piezoelectricity *ACS Appl. Mater. Interfaces* **10** 33105–11

[245] Dudem B, Kim D H, Bharat L K and Yu J S 2018 Highly-flexible piezoelectric nanogenerators with silver nanowires and barium titanate embedded composite films for mechanical energy harvesting *Appl. Energy* **230** 865–74

[246] Park S, Kim Y, Jung H, Park J-Y, Lee N and Seo Y 2017 Energy harvesting efficiency of piezoelectric polymer film with graphene and metal electrodes *Sci. Rep.* **7** 17290

[247] He Z, Gao B, Li T, Liao J, Liu B, Liu X, Wang C, Feng Z and Gu Z 2018 Piezoelectric-driven self-powered patterned electrochromic supercapacitor for human motion energy harvesting *ACS Sustain. Chem. Eng.* **7** 1745–52

[248] Guan M and Liao W 2007 On the efficiencies of piezoelectric energy harvesting circuits towards storage device voltages *SMS* **16** 498

[249] Chen S, Huang T, Zuo H, Qian S, Guo Y, Sun L, Lei D, Wu Q, Zhu B and He C 2018 A single integrated 3D-printing process customizes elastic and sustainable triboelectric nanogenerators for wearable electronics *Adv. Funct. Mater.* **28** 1805108

[250] Valentine A D, Busbee T A, Boley J W, Raney J R, Chortos A, Kotikian A, Berrigan J D, Durstock M F and Lewis J A 2017 Hybrid 3D printing of soft electronics *Adv. Mater.* **29** 1703817

[251] Zhou X, Parida K, Halevi O, Liu Y, Xiong J, Magdassi S and Lee P S 2020 All 3D-printed stretchable piezoelectric nanogenerator with non-protruding kirigami structure *Nano Energy* **72** 104676

[252] Wei P, Leng H, Chen Q, Advincula R C and Pentzer E B 2019 Reprocessable 3D-printed conductive elastomeric composite foams for strain and gas sensing *ACS Appl. Polym. Mater.* **1** 885–92

[253] Ali A, Baheti V and Militky J 2019 Energy harvesting performance of silver electroplated fabrics *Mater. Chem. Phys.* **231** 33–40

[254] Somkuwar V U, Pragya A and Kumar B 2020 Structurally engineered textile-based triboelectric nanogenerator for energy harvesting application *J. Mater. Sci.* **55** 5177–89

[255] Lee S and Lim Y 2018 Generating power enhancement of flexible PVDF generator by incorporation of CNTs and surface treatment of PEDOT: PSS electrodes *Macromol. Mater. Eng.* **303** 1700588

[256] Ng K E, Ooi P C, Haniff M A S M, Goh B T, Dee C F, Chang W S, Wee M M R and Mohamed M A 2020 Performance of all-solution-processed, durable 2D MoS_2 flakes–$BaTiO_3$ nanoparticles in polyvinylidene fluoride matrix nanogenerator devices using N-methyl-2-pyrrolidone polar solvent *J. Alloys Compd.* **820** 153160

IOP Publishing

Energy Harvesting Properties of Electrospun Nanofibers (Second Edition)

Jian Fang and Tong Lin

Chapter 10

Piezoelectric and triboelectric nanogenerators based on electrospun PVDF-nanofiller composites

T Sathies, Govind S Ekbote and S Anandhan

10.1 Introduction

The interest towards nanogenerators (NGs) and self-powered devices has surged in the 21st century because the pollution and global warming issues arising out of non-renewable energy sources compelled researchers to look into environment-friendly energy sources and energy harvesting methods. Piezoelectric and triboelectric NGs (PENGs and TENGs) are devices designed to harness electrical energy from various sources of mechanical energy, which are freely available in the environment. NGs are expected to be compact, portable, durable, energy-efficient and should also meet the power requirement of microelectronic devices. Flexibility also comes into the picture in the case of wearable applications. Materials with ferroelectric and electroactive properties can harness the unused mechanical energy available in the environment [1]. Numerous piezoelectric and triboelectric materials have been explored by the scientific community for the construction of efficient and durable energy harvesters. Polymers and ceramics are well-known piezoelectric materials, which have the potential to generate electrical energy from unused omnipresent energy sources. Ferroelectric ceramics usually have a high electroactive potential and exhibit a higher piezoelectric coefficient than polymers. However, the wide usage of ceramics for the development of energy harvesters and self-powered devices is limited as the ceramics lack flexibility and pose problems during the synthesis. Hence the research activities on piezoelectric polymers are constantly increasing and interesting outcomes have been reported. Even though piezoelectric coefficients of polymers are inferior to those of ceramics, polymer character-istics such as flexibility, biocompatibility, light weight, affordability, and enhanced mechanical strength are favourable for the construction of energy harvesters. Numerous

doi:10.1088/978-0-7503-5487-5ch10

natural (collagen, cellulose, keratin) and synthetic polymers (polyamides, poly(L-lactic acid), poly(vinylidene fluoride) and co-polymers of vinylidene fluoride) have been tested as piezoelectric energy harvesting materials and their energy harvesting potential mainly depends on the electroactive phases present and the synthesis methods involved [2, 3]. Since pure polymers have lower piezoelectric coefficients, the energy generated by the pure polymer-based energy harvesters is not able to meet the power demand of miniature electronic devices. Hence, numerous nanoparticle (NP)-incorporated polymer nano-composites with improved energy harvesting capability are reported by researchers. Even though several polymers demonstrate piezoelectric properties, vinylidene fluoride's homopolymer and co-polymers are considered potential candidates for piezoelectric energy harvesting owing to their high piezoelectric coefficient (among polymers), high thermal stability, flame retardancy, biocompatibility and flexibility. Poly(vinylidene fluoride) (PVDF) is a semicrystalline polymer that exhibits five different polymorphs namely α, β, γ, δ, and ε. The most commonly observed phases are α (TGTG'), β (TTTT), γ (T3GT3G') and the phases formed depend on the processing technique. The α-phase is non-polar and thermodynamically stable and it is usually formed during the crystallization from melt. PVDF films with a high electroactive phase (β- or γ-phase) are preferred for energy harvesting applications. The β-phase is recognized as a highly polar phase that exhibits unidirectional polarization. PVDF films with non-polar phases are converted into electroactive phases by subjecting them to electrical poling, mechanical stretching, and thermal annealing. Another possible way is to incorporate nanofillers during synthesis which can assist in the conversion of the non-polar α polymorph to its β counterpart [4]. Techniques such as solution casting, injection moulding, spin coating, electrospinning, and 3D printing are involved in the processing of PVDF-based polymer nanocomposites for energy harvesting applications [2]. Among the mentioned techniques, lightweight and flexible PVDF nanofiber mats with high electroactive phase can be fabricated through the electrospinning technique. The simultaneous application of high voltage and stretching effects during electrospinning facilitates the formation of electroactive phases in PVDF. The need for additional electrical poling, thermal annealing and uni- or bi-axial stretching is eliminated in the case of the electrospinning technique. The solution preparation strategy and the electrospinning process parameters influence the formation of the β-phase. The morphology of nanofibers produced through electrospinning depends on solution viscosity, applied voltage, tip-to-collector distance, flow rate, solution conductivity and collector type. The concentration, and molecular weight of PVDF and the solvents employed for solution preparation determine the solution viscosity [5, 6]. PVDF nano-composites with high surface area and surface charge density can also be developed through electrospinning for constructing TENGs. The focus of this chapter is on the advancements in electrospun PVDF nanocomposites for piezoelectric and triboelectric energy harvesting applications.

10.2 Electrospun PVDF nanocomposite-based piezoelectric nanogenerators

Electrospun PVDF-based nanocomposites with high piezoresponsive properties are promising candidates for constructing high-performance and flexible PENGs.

The piezoelectric properties of PVDF are generally modified through the addition of nanofillers, copolymerization and blending. Nanofillers are infused with an objective of promoting the dielectric property and electroactive phases of PVDF without compromising its thermal, chemical and mechanical properties. The addition of nanofillers has a significant impact on the physico-chemical properties of PVDF in comparison with blending and copolymerization and it is attributed to the high surface-to-volume area of nanofillers. The nanofillers interact with the electronegative fluorine atoms or electropositive hydrogen atoms of PVDF through physical forces of attraction such as hydrogen bonding, electrostatic interaction, ion–dipole interaction and π cloud dipole interaction and improve the electroactive β-phase of PVDF. Figure 10.1 summarizes the different nanofillers capable of improving the electroactive phases of PVDF and their plausible mechanism of interaction with the PVDF matrix. The NPs-filled PVDF composite also undergoes surface charge-induced polarization or stress-induced polarization and it favours β-phase nucleation in PVDF nanocomposite. The interaction between the nanofiller and PVDF matrix depends on the composition, and morphology of the nanofiller. Other factors affecting the properties of PVDF nanocomposites are nanofiller dispersion, relative fraction of nanofiller, and processing conditions. The nanofillers considered for improving the piezoelectric characteristics of electrospun PVDF nanofabrics can be categorized as organic and inorganic nanofillers. The incorporation of inorganic nanofillers such as metallic NPs, ceramic NPs, carbon-based NPs and hybrid nanofillers induces electroactive phases and enhances the energy harvesting capability of PVDF. In addition to β-phase nucleation, inorganic nanofiller

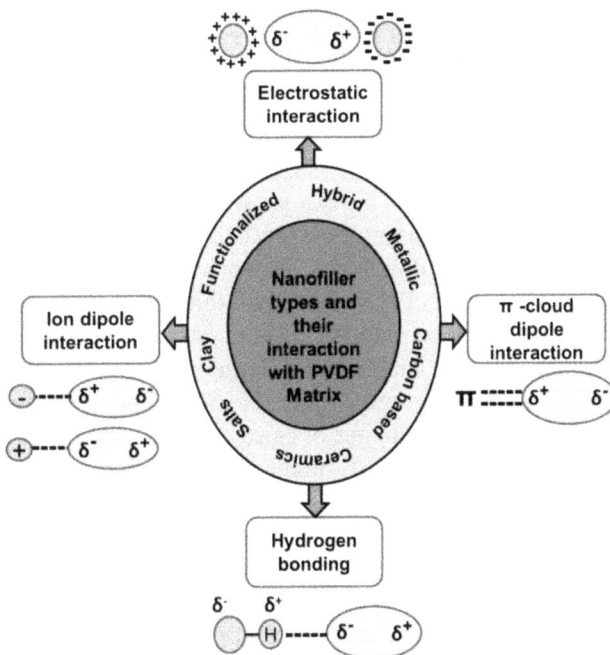

Figure 10.1. Nanofillers and their plausible mechanism of interaction with PVDF chains.

addition also affects the dielectric properties, degree of crystallinity, flexibility and thermal and mechanical properties of PVDF.

The piezoelectric performance of electrospun PVDF nanocomposite is attributed to the change in dipole moment upon the application of dynamic mechanical stimulation resulting in the generation of electrical energy. PVDF films dispersed with ferroelectric ceramic NPs, metal oxides, and conductive NPs and their hybrids usually exhibit higher dielectric properties and piezoelectric coefficients and it is beneficial for energy harvesting applications. Piezoelectric nanofillers (ferroelectric ceramics) in various forms namely NPs, nanosheets, nanorods (NRs), nanoflakes, nanowires (NWs) and nanofibers have been utilized to improve the piezoelectricity of electrospun PVDF nanofibers. Lead-free piezoceramic nanofillers are preferred over lead-based nanofillers owing to the toxicity associated with lead and some of the widely adopted lead-free piezoceramic nanofillers are potassium sodium niobate (KNN) [7], barium titanate ($BaTiO_3$) [8], zinc oxide (ZnO) [9] and boron nitride. The piezoelectric nanofillers not only impart an additional piezoelectric effect to PVDF but serve as a nucleation agent and favour electroactive phase formation through electrostatic interaction. Additionally, piezoelectric fillers act as stress-reinforcing agents and improve the force transfer to the internal crystalline phases of the piezoelectric strips during the piezo test. A wireless piezoelectric device developed from electrospun PVDF/$BaTiO_3$ nanocomposite was able to sense various human motions and functioned as a self-powered vibration-sensing device [8]. Dehaghi *et al* utilized a co-axial electrospinning process to develop hollow nanofibers from PVDF solution containing $BaTiO_3$ NPs dispersion. For the preparation of hollow nanofibers, argon gas was supplied through the core needle and polymer solution was fed through the shell needle. The PVDF/8 wt.% $BaTiO_3$ hollow nanofibers functioned as piezoelectric actuators and exhibited an actuation deflection of 28 μm when they were subjected to a voltage of 480 V [10]. PENGs fabricated from PVDF/2 wt.% Ce–Co_3O_4 and PVDF/2 wt.% Ce–Fe_2O_3 composite nanofiber mats exhibited higher electroactive phase and superior energy harvesting potential in comparison to pure PVDF and developed power from human wrist movements. The observed results were attributed to ion–dipole interactions between the positively charged magnetic particles and –CF_2– dipoles of PVDF [11]. Piezoelectric nanofiber mat synthesized from 5 wt.% ZnO NP dispersed PVDF solution exhibited higher electroactive phase and mechanical strength than pure PVDF. The crystallinity of composite nanofiber mat increased because of the inclusion of large aspect ratio ZnO NRs and it contributed to the performance improvement. The morphology of the nanofiller also influenced the piezoelectric performance of electrospun PVDF nanocomposites. For instance, ZnO NRs dispersed PVDF mats possessed higher energy harvesting potential than ZnO NPs dispersed PVDF mats. ZnO NRs with a high aspect ratio aligned along the nanofiber and functioned as nucleating agents for β-phase formation [12]. The effect of adding nanofillers with a high aspect ratio on the crystallization of PVDF during electrospinning is schematically represented in figure 10.2. The thermal stability, degree of crystallinity and tensile strength of electrospun PVDF nanofibers increased upon the incorporation of propylammonium lead iodide ($C_3H_7NH_3PbI_3$). A PENG constructed from PVDF/4.76 wt.%

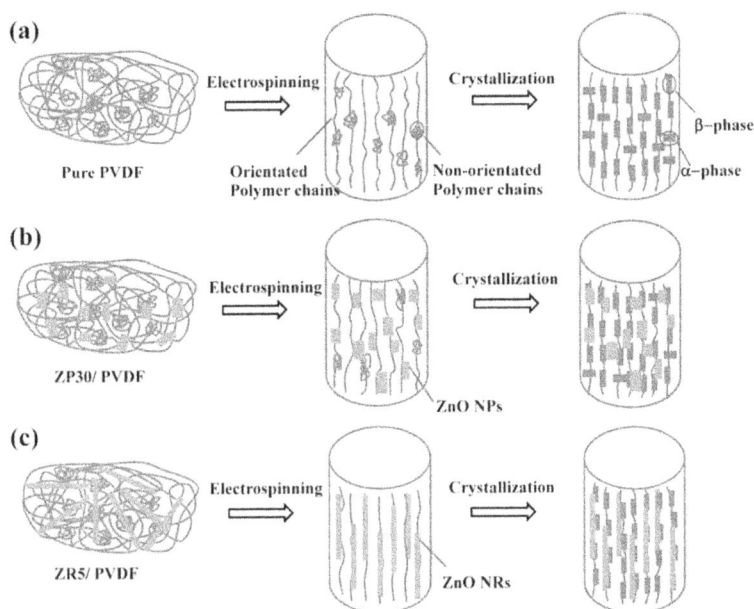

Figure 10.2. Schematic showing the effect of electrospinning and nanofiller aspect ratio on the crystallization of PVDF. Reprinted with permission from [12], Copyright 2019, American Chemical Society.

$C_3H_7NH_3PbI_3$ composite nanofibers generated an open circuit voltage (V_{OC}) of 60 V and a short-circuit current (I_{SC}) of 27.5 μA. Further, the $C_3H_7NH_3PbI_3$ incorporated PVDF nanofibers possessed excellent mechano-sensitivity and developed 12.12 $V N^{-1}$ with an energy conversion efficiency of 42.03%. The interaction between the –CH_2– dipoles of the PVDF and surface charges of $C_3H_7NH_3PbI_3$ contributed to the piezoelectric performance improvement [13].

Hybrid nanofillers are introduced into polymer solutions to garner the simultaneous benefits offered by two or more nanofillers. The output voltage and current generated by hybrid nanofillers-reinforced PVDF nanocomposite fibers are better than single nanofiller-incorporated composite. The synergetic interaction between the hybrid nanofillers and polymer matrix develops a path for additional interfacial polarization and enables electrospun piezoelectric PVDF nanocomposite patches to be highly effective. Conductive and semiconductive NPs are used as a constituent of hybrid nanofillers so that the conductivity of the hybrid nanofillers dispersed precursor polymer solution increases and the fibers undergo better mechanical stretching during electrospinning. Some of the hybrid nanofillers suggested by researchers for boosting the piezoelectric energy harvesting potential of electrospun PVDF nanofibers are KNN/ZnO [14], KNN/carbon nanotube (CNT) [15], ZnO/graphene [16], ZnO/iron oxide [17], MXene/manganese dioxide (MnO_2) [18], ZnO/single-walled carbon nanotube (SWCNT) [19], bismuth ferrite ($BiFeO_3$)/$BaTiO_3$ [20], zinc stannate ($ZnSnO_3$)/molybdenum disulphide (MoS_2) [21], multi-walled carbon nanotube (MWCNT)/copper (Cu) [22], MWCNT/$BaTiO_3$ [23], bismuth trichloride ($BiCl_3$)/ZnO [24].

Numerous clay minerals (phyllosilicate group) with high dielectric constant and specific surface area occur naturally and as reinforcement, with the capability to induce to electroactive β-phase in PVDF. Piezoelectric patches developed from electrospun PVDF/15 wt.% nano clay generated an output voltage of 70 V with a power density of 68 mW cm^{-2}. The two-dimensional clay platelets improved the thermal stability, and toughness and induced electroactive β-phase, thereby the energy harvesting capability of the composite nanofiber mat increased [25]. Similarly, electrospun PVDF nano-composite fabrics with 0.50 wt.% of talc possessed superior energy harvesting potential than pure PVDF fabrics and it was attributed to the hydrogen bonding interaction between electronegative fluorine atoms of PVDF and hydroxyl groups present in talc [26]. Mica nanosheet (MNS) dispersed PVDF composite nanofabrics functioned as a PENG and were able to generate a V_{OC} of ∼8.4 V when a load of 8 N was applied. The plausible mechanism of interaction between PVDF chains and MNS is shown in figure 10.3. The incorporation of 0.75 wt.% MNS enhanced the dielectric constant and promoted the β-phase content of composite nanofabrics. Further, the interaction between CH$_2$ groups of PVDF chains and O-atoms of MNS and the establishment of hydrogen bonding between the CF$_2$ groups of the polymer and hydroxyl groups of MNS contributed to the enhancement of electroactive phases of PVDF/MNS composite nanofabrics [27].

Conductive nanofillers offer better electrical properties, high surface area and aspect ratio and can augment the piezoelectric performance of electrospun PVDF

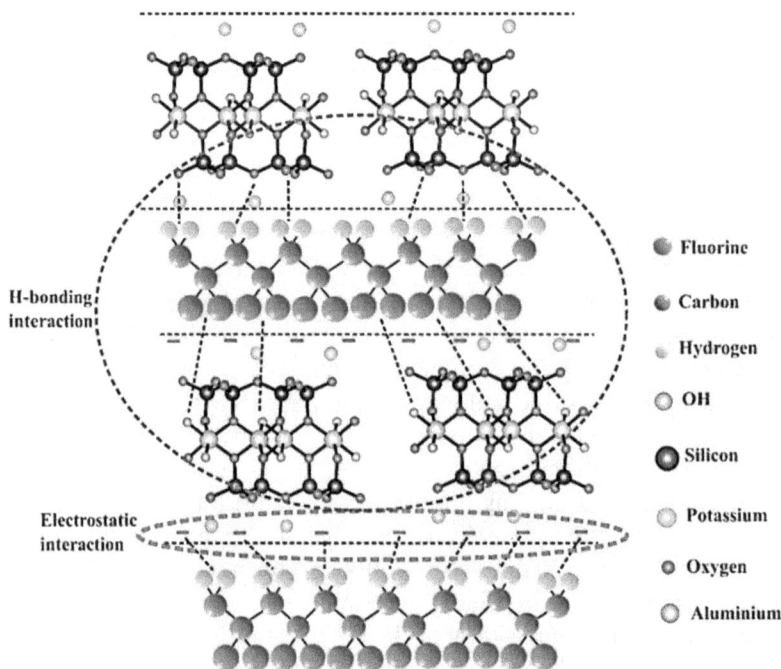

Figure 10.3. Plausible mechanism of interaction between PVDF chains and mica nanosheets. Reprinted from [27] with permission from the Royal Society of Chemistry. CC BY 3.0.

nanofibers. The conductive NPs serve as a nucleating agent for β-phase formation and assist in dipole alignment and charge mobility. Moreover, several micro-capacitors are formed within conductive NP-reinforced PVDF film and it helps in enhancing the piezoelectric response. Conductive nanofillers such as graphene [28], graphene oxide [29], carbon nanofibers [30], MWCNTs [31], silver NPs [32], and MXene [33] have been employed as reinforcements for augmenting the piezoelectric properties of electrospun PVDF nanofibers. The incorporation of conductive nanoparticles improves the conductivity of the solution and enhances the stretching effect during electrospinning, resulting in the formation of the β-phase. In addition to β-phase improvement, the electrospun PVDF/conductive filler nanocomposites also benefit from the improved volume and surface conductivities of the nanofiber mats. The piezoelectric mats with the optimum volume and surface conductivity facilitate the efficient charge transfer during the piezo test. However, excessive addition of conductive nanofillers beyond the percolation threshold limit may result in dielectric loss. As a result, the leakage current from the device increases and lowers the piezoelectric response of the developed NG. Functionalized nanofillers and surface-modified nanofillers are another class of reinforcements considered for strengthening the energy harvesting potential of electrospun PVDF nanofiber mats. The physical and chemical functionalization avoids the aggregation of nanofillers and creates an additional site for interaction between the nanofillers and polymer matrix. Researchers have developed high-performance PVDF nanocomposite piezo-electric patches by embedding functionalized and surface-modified nanofillers such as phenyl-isocyanate functionalized GO [34], oleylamine–functionalized boron nitride (BN) nanosheets [35], amino-functionalised BN nanosheets [36], carboxy-lated functionalized hBN [37], silane-modified KNN [38], poly(methyl methacrylate) (PMMA)-coated hyperbranched $BaTiO_3$ [39], COOH-functionalized graphene nanosheet [40], polyaniline-coated halloysite nanotube [41], and polyaniline (PANI)-coated graphitic carbon nitride (g-C_3N_4) [42]. Table 10.1 summarizes the electrical output generated by electrospun PVDF nanocomposite-based PENGs.

The findings reported in published works confirm that electrospinning is an effective method for fabricating PVDF nanocomposite fabrics with enhanced electroactive phases for energy harvesting applications. Advancements in the Internet of Things and flexible NG technology have resulted in extensive usage of wearable electronics in recent years. The electrospun piezoelectric patches can be integrated with conventional textiles and used as flexible NGs to generate electricity from biomechanical movements and power wearable smart devices. The piezo-electric NGs developed from electrospun PVDF nanocomposites can also serve as power sources for various sensors and self-powered devices. The source of mechanical energy for actuating piezoelectric NGs can be from human movements, wind flow, fluid flow or vibrations from automobiles, and industrial machinery. The incorporation of nanofillers endows electrospun PVDF nanofiber mats with improved electroactive phases, crystallinity, and mechanical strength. Additionally, noticeable changes were observed in the surface morphology of the nanofiber mat.

Table 10.1. Summary of piezoelectric output achieved by electrospun PVDF nanocomposite-based PENGs.

Nanofiller	Optimum filler loading (wt.%)	Maximum β phase %	Output performance and testing conditions	References
KNN	5	68	1.9 V	[7]
ZnO	5	—	4.4 V and 160 nA@5 N, 4 Hz	[9]
BaTiO$_3$	10	91	50 V, 0.312 mA m^{-2}, and 4.07 mW m^{-2} @ ~3 N, 4 Hz	[8]
Ce–Fe$_2$O$_3$	2	75	20 V and 0.010 μA cm^{-2} @ 2.5 N	[11]
ZnO NRs	5	90.7	85 V and 2.2 μA @ 4 Hz repetitive bending and relaxation	[12]
KNN/ZnO	3 wt.% KNN/ 2 wt.% ZnO	94	25 V, 1.81 μA and 11.31 μW cm^{-2} @ finger tapping	[14]
KNN/CNTs	3 wt.% KNN/ 0.1 wt.% CNTs	84	23.24 V, 9 mA and 52.29 mW cm^{-2}	[15]
Graphene–ZnO nanocomposite	1	62.36	840 mV @ 1 N	[16]
FeO–ZnO	3	—	5.9 V @2.5 N	[17]
ZnO–decorated SWCNT	0.75 wt.% ZCNT	95	15.5 V and 8.1 μW cm^{-2} @ heel pressing (0.5–0.6 MPa)	[19]
0.67BiFeO$_3$– 0.33BaTiO$_3$ nanocomposite	30	86.4	83 V, 1.62 μA and 142 mW m^{-2} @ 0.1 kgf at 3 Hz	[20]
ZnSnO$_3$–MoS$_2$	40 wt.% ZnSnO$_3$ and 6 wt.% MoS$_2$	87	26 V, 0.5 μA, and 28.9 mW m^{-2} @ finger tapping	[21]
MWCNTs–Cu	5 wt.% Cu and 0.1 wt.% MWCNT	64.7	18 to 20 V @ foot pressure	[22]
Cloisite 30B	15	79	70 V and 68 μW cm^{-2} @ finger tapping.	[25]
Talc	0.50	89.6	9.1 V and 1.12 μW cm^{-2} @ finger tapping	[26]
Mica nanosheet	0.75	84.3	8.4 V and ~3 μW cm^{-2} @ 8 N force	[27]
Graphene	0.1	83	7.9 V @ finger tapping	[28]
Graphene oxide (GO)	2	87	700 nA and 16 V @ finger tapping	[29]
Functionalized CNF	1	85	44 V, 1.3 μA and 57.2 μW @ finger tapping mode	[30]

Phenyl-isocyanate functionalized GO	0.05	97.1	6.8 V @ 1 N	[34]
Functionalized hexagonal boron nitride nanosheets	1	90	23 V @ 1.5 N and 4 Hz tapping force	[37]
Surface-modified KNN	3	95	$\sim$21 V, $\sim$22 mA, $\sim$5.5 mA cm^{-2} and $\sim$115.5 mW cm^{-2} @1.1 kPa pressure	[38]
Polyaniline-coated halloysite nanotube	1	87.7	30 V, 600 nA and 41.6 1W @ 5 N, 3 Hz	[41]
Polyaniline-coated graphitic carbon nitride nanosheets	20	97	30 V and 3.7 µA @ punching	[42]

10.3 Electrospun PVDF nanocomposite-based triboelectric nanogenerators

A triboelectric energy harvesting system uses triboelectrification and electrostatic induction principles to scavenge electrical energy from the unused kinetic energy available in the environment. A TENG comprising tribopositive and tribonegative layers generates electrical energy once the periodic contact between tribolayers occurs. Common operating modes of a TENG are vertical contact separation mode, single-electrode mode, lateral sliding mode and free-standing mode (figure 10.4) [43]. Robust tribolayers with high surface roughness, surface area, mechanical strength and electronegativity difference are preferred for the effective harnessing of ambient energy available in the environment. Various factors affecting the energy harvesting performance of TENGs are summarized in figure 10.5 [44]. The electrospun PVDF composite NFs exhibit high flexibility, specific surface area, biocompatibility, surface wettability and improved dielectric properties. Hence, the TENGs based on composite nanofibers have the potential to be used as a power source for a wide range of electronic devices. The performance of the TENGs is influenced by the device structure, functional materials and surface microstructure of tribolayers. To achieve a large difference in electronegativity between tribolayers, materials that are far way in the tribo series must be selected for the construction of TENGs. Different materials according to their triboelectric charge densities are summarized in figure 10.6 [45]. When electrospun PVDF composite nanofiber mat is used as a functional material for the construction of TENG, the electronegativity difference and the effective contact area between the tribolayers are enhanced because of the coupling effect between porous structure and nano-network structure of electrospun composite nanofiber mat. Electrospun PVDF nanocomposite mats with a high dielectric constant are used as a tribonegative layer of TENG and the tribopositive

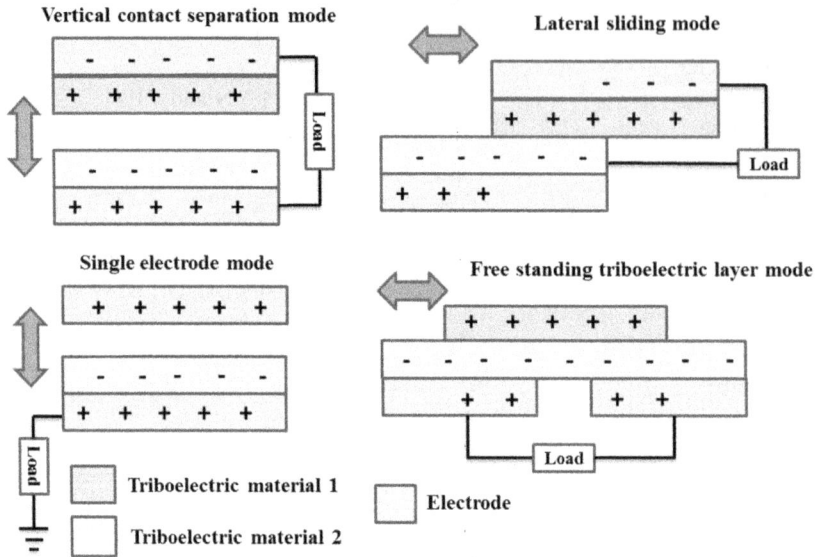

Figure 10.4. Different operating modes of TENGs. Reprinted from [43], copyright 2022, with permission from Elsevier.

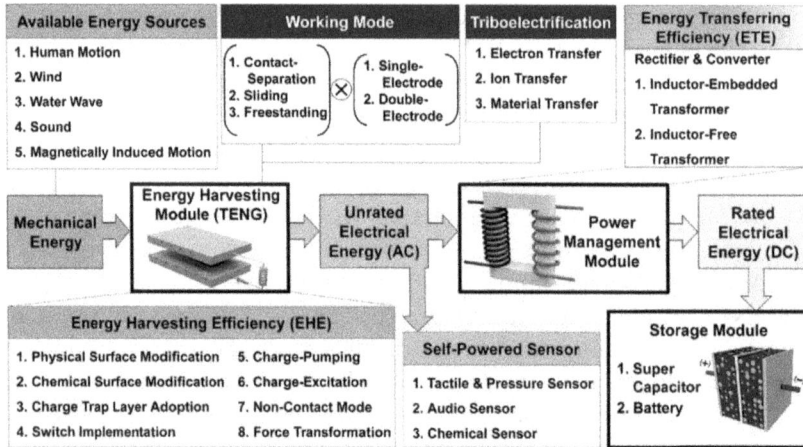

Figure 10.5. Various factors influencing the energy harvesting performance of TENGs. Reprinted with permission from [44], Copyright 2021, American Chemical Society.

layer may be an electrospun nanofiber mat or readily available materials such as copper, aluminium, silk, nylon, etc [46].

Numerous PVDF-based nanocomposite mats with high dielectric properties, mechanical strength and flexibility have been reported by researchers for constructing TENGs with high power density. Conductive, semiconductive and high dielectric constant nanofillers are adopted by researchers for improving the energy harvesting performance of electrospun nanofiber-based TENGs. The addition of

(+)	Tribopositive	Tribonegative	
	Borosilicate Glass	Poly(ethylene terephthalate)	
	Quartz	Natural rubber	
	Mica	Poly(acrylonitrile)	
	Polyamide 11, Polyamide (Nylon 6,6)	High impact polystyrene	
	Polyurethane foam	High temperature silicone rubber	
	Human skin	Polyethylene	
	Leather	Poly(ether ether ketone)	
	Wool	Poly(vinylidene fluoride)	
	Fur	Polyimide	
	Silk	Polyester fabric	
	Poly(ethylene glycol succinate)	Poly(dimethylsiloxane)	
	Cellulose	Poly(etherimide)	
	Paper	Polystyrene	
	Cotton	Polycarbonate	
	Wood	Acrylonitrile butadiene styrene	
	Poly(methyl methacrylate)	Poly(vinyl chloride)	
	Poly(vinyl alcohol)	Poly(trifluorochloroethylene)	
	Aluminium, steel	Fluoro elastomer rubber	
	Nickel, Copper	Poly(tetrafluoroethylene)	
	Brass, Silver	Fluorinated ethylene propylene	(-)

Figure 10.6. Triboelectric material series [44, 45].

conductive nanofiller creates charge-trapping sites and enhances the charge transfer efficiency of nanofiber mats. The PVDF nanofiber mat incorporated with semi-conductive nanofillers undergoes polarization and the micro-electric field developed by the polarized charge contributed to the high surface charge density. Similarly, nanofillers with high permittivity enhance the output performance of nanofiber mat by improving its polarization strength. Varghese *et al* constructed a flexible TENG using an electrospun PVDF nanofiber mat as tribonegative material and paper as counter material for biomechanical energy harvesting. The proposed TENG finds application in telerehabilitation by harnessing energy from different human motions namely elbow bending, wrist flexion and finger tapping. A maximum V_{OC} of 160 V with a short-circuit current density of 270 μA m^{-2} was generated from finger tapping [47]. Electrospun PVDF/natural sepiolite nanocomposite with improved tensile strength and surface roughness was reported by Yin *et al*. The morphology and diameter of the fiber are also affected by the nanofiller addition. The TENG comprising PVDF/15 wt.% sepiolite nanocomposite mat (tribonegative) and glass fiber fabric (tribopositive) layer was able to generate a V_{OC} of 740 V and maintain the performance over 1000 cycles [48]. Fullerene (C_{60}) finds application in numerous electroactive systems because of its exceptional electron-accepting characteristics. Further, the presence of π-bonds and sp^2 hybridization enables C_{60} NPs to exhibit high electronegativity and charge trap ability.

The incorporation of electronegative C_{60} additives made electrospun PVDF nano-composite fibers more electronegative by providing a greater number of charge trap sites and augmenting the surface tribo-charge density of composite nanofibers. As a result, TENGs fabricated from aluminium contact electrodes and PVDF/0.025 wt.% C_{60} composite nanofibers delivered a maximum output power of 282 μW [49]. A ZnO nanowire dispersed PVDF and nylon composite nanofiber mat with high polar crystalline β-phase and δ'-phase (nylon) was electrospun by Pu et al to construct a TENG. The addition of ZnO NWs enhanced the mechanical properties and thermal stability of nylon and PVDF fibrous mats. The electrostatic interaction between the dipoles of the PVDF polymer chains and the surface charges of the ZnO NWs leads to the alignment of ZnO NWs within fibers and promotes the formation of polar phases. The triboelectric performance of the nanofiber mat was significantly enhanced by the addition of 5 wt.% ZnO NWs and a maximum V_{OC} of 330 V was generated by the proposed TENG system [50]. Li et al used carbon-coated zinc oxide (ZnO@C) NPs as fillers for modifying the piezo and triboelectric properties of an electrospun PVDF nanofiber mat. When 5 wt.% of ZnO@C NPs were added, the PVDF's surface potential was shifted to −740 mV, thereby improving the triboelectric property of the composite nanofiber mat. Wearable sensors developed from ZnO@C/PVDF nanofabric exhibited a high sensitivity of 0.98 V kPa^{-1} and served as body motion sensors [51].

Guo et al incorporated a fluorinated metal–organic framework (F-MOF) with high charge-trapping and charge-inducing capabilities into PVDF to enhance its triboelectric performance. The peak power density produced by the PVDF@F-MOF NFs and Al foil pairs reached a maximum of 121 μW cm^{-2} and it was attributed to the surge in the hydrophobicity and surface roughness of PVDF@F-MOF composite NF mat [52]. Rahman et al used cobalt-based nanoporous carbon (Co-NPC) as a nanofiller for modifying the triboelectric properties of PVDF. The Co-NPC derived from MOF possessed ultrahigh surface area with fine pore-size distribution. The nanoporous carbon acted as a nucleating agent for the formation of electroactive β-phase and improved the dielectric constant of the composite nanofiber mat. Further, the charge trapping capability and surface potential of the PVDF composite nanofibers increased upon the incorporation of nanoporous carbon resulting in performance improvement of Co-NPC/PVDF-based TENG [53]. Sohn et al used MOFs with MIL-101(Cr) structure as reinforcement for improving negative surface potential and electroactive β-phase of electrospun PVDF nanofiber mat. The addition of MIL-101(Cr) increased the negative surface potential and the PVDF/MIL-101(Cr) composite nanofiber mat undergoes self-poling. During electrospinning, pores present in the MIL-101(Cr) filler retained solvent within the nanofiber, and the retained solvent evaporated gradually from the nanofiber mat during the ageing process, resulting in β-phase improvement through the self-poling effect. The TENG fabricated from self-poled PVDF/0.8 wt.% MIL-101(Cr) composite nanofiber generated power with a maximum power density of 871.2 μW cm^{-2} and it was 13.7 times higher than power developed by pure PVDF nanofibers [54].

The surface charge density, dielectric property, surface roughness and effective surface area of electrospun PVDF nanofiber mat increased upon the incorporation of

10 wt.% MXene ($Ti_3C_2T_x$). The formation of a micro-capacitor network and microscopic dipole contributed to the dielectric constant increment. The TENG based on electrospun Nylon 6/6 nanofiber mat and PVDF/MXene composite nanofibers delivered a peak power of 4.6 mW and was used as a power source for commercial LEDs and low-power electronics [55]. A book-shaped TENG made from poly(3-hydroxybutyrate-co-3-hydroxyvalerate) (PHBV) nanofiber mat and PVDF/0.7 wt.% graphene oxide (GO) composite nanofibers produced peak-to-peak voltage of 340 V and was stable over 18 000 cycles. The GO served as a charge trapping site and increased the charge storage ability and output performance of PVDF/GO composite nanofibers [56]. In a study by Dai *et al*, PVDF/GO nanosheets composite nanofibers and cotton fibers were used to fabricate high-performance TENGs. A reduction in the average fiber diameter was observed upon the addition of GO and the triboelectric potential difference between the PVDF/GO layer and cotton layer increased. The higher triboelectric potential differences between layers have a positive effect on TENG performance and a maximum output voltage of 450 V and I_{SC} of 35μA was generated by the proposed setup [57]. Similarly, TENG comprising PA6 films and PVDF/1.5 wt. % graphene nanosheet composite nanofibers generated a maximum output voltage of ~1511 V with a peak power density of ~130.2 Wm^{-2}. The graphene nanosheet acted as a charge-trapping site and improved the surface potential and charge density of the composite nanofibers [58]. A lightweight TENG constructed from copper mesh (tribopositive) and polyacrylonitrile (PAN)/PVDF/1 wt.% MWCNTs composite nanofiber mat functioned as an acoustic energy harvester and developed a V_{OC} of 120 V from incident acoustic sound waves of 116 dB at 200 Hz. The charged voltage of the 2.2 μF capacitor reached 26.5 V after 120 s of charging by the TENG [59]. In the presence of carbon-based nanofillers, the interfacial polarization and the dielectric constant of the electrospun PVDF composite increased, resulting in an enhancement of triboelectric charge density and the output voltage and current.

Magnetic magnetite (Fe_3O_4) NPs dispersed PVDF composite nanofibers with improved energy harvesting capability and EMI shielding performance were proposed by Im *et al*. The PVDF nanofibers with 11.3 wt.% of Fe_3O_4 showed better output and the performance improvement was ascribed to the presence of NPs on the fiber surface and the electret doping effect induced by NP [60]. A TENG composed of 0.5 wt.% hexagonal boron nitride nanosheets (hBNNSs) incorporated PVDF nanofibers and nylon-11 nanofibers showed high flexibility and generated V_{OC} as high as 500 V. The large specific surface area and piezoelectric properties of hBNNSs enabled the composite nanofiber mat to efficiently convert the mechanical energy to electric energy [61]. Govind *et al* fabricated arc-shaped and fluttering-driven TENG from barium tungstate NRs (BWN) dispersed PVDF nanofabrics and used them for harvesting electrical energy from biomechanical movements and wind energy. PVDF-3wt.% BWN composite nanofabrics benefit from the higher dielectric constant of BWN and the interfacial/electrostatic interaction of BWN with PVDF chains enhanced the electroactive phases and dielectric constant of the composite nanofabrics. The TENG produced maximum voltage output of 200 V and 84 V from single finger tapping and a wind speed of 7 m s^{-1}, respectively [62]. Zhou *et al* fabricated a microwaved patterned nanofiber mat from silicon nitride (Si_3N_4) doped

PVDF and TPU solution and used it as a tribonegative and tribopositive layer for constructing TENG. Further, the PVDF/1.5 wt.% Si_3N_4 composite nanofiber mat was subjected to electret doping to improve the tribo performance. The micro-waveform structured nanofiber mat demonstrated better power output in comparison to plain nanofiber mat and a maximum current and output voltage of 1.02 μA and 102 V was generated by the system [63]. Pandey *et al* synthesized a PVDF composite nanofiber mat with enhanced negative surface potential by incorporating Nafion™ functionalized $BaTiO_3$ NPs as reinforcement. Nafion™ undergoes hydrophobic interaction with PVDF chains and hydrophilic interaction with $BaTiO_3$ NPs because of which the dispersion of NPs within the PVDF matrix is promoted. As a result, the effective stress transfer happened between the interface of PVDF and $BaTiO_3$ and a TENG based on PVDF/5 wt.% $BaTiO_3$ composite nanofiber mat produced a maximum voltage of 309 V with a current density of 1.8 μA cm^{-2} [64].

The discussed works confirm that electrospun nanofiber mats can meet the requirements of TENGs and be used as functional tribolayers. The TENGs based on electrospun nanofiber mats find application in numerous fields including human energy harvesting, self-powered wearable devices, flexible electronics, wind energy harvesting, and acoustic energy harvesting. Flexible TENGs can function as pressure/touch sensors, acoustic sensors, biomedical sensors, chemical sensors, and vehicle and industrial machinery monitoring sensors. The output current, voltage and power developed by TENGs depends on the triboelectric charge density of the tribolayers. The material composition, effective contact area and environmental conditions determine the triboelectric charge density. An electrospun nanofiber mat enjoys the benefit of high specific surface area and its tribo properties are tailored through the incorporation of nanofillers. Self-powered devices and sensors constructed based on electrospun PVDF nanocomposites are discussed in the forthcoming section.

10.4 Electrospun PVDF nanocomposite-based hybrid piezo and triboelectric nanogenerators

A hybrid NG is a unified energy harvesting device developed by merging various energy harvesting principles into a single unit. It is a typical integration of different energy harvesting mechanisms such as piezoelectric, pyroelectric, photoelectric, triboelectric and electromagnetic. This integration allows the utilization of different energy sources either independently or simultaneously, providing flexibility in capturing diverse forms of ambient energy. Additionally, the strategic integration of two or more harvesters enhances the overall electric power output [65–67]. The need for hybrid NGs arises from the requirement to efficiently extract energy from a variety of environmental resources. Hybridization provides various benefits such as improved efficiency, continuous power supply, and flexibility. Hybrid NGs provide better overall efficiency than standalone systems by integrating several energy collecting techniques and also, the capacity to tap into numerous kinds of energy guarantees a more consistent and dependable power supply [68].

The piezoelectric effect refers to the phenomenon in certain materials where mechanical stress induces a change in polarization, leading to the generation of an electric field. PENGs use the principle of direct piezoelectric effect to generate electricity [69]. In contrast, a TENG generates electricity through the combination of triboelectric effect and electrostatic induction. The triboelectric effect is a phenomenon that arises when two dissimilar materials make contact, leading to the generation of electricity [70]. Because PENG and TENG share similar impedance and structural characteristics, and their shared capability to convert mechanical energy into electricity, combining them into a single system is highly plausible to enhance the electrical output from their unique mechanisms. The primary limitation of PENGs lies in their lower electrical outputs, which narrow their applicability in various applications, whereas TENGs exhibit humidity-sensitive electrical responses. Therefore, adopting a hybrid approach between PENG and TENG can mitigate these constraints and enhance the unique strength of each NG [66]. The alignment of piezoelectrically polarized charges and tribo-electrically induced charges in the same direction enhances power outputs, whereas their opposite alignment significantly reduces electricity generation [65].

Based on the structural design featuring various shared components, three hybrid coupling modes exist for hybrid piezoelectric and triboelectric nanogenerators (HPTNGs). Firstly, the two-electrode mode (Mode-I) involves the use of only two electrodes in the fabrication of HPTNGs. Secondly, the three-electrode mode (Mode-II) employs three electrodes, with one electrode serving as a common element for both PENG and TENG. Finally, the four-electrode mode (Mode-III) in which the device consists of four electrodes allows the independent operation of PENG and TENG [67]. In mode-I, the NG is still considered a HPTNG, even if the dielectric layer is absent (figure 10.7).

Guo et al have developed an all-fiber HPTNG using electrospun PVDF and silk/PEO nanofibers. The hybrid NG was configured as mentioned in mode-I (figure 10.8). The HPTNG demonstrated exceptional electrical performance, harnessing the synergistic effects of both piezoelectric and triboelectric mechanisms. The V_{OC}, I_{SC} and instantaneous power density of 500 V, 12 μA and 0.31 mW cm^{-2}, respectively were generated. Additionally, the HPTNG was applied for wearable gesture monitoring [71]. Similarly, Ünsal et al synthesized porous PVDF and TPU nanofiber through electrospinning and utilized them for constructing a HPTNG. For improving output performance, electrospun nanofibers were decorated with ZnO NWs and rGO NPs by the application of hydrothermal growth and

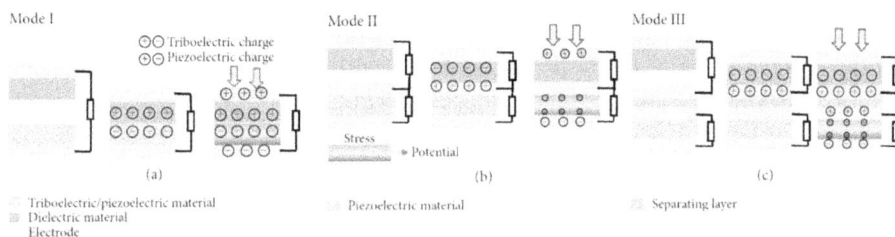

Figure 10.7. Operational modes of HPTNG are (a) two-electrode mode, (b) three-electrode mode, and (c) four-electrode mode. Reprinted with permission from [67], Copyright 2021 Lingjie Xie et al. CC BY 3.0.

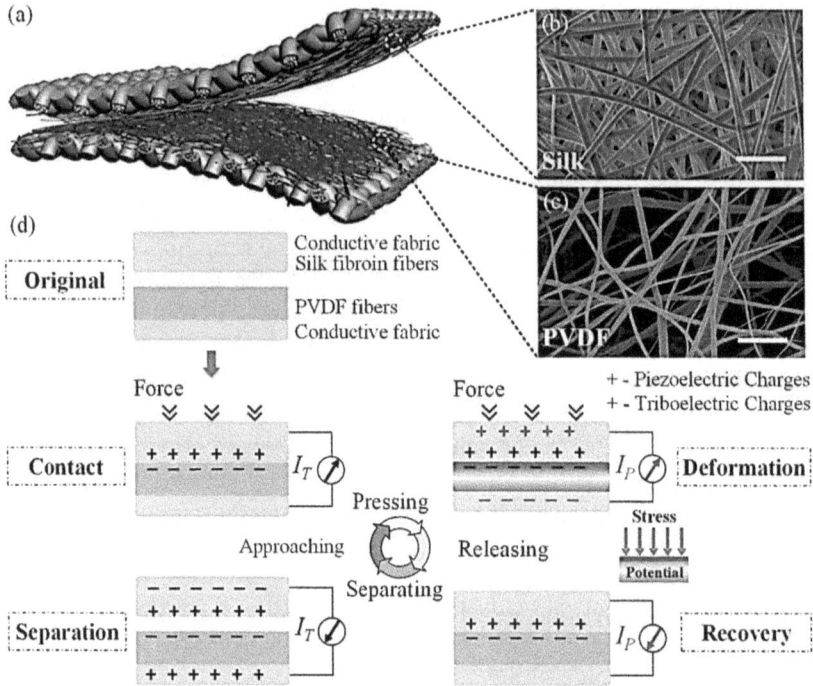

Figure 10.8. (a) Graphical illustration of HPTNG, SEM image of electrospun (b) PVDF nanofibers, (c) silk nanofibers, (d) operating modes of electrospun nanofiber-based HPTENG. Reprinted from [71], copyright 2018, with permission from Elsevier.

electrospraying techniques. The HPTNG constructed from NP-coated nanofiber generated the maximum power with a voltage density of 2.35 kV m^{-2} and current density of 3.40 mA m^{-2} [72]. Song *et al* fabricated flexible textile-based hybrid NG consisting of electrospun PVDF/CNT/BaTiO$_3$ composite nanofabrics as a piezo-electric layer and poly(dimethylsiloxane) (PDMS)/MWCNT/graphite composite film as a triboelectric layer. Both the layers share a common electrode, representing the mode-II configuration. The HPTNG generated the rectified average peak V_{OC} of 161.66 V, an instantaneous power output of 2.22 W m^{-2} and lit up 150 LEDs [73].

In summary, the combined potential of piezoelectric and TENGs opens up exciting possibilities across various applications for HPTNGs. Their capability to harvest ambient vibrational energy efficiently makes them versatile and reliable sources of power. The ongoing exploration and optimization of NGs indicate their potential to revolutionize methods of energy harvesting and contribute significantly to a more sustainable future.

10.5 Self-powered devices based on electrospun PVDF nanofabrics

Self-powered technology enables a device to meet its power requirement by harnessing the free energy available in the environment and eliminating the need for an external power source. PENG- and TENG-based self-powered devices and sensors could be potential candidates for numerous applications including human motion sensing, pressure-sensing, gas sensing, air filtering, tactile sensing, structural health monitoring, acoustic energy harvesting, etc. Zhang *et al* presented a liquid metal-based TENG for testing vibrations of different accelerations. The proposed self-powered acceleration sensor encompassed mercury droplets as a tribopositive layer and electrospun PVDF nanofabrics as tribonegative material. The performance of the device was found to be a function of the volume ratio of the mercury droplet with respect to nanofiber volume. The triboelectric sensor with a volume ratio of 0.25:1 exhibited better performance and generated a V_{OC} of 15.5 V and I_{SC} of 300 nA at an acceleration of 60 m s^{-2}. Further, the sensor was able to identify the operational status of a car engine and air compressor. The sensor affixed to the car engine generated a voltage of 2.5 V as soon as the ignition started, thereafter the V_{OC} dropped to 0.25 V and maintained the same level till the engine ran. The sensor stops generating the response once the car engine stops. The authors also employed the sensor for human gait analysis by attaching it to shoes. The sensor detected the changes in the vibration arising from human movements i.e., walking and running [74].

Fuh *et al* fabricated a self-powered piezoelectric pressure sensor by integrating 3D printing and a near-field electrospinning process (NFES). PVDF nanofibers were deposited directly on the 3D-printed thermoplastic elastomer (TPE) wavy substrate and used for sensing applications. The TPE substrate with different surface topologies namely planar, square and sinusoidal was printed and the effect of surface topology on the piezoelectric output was determined. Among the topologies considered, the sinusoidal surface-based substrate exhibited higher piezoelectric output, and it was attributed to the larger strain developed by the sinusoidal pattern. The developed wavy sensor functioned as a self-powered foot pressure mapping sensor and detected the pressure distribution across the foot by generating electrical responses. A relationship existed between foot pressure and the electrical response of the sensor proving the ability of the sensor for human health monitoring [75]. Zhang *et al* introduced self-powered friction piezoelectric sensors for pressure-sensing applications. Core–shell nanofibers were electrospun from PVDF/2 wt.% ZnO and PVDF/3 wt.% MXene precursor solution and utilized for sensing. The sensor recognized various human motions and had a maximum sensitivity of 0.751 V N^{-1} [76].

Garcia *et al* developed a self-powered triboelectric sensor to measure and detect the impact energy applied on composite structures. Electrospun poly(vinyl pyrrolidone) (PVP) and PVDF nanofibers served as tribolayers of the sensor. The triboelectric sensor sensitivity was reported in terms of voltage as well as current under the application of impact energy of different amplitude. The sensor's electric response was affected by the amplitude of the impact energy, and a linear relationship existed between the amplitude of the impact energy and the sensitivity of the triboelectric sensor. Notably, the sensor exhibited an excellent impact sensitivity of 160 mV J^{-1} and 13 nA J^{-1} in the energy range of 2–30 J. Further, a sensitivity of 0.4 mV N^{-1} and 0.03 nA N^{-1} was observed when the impact forces were varied from 2000 to 15 000 N. The finding suggests that the developed sensor can accurately measure impacts in real time with minimal delay, making it a potential candidate for monitoring impacts in structures such as aircraft, wind turbines, and bridges [77].

Chen *et al* introduced a unique design for a TENG that can harvest acoustic energy from the environment and measure the velocity of objects in motion. For harvesting acoustic energy, a ring-shaped TENG containing electrospun PVDF nanofabric and conductive fabric electrode was housed inside a polymer tube. The construction and working mechanism of sound-driven TENG are shown in figure 10.9. The poly(vinylidene chloride) (PVC) tubing helped in enhancing the pressure of sound waves hitting the TENG. The sound-driven TENG generated an open circuit voltage of 400 V when it was subjected to sound pressure of 115 dB at a frequency of 170 Hz. The velocity sensing system devised by the authors had a larger PVC pipe in which two sound-driven TENGs were placed at a distance of 70 mm from each other. The velocity sensing system was positioned near the path of moving sound objects and the sound-driven TENGs generated an electrical signal when the linear motor platform with speaker moved at specific velocity. The time difference between two maximum peaks and the distance between the two sound-driven TENGs was utilized by the authors for calculating the velocity of the speaker. Further, the proposed system was able to record sound and detect the sources of sound [78]. Similarly, Sultana *et al* synthesized a piezoelectric acoustic sensor from 0.5 wt.% zinc sulphide nanorods (ZnS-NRs) infused PVDF nanofiber mat and evaluated its energy harvesting and sensing capabilities. The acoustic energy harvester was able to produce a maximum V_{OC} of ~6 V with an acoustic sensitivity of around ~3 V Pa^{-1} when subjected to a sound pressure level of 100 dB. Additionally, the acoustic energy harvester demonstrated a power density of ~0.15 µW cm^{-2} across a load resistance of ~0.7 MΩ. The piezoelectric acoustic

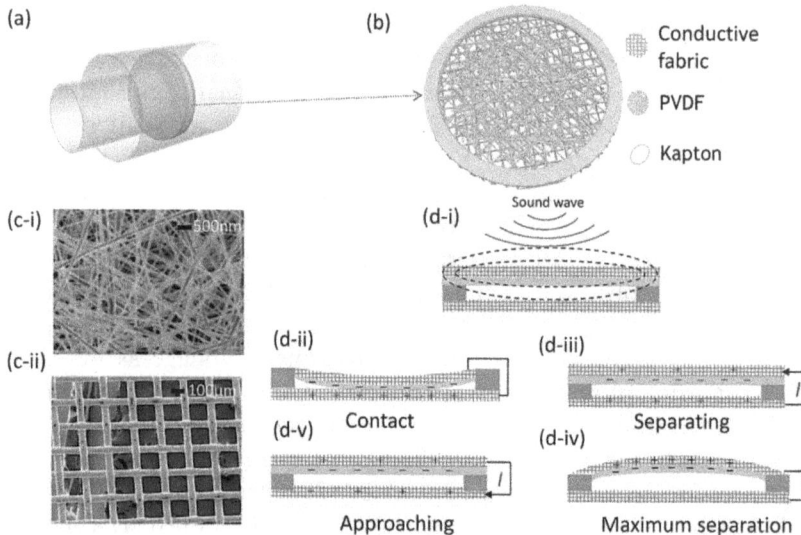

Figure 10.9. Schematic of sound-driven TENG and its working mode, (a) Construction of sound-driven TENG, (b) TENG comprising PVDF nanofiber mat and conductive fabric, (c) SEM image of PVDF nanofiber mat and conductive fabric, (d) Different operating modes of sound-driven TENG. Reprinted from [78], copyright 2019, with permission from Elsevier.

sensor was also examined as a transducer/microphone and the acoustic sensor generated electrical signals from the sound pressure of a human voice. The acoustic sensor's response waveform corresponding to the human voice was processed using short-time Fourier transforms (STFTs), and the resulting spectrogram was compared with the output profile of a commercial microphone. Both the acoustic sensor and the commercial microphone exhibited similar histograms for a particular sound level, which indicates that the acoustic sensor can serve as a self-powered microphone for sound recording [79].

Azimi *et al* utilized an electrospun PVDF/ZnO/GO nanocomposite mat for developing a self-powered cardiac pacemaker. The PENG constructed from a nanofiber mat containing 0.1 wt.% of ZnO/GO (90/10) generated a V_{OC} of ~6 V, I_{SC} of ~3.46 μA, and power density of ~138 μW cm^{-3}. For performing *in vitro* analysis, the developed PENG was encapsulated inside the PDMS structure and implanted into a dog so that it could function as a power source for the heart pacemaker. Biocompatibility studies indicated favourable cytocompatibility for both the designed PENG and the composite fiber mat. The authors sutured the proposed PENG to a dog's heart and studied the piezoelectric response resulting from the biomechanical movement of the heart. The electrical voltage generated by the implanted PENG synchronized with the heartbeat, and generated a V_{OC} of 3.9 V and I_{SC} of 2.33 μA for every cycle. An electrical energy of 0.487 J was produced by the implanted PENG device, which was higher than the energy required by a commercial pacemaker (0.377 J). The durability and mechanical integrity of the device were tested over 6 h, and no significant reduction in piezoelectric performance was observed [80].

Kim *et al* produced functional electroactive PVDF/silver (Ag) composite nanofibers through electrospinning for particulate matter absorption. The friction between the flowing air and the nanofibers generated tribocharges and it helped in particulate matter absorption. The composite nanofiber mat attached to the commercially available particulate absorbing filter demonstrated improved particle absorption efficiency and the pressure drop and power consumption of the nanofiber affixed filter were reduced [81]. A breathable nanofiber mat with improved UV resistance was developed by Sun *et al* by combining electrospinning and electrospraying techniques. The 4 wt.% titanium dioxide (TiO$_2$) NPs dispersed in DMF were electrosprayed during the electrospinning of the PVDF/PDMS blend. The TiO$_2$ nanoparticles embedded in PVDF/PDMS nanofiber exhibited higher surface roughness and dielectric constant and functioned as TENG in single-electrode mode. The proposed TENG involved human skin as tribopositive material and generated electrical signals from various physical activities including breathing [82]. A breathable mask with continuous electrostatic charge replenishing capacity was proposed by Peng *et al*. Figure 10.10(a) shows a schematic of the proposed mask, and the electrical responses generated by the TENG-based mask from human breathing is depicted in figure 10.10(d). The mask containing PVDF nanofiber mat and nylon fabric continuously generated electrostatic charges from human breathing and demonstrated a filtration efficiency of 95.8% for a particle size of 0.3 μm [83]. Cao *et al* fabricated a breathable and skin-friendly self-powered triboelectric sensor

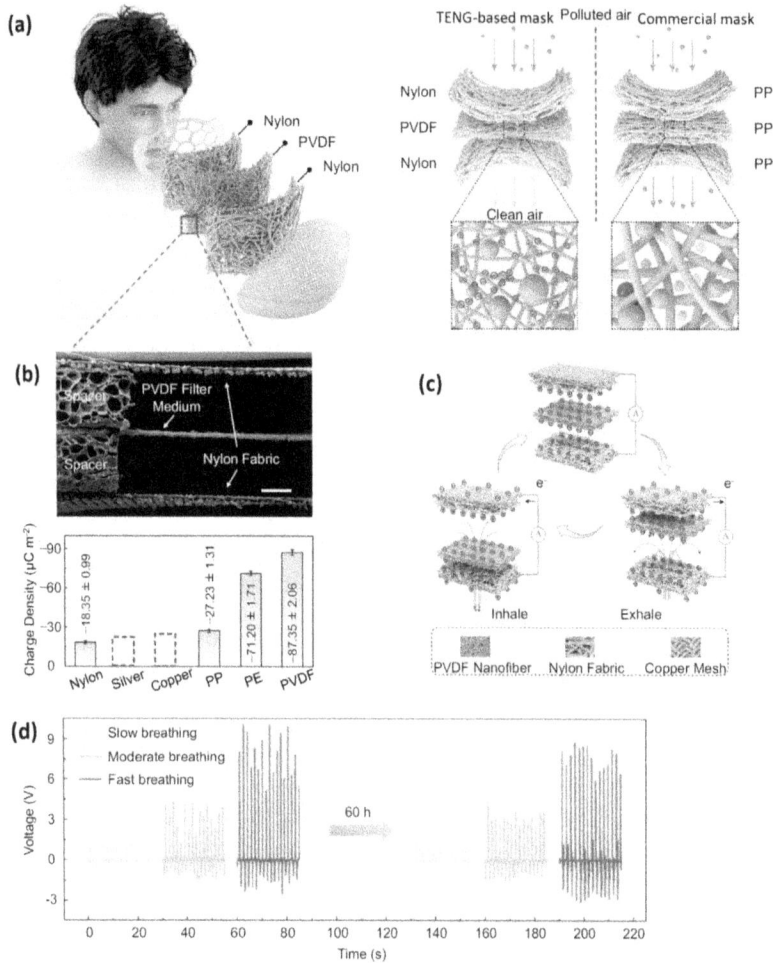

Figure 10.10. (a) Schematic of electrospun PVDF nanofiber-based wearable mask, (b) Cross-sectional SEM images of nanofiber mat, (c) Operating modes of TENG-based wearable mask, (d) Output voltage generated by mask in response to human breathing. Reprinted with permission from [83], Copyright 2022, Zehua Peng *et al.* CC BY 4.0.

for monitoring human respiration. The sensor contained electrospun PVDF nano-fabric as a tribolayer and screen-printed Ag NPs as an electrode. The screen-printed Ag electrode ensured the air permeability of the sensor. The sensor integrated with the breather valve of the mask produced an electrical response proportional to the human breathing rate and functioned as a respiration monitoring sensor [84]. The interesting outcomes reported by the researchers prove that electrospun nanofiber-based PENGs and TENGs can function as sensor and self-powered devices. The nanofibers with optimum filler loading exhibited excellent power generation capabilities and eliminated the external power source requirement for sensors and miniature electronic devices.

10.6 Challenges and future outlook

Even though the piezoelectric and triboelectric energy harvesters are capable of harvesting electrical energy from unused energy sources, the widespread adoption of electrospun polymer composite-based PENGs and TENGs is limited because of certain constraints. Sustainable output and a stable performance are the main prerequisites for energy harvesters. Uninterrupted external excitation is essential for the NGs to generate power continuously. The maximum instantaneous output power delivered by the electrospun PVDF nanocomposite-based PENG and TENG is lower than the power requirement of several electronic devices and a strong power management and power storage system is needed. Environmental factors such as atmospheric temperature and humidity may impact the performance of PENG and TENG. So the device encompassing the NGs must be resistant to atmospheric conditions. The power output of pure polymer nanofiber-based NGs is lower, and therefore great care must be taken when selecting the nanofiller for electrospinning composite nanofibers. The power generation capability of a functional layer in an NG is also influenced by the device structure, and the NG design that can efficiently transfer external excitation to the functional layer is recommended for effective energy harvesting. It can be concluded from the observations that the major factors influencing the efficiency of nanofiber-based PENG and TENG are functional material composition, device structure and power management circuit, and the NG with an optimum combination of the above-mentioned factors can meet the power requirement of miniature and self-powered electronic devices.

References

[1] Lu L, Ding W, Liu J and Yang B 2020 Flexible PVDF based piezoelectric nanogenerators *Nano Energy* **78** 105251
[2] Mohammadpourfazeli S, Arash S, Ansari A, Yang S, Mallick K and Bagherzadeh R 2023 Future prospects and recent developments of polyvinylidene fluoride (PVDF) piezoelectric polymer; fabrication methods, structure, and electro-mechanical properties *RSC Adv.* **13** 370–87
[3] Yan J, Liu M, Jeong Y G, Kang W, Li L, Zhao Y, Deng N, Cheng B and Yang G 2019 Performance enhancements in poly (vinylidene fluoride)-based piezoelectric nanogenerators for efficient energy harvesting *Nano Energy* **56** 662–92
[4] Purushothaman S M, Tronco M F, Kottathodi B, Royaud I, Ponçot M, Kalarikkal N, Thomas S and Rouxel D 2023 A review on electrospun PVDF-based nanocomposites: recent trends and developments in energy harvesting and sensing applications *Polymer* **283** 126179
[5] Shetty S and Anandhan S 2021 Electrospun PVDF-based composite nanofabrics: an emerging trend toward energy harvesting *Nano Tools and Devices for Enhanced Renewable Energy* (Elsevier) ch 10 pp 215–36
[6] Zhang M, Liu C, Li B, Shen Y, Wang H, Ji K, Mao X, Wei L, Sun R and Zhou F 2023 Electrospun PVDF-based piezoelectric nanofibers: materials, structures, and applications *Nanoscale Adv.* **5** 1043–59
[7] Teka A, Bairagi S, Shahadat M, Joshi M, Ziauddin Ahammad S and Wazed Ali S 2018 Poly (vinylidene fluoride)(PVDF)/potassium sodium niobate (KNN)–based nanofibrous web: a

unique nanogenerator for renewable energy harvesting and investigating the role of KNN nanostructures *Polym. Adv. Technol.* **29** 2537–44

[8] Athira B S, George A, Vaishna Priya K, Hareesh U S, Gowd E B, Surendran K P and Chandran A 2022 High-performance flexible piezoelectric nanogenerator based on electrospun PVDF-BaTiO$_3$ nanofibers for self-powered vibration sensing applications *ACS Appl. Mater. Interfaces* **14** 44239–50

[9] Li G Y, Zhang H D, Guo K, Ma X S and Long Y Z 2020 Fabrication and piezoelectric-pyroelectric properties of electrospun PVDF/ZnO composite fibers *Mater. Res. Express* **7** 095502

[10] Dehaghi N G and Kokabi M 2023 Polyvinylidene fluoride/barium titanate nanocomposite aligned hollow electrospun fibers as an actuator *Mater. Res. Bull.* **158** 112052

[11] Parangusan H, Ponnamma D and AlMaadeed M A A 2019 Toward high power generating piezoelectric nanofibers: influence of particle size and surface electrostatic interaction of Ce–Fe$_2$O$_3$ and Ce–Co$_3$O$_4$ on PVDF *ACS Omega* **4** 6312–23

[12] Li J, Chen S, Liu W, Fu R, Tu S, Zhao Y, Dong L, Yan B and Gu Y 2019 High performance piezoelectric nanogenerators based on electrospun ZnO nanorods/poly (vinylidene fluoride) composite membranes *J. Phys. Chem.* C **123** 11378–87

[13] Sengupta P, Sadhukhan P, Saha S, Das S and Ray R 2023 Improved energy harvesting ability of C$_3$H$_7$NH$_3$PbI$_3$ decorated PVDF nanofiber based flexible nanogenerator *Nano Energy* **109** 108277

[14] Bairagi S and Ali S W 2020 A hybrid piezoelectric nanogenerator comprising of KNN/ZnO nanorods incorporated PVDF electrospun nanocomposite webs *Int. J. Energy Res.* **44** 5545–63

[15] Bairagi S and Ali S W 2020 Investigating the role of carbon nanotubes (CNTs) in the piezoelectric performance of a PVDF/KNN-based electrospun nanogenerator *Soft Matter* **16** 4876–86

[16] Hasanzadeh M, Ghahhari M R and Bidoki S M 2021 Enhanced piezoelectric performance of PVDF-based electrospun nanofibers by utilizing *in situ* synthesized graphene-ZnO nanocomposites *J. Mater. Sci., Mater. Electron.* **32** 15789–800

[17] AlAhzm A M, Alejli M O, Ponnamma D, Elgawady Y and Al-Maadeed M A A 2021 Piezoelectric properties of zinc oxide/iron oxide filled polyvinylidene fluoride nanocomposite fibers *J. Mater. Sci., Mater. Electron.* **32** 14610–22

[18] Yang L, Sun J, Zhang D, Bao H, Zhang R, Zhao Q, Bie Y, He H, Huang H and Xu Y 2023 A novel topographically patterned MXene@ MnO$_2$/PVDF piezo-active hybrid for flexible real-time and sensitive force sensor *Compos. Sci. Technol.* **241** 110127

[19] Khalifa M, Peravali S, Varsha S and Anandhan S 2022 Piezoelectric energy harvesting using flexible self-poled electroactive nanofabrics based on PVDF/ZnO-decorated SWCNT nanocomposites *JOM* **74** 3162–71

[20] Muduli S P, Veeralingam S and Badhulika S 2022 Multilayered piezoelectric nanogenerator based on lead-free poly (vinylidene fluoride)-(0.67 BiFeO$_3$-0.33 BaTiO$_3$) electrospun nanofiber mats for fast charging of supercapacitors *ACS Appl. Energy Mater.* **5** 2993–3003

[21] Muduli S P, Veeralingam S and Badhulika S 2021 Interface induced high-performance piezoelectric nanogenerator based on a electrospun three-phase composite nanofiber for wearable applications *ACS Appl. Energy Mater.* **4** 12593–603

[22] Selvan R T, Jia C Y, Jayathilaka W A D M, Chinappan A, Alam H and Ramakrishna S 2020 Enhanced piezoelectric performance of electrospun PVDF-MWCNT-Cu nanocomposites for energy harvesting application *Nano* **15** 2050049

[23] Lin X, Yu F, Zhang X, Li W, Zhao Y, Fei X, Li Q, Yang C and Huang S 2023 Wearable piezoelectric films based on MWCNT-BaTiO$_3$/PVDF composites for energy harvesting sensing, and localization *ACS Appl. Nano Mater.* **6** 11955–65

[24] Zhang D, Zhang X, Li X, Wang H, Sang X, Zhu G and Yeung Y 2022 Enhanced piezoelectric performance of PVDF/BiCl$_3$/ZnO nanofiber-based piezoelectric nanogenerator *Eur. Polym. J.* **166** 110956

[25] Tiwari S, Gaur A, Kumar C and Maiti P 2019 Enhanced piezoelectric response in nanoclay induced electrospun PVDF nanofibers for energy harvesting *Energy* **171** 485–92

[26] Shetty S, Mahendran A and Anandhan S 2020 Development of a new flexible nanogenerator from electrospun nanofabric based on PVDF/talc nanosheet composites *Soft Matter* **16** 5679–88

[27] Ekbote G S, Khalifa M, Perumal B V and Anandhan S 2023 A new multifunctional energy harvester based on mica nanosheet-dispersed PVDF nanofabrics featuring piezo-capacitive, piezoelectric and triboelectric effects *RSC Appl. Polym.* **1** 266–80

[28] Abolhasani M M, Shirvanimoghaddam K and Naebe M 2017 PVDF/graphene composite nanofibers with enhanced piezoelectric performance for development of robust nanogenerators *Compos. Sci. Technol.* **138** 49–56

[29] Yang J, Zhang Y, Li Y, Wang Z, Wang W, An Q and Tong W 2021 Piezoelectric nanogenerators based on graphene oxide/PVDF electrospun nanofiber with enhanced performances by *in situ* reduction *Mater. Today Commun.* **26** 101629

[30] Tiwari S, Dubey D K, Prakash O, Das S and Maiti P 2023 Effect of functionalization on electrospun PVDF nanohybrid for piezoelectric energy harvesting applications *Energy* **275** 127492

[31] Eun J H, Sung S M, Kim M S, Choi B K and Lee J S 2021 Effect of MWCNT content on the mechanical and piezoelectric properties of PVDF nanofibers *Mater. Des.* **206** 109785

[32] Thotadara Shivalingappa R, Narasimha Murthy H N R, Purushothaman P, Badiger P, Savarn S, Majumdar U and Angadi G 2021 Development of poly (vinylidene fluoride)/silver nanoparticle electrospun nanofibre mats for energy harvesting *Polym. Polym. Compos.* **29** 1084–91

[33] Pan C T, Dutt K, Kumar A, Kumar R, Chuang C H, Lo Y T, Wen Z H, Wang C S and Kuo S W 2023 PVDF/AgNP/MXene composites-based near-field electrospun fiber with enhanced piezoelectric performance for self-powered wearable sensors *Int. J. Bioprint.* **9** 647

[34] Ramasamy M S, Rahaman A and Kim B 2021 Effect of phenyl-isocyanate functionalized graphene oxide on the crystalline phases, mechanical and piezoelectric properties of electrospun PVDF nanofibers *Ceram. Int.* **47** 11010–21

[35] Ramasamy M S, Rahaman A and Kim B 2021 Influence of oleylamine–functionalized boron nitride nanosheets on the crystalline phases, mechanical and piezoelectric properties of electrospun PVDF nanofibers *Compos. Sci. Technol.* **203** 108570

[36] Zhang J, Liu D, Han Q, Jiang L, Shao H, Tang B, Lei W, Lin T and Wang C H 2019 Mechanically stretchable piezoelectric polyvinylidene fluoride (PVDF)/Boron nitride nanosheets (BNNSs) polymer nanocomposites *Compos. Part B Eng.* **175** 107157

[37] Eslami R, Malekkhouyan A, Santhirakumaran P, Mehrvar M and Zarrin H 2023 High-performance flexible nanogenerators based on piezoelectric hBN-induced polyvinylidene fluoride nanofibers *Mater. Today Chem.* **30** 101609

[38] Bairagi S and Ali S W 2020 Flexible lead-free PVDF/SM-KNN electrospun nanocomposite based piezoelectric materials: significant enhancement of energy harvesting efficiency of the nanogenerator *Energy* **198** 117385

[39] Zhao B, Chen Z, Cheng Z, Wang S, Yu T, Yang W and Li Y 2022 Piezoelectric nanogenerators based on electrospun PVDF-coated mats composed of multilayer polymer-coated $BaTiO_3$ nanowires *ACS Appl. Nano Mater.* **5** 8417–28

[40] Shetty S, Shanmugharaj A M and Anandhan S 2021 Physico-chemical and piezoelectric characterization of electroactive nanofabrics based on functionalized graphene/talc nanolayers/PVDF for energy harvesting *J. Polym. Res.* **28** 1–17

[41] Qi Z, Zhang S, Huang J, Li J, Jiang J, Fan P and Yang J 2023 Effect of HNT@ PANI hybrid nanoparticles on performance enhancement of electrospun PVDF nanofiber mat for flexible piezoelectric nanogenerators *J. Mater. Sci., Mater. Electron.* **34** 1352

[42] Khalifa M and Anandhan S 2019 PVDF nanofibers with embedded polyaniline–graphitic carbon nitride nanosheet composites for piezoelectric energy conversion *ACS Appl. Nano Mater.* **2** 7328–39

[43] Sriphan S and Vittayakorn N 2022 Hybrid piezoelectric-triboelectric nanogenerators for flexible electronics: recent advances and perspectives *J. Sci.: Adv. Mater. Devices.* **7** 100461

[44] Kim W G, Kim D W, Tcho I W, Kim J K, Kim M S and Choi Y K 2021 Triboelectric nanogenerator: structure, mechanism, and applications *ACS Nano* **15** 258–87

[45] Zou H *et al* 2019 Quantifying the triboelectric series *Nat. Commun.* **10** 1427

[46] Li Y, Xiao S, Luo Y, Tian S, Tang J, Zhang X and Xiong J 2022 Advances in electrospun nanofibers for triboelectric nanogenerators *Nano Energy* **104** 107884

[47] Varghese H, Athira B S and Chandran A 2023 Highly flexible triboelectric nanogenerator based on PVDF nanofibers for biomechanical energy harvesting and telerehabilitation via human body movement *IEEE Sens. J.* **23** 13

[48] Yin H, Zheng Z, Yu D, Chen Y, Liu H and Guo Y 2023 Natural sepiolite modified PVDF electrospun films for mechanically robust and high-performance triboelectric nanogenerators *Appl. Clay Sci.* **233** 106819

[49] Sim D J, Choi G J, Sohn S H and Park I K 2022 Electronegative polyvinylidene fluoride/C60 composite nanofibers for performance enhancement of triboelectric nanogenerators *J. Alloys Compd.* **898** 162805

[50] Pu X, Zha J W, Zhao C L, Gong S B, Gao J F and Li R K 2020 Flexible PVDF/nylon-11 electrospun fibrous membranes with aligned ZnO nanowires as potential triboelectric nanogenerators *Chem. Eng. J.* **398** 125526

[51] Li X, Ji D, Yu B, Ghosh R, He J, Qin X and Ramakrishna S 2021 Boosting piezoelectric and triboelectric effects of PVDF nanofiber through carbon-coated piezoelectric nanoparticles for highly sensitive wearable sensors *Chem. Eng. J.* **426** 130345

[52] Guo Y, Cao Y, Chen Z, Li R, Gong W, Yang W, Zhang Q and Wang H 2020 Fluorinated metal-organic framework as bifunctional filler toward highly improving output performance of triboelectric nanogenerators *Nano Energy* **70** 104517

[53] Rahman M T, Rana S S, Zahed M A, Lee S, Yoon E S and Park J Y 2022 Metal-organic framework-derived nanoporous carbon incorporated nanofibers for high-performance triboelectric nanogenerators and self-powered sensors *Nano Energy* **94** 106921

[54] Sohn S H, Choi G J and Park I K 2023 Metal-organic frameworks-induced self-poling effect of polyvinylidene fluoride nanofibers for performance enhancement of triboelectric nanogenerator *Chem. Eng. J.* **475** 145860

[55] Bhatta T, Maharjan P, Cho H, Park C, Yoon S H, Sharma S, Salauddin M, Rahman M T, Rana S S and Park J Y 2021 High-performance triboelectric nanogenerator based on MXene functionalized polyvinylidene fluoride composite nanofibers *Nano Energy* **81** 105670

[56] Huang T, Lu M, Yu H, Zhang Q, Wang H and Zhu M 2015 Enhanced power output of a triboelectric nanogenerator composed of electrospun nanofiber mats doped with graphene oxide *Sci. Rep.* **5** 13942

[57] Dai Y, Zhong X, Xu T, Li Y, Xiong Y and Zhang S 2023 High-performance triboelectric nanogenerator based on electrospun polyvinylidene fluoride-graphene oxide nanosheet composite nanofibers *Energy Technol.* **11** 2300426

[58] Shi L *et al* 2021 High-performance triboelectric nanogenerator based on electrospun PVDF-graphene nanosheet composite nanofibers for energy harvesting *Nano Energy* **80** 105599

[59] Sun W, Ji G, Chen J, Sui D, Zhou J and Huber J 2023 Enhancing the acoustic-to-electrical conversion efficiency of nanofibrous membrane-based triboelectric nanogenerators by nano-composite composition *Nano Energy* **108** 108248

[60] Im J S and Park I K 2018 Mechanically robust magnetic Fe_3O_4 nanoparticle/polyvinylidene fluoride composite nanofiber and its application in a triboelectric nanogenerator *ACS Appl. Mater. Interfaces* **10** 25660–5

[61] Yang Z, Zhang X and Xiang G 2022 2D boron nitride nanosheets in polymer nanofibers for triboelectric nanogenerators with enhanced performance and flexibility *ACS Appl. Nano Mater.* **5** 16906–11

[62] Ekbote G S, Khalifa M, Perumal B V and Anandhan S 2023 Development of a flexible piezoelectric and triboelectric energy harvester with piezo capacitive sensing ability from barium tungstate nanorod-dispersed PVDF nanofabrics *Flex. Print. Electron.* **8** 025011

[63] Zhou Y, Tao X, Wang Z, An M, Qi K, Ou K, He J, Wang R, Chen X and Dai Z 2022 Electret-doped polarized nanofiber triboelectric nanogenerator with enhanced electrical output performance based on a micro-waveform structure *ACS Appl. Electron. Mater.* **4** 2473–80

[64] Pandey P, Jung D H, Choi G J, Seo M K, Lee S, Kim J M, Park I K and Sohn J I 2023 Nafion-mediated barium titanate-polymer composite nanofibers-based triboelectric nano-generator for self-powered smart street and home control system *Nano Energy* **107** 108134

[65] Dong K, Peng X and Wang Z L 2020 Fiber/fabric-based piezoelectric and triboelectric nanogenerators for flexible/stretchable and wearable electronics and artificial intelligence *Adv. Mater.* **32** 1–43

[66] Kumar A, Kumar S, Pathak A K, Ansari A A, Rai R N, Lee Y, Kim S Y, Le Q V and Singh L 2023 Recent progress in nanocomposite-oriented triboelectric and piezoelectric energy generators: an overview *Nano-Struct. Nano-Objects* **36** 101046

[67] Xie L, Zhai N, Liu Y, Wen Z and Sun X 2021 Hybrid triboelectric nanogenerators: from energy complementation to integration *Research* **2021** 9143762

[68] Zhao X, Li C, Wang Y, Han W and Yang Y 2021 Hybridized nanogenerators for effectively scavenging mechanical and solar energies *iScience* **24** 102415

[69] Mishra S, Unnikrishnan L, Nayak S K and Mohanty S 2019 Advances in piezoelectric polymer composites for energy harvesting applications: a systematic review *Macromol. Mater. Eng.* **304** 1800463

[70] Wang Z L, Long L, Chen J, Niu S and Zi Y 2016 Triboelectric nanogenerators *Green Energy and Technology* (Springer International Publishing)

[71] Guo Y, Zhang X S, Wang Y, Gong W, Zhang Q, Wang H and Brugger J 2018 All-fiber hybrid piezoelectric-enhanced triboelectric nanogenerator for wearable gesture monitoring *Nano Energy* **48** 152–60

[72] Faruk Ünsal O and Çelik Bedeloğlu A 2023 Three-dimensional piezoelectric–triboelectric hybrid nanogenerators for mechanical energy harvesting *ACS Appl. Nano Mater.* **6** 14656–68

[73] Song J, Yang B, Zeng W, Peng Z, Lin S, Li J and Tao X 2018 Highly flexible, large-area, and facile textile-based hybrid nanogenerator with cascaded piezoelectric and triboelectric units for mechanical energy harvesting *Adv. Mater. Technol.* **3** 1800016

[74] Zhang B *et al* 2017 Self-powered acceleration sensor based on liquid metal triboelectric nanogenerator for vibration monitoring *ACS Nano* **11** 7440–6

[75] Fuh Y K, Wang B S and Tsai C Y 2017 Self-powered pressure sensor with fully encapsulated 3D printed wavy substrate and highly-aligned piezoelectric fibers array *Sci. Rep.* **7** 1–7

[76] Zhang M, Tan Z, Zhang Q, Shen Y, Mao X, Wei L, Sun R, Zhou F and Liu C 2023 Flexible self-powered friction piezoelectric sensor based on structured PVDF-based composite nanofiber membranes *ACS Appl. Mater. Interfaces* **15** 30849–58

[77] Garcia C, Trendafilova I and Sanchez del Rio J 2019 Detection and measurement of impacts in composite structures using a self-powered triboelectric sensor *Nano Energy* **56** 443–53

[78] Chen F, Wu Y, Ding Z, Xia X, Li S, Zheng H, Diao C, Yue G and Zi Y 2019 A novel triboelectric nanogenerator based on electrospun polyvinylidene fluoride nanofibers for effective acoustic energy harvesting and self-powered multifunctional sensing *Nano Energy* **56** 241–51

[79] Sultana A, Alam M M, Ghosh S K, Middya T R and Mandal D 2019 Energy harvesting and self-powered microphone application on multifunctional inorganic-organic hybrid nano-generator *Energy* **166** 963–71

[80] Azimi S, Golabchi A, Nekookar A, Rabbani S, Amiri M H, Asadi K and Abolhasani M M 2021 Self-powered cardiac pacemaker by piezoelectric polymer nanogenerator implant *Nano Energy* **83** 105781

[81] Kim H J, Yoo S, Chung M H, Kim J and Jeong H 2022 Energy-efficient PM adhesion method using functional electroactive nanofibers *Energy Reports* **8** 7780–8

[82] Sun N, Zhang X N, Li J Z, Cai Y W, Wei Z, Ding L and Wang G G 2023 Waterproof, breathable, and UV-protective nanofiber-based triboelectric nanogenerator for self-powered sensors *ACS Sustain. Chem. Eng.* **11** 5608–16

[83] Peng Z *et al* 2022 Self-charging electrostatic face masks leveraging triboelectrification for prolonged air filtration *Nat. Commun.* **13** 7835

[84] Cao *et al* 2018 Self-powered nanofiber-based screen-print triboelectric sensors for respiratory monitoring *Nano Res.* **11** 3771–9

www.ingramcontent.com/pod-product-compliance
Lightning Source LLC
Chambersburg PA
CBHW082131210326
41599CB00031B/5943